IMPORTANT FIGURES IN THE NUMBER THE

(Chronological)

Pythagoras   585 - 501 B.C.
Euclid   323 - 285 B.C.
Archimedes   287 - 212 B.C.
Eratosthenes   c.230 B.C.

Diophantus   c.250 A.D.
Sun-Tzu   c.250
Aryabhata   c.475 - 550
Brahmagupta   c.625

al-Karaji   c.1090
Bhaskhara   1114 - c.1185
Fibonacci   c.1175 - 1250
Ch'in Chiu-shao   c.1202 - 1261

Rafael Bombelli   1526 - 1572
Claude Bachet   1587 - 1638
Marin Mersenne   1588 - 1648

Pierre de Fermat   1601 - 1665
Bernard Frenicle de Bessy   c.1602 - 1675
John Pell   1610 - 1685
John Wallis   1610 - 1703
William Brouncker   1620 - 1684
Christian Goldbach   1690 - 1764

Leonhard Euler   1707 - 1783
Edward Waring   1734 - 1793
Joseph Louis Lagrange   1736 - 1813
John Wilson   1741 - 1793
Adrien-Marie Legendre   1752 - 1833
Carl Friedrich Gauss   1777 - 1855
Gabriel Lamé   1795 - 1870

Carl Gustav Jacobi   1804 - 1851
Peter Lejeune Dirichlet   1805 - 1859
Ernst Kummer   1810 - 1893
Pavnuty Chebyshev   1821 - 1894
Ferdinand Eisenstein   1823 - 1852
Edouard Lucas   1842 - 1891
David Hilbert   1862 - 1943
Axel Thue   1863 - 1922
Jacques Hadamard   1865 - 1963
Charles de la Vallée-Poussin   1866 - 1962
G. H. Hardy   1877 - 1947
I. M. Vinogradov   1891 - 1983

The Theory of Numbers

A TEXT AND SOURCE BOOK OF PROBLEMS

Andrew Adler   John E. Coury

The University of British Columbia

Jones and Bartlett Publishers
Sudbury, Massachusetts

Boston   London   Singapore

Editorial, Sales, and Customer Service Offices

Jones and Bartlett Publishers
40 Tall Pine Drive
Sudbury, MA 01776
1-508-443-5000
1-800-832-0034
info@jbpub.com
http://www.jbpub.com

Jones and Bartlett Publishers International
Barb House, Barb Mews
London W6 7PA
UK

Copyright © 1995 by Jones and Bartlett Publishers, Inc.

All rights reserved. No part of the material protected by this copyright notice may be reproduced or utilized in any form, electronic or mechanical, including photocopying, recording, or by any information storage and retrieval system, without written permission from the copyright owner.

Library of Congress Cataloging-in-Publication Data
Adler, Andrew.
   The theory of numbers: a text and source book of problems / Andrew Adler, John E. Coury.
      p. cm.
   Includes bibliographical references and index.
   ISBN 0-86720-472-9
   1. Number theory. I. Coury, John E. II. Title.
QA241.A244   1995
$512'.7$–dc20   94-41865
                                                                                  CIP

Printed in the United States of America
98 97 96   10 9 8 7 6 5 4 3 2

For
Marie, Bob, and Michael
J.C.

For
Alan, Lissett, Rachel, and Sara
A.A.

Contents

Preface   ix

Introduction   1

Chapter One: Divisibility, Primes, and the Euclidean Algorithm   6
    Results   7
        Divisibility   7
        Primes   10
        The Euclidean Algorithm   13
        The Equation $ax + by = c$ 15
    Problems and Solutions   17
    Exercises   32
    Notes, Biographical Sketches, References   35

Chapter Two: Congruences   39
    Results   40
        Divisibility Tests   42
        Linear Congruences   43
        Techniques for Solving $ax \equiv b \pmod{m}$ 44
        The Chinese Remainder Theorem   46
        An Application: Finding the Day of the Week   47
    Problems and Solutions   48
    Exercises   64
    Notes, Biographical Sketches, References   66

Chapter Three: The Theorems of Fermat, Euler, and Wilson   71
    Results   72
        Fermat's Theorem and Wilson's Theorem   72
        Euler's Theorem and the Euler ϕ-function   75
    Problems and Solutions   78

Exercises   94
Notes, Biographical Sketches, References   96

Chapter Four: Polynomial Congruences   **101**
Results   101
   General Polynomial Congruences   101
   Solutions of $f(x) \equiv 0 \pmod{p^k}$ 106
   The Congruence $x^2 \equiv a \pmod{p^k}$ 109
Problems and Solutions   110
Exercises   122
Notes, References   124

**Chapter Five: Quadratic Congruences and the Law
         of Quadratic Reciprocity**   **125**
Results   126
   General Quadratic Congruences   126
   The Congruence $x^2 \equiv a \pmod{m}$ 127
   Quadratic Residues   128
   The Law of Quadratic Reciprocity   131
Problems and Solutions   137
Exercises   153
Notes, Biographical Sketches, References   155

Chapter Six: Primitive Roots and Indices   **158**
Results   158
   The Order of an Integer   158
   Primitive Roots   160
   Power Residues and Indices   164
   The Existence of Primitive Roots   167
Problems and Solutions   169
Exercises   189
Notes, Biographical Sketches, References   191

Chapter Seven: Prime Numbers   **194**
Results   195
   The Sieve of Eratosthenes   195
   Perfect Numbers   196
   Mersenne Primes   197
   Fermat Numbers   198
   The Prime Number Theorem   200
   Dirichlet's Theorem   202
   Goldbach's Conjecture   203
   Other Open Problems   204

Problems and Solutions   205
Exercises   216
Notes, Biographical Sketches, References   217

**Chapter Eight: Some Diophantine Equations and
Fermat's Last Theorem**   221

Results   222
   The Equation $x^2 + y^2 = z^2$ 222
   Fermat's Last Theorem   224
   Sums of Two Squares   226
   Sums of Two Relatively Prime Squares   229
   Sums of Four Squares   233
   Sums of Three Squares   235
   Waring's Problem   236
Problems and Solutions   237
Exercises   263
Notes, Biographical Sketches, References   265

Chapter Nine: Continued Fractions   270

Results   271
   Finite Continued Fractions   271
   An Application: Solutions of $ax + by = c$ 274
   Infinite Continued Fractions   275
   The Infinite Continued Fraction of an Irrational Number   276
   Periodic Continued Fractions   278
   Purely Periodic Continued Fractions   281
   Rational Approximations to Irrational Numbers   282
   An Application: Calendars   285
Problems and Solutions   286
Exercises   308
Notes, Biographical Sketches, References   310

Chapter Ten: Pell's Equation   314

Results   315
   Pell's Equation $x^2 - dy^2 = 1$ 315
   The Equation $x^2 - dy^2 = -1$ 322
   The Equation $x^2 - dy^2 = N$ 324
   Pell's Equation and Sums of Two Squares   325
   An Application: Factoring Large Numbers   327
Problems and Solutions   329
Exercises   352
Notes, Biographical Sketches, References   354

Chapter Eleven: The Gaussian Integers and Other Quadratic Extensions 357

Results 358
  The Gaussian Integers 358
  Unique Factorization for Gaussian Integers 361
  The Gaussian Primes 362
  An Application: Gaussian Integers and Sums of Two Squares 363
  Applications of Gaussian Integers to Diophantine Equations 364
  The Integers of $Q(\sqrt{d})$ 365
  Primes of $Q(\sqrt{d})$ and Diophantine Equations 369
  Units of $Q(\sqrt{d})$ 370
Problems and Solutions 372
Exercises 387
Notes, Biographical Sketches, References 388

Appendix 391
  Table of Primes and Their Least Primitive Root 392
  Table of Continued Fraction Expansion of $\sqrt{d}$ 393

General References 394

Index 398

Preface

This book presents the principal ideas of classical elementary number theory, emphasizing the historical development of these results and the important figures who worked on them. The book is also intended to introduce students to mathematical proofs by presenting them in a clear and simple way and by providing complete, step-by-step solutions to the problems with as much detail as students would be expected to provide themselves. Throughout, we have tried to indicate the important ideas in a proof or numerical technique and to show the students computational shortcuts whenever possible.

We feel that the historical background and comments are important not only for putting the various results in perspective, but also because they capture the interest and imagination of students and make the material more relevant. We had three goals in mind as we wrote this book. First of all, we wanted to make the material interesting and as easy to learn as possible; to this end, we have included applications of the theoretical material (for example, a discussion of calendars, how to find the day of the week for a given date, determining "best" rational approximations to π, and an algorithm for factoring large numbers). Second, the topics have been organized and the proofs and solutions written in such a way that it is very easy for instructors to teach from the book. Finally, we wanted to make the proofs sufficiently transparent and motivated so that students understand the nature of a mathematical proof – both simple and more complicated arguments – and eventually learn how to construct rigorous and logical proofs of their own.

Since many of the basic concepts (primes, divisibility, factoring) are already familiar to students, number theory is an ideal way to introduce students to mathematical proofs – better, in fact, than a course in elementary analysis, where the concepts are much less familiar and the proofs (ϵ-δ arguments, for example) are more difficult to understand. But students need clear models of how to write out proofs and solutions. A unique feature of this book is that it provides detailed solutions for almost 800 problems, with complete references to the results used so that the student can follow each step of the argument. There is also a large collection of problems *without* solutions at the end of each chapter that may be used for homeworks or exams. In our experience, students at this level do not get enough practice doing problems, especially when the course runs for only one term (as is now the case at most universities). This is particularly true for problems involving proofs, even simple proofs. However, if students can see a large number of problems on each topic worked out in *detail*, they have a much better chance of doing similar problems. When only a numerical answer or sketch of a solution is given, students often do not realize what the important points in the proof are. Consequently, they may not be able to solve other problems of this type that do not closely resemble the few problems worked out in the text.

The solved problems, in fact, provide a clear model to follow, showing the student how to put together previous results to solve a new problem and indicating what should – and what need not – be included, as well as the level of detail expected. Sometimes, several solutions to the same problem are given to emphasize that there are often many ways to arrive at a solution.

A word about the proofs in this book. We have taken special pains to make the proofs as straightforward and clear as possible, preferring clarity to either a short proof or an "elegant" proof. In a number of cases, the proofs given are nonstandard and considerably more transparent than the usual arguments (for example, the Four-Squares Theorem and the Law of Quadratic Reciprocity). Our guiding philosophy throughout has been to make the proofs of the theorems and the solutions of the problems easy to present in class. We have devoted a great deal of time to organizing the material, and particular attention has been paid to motivating the results, often by looking at concrete examples or applications.

Our book is intended as a text for either a one-term or two-term course in elementary number theory, usually given at American and Canadian colleges and universities in the third year. Typically, such courses are taken by juniors and seniors, but increasingly, second-year students (including those at two-year colleges) take such a course, and the material is very accessible to them. There are few formal prerequisites for the material in this book; in particular, no previous course in abstract algebra is necessary. Students should be familiar with proofs by mathematical induction, and frequent use will be made of the fact that *any nonempty set of positive integers contains a smallest element*. A few of the proofs use basic properties of limits of sequences of real numbers; for example, students should know that *an increasing sequence of real numbers which is bounded above is convergent*. Finally, the material in Chapter 11 – which is seldom covered nowadays in a course in elementary number theory – requires some familiarity with complex numbers.

FEATURES

This book is intended to be a self-contained text for a course in elementary number theory, as well as a source book of solved problems. *It is the only source book of problems in number theory that has detailed, step-by-step solutions to all of the problems.* As such, it is a valuable reference even if it is not used as the principal text; the solved problems are ideal for supplementing class lectures, as well as for homework assignments, exams, and review. All of the standard topics are presented, along with a number of topics that are not found in many current books: polynomial congruences, factorization of large numbers, Gaussian integers and the integers of other quadratic fields, to name a few. This last topic is seldom taught nowadays in an introductory course in number theory, but it ties together nicely a number of topics in the previous chapters and provides simple proofs for many of the results.

We have also included a very complete treatment of quadratic reciprocity, primitive roots, representations of integers as sums of two squares (including a derivation of the formula for the number of such representations), rational approximation of irrational numbers, and Pell's Equation. Thus instructors can customize a course to reflect their own interests as well as the background of their students; for example, more computational topics can be added to the standard material.

Chapters 1 and 2 contain the basic concepts that will be used throughout the book; the more advanced topics (including optional material) appear in subsequent chapters. With few exceptions, each chapter begins with a historical introduction, and historical comments appear throughout the text, including the dates particular results were proved and by whom. Where relevant, we also mention unsolved problems or open questions. In particular, each chapter is arranged as follows:

Basic Results and Proofs. In many cases, we have provided new or much simpler proofs. There are also detailed worked examples, applications, computational notes, and a discussion of algorithms for efficient computation in numerical problems.

Solved Problems. Each chapter contains approximately 50 to 100 such problems with complete and detailed solutions, fully referenced to the results in the text. This feature is unique to this book. The problems are arranged according to the sections within the chapter and cover a wide range of difficulty and computational skill, from straightforward (a numerical computation or one-step argument using a theoretical result from the text) to more challenging (requiring several steps in the proof or drawing on other results). The most challenging problems are denoted in the margin by a ▷ and are intended for the better students. These problems develop other areas of the chapter material, provide additional theory (for example, the Jacobi symbol in Chapter 5 and secondary convergents in Chapter 9), or give a new or unusual proof of a standard result. *These problems are not needed for later chapters*; indeed, with few exceptions, the results in the text are independent of the Solved Problems.

Finally, many problems appear with a hint, and a number are stated as "Prove or disprove" to encourage students to experiment and think about whether the statement seems reasonable.

Exercises. Each chapter contains a large number of problems with no solutions, although many appear with hints. The Exercises are very similar to the Solved Problems and are ideal for homework assignments, quizzes, and exams.

Chapter Notes. These expand on the text material, indicating different approaches, additional results, and open questions. Some of the Notes present discussions that relate the material to other areas of mathematics.

Biographical Sketches. A brief summary is given of the lives and work of the more important mathematicians who worked in number theory.

Annotated References. These provide a source of additional material for interested students, indicating the distinctive features of each book.

HOW TO USE THIS BOOK

There is more than enough material in this text for two one-term courses in elementary number theory; in fact, some selection of topics will have to be made even if two courses are offered. The material can be split up in many different ways. At the University of British Columbia, our original full-year course in number theory has been redesigned into two one-term courses, each lasting about 13 weeks. The first course is a prerequisite for the second and may be taken by itself, although most students take both.

A one-term course can be designed as follows. Chapters 1, 2, and 3 should be covered, since this basic material – on divisibility, primes, the Euclidean Algorithm, congruences, the Chinese Remainder Theorem, and the theorems of Fermat, Euler, and Wilson – is used in the subsequent chapters. The topics in Chapter 4 (Polynomial Congruences) are optional and are generally not covered in most introductory courses in number theory. However, they are a very nice source of computational problems. (At our university, this material is usually skipped in favor of the material in Chapter 5.) In Chapter 5, the theory of quadratic residues and Gauss's Law of Quadratic Reciprocity should be covered, even if the proof of the Quadratic Reciprocity Law is not presented. The proof we give is not the standard one (due to Eisenstein), and we believe that it is very accessible to a class of third-year and fourth-year mathematics students. The Law of Quadratic Reciprocity is one of the most important results in the classical theory of numbers, and this material is an excellent source of numerical as well as theoretical problems.

At our university, a second one-term course would normally cover the following topics. We begin with the material in Chapter 6 on primitive roots and indices, but some of the existence proofs could easily be left out or given as reading assignments. This material is a nice blend of theory and computational techniques. Chapter 7 is optional, although the general discussion of primes and the material on perfect numbers and Mersenne primes are included in the course we offer. The section on Fermat numbers is short enough to include as well. The material in Chapter 8 on Pythagorean triples and sums of two squares, as well as the statement of the Four-Squares Theorem, is also covered, with the discussion on primitive representations, the proof of the Four-Squares Theorem, and Waring's Problem left as optional topics. In Chapter 9, most of the basic results on finite and infinite continued fractions are presented, but few of the proofs need be given, since many are proved by using induction and are quite repetitive. The material on rational approximations could be skipped, but both authors do include it since it is a nice application of the theory as well as a good source of numerical problems. Chapter 10 (Pell's Equation) is an important application of continued fractions and also gives rise to many numerical problems. We would generally expect to cover both of the equations $x^2 - dy^2 = 1$ and $x^2 - dy^2 = -1$, although the latter could be omitted in the interest of time or to present the material on factoring large numbers, a topic that students really seem to find interesting. Finally, Chapter 11 is optional and would seldom be covered unless other topics (for example, primitive roots or Pell's Equation) are omitted. In this case, the material on Gaussian primes could be presented, which allows for some elegant proofs of results in previous chapters, among them the formula for Pythagorean triples and the number of ways to represent an integer as a sum of two squares.

A final comment about the Solved Problems and Exercises. A collection of Solved Problems could be assigned weekly – or even daily – for students to read, with a 20- or 25-minute quiz every week or so using either other Solved Problems or questions from the Exercises. In this way, valuable class time need not be taken to go over solutions of a number of problems in detail (which can be very time-consuming).

CLOSING REMARKS AND ACKNOWLEDGMENTS

Over the past twenty-five years, both of us have taught courses in number theory many times. This book, and the problems in it, are the result of our experience and long-standing interest in the subject. We both feel that the best way to teach number theory is to complement the theoretical results with a large number of problems that have detailed solutions, so students understand the various techniques for writing out their own proofs. For a number of years, we have used this approach in our teaching. Since there was no source book of problems with step-by-step solutions, we decided to organize the many hundreds of problems we have collected over the years and write our own book.

We would like to acknowledge our debt to the books written by G. H. Hardy and E. M. Wright, Ivan Niven and Herbert Zuckerman, and Harold

Davenport. Each in its own way has played an important role in our approach to number theory and our continuing interest in the subject.

We wish to thank Professors Stephen Chase and Don Redmond for reading the manuscript and making a number of helpful suggestions. We also express our appreciation to the people at Jones and Bartlett for their support and cooperation. Our special thanks go to Carl Hesler, who encouraged us in the project and has always been available to provide assistance.

Finally, Professor Coury would like to express his gratitude to Edwin Hewitt and Herbert Zuckerman, the one for guiding his early mathematical career and the other for revealing the beauty of number theory. He is greatly indebted to both of them.

Vancouver, Canada
October 1994

Andrew Adler
John E. Coury

Introduction

The *theory of numbers*, sometimes called the *higher arithmetic*, is one of the oldest areas of mathematics, dating back several thousand years. The earliest problems considered were based on the notion of counting and the elementary concepts of arithmetic, ideas that are even older and appeared in Babylonian tablets some 4000 years ago. Mathematical puzzles and word problems, dating from antiquity, have been another source of investigation in number theory. Word problems appear in Greek mathematics beginning in the first century A.D. and in Chinese mathematics in the fifth century A.D. They also occur in the writings of the Indian mathematicians Brahmagupta (seventh century) and Bhaskara (twelfth century), as well as the work of the Italian mathematician Fibonacci (early thirteenth century).

In a broad sense, number theory is concerned with the properties of the positive integers (or natural numbers), including divisibility, the greatest common divisor of two integers, and the study of primes and composite numbers. The problems and conjectures in number theory are, by and large, easy to state but often quite difficult to prove. A good illustration of this is *Goldbach's Conjecture*, which asserts that every even integer greater than 2 is the sum of two primes. Much work has been done on this problem since it first appeared in 1742, but it remains unsolved. Another example concerns the representation of certain positive integers as a sum of two squares. While Diophantus, in the third century, treated this question in his *Arithmetica*, it was some 1500 years later that the question was finally resolved.

Early Greek mathematics dealt with the problems of primes and divisibility, finding right triangles with sides of integral length, and investigating perfect numbers (that is, numbers which are equal to the sum of their proper positive divisors). Beginning with Pythagoras in the sixth century B.C., these problems were studied in some detail, with the results usually of a theoretical nature. In the third century B.C., Euclid compiled much of the mathematics known to the ancient Greeks in his *Elements*, arguably the most important mathematical treatise ever written. In addition to a detailed development of geometry, the *Elements* contains a discussion of prime numbers, including a proof that there are infinitely many primes; a method for generating perfect numbers; the well-known *Euclidean Algorithm* for finding the greatest common divisor of two integers; and the tools for proving the *Fundamental Theorem of Arithmetic*, which asserts that every integer greater than 1 can be expressed in just one way, apart from the order of the factors, as a product of primes. (This theorem was first stated and proved by Carl Friedrich Gauss in 1801.) Somewhat later, Eratosthenes developed an interesting technique, called the *Sieve of Eratosthenes*, for determining all of the primes less than a given positive integer. And in the third century A.D., Diophantus of Alexandria gave the first systematic treatment of what are now known as Diophantine equations, that is, algebraic equations for which integer solutions are sought (or, in the case of Diophantus, rational solutions). His *Arithmetica*, which for the first time used symbols rather than words to express equations, contains over 250 such problems and solutions.

From the time of Diophantus to the thirteenth century, Indian, Chinese, and Arab mathematicians produced various algorithms (such as the *Chinese Remainder Theorem*) and studied certain Diophantine equations, including the linear equation $ax + by = c$ and the quadratic equation $x^2 - dy^2 = 1$, which eventually became known as *Pell's Equation*. Brahmagupta and Bhaskara examined this latter equation in detail and obtained results that would not be matched in Europe until the seventeenth century. Particular versions of Pell's Equation had also been studied by the Greeks, since these and the closely related idea of continued fractions arise in the problem of finding good rational approximations to the irrational number $\sqrt{d}$. In the third century B.C., for example, Archimedes approximated $\sqrt{3}$ by $265/153$ and $1351/780$, which are accurate to four and six decimal places, respectively. One of the earliest appearances of Pell's Equation occurs in the third century in connection with the *Cattle Problem of Archimedes*, which leads to the equation $x^2 - 4729494y^2 = 1$, the least positive solution (found in 1880) having a y-value that is 41 digits long.

Until the twelfth century, there was very little mathematical development in medieval Europe. The most gifted mathematician in Europe during the Middle Ages was Leonardo of Pisa (c. 1175–1250), better known as Fibonacci. He introduced the use of Arabic numerals in his book *Liber Abaci* ("Book of Calculation"), and in *Liber Quadratorum* ("Book of Squares"), Fibonacci investigated the solution of certain Diophantine equations involving squares. *Liber Abaci* contains many word problems, including Fibonacci's famous "rabbit problem": Beginning with a single pair of rabbits, how many pairs will be produced in one year if every month each pair bears a new pair that becomes productive from the second month on? The answer is 377, the twelfth term in the *Fibonacci sequence* 2, 3, 5, 8, 13, 21, ... , 377, ... , where each term, beginning with the third, is the sum of the two preceding terms. This sequence has many interesting properties. For example, any two successive terms have no divisor in common except 1, and the ratio of sufficiently large successive terms is arbitrarily close to the "golden ratio" $(\sqrt{5}-1)/2$, which was of interest to the ancient Greeks.

Beginning in the twelfth century in Western Europe, Euclid's *Elements* and a number of Arabic texts were translated into Latin, although the first printed edition of the *Elements* did not appear until 1482. Almost a hundred years later, a Latin translation of Diophantus's *Arithmetica* was published, followed in 1621 by a greatly improved edition. With the availability of these books, the quality of mathematics in Europe advanced significantly.

As late as the seventeenth century, it was common for mathematicians to work alone, conveying their results by letter to one another. The Franciscan monk Marin Mersenne (1588–1648) corresponded with many of the scholars of the day and acted as a clearinghouse for their scientific work. One of the mathematicians with whom Mersenne exchanged ideas regularly was Pierre de Fermat (1601–1665). Bachet's 1621 translation of Diophantus's *Arithmetica* introduced Fermat to the problems of number theory. Later called the "Prince of Amateurs" (he was a magistrate by profession), Fermat was the last great mathematician for whom mathematics was essentially a hobby. After Fermat, mathematical research would be conducted predominantly by professional mathematicians at universities and scientific academies.

As a systematic area of study, the theory of numbers really begins with the work of Fermat in the seventeenth century. Many mathematicians since the time of Pythagoras had made contributions to this field, but it was Fermat who highlighted the problems and themes in number theory that would be studied for the next 150 years. Fermat was interested in the theoretical ideas that bound together individual numerical results, and his work covered a wide range of problems: perfect numbers, divisibility, primes, and various Diophantine equations, including the first serious treatment of Pell's Equation. Fermat stated, without proof, that every prime of the form $4k+1$ has a unique representation as a sum of two squares, a question that arises from the work of Diophantus. During his career, Fermat offered proofs for very few of his assertions; most of his work appears without proof in correspondence with other mathematicians, often in the form of a challenge to solve particular problems. However, Fermat did use what he called his *method of infinite descent* to prove some of his results; this technique is essentially equivalent to the principle of mathematical induction. Fermat's most famous unproved assertion, and one of the best-known unsolved problems in all of mathematics, states that the equation $x^n + y^n = z^n$ has no solution in nonzero integers if $n \geq 3$. Known as *Fermat's Last Theorem*, this conjecture defied proof for over three and a half centuries; a complete proof was finally given in October, 1994.

Fermat, the foremost figure in number theory in the seventeenth century, was succeeded by Leonhard Euler (1707–1783) and Joseph Louis Lagrange (1736–1813). Euler was the most prominent mathematician of the eighteenth century and also one of the most prolific in history, publishing an enormous number of papers in his lifetime. He proved many of the results that Fermat had only stated, including the fact that a prime of the form $4k + 1$ is a sum of two squares in just one way. He generalized a number of Fermat's results and formulated, in 1746, a version of the famous *Law of Quadratic Reciprocity*, which would be proved some 50 years later by Gauss. Euler also introduced what is now known as the Euler ϕ-function, a concept of great importance in number theory, as well as the idea of congruence and residue classes, which was refined by Gauss at the end of the century.

Joseph Louis Lagrange, second only to Euler in mathematical prominence in the eighteenth century, succeeded Euler at the Academy of Berlin when Euler accepted a post in St. Petersburg. While much of Lagrange's work was outside number theory, he was the first to prove, in 1770, that every positive integer can be expressed as a sum of no more than four squares, a result that had eluded even Euler. Lagrange also gave the first published proof of Wilson's Theorem in 1771 and proved an important theorem on the number of roots of certain polynomial congruences. And in a series of papers presented to the Berlin Academy around 1770, Lagrange gave the first rigorous treatment of Pell's Equation using continued fractions (a connection that Euler had noted some ten years earlier).

The foremost number theorist in the nineteenth century was Carl Friedrich Gauss (1777–1855). Called the "Prince of Mathematicians" by his contemporaries, Gauss is generally considered to be the founder of modern number theory and one of the three greatest mathematicians in history, along with Archimedes and Isaac Newton. With the publication, in 1801, of his landmark book on the theory of numbers, *Disquisitiones Arithmeticae* ("Investigations in Arithmetic"), Gauss put the theory of numbers on a sound mathematical basis. By arithmetic, Gauss meant number theory; in fact, in the preface to his book, Gauss coined the phrase "the higher arithmetic," which includes more general inquiries concerning the integers, to distinguish it from what he called "elementary arithmetic." In *Disquisitiones*, Gauss presented most of the concepts and notation that are still used today. He introduced the modern definition of congruence and residues, which greatly simplified computations involving integers, as well as the notation $\equiv$ for congruence that has been used ever since. Gauss's book also contains the first complete proof of the *Law of Quadratic Reciprocity* (he would eventually give six proofs of this result), a detailed treatment of linear congruences, and a comprehensive discussion of primitive roots. In addition, *Disquisitiones* includes the first statement and proof of the Fundamental Theorem of Arithmetic.

Gauss formulated, but did not prove, the celebrated *Prime Number Theorem* (the first proof was not given until 1896, some 40 years after his death), and later in his career he made a detailed study of the properties of what are now called *Gaussian integers* (that is, complex numbers $a + bi$, where a and

b are integers). Gauss generalized the notion of primes to Gaussian integers and proved that these integers, like the ordinary integers, can also be factored in an essentially unique way as a product of "Gaussian" primes.

Throughout his long and distinguished career in many areas of mathematics and science, Gauss always had a special fondness for number theory. He once described mathematics as the queen of sciences and the theory of numbers as the queen of mathematics. One reason that number theory has held the interest of mathematicians since ancient times is that the ideas and concepts (for example, divisibility, prime numbers, and factoring) are so familiar. Many of the conjectures are easy to formulate and understand, even those, such as Fermat's Last Theorem, that resisted proof for centuries.

Referring to the difficulty in trying to prove results that seem quite evident on the basis of numerical observations, Gauss once said: "It is precisely this which gives the higher arithmetic the magical charm that has made it the favorite science of the greatest mathematicians, not to mention its inexhaustible wealth, wherein it so greatly surpasses other parts of mathematics."

CHAPTER ONE

Divisibility, Primes, and the Euclidean Algorithm

The first systematic development of the theory of divisibility can be found in Books VII–IX of Euclid's *Elements* (c. 300 B.C.). There were systematic treatments of basic number theory before Euclid, for example, by Archytas and by the great mathematician Eudoxus. Although these have been lost, there is reason to believe that a great deal of Euclid's number theory comes from earlier sources.

Much of the theoretical content of this chapter can be found in Euclid. He did not state the Unique Factorization Theorem, but some have argued that it is essentially contained in his *Elements*. Euclid did not consider the question of solving the equation $ax+by = c$ in integers, even though the solution comes fairly simply from his algorithm for finding the greatest common divisor of two numbers.

Methods for finding integer solutions of $ax + by = c$ were obtained in sixth-century India by Aryabhata and refined in the seventh century by Brahmagupta. Their method, called *kuttaka* (the pulverizer), continued to play an important role in Indian mathematics for several centuries. It is closely related to the back substitution method described after Theorem 1.23.

In Western Europe, a thorough understanding of the equation $ax+by = c$ seems to have been reached only in the early seventeenth century. Claude Bachet de Méziriac (1587–1638) gave a full discussion in 1612. His method is again closely related to the Euclidean Algorithm. Like earlier mathematicians, he was hampered by a reluctance to use negative numbers. In the eighteenth century, Leonhard Euler (1707–1783) and Joseph Louis Lagrange (1736–1813) reached full technical mastery of the subject. In 1801, Carl Friedrich Gauss (1777–1855) gave number theory a proper theoretical framework in his *Disquisitiones Arithmeticae*.

RESULTS FOR CHAPTER 1

Divisibility

(1.1) Definition. Let a and b be integers, with a nonzero. We say that a *divides* b, or that b is a *multiple* of a, if there is an integer q such that $b = qa$. In this case, we write $a|b$ and say that a is a *divisor* of b. If a does not divide b, we write $a \nmid b$.

The proof of the next result is a direct consequence of this definition.

(1.2) Theorem. *Let a, b, and c be integers.*

(i) *If $a|b$, then $a|kb$ for any integer k.*

(ii) *If $a|b$ and $b|a$, then $a = \pm b$.*

(iii) *If $a|b$ and $b|c$, then $a|c$.*

(iv) *If $a|b$ and $a|c$, then $a|sb + tc$ for any integers s and t.*

(v) *For any nonzero integer k, $a|b$ if and only if $ka|kb$.*

The following familiar result, known as the *Division Algorithm*, is an important tool in number theory. Roughly speaking, it states that we can divide an integer b by the integer a and leave a remainder smaller than a. The proof appeals to the *Well-Ordering Property*, a fact that will be used frequently in the book: *Every nonempty set of positive integers contains a smallest element*.

(1.3) Theorem (Division Algorithm). *Let a and b be integers, with a positive. Then there exist unique integers q and r such that $b = qa + r$ and $0 \leq r < a$.*

Proof. Let S be the set of positive integers that are greater than b/a. By the Well-Ordering Property, S contains a smallest element t; thus $t - 1 \leq b/a < t$. Let $q = t - 1$; then $qa \leq b < (q+1)a$. If we set $r = b - qa$, then $b = qa + r$ and $0 \leq r < a$.

The integer q is called the *quotient* and r is called the *remainder*.

Note. The word *algorithm* derives from the name of the ninth-century mathematician al-Khwarizmi, who wrote a book explaining the Indian number system (namely, the now-universal decimal representation of integers). In the late Middle Ages, people who knew the art of computing using decimal representation were called *algorists*. Gradually they supplanted the *abacists*, who computed with the traditional abacus.

An algorithm has come to mean any mechanical computing procedure. In that sense, there is an algorithm implicit in the proof of Theorem 1.3. For example, if $b \geq 0$, keep subtracting a from b until what is left becomes less than a. It is clear that what remains is r, and the number of times we have subtracted a is q. In general, this algorithm is inefficient – the familiar "long division" procedure is far better.

(1.4) Definition. *The largest positive integer that divides both a and b is called the greatest common divisor (or gcd) of a and b. We denote it by (a,b).*

Definition 1.4 contains implicitly an algorithm for computing the gcd. If a and b are both 0, the gcd does not exist. If $a = 0$ and $b \neq 0$, the gcd is $|b|$. If a and b are both nonzero, with $|a| \leq |b|$, we list all the positive divisors of a. The largest of these that also divides b is the gcd. This algorithm is in general very inefficient.

The next theorem gives a very useful characterization of the greatest common divisor of a and b in terms of their *linear combinations*, that is, sums of the form $sa+tb$, where s and t are *integers*. The result will be used frequently in subsequent proofs in this chapter.

(1.5) Theorem. *Suppose a and b are not both 0, and let $d = (a,b)$. Then d is the smallest positive integer that can be expressed as a linear combination of a and b.*

Proof. Since the set of all linear combinations of a and b clearly contains positive integers (as well as negative integers and 0), it contains a smallest positive element m, say, $m = sa+tb$. Use the Division Algorithm to write $a = qm+r$, where $0 \leq r < m$. Then $r = a-qm = a-q(sa+tb) = (1-qs)a+(-qt)b$, and hence r is also a linear combination of a and b. But $r < m$, so it follows from the definition of m that $r = 0$. Thus $a = qm$, that is, $m \mid a$; similarly, $m \mid b$. Therefore m is a common divisor of a and b.

Since d divides a and b, d divides any linear combination of them, by (1.2.iv). Hence d divides m and thus $d \leq m$. Since d is the *greatest* common divisor, we must have $d = m$.

(1.6) Corollary. *If c is any common divisor of a and b, then c divides (a,b).*

Proof. Let $d = (a,b)$ and write $d = sa+tb$. Since $c \mid a$ and $c \mid b$, (1.2.iv) implies that c divides d.

Note. The greatest common divisor d of a and b can also be characterized as follows: d is the unique positive integer such that (i) $d \mid a$ and $d \mid b$, and (ii) if $e \mid a$ and $e \mid b$, then $e \mid d$.

(1.7) Theorem. *Let a and b be integers. Then*

(i) $(ca, cb) = c(a,b)$ *for any positive integer c;*

(ii) $(a/d, b/d) = 1$ if $d = (a,b)$.

Proof. To prove (i), note that the smallest positive integer which is a linear combination of ca and cb is simply c times the least positive integer which is a linear combination of a and b. Now apply (1.5). Part (ii) follows from (i) by using a/d, b/d, and $c = d$.

(1.8) Definition. The integers a and b are said to be *relatively prime* if $(a,b) = 1$, that is, if they have no (positive) factor in common except 1. The integers $m_1, m_2, \ldots, m_k$ are *relatively prime in pairs* if $(m_i, m_j) = 1$ whenever $i \neq j$.

It follows from (1.5) that if $sa + tb = 1$ for some choice of integers s and t, then a and b must be relatively prime. Also, if a and b are each divided by their greatest common denominator d, then a/d and b/d are relatively prime (see (1.7.ii)).

The following property of relatively prime integers is very useful and plays an important role in the proof of the Fundamental Theorem of Arithmetic (Theorem 1.16).

(1.9) Theorem (Euclid). *If a divides bc and $(a,b) = 1$, then a divides c.*

Proof. Theorem 1.7 implies that $(ac, bc) = c(a,b) = c$. Note that a is a divisor of ac (clearly) and bc (by hypothesis). Thus it follows from (1.6) that a divides c.

(1.10) Theorem (Euclid). *Let a, b, and c be integers.*

(i) *If $(a,b) = (a,c) = 1$, then $(a, bc) = 1$.*

(ii) *If $a|c$, $b|c$, and $(a,b) = 1$, then $ab|c$.*

Proof. (i) Use (1.5) to write $sa + tb = 1$ and $ua + vc = 1$ for integers s, t, u, and v. Then $tb \cdot vc = (1 - sa)(1 - ua) = 1 - ma$, where $m = s + u - sua$. Hence $ma + tv(bc) = 1$, and the result follows from (1.5).

(ii) Let $c = mb$. Since $a|mb$ and $(a,b) = 1$, it follows from (1.9) that $a|m$. If $m = na$, then $c = nab$ and hence $ab|c$.

We now define a concept complementary to that of the greatest common divisor of two integers. Theorem 1.13 below indicates the connection between them.

(1.11) Definition. Suppose a and b are integers, not both zero. The smallest positive integer that is a multiple of a and b is called the *least common multiple* of a and b. It is denoted by $[a,b]$. The least common multiple of $a_1, a_2, \ldots, a_r$ is defined analogously and is denoted by $[a_1, a_2, \ldots, a_r]$.

(1.12) Theorem. *If k is a common multiple of a and b, then $[a,b]$ divides k. Thus every common multiple of a and b is of the form $t[a,b]$ for some integer t.*

Proof. Let $m = [a,b]$ and use the Division Algorithm to write $k = qm + r$, where $0 \le r < m$. Since $a\,|\,k$ and $a\,|\,m$, we have $a\,|\,r$; similarly, $b\,|\,r$. Thus r is a common multiple of a and b which is less than m. Since m is the *least* common multiple, it follows that $r = 0$, and hence m divides k.

(1.13) Theorem. *Let a and b be positive integers. Then $(a,b)[a,b] = ab$.*

Proof. Let $d = (a,b)$ and use (1.5) to write $d = sa + tb$. Since d divides ab, $m = ab/d$ is an integer. It suffices to prove that $m = [a,b]$.

Clearly, m is a common multiple of a and b. Also, if n is any positive common multiple of a and b, then $n/m = nd/ab = n(sa+tb)/ab = (\hat{n}/b)s + (n/a)t$ is an integer. Hence m divides n and thus $m \le n$. It follows that m is the least common multiple of a and b.

Primes

The study of primes has always been an important part of number theory, and early results in this area date back to the Greeks in the fourth century B.C. In this section we present basic facts about prime numbers. The deeper properties of primes will be investigated in Chapter 7. We begin with the definition of a prime number.

(1.14) Definition. A *prime number* is an integer $p > 1$ that has no positive divisors other than 1 and itself. (In other words, p has no *proper divisors*.) An integer greater than 1 that is not prime is called *composite*. (The integer 1 is neither prime nor composite.)

In a certain sense, prime numbers are the building blocks for the integers. The Fundamental Theorem of Arithmetic asserts that every integer greater than 1 can be expressed in an essentially unique way as a product of prime numbers (possibly with repetition). All the tools needed to prove this theorem are present in Euclid's *Elements*, but Gauss, in his *Disquisitiones Arithmeticae* of 1801, was the first to state and prove the theorem. We require the following lemma.

(1.15) Lemma. *If p is prime and $p\,|\,ab$, then $p\,|\,a$ or $p\,|\,b$. In general, if p divides the product $a_1 a_2 \cdots a_r$, then p divides at least one of the a_i.*

Proof. If $p\,|\,ab$ and $p \nmid a$, then $(p,a) = 1$ and so (1.9) implies that $p\,|\,b$. Now suppose that p divides $a_1 a_2 \cdots a_r$. If $p \nmid a_1$, then $p\,|\,a_2 \cdots a_r$. If $p\,|\,a_2 \cdots a_r$ and $p \nmid a_2$, then $p\,|\,a_3 \cdots a_r$, and so on. Thus if p does not divide any of the integers $a_1, \ldots, a_{r-1}$, then p must divide a_r.

(1.16) The Fundamental Theorem of Arithmetic. *Every integer $n > 1$ is a product of primes. The representation is unique, except for the order of the factors.*

Proof. We use proof by contradiction to show that n has at least one such representation. If there is an integer greater than 1 that is not the product of primes, then there must be a smallest such integer, say m; clearly, m is not prime. Thus we can write $m = rs$ with $1 < r < m$ and $1 < s < m$. Since r and s are smaller than m, each must be a product of primes, and therefore m is also, contradicting the definition of m. Thus we conclude that every $n > 1$ is a product of primes (not necessarily distinct).

Now suppose there exist integers greater than 1 with two different factorizations; then there is a smallest such integer, say n, and clearly n is not prime. Assume that n has two essentially different factorizations $n = p_1^{a_1} p_2^{a_2} \cdots p_r^{a_r} = q_1^{b_1} q_2^{b_2} \cdots q_s^{b_s}$, where the p_i are distinct primes and the q_j are distinct primes. Since p_1 divides the right side, the preceding lemma implies that $p_1 \mid q_k$ for some k; hence $p_1 = q_k$, since both are prime. Thus we may divide each side by p_1 to obtain two different factorizations of n/p_1, which contradicts the definition of n since $1 < n/p_1 < n$. Thus we conclude that, apart from the order of the factors, the representation of any integer greater than 1 as a product of primes is unique.

The Fundamental Theorem of Arithmetic allows us to write any integer $n > 1$ in the form $p_1^{a_1} p_2^{a_2} \cdots p_r^{a_r}$, where the primes p_i are distinct and the exponents are positive. This is usually called the *prime factorization* of n. We will often use the notation $\prod_1^r p_i^{a_i}$, or more simply $\prod p_i^{a_i}$, to indicate $p_1^{a_1} p_2^{a_2} \cdots p_r^{a_r}$.

Theorem 1.16 also provides a way of finding the greatest common divisor and least common multiple of two integers. By taking some of the exponents to be zero if necessary, we may use the same primes in the factorization of the two integers, as in the next result.

(1.17) Theorem. *Let $a = p_1^{a_1} p_2^{a_2} \cdots p_r^{a_r}$ and $b = p_1^{b_1} p_2^{b_2} \cdots p_r^{b_r}$, where the a_i and the b_i are nonnegative. For $i = 1, 2, \ldots, r$, define m_i to be the minimum of a_i and b_i, and let M_i denote the maximum of a_i and b_i. Then*

$$(a, b) = p_1^{m_1} p_2^{m_2} \cdots p_r^{m_r} \quad \text{and} \quad [a, b] = p_1^{M_1} p_2^{M_2} \cdots p_r^{M_r}.$$

Theorem 1.17 provides a very easy proof of (1.13): Simply note that $\min(m, n) + \max(m, n) = m + n$. In general, a problem that involves only multiplication (this includes the notions of divisibility, greatest common divisor, and least common multiple) can usually be settled in a straightforward way by using the Fundamental Theorem of Arithmetic.

Having shown that every integer greater than 1 has a prime divisor, we are now in a position to prove that the number of primes is infinite. The proof is extremely simple and appears in Book IX of Euclid's *Elements*.

(1.18) Theorem (Euclid). *There exist infinitely many primes.*

Proof. We will show that given any finite collection of primes, we can always find a prime q that is *not* in the collection. Let $p_1, p_2, \ldots, p_n$ be given primes, and let $N = p_1 p_2 \cdots p_n + 1$. By (1.16), N has a prime divisor q (which could be N itself). If q is one of the p_i, then q divides the product $p_1 p_2 \cdots p_n$, and since q divides N, it follows that q divides their difference, that is, $q \mid 1$. This contradiction establishes the result.

While there are infinitely many primes, it is easy to show that the gap between consecutive primes can be arbitrarily large. (See Problem 1-28.)

We show next how the prime factorization of a positive integer can be used to determine the number of its positive divisors and the sum of these divisors.

(1.19) Definition. If n is a positive integer, let $\tau(n)$ denote the *number of positive divisors of* n, and let $\sigma(n)$ denote the *sum of all of the positive divisors of* n.

In the next theorem, we obtain formulas for $\tau(n)$ and $\sigma(n)$ in terms of the prime factorization of n.

(1.20) Theorem. *Let $n > 1$ and suppose $n = p_1^{n_1} p_2^{n_2} \cdots p_r^{n_r}$. Then*

$$\tau(n) = (n_1 + 1)(n_2 + 1) \cdots (n_r + 1)$$

and

$$\sigma(n) = \frac{(p_1^{n_1+1} - 1)}{p_1 - 1} \frac{(p_2^{n_2+1} - 1)}{p_2 - 1} \cdots \frac{(p_r^{n_r+1} - 1)}{p_r - 1}.$$

Proof. Let $d = p_1^{d_1} p_2^{d_2} \cdots p_r^{d_r}$ be a positive divisor of n; then $d_i \leq n_i$ for each i. There are $n_i + 1$ choices for d_i (namely, $0, 1, \ldots, n_i$), and hence the exponents $d_1, d_2, \ldots, d_r$ can be chosen in precisely $(n_1 + 1)(n_2 + 1) \cdots (n_r + 1)$ ways.

To derive the expression for $\sigma(n)$, note that the product

$$P = (1 + p_1 + p_1^2 + \cdots + p_1^{n_1})(1 + p_2 + p_2^2 + \cdots + p_2^{n_2}) \cdots (1 + p_r + p_r^2 + \cdots + p_r^{n_r}),$$

when multiplied out, is the sum of all possible products $p_1^{a_1} p_2^{a_2} \cdots p_r^{a_r}$, where $0 \leq a_i \leq n_i$ for each i. But the collection of all such products is precisely the set of all positive divisors of n, and thus $\sigma(n) = P$. To complete the argument, note that by the usual formula for the sum of a geometric progression, $1 + p + p^2 + \cdots + p^k = (p^{k+1} - 1)/(p - 1)$. (To prove this, multiply $1 + p + p^2 + \cdots + p^k$ by $p - 1$.)

Note. The functions τ and σ are examples of *number-theoretic functions*, and they have an important property in common: Both τ and σ are *multiplicative functions*, that is,

$$\sigma(mn) = \sigma(m)\sigma(n) \quad \text{and} \quad \tau(mn) = \tau(m)\tau(n)$$

for every pair of relatively prime integers m and n. More generally, a function f defined on the positive integers is *multiplicative* if $f(mn) = f(m)f(n)$ whenever $(m, n) = 1$. The fact that τ and σ are multiplicative can be proved in a straightforward way from the formulas obtained in Theorem 1.20. (Another important example of a multiplicative function is Euler's ϕ-function, which will be introduced in Chapter 3.)

We conclude this section with a useful fact about the prime factorization of factorials.

(1.21) Theorem (Legendre). *Let p be a prime. Then the largest power of p that divides $n!$ is $[n/p] + [n/p^2] + [n/p^3] + \cdots$, where $[x]$ denotes the greatest integer not exceeding x.*

Proof. We want to count the number of factors of p that occur in $n!$. The number of integers among $1, 2, \ldots, n$ that are divisible by p is simply $[n/p]$. Some of these are also divisible by p^2 and therefore contribute an extra factor of p. In particular, there are exactly $[n/p^2]$ positive integers not exceeding n that are divisible by p^2. Similarly, the numbers divisible by p^3 contribute another extra factor of p, and so on. Hence the sum $[n/p] + [n/p^2] + \cdots$ gives the total number of factors of p in $n!$. Note that this sum has only a finite number of nonzero terms, since for a given n, $n/p^k < 1$ for all sufficiently large k and thus $[n/p^k] = 0$.

The Euclidean Algorithm

We now describe the *Euclidean Algorithm*, a systematic and efficient procedure for computing the greatest common divisor of two numbers. Proposition VII.2 of Euclid's *Elements* describes the algorithm and proves that it calculates the greatest common divisor, but the method probably predates Euclid.

The following lemma is the key to understanding the Euclidean Algorithm.

(1.22) Lemma. *Let m and n be integers, not both zero. Then $(m, n) = (n, m - tn)$ for any integer t.*

Proof. If e is any common divisor of m and n, then by (1.2), e divides $m - tn$, and so e is a common divisor of n and $m - tn$. Similarly, if e is any common divisor of n and $m - tn$, then e divides m, and hence e is a common divisor of m and n. It follows, in particular, that the *greatest* common divisor of m and n is the same as the greatest common divisor of n and $m - tn$.

We illustrate the usefulness of the lemma with the following example.

Example. Suppose we want to find $(996, 234)$. By the preceding lemma, with $t = 1$, we have $(996, 234) = (234, 996 - 234)$; the gcd has not changed, but one of the numbers is smaller. We could continue like this, continually subtracting the smaller number from the larger (this was Euclid's wording), but there is a quicker way. Divide 996 by 234, obtaining the quotient 4 and remainder 60; thus $996 = 4 \cdot 234 + 60$. By (1.22), with $t = 4$, we then have $(996, 234) = (234, 996 - 4 \cdot 234) = (234, 60)$. Now apply the lemma again, to $(234, 60)$. Divide 234 by 60 to obtain the quotient 3 and remainder 54; thus $(234, 60) = (60, 54)$. Applying (1.22) once more, with $t = 1$, we find that $(60, 54) = (54, 6)$. We can either apply (1.22) again to get $(54, 6) = (6, 0) = 6$, or simply note that $(54, 6) = 6$ since $6 \mid 54$. (If we are doing calculations by hand, we can stop as soon as the gcd becomes obvious.)

What makes the preceding computation work is that while the numbers that we are considering change (indeed, decrease quite rapidly), their greatest common divisor *does not change*. Now consider the problem in general. For any integer c, the divisors of c are precisely the same as the divisors of $-c$; hence we always have $(a, b) = (|a|, |b|)$. Thus, in calculating (a, b), we may assume that a and b are nonnegative. Without loss of generality, we will suppose that $a \geq b > 0$.

Apply repeatedly the reduction procedure illustrated in the preceding example. If at a certain stage we are trying to find (m, n), where $m \geq n$ and $n \neq 0$, let r be the remainder when m is divided by n; thus $m = qn + r$ for some quotient q. Then $r = m - qn$ and therefore $(m, n) = (n, r)$, by (1.22). If $r \neq 0$, apply the procedure again to the pair n, r. If $r = 0$, we stop; the greatest common divisor of m and n is in this case equal to n.

It is clear that the procedure described above must terminate: At each step, the smaller of the two numbers we are considering decreases by at least one and thus must reach zero in at most b steps. In fact, the Euclidean Algorithm terminates much faster than that.

(1.23) Euclidean Algorithm. Suppose a and b are positive, with $a \geq b$. To find (a, b), first set $m = a$ and $n = b$, and let r be the remainder when m is divided by n. If $r \neq 0$, replace m by n and n by r, then repeat the process. If $r = 0$, then $(a, b) = n$.

We show next how to use the Euclidean Algorithm to compute integers x and y such that $ax + by = (a, b)$. For notational convenience, let $r_0 = a$ and $r_1 = b$. The Euclidean Algorithm can then be described as follows. Let r_2 be the remainder when r_0 is divided by r_1, r_3 the remainder when r_1 is divided by r_2, and so on. For some k, $r_{k+1} = 0$ and the computation terminates. Then $(a,b) = r_k$, the last nonzero remainder. If q_i is the quotient when r_{i-1} is divided by r_i, we have

$$a = r_0 = q_1 r_1 + r_2, \qquad 0 < r_2 < r_1,$$
$$b = r_1 = q_2 r_2 + r_3, \qquad 0 < r_3 < r_2,$$
$$r_2 = q_3 r_3 + r_4, \qquad 0 < r_4 < r_3,$$
$$\vdots$$
$$r_{k-2} = q_{k-1} r_{k-1} + r_k, \quad 0 < r_k < r_{k-1},$$
$$r_{k-1} = q_k r_k.$$

Start with the next-to-last equation above and solve for r_k in terms of r_{k-1} and r_{k-2}. Now solve the preceding equation for r_{k-1} and substitute. Continue this way, successively eliminating the remainders $r_{k-1}, r_{k-2}, \ldots, r_2$. The final equation then gives r_k, the greatest common divisor of a and b, as a linear combination of a and b. (See Problem 1-3.)

Computational Note. There is a variant of the Euclidean Algorithm, called the *Extended Euclidean Algorithm*, that simultaneously finds integers x and y such that $ax + by = (a,b)$. The procedure is efficient and easy to implement on a computer or programmable calculator.

We apply the algorithm as described in (1.23). Clearly, a and b are linear combinations of a and b. Now suppose that we have expressed m and n as linear combinations of a and b, say $m = as + bt$ and $n = au + bv$. Write $m = qn + r$, where $0 \leq r < n$. If $r = 0$, then $n = (a,b)$, and thus we have found a representation of (a,b) as a linear combination of a and b. If $r \neq 0$, then $r = m - nq = as + bt - q(au + bv) = a(s - qu) + b(t - qv)$ is a linear combination of a and b. Since (a,b) is the last nonzero remainder, we eventually obtain a representation of (a,b) as $ax + by$.

The Equation $ax + by = c$

Systematic methods for producing rational solutions of certain equations appear in the work of Diophantus of Alexandria (c. 250 A.D.). Originally, the term *Diophantine equation* referred to an equation where solutions were restricted to rational numbers, but it now commonly applies to equations where integer solutions are sought. Such equations will be studied in detail in Chapter 8.

In this section, we analyze the integer solutions of the equation $ax + by = c$, since the existence and form of its solutions follow from the previous results of this chapter. In particular, we will show that if such an equation has a solution in integers, then it has infinitely many.

(1.24) Theorem. *The equation $ax + by = c$ has a solution in integers if and only if $d \mid c$, where $d = (a,b)$. If x^*, y^* is any particular solution of the equation, then all solutions are given by*

$$x = x^* + (b/d)t \quad \text{and} \quad y = y^* - (a/d)t, \tag{1}$$

where t is an arbitrary integer.

Proof. Suppose $ax + by = c$ has solutions; since $d \mid a$ and $d \mid b$, we must also have $d \mid c$, by (1.2.iv). Hence there are no solutions in integers unless $d \mid c$.

Now suppose $d \mid c$, and write $c = kd$. By (1.5), there are integers r and s such that $ra + sb = d$, and hence $x^* = rk$, $y^* = sk$ is a solution of $ax + by = c$. It is clear that if x and y are defined as in (1), then $ax + by = c$. Conversely, if x, y is any solution of the equation, then $ax + by = ax^* + by^* = c$, and thus $a(x-x^*) = b(y^*-y)$. Divide each side by d to get $(a/d)(x-x^*) = (b/d)(y^*-y)$. Since a/d divides the right side and $(a/d, b/d) = 1$ (from (1.7.ii)), a/d must divide $y^* - y$, by (1.9). Thus $y^* - y = (a/d)t$ for some integer t, that is, $y = y^* - (a/d)t$. Substituting this in $(a/d)(x - x^*) = (b/d)(y^* - y)$ then gives $x - x^* = (b/d)t$, and so $x = x^* + (b/d)t$.

Note. Define a *lattice point* in the plane to be a point (x,y) where x and y are both integers. Then (1.24) implies that if the straight line $ax + by = c$ (with a, b, c integers) passes through one lattice point, it must pass through infinitely many.

We look finally at conditions on a, b, and c that guarantee that a *positive solution* exists, that is, a solution with both x and y positive. In considering the equation $ax + by = c$, we may suppose that the greatest common divisor d of a and b is 1. (Otherwise, consider the equivalent equation $(a/d)x + (b/d)y = c/d$, where $(a/d, b/d) = 1$.)

(1.25) Theorem. *Let a, b, and c be positive integers, with $(a,b) = 1$, and suppose that x^*, y^* is any solution of $ax + by = c$. Then the number of positive solutions of $ax + by = c$ is the number of integers t for which $-x^*/b < t < y^*/a$. In particular, $ax + by = c$ has at least n positive solutions if $c > nab$.*

Proof. Setting $x > 0$ and $y > 0$ in the general form of the solution yields the inequalities $-x^*/b < t < y^*/a$. Thus the number of positive solutions is the number of integers in this interval. It follows that there will be at least n positive solutions of $ax + by = c$ if $y^*/a - (-x^*/b) > n$. This last inequality holds if and only if $by^* + ax^* > nab$. Since $by^* + ax^* = c$, the result follows.

Let N be the number of positive solutions of $ax + by = c$. If c/ab is an integer, then N is $(c/ab) - 1$. If c/ab is not an integer, then N is either $[c/ab]$ or $[c/ab] + 1$. Thus the number of positive solutions is almost, but not entirely, determined by the quotient c/ab. (Consider, for example, the equations $x + 15y = 23$ and $3x + 5y = 23$. The first equation has one positive solution and the second has two, but $c/ab = 23/15$ in each case.)

PROBLEMS AND SOLUTIONS

Divisibility, Greatest Common Divisor, Least Common Multiple, Euclidean Algorithm

1-1. *Determine the greatest common divisor of 210 and 495, and express it as an integral linear combination of 210 and 495.*

Solution. Use the Euclidean Algorithm: $495 = 2 \cdot 210 + 75$, $210 = 2 \cdot 75 + 60$, $75 = 1 \cdot 60 + 15$, $60 = 4 \cdot 15$. Thus $(495, 210) = 15$, the last nonzero remainder. Also, $15 = 75 - 1 \cdot 60 = 75 - 1(210 - 2 \cdot 75) = 3 \cdot 75 - 1 \cdot 210 = 3(495 - 2 \cdot 210) - 1 \cdot 210 = 3 \cdot 495 - 7 \cdot 210$.

1-2. *Use the Euclidean Algorithm to find the greatest common divisor of (a) 271 and 337; (b) 1128 and 1636; (c) 519 and 1730.*

Solution. (a) $337 = 1 \cdot 271 + 66$, $271 = 4 \cdot 66 + 7$, $66 = 9 \cdot 7 + 3$, $7 = 2 \cdot 3 + 1$; thus $(271, 337) = 1$, the last nonzero remainder.
    (b) $1636 = 1 \cdot 1128 + 508$, $1128 = 2 \cdot 508 + 112$, $508 = 4 \cdot 112 + 60$, $112 = 1 \cdot 60 + 52$, $60 = 1 \cdot 52 + 8$, $52 = 6 \cdot 8 + 4$, $8 = 2 \cdot 4$; thus $(1128, 1636) = 4$.
    (c) $1730 = 3 \cdot 519 + 173$, $519 = 3 \cdot 173$, and so $(519, 1730) = 173$.

1-3. *Find the greatest common divisor of 1769 and 2378, and express it as a linear combination of these two numbers.*

Solution. $2378 = 1 \cdot 1769 + 609$, $1769 = 2 \cdot 609 + 551$, $609 = 1 \cdot 551 + 58$, $551 = 9 \cdot 58 + 29$, and $58 = 2 \cdot 29$. So $(1769, 2378) = 29$, the last nonzero remainder. Then $29 = 551 - 9 \cdot 58 = 551 - 9(609 - 1 \cdot 551) = 10 \cdot 551 - 9 \cdot 609 = 10(1769 - 2 \cdot 609) - 9 \cdot 609 = 10 \cdot 1769 - 29 \cdot 609 = 10 \cdot 1769 - 29(2378 - 1 \cdot 1769) = 39 \cdot 1769 - 29 \cdot 2378$.

1-4. *Use the Binary GCD Algorithm described in the Notes at the end of the chapter to find the greatest common divisors of the three pairs of numbers in Problem 1-2.*

Solution. (a) $(271, 337) = (271, 337 - 271) = (271, 66) = (271, 33) = (33, 271 - 33) = (33, 238) = (33, 119) = (33, 119 - 33) = (33, 86) = (33, 43) = 1$. (We stopped computing when the answer became obvious.)
    (b) $(1128, 1636) = 2(564, 818) = 4(282, 409) = 4(141, 409) = 4(141, 268) = 4(141, 134) = 4(141, 67) = 4(74, 67) = 4(37, 67) = 4$.
    (c) $(519, 1730) = (519, 865) = (519, 346) = (519, 173) = (346, 173) = (173, 173) = 173$.

1-5. *Do there exist integers a and b that add to 500 and whose greatest common divisor is 7?*

Solution. No. If $(a,b) = 7$, then $7 \mid a$ and $7 \mid b$, and hence $7 \mid a+b$. But 500 is not divisible by 7.

1-6. *Let a, b, c, and d be positive integers, with $b \neq d$. Show that if a/b and c/d are two fractions in lowest terms (i.e., $(a,b) = 1$ and $(c,d) = 1$), then $a/b + c/d$ cannot be an integer.*

Solution. Suppose to the contrary that $a/b + c/d = n$, where n is an integer. Then $ad + bc = bdn$, i.e., $ad = b(dn - c)$. Thus $b \mid ad$, and hence $b \mid d$ since $(a,b) = 1$. Similarly, we can show that $d \mid b$. Hence $b = d$, a contradiction.

1-7. *Prove that n and $n+1$ are always relatively prime.*

Solution. Any common divisor of n and $n+1$ must divide $(n+1) - n = 1$.

1-8. *Show that $n! + 1$ and $(n+1)! + 1$ are relatively prime. (Hint. Multiply the first number by $n+1$.)*

Solution. If $d > 0$ is a common divisor of the two numbers, then d divides the linear combination $(n+1)(n!+1) - ((n+1)!+1)$, which equals n. But if $d \mid n$ and $d \mid n!+1$, then $d \mid 1$. Hence $d = 1$.

1-9. *Prove that if n is odd, then n and $n-2$ are relatively prime.*

Solution. If $d = (n, n-2)$, then d divides $n - (n-2) = 2$. But since n is odd, $d \neq 2$, so $d = 1$.

1-10. *If $(a,b) = 1$, prove that $(a+b, a-b) = 1$ or 2.*

Solution. Let $d = (a+b, a-b)$; then d divides $(a+b) \pm (a-b)$, i.e., $d \mid 2a$ and $d \mid 2b$. If exactly one of a and b is odd, then $a+b$ and $a-b$ are both odd, so d is odd. Hence $(d, 2) = 1$ and thus $d \mid a$ and $d \mid b$. Since $(a,b) = 1$, we conclude that $d = 1$. If a and b are both odd, then $a+b$ and $a-b$ are even; hence d is even, say, $d = 2e$. Then $d \mid 2a$, $d \mid 2b$ imply that $e \mid a$, $e \mid b$. Hence $e = 1$, and so $d = 2$.

1-11. *Prove or disprove: For every $k \geq 1$, the integers $6k + 5$ and $7k + 6$ are relatively prime.*

Solution. This is true, since $6(7k+6) - 7(6k+5) = 1$. Thus any common divisor of $6k+5$ and $7k+6$ must divide 1.

1-12. *If $(a,b) = 1$ and c divides $a+b$, prove that $(a,c) = (b,c) = 1$.*

Solution. Let $d = (a,c)$; then $d \mid c$ implies $d \mid a+b$. Since $d \mid a$, we also have $d \mid b$. Thus $d = 1$. A similar argument shows that $(b,c) = 1$.

1-13. Show that if $(b,c) = 1$ and $m \mid b$, then $(m,c) = 1$.

Solution. Let $d = (m,c)$; then $d \mid c$ and $d \mid m$. Since $m \mid b$, we also have $d \mid b$, and so d is a common divisor of b and c. Since the greatest common divisor of b and c is 1, it follows that $d = 1$.

Another proof: By (1.5), there exist integers r and s such that $rb + sc = 1$. Let $b = mk$. Then $(rk)m + sc = 1$, and hence $(m,c) = 1$.

1-14. Show that if b is positive, then exactly (b,n) of the numbers $n, 2n, 3n, \ldots, bn$ are multiples of b.

Solution. Let $d = (b,n)$, and write $n = md$, $b = ad$. Then kn is a multiple of b if and only if km is a multiple of a. But since $(a,m) = 1$, this holds if and only if k is a multiple of a. There are $b/a = d$ such k with $1 \le k \le b$.

▷ **1-15.** The sum of two positive numbers is 5432 and their lowest common multiple is 223020. Find the numbers.

Solution. Let a and b be the two numbers; then $a + b = 5432$ and $[a,b] = 223020$. If p is any prime that divides $[a,b]$ and $a + b$, then since p divides $[a,b]$, it divides at least one of a and b; since p also divides $a + b$, it must divide *both* a and b. Now we can divide a and b by p, thus dividing $[a,b]$ by p. We keep doing this until we obtain two relatively prime numbers A and B. In our case, since $(5432, 223020) = 28$, we obtain $a = 28A$, $b = 28B$, $A + B = 5432/28 = 194$, $[A,B] = 223020/28 = 7965$.

Since $(A,B) = 1$, it follows from (1.13) that $[A,B] = AB$. Now solve the equations $A + B = 194$, $AB = 7965$ for A and B. Since $B = 194 - A$, substitution in the second equation gives $A^2 - 194A + 7965 = 0$. This equation has solutions $A = 59$ and $A = 135$, giving $B = 135$ or $B = 59$. Hence our two numbers are $28 \cdot 59 = 1652$ and $28 \cdot 135 = 3780$.

1-16. Prove that $[ma, mb] = m[a,b]$ if $m \ge 1$.

Solution. Let $s = [ma, mb]$ and $t = [a,b]$; then mt is a multiple of ma and mb, and hence $mt \ge s$. Since s is also a multiple of ma and mb, s/m is a multiple of a and b and so $s/m \ge t$, i.e., $s \ge mt$. Thus $s = mt$.

Another proof: By (1.7), $(ma, mb) = m(a,b)$, and hence (1.13) implies that $[ma, mb] = (ma)(mb)/(ma, mb) = mab/(a,b) = m[a,b]$. (We could also prove the result by looking at the prime factorizations.)

1-17. Show that if d and M are positive integers, then there exist integers a, b such that $d = (a,b)$ and $M = [a,b]$ if and only if $d \mid M$.

Solution. Since any common divisor of two numbers divides their least common multiple, the condition $d \mid M$ is necessary. Suppose then that $d \mid M$. Let $a = d$ and $b = M$. It is clear that $(a,b) = d$ and $[a,b] = M$.

1-18. What is the smallest positive rational number that can be expressed in the form $x/30 + y/36$ with x and y integers?

Solution. Let $x/30 + y/36 = r$. Then $36x + 30y = (30 \cdot 36)r$. To make r positive and as small as possible, we make $36x + 30y$ positive and as small as posssible. The smallest positive value of $36x + 30y$ is $(36, 30) = 6$. Hence the smallest positive value of $x/30 + y/36$ is $6/(30 \cdot 36) = 1/180$. (The same argument shows that the smallest positive value of $x/a + y/b$ is $1/N$, where N is the least common multiple of a and b.)

▷ **1-19.** *Across an eleven-inch-high piece of paper, 21 parallel blue lines are drawn, dividing the paper into 22 strips of equal height. Now 37 parallel red lines are drawn, dividing the paper into 38 strips of equal height. What is the shortest distance between a blue line and a red line?*

Solution. Let $a = 11/22$ and let $b = 11/38$. We want to find positive integers $x \leq 21$ and $y \leq 37$ such that $|x(11/22) - y(11/38)|$ is as small as possible. This will be accomplished if $|19x - 11y|$ is as small as possible. The smallest possible value of $|19x - 11y|$ is clearly 1, since 11 and 19 are relatively prime; it is reached, for example, when $x = 4$ and $y = 7$. This gives a minimum distance of $1/38$.

Primes and Prime Factorization

1-20. *Find the greatest common divisor and least common multiple of $a = 2^3 \cdot 3^2 \cdot 11^4 \cdot 37^3$ and $b = 2^2 \cdot 3 \cdot 5^2 \cdot 7 \cdot 11 \cdot 29 \cdot 37^4$.*

1-21. *What is the least common multiple of the numbers $1, 2, 3, \ldots, 30$?*

Solution. For any prime p, the largest power of p that divides the least common multiple of $1, 2, \ldots, 30$ is the largest power of p dividing at least one of $1, 2, \ldots, 30$. So the answer is $2^4 \cdot 3^3 \cdot 5^2 \cdot 7 \cdot 11 \cdot 13 \cdot 17 \cdot 19 \cdot 23 \cdot 29$.

1-22. *Prove that if $a^3 | b^2$, then $a | b$. Does $a^2 | b^3$ imply $a | b$?*

Solution. For any prime p, let p^m and p^n be the highest powers of p that divide a and b, respectively. Then $a^3 | b^2$ implies that $3m \leq 2n$, and hence $m \leq n$. It follows that $a | b$. If $a^2 | b^3$, then a need not divide b; for example, take $a = 8$, $b = 4$.

1-23. *Let $(a, b) = 10$. Find all possible values of (a^3, b^4).*

Solution. Use (1.7.ii) to write $a = 10A$, $b = 10B$, where $(A, B) = 1$. Thus $a^3 = 1000A^3$, $b^4 = 10000B^4$, and $(a^3, b^4) = 1000(A^3, 10B^4)$. The numbers A^3 and B^4 are relatively prime, for if p is a prime such that $p | A^3$ and $p | B^4$, then $p | A$ and $p | B$, contradicting $(A, B) = 1$. Thus any common divisor of A^3 and $10B^4$ must divide 10. It is easy to arrange for $(A^3, 10B^4)$ to be any of 1, 2, 5, or 10. Thus the possibilities for (a^3, b^4) are 1000, 2000, 5000, and 10000.

1-24. *Prove that $(a^n, b^n) = (a, b)^n$ for any $n \geq 1$. In particular, if $(a, b) = 1$, then $(a^n, b^n) = 1$.*

Solution. Let $a = \prod p_i^{a_i}$ and $b = \prod p_i^{b_i}$. Then by (1.17), $(a, b) = \prod p_i^{m_i}$, where $m_i = \min(a_i, b_i)$; similarly, $(a^n, b^n) = \prod p_i^{k_i}$, where $k_i = \min(na_i, nb_i)$. Since $k_i = n \cdot \min(a_i, b_i) = nm_i$, it follows that $(a^n, b^n) = \prod p_i^{nm_i} = (a, b)^n$.

1-25. *If a^n divides b^n, must a divide b? (Hint. Use the preceding problem.)*

Solution. Yes, because $a^n \mid b^n$ implies $(a^n, b^n) = a^n$. Since $(a^n, b^n) = (a, b)^n$ by the preceding problem, it follows that $(a, b) = a$, that is, $a \mid b$. (We can also prove this by writing $a = \prod p_i^{a_i}$ and $b = \prod p_i^{b_i}$, noting that $a^n \mid b^n$ implies $na_i \leq nb_i$, i.e., $a_i \leq b_i$, for each i, whence $a \mid b$.)

▷ **1-26.** *Let $n > 0$, and suppose n has r distinct prime divisors. Show that there are 2^r ordered pairs (x, y) of relatively prime positive integers such that $xy = n$.*

Solution. We calculate the number of choices for x; once x is chosen, y is determined. We find x by constructing its prime factorization. Consider one by one the r primes that divide n. For such a prime p, we cannot have $p \mid x$ and also $p \mid y$, so either x contains the largest power of p that divides n, or it has no factor of p at all. This gives two choices for each prime and hence 2^r choices in all. (Equivalently, we could say that x is characterized by the set of primes it contains. But any set of r elements has 2^r subsets.)

1-27. *Find all primes p such that $17p + 1$ is a square.*

Solution. Suppose that $17p + 1 = x^2$. Since 17 and p are primes and $17p = x^2 - 1 = (x - 1)(x + 1)$, we must have $x - 1 = 17$, giving $p = x + 1 = 19$. (We cannot have $x + 1 = 17$ since 15 is not prime.)

1-28. *Show that if $n > 1$, then the numbers $n! + 2, n! + 3, \ldots, n! + n$ are all composite. (This shows that there are arbitrarily long sequences of composite numbers.)*

Solution. If $2 \leq i \leq n$, then i divides $n!$, and therefore i divides $n! + i$. Since $n! + i > i$, $n! + i$ is composite.

1-29. *Suppose that p and $p + 2$ are both primes, with $p > 3$. Show that their sum $2p + 2$ is divisible by 12.*

Solution. Since $2p + 2 = 2(p + 1)$, it is enough to show that $p + 1$ is divisible by 6. Since p is odd, $p + 1$ is even and hence divisible by 2. Also, p is of the form $3k + 1$ or $3k + 2$; but if $p = 3k + 1$, then $p + 2 = 3(k + 1)$ is divisible by 3 and hence not prime. We conclude that $p = 3k + 2$ and so $p + 1$ is divisible by 3. Since 2 and 3 divide $p + 1$ and $(2, 3) = 1$, it follows that 6 divides $p + 1$.

1-30. *Prove that any positive integer of the form $4k + 3$ has a prime factor of the same form.*

Solution. Every integer can be written as $4k$, $4k + 1$, $4k + 2$, or $4k + 3$ (by the Division Algorithm), and hence every prime different from 2 must be of the form $4k + 1$ or $4k + 3$. Suppose $N = q_1 q_2 \cdots q_r$, where the q_i are (not necesssarily distinct) odd primes. The product of two numbers of the form $4k + 1$ is also of that form, since $(4m + 1)(4n + 1) = 4(4mn + m + n) + 1$. Hence, if all the q_i were of the form $4k + 1$, their product would also be of that form, contradicting the fact that N is of the form $4k + 3$.

22   CHAPTER 1: DIVISIBILITY AND PRIMES

▷ **1-31.** *Prove that there are infinitely many primes of the form $4k + 3$. (Hint. Consider $N = 4p_1 p_2 \cdots p_n - 1$, where $p_1, p_2, \ldots, p_n$ are primes of this form, and use the preceding problem.)*

Solution. Note that every odd prime is of the form $4k+1$ or $4k+3$. Define N as in the hint. Since $N = 4(p_1 p_2 \cdots p_n - 1) + 3$, N must then have a prime factor q of the form $4k+3$, by the preceding problem. The prime q is not one of the p_i, for otherwise, since $q \mid N$ and $q \mid 4p_1 p_2 \cdots p_n$, we would have $q \mid 1$, a contradiction. Thus we have shown that given any finite set of primes of the form $4k + 3$, we can always find a different prime of this form. Hence there are infinitely many primes of the form $4k + 3$.

Note. The same type of argument can be used to show that there are infinitely many primes of the form $3k + 2$, but it will not show, for example, that there exist infinitely many primes of the form $3k + 1$ or $4k + 1$. These cases will be dealt with in Chapter 5.

1-32. *Let $n = \prod p_i^{n_i}$ be the prime factorization of n. Prove that n is a perfect square if and only if each n_i is even.*

Solution. If each n_i is even, say, $n_i = 2c_i$, then $n = \left(\prod p_i^{c_i}\right)^2$. Now suppose n is a square, say, $n = m^2$. If $m = \prod p_i^{m_i}$, then $n_i = 2m_i$ for each i.

1-33. *Prove that if $(a,b) = 1$ and ab is a kth power, then a and b are each kth powers.*

Solution. Let $a = p_1^{a_1} p_2^{a_2} \cdots p_r^{a_r}$ and $b = q_1^{b_1} \cdots q_s^{b_s}$ be the prime factorizations of a and b; since $(a, b) = 1$, no p_i is a q_j. If $ab = n^k$, the prime divisors of n are clearly just the p_i and q_j. Write $n = p_1^{c_1} \cdots p_r^{c_r} \cdot q_1^{d_1} \cdots q_s^{d_s}$; then $ab = n^k$ implies that $a_i = kc_i$ and $b_i = kd_i$ for each i. Thus $a = p_1^{kc_1} \cdots p_r^{kc_r} = (p_1^{c_1} \cdots p_r^{c_r})^k$ and $b = (q_1^{d_1} \cdots q_s^{d_s})^k$.

1-34. *(a) Let a, b, c be positive integers. Show that if ab, ac, and bc are perfect cubes, then a, b, and c must be perfect cubes.*
(b) Discuss what happens if we replace "perfect cube" by "perfect kth power."

Solution. (a) We use the Unique Factorization Theorem. For any prime p, let p^{a_p} be the largest power of p that divides a, and define b_p and c_p analogously. Then for any prime p, the numbers $a_p + b_p$, $a_p + c_p$ and $b_p + c_p$ are all divisible by 3. Thus $a_p - c_p$ is a multiple of 3; since $3 \mid a_p + c_p$, it follows that $3 \mid 2a_p$ and hence $3 \mid a_p$. Therefore a is a perfect cube, and by symmetry, so are b and c.

(b) The argument of (a) works if we replace "cube" by "kth power," where k is odd. For k even, the argument breaks down, since we cannot conclude that k divides a_p from the fact that $k \mid 2a_p$. In fact, the result is false for k even. For example, let $k = 2m$ and $a = b = c = 2^m$.

1-35. *Let d and k be positive integers. Using the Unique Factorization Theorem, show that if $\sqrt[k]{d}$ is a rational number, then $d = b^k$ for some positive* integer b. In particular, $\sqrt{d}$ is irrational if d is a positive integer that is not a perfect square.

Solution. Suppose that $\sqrt[k]{d} = r/s$, where r and s are positive integers. By taking the kth power of both sides, we obtain $ds^k = r^k$. For any prime p, let p^{d_p} be the largest power of p that divides d. Define similarly s_p and r_p. Matching powers of p in the equation $ds^k = r^k$, we obtain $d_p + ks_p = kr_p$. It follows that d_p is divisible by k for any p, and hence d is a perfect kth power.

The Equation $ax + by = c$

1-36. (a) Find all solutions in integers of $15x + 7y = 210$. (b) Determine the number of solutions in positive integers.

Solution. (a) By inspection, $x = 0$, $y = 30$ is a solution. By (1.24), since 15 and 7 are relatively prime, all solutions are given by $x = 7t$, $y = 30 - 15t$, where t ranges over the integers.

(b) Since $x > 0$, we must have $t \geq 1$; since $y > 0$, we must have $t < 2$. Thus $t = 1$, and there is only one solution in positive integers.

1-37. Find the solutions of the equation $91x + 221y = 1053$. Are there solutions in positive integers?

Solution. Since each coefficient is divisible by 13, the equation is equivalent to $7x + 17y = 81$. By inspection, one solution is $x = 14$, $y = -1$. The general solution is therefore $x = 14 + 17t$, $y = -1 - 7t$. To make y positive, t must be negative, but then x is negative. Thus there are no solutions in positive integers.

1-38. Find all solutions in positive integers of $11x + 7y = 200$.

Solution. Since $(11, 7) = 1$, (1.24) guarantees that integer solutions exist. Note that $11 \cdot 2 - 7 \cdot 3 = 1$, so $11(2 \cdot 200) - 7(3 \cdot 200) = 200$. Hence $x = 400$, $y = -600$ is one solution of $11x + 7y = 200$, and thus, by (1.24), all solutions are given by $x = 400 + 7t$, $y = -600 - 11t$. Setting $x > 0$ and $y > 0$ gives $-400/7 < t < -600/11$, and hence positive solutions occur only for $t = -55, -56$, and -57. Therefore the only positive solutions are $x = 15$, $y = 5$; $x = 8$, $y = 16$; $x = 1$, $y = 27$. (Note that for decreasing values of t, the x-values decrease by 7, which is the coefficient of t in $x = 400 + 7t$, and the y-values increase by 11, the negative of the coefficient of t in $y = -600 - 11t$.)

1-39. Do there exist infinitely many positive integer solutions of $10x - 7y = -17$? Explain.

Solution. Yes. By inspection, $10(-1) - 7 \cdot 1 = -17$, so $x = -1$, $y = 1$ is one solution of the equation. Hence all solutions are given by $x = -1 - 7t$, $y = 1 - 10t$. If $t < -1/7$, then $x > 0$, and if $t < 1/10$, then $y > 0$, and therefore any integer $t \leq -1$ yields a positive solution.

1-40. *Find the smallest positive integer b such that the linear Diophantine equation* $1111x + 704y = 15000 + b$ *has a solution.*

Solution. Since $(1111, 704) = 11$, it follows from (1.24) that solutions exist if and only if 11 divides $15000 + b$. The smallest positive value of b is thus 4.

1-41. *Find the smallest number n such that the equation* $10x + 11y = n$ *has exactly nine solutions in nonnegative integers.*

Solution. By inspection, $x = -n$, $y = n$ is a solution for any n, so the general solution is $x = -n + 11t$, $y = n - 10t$. Setting $x \geq 0$ and $y \geq 0$ gives $n/11 \leq t \leq n/10$. The interval $[n/11, n/10]$ has length $n/110$. So if $n = 880$, the allowed values of t go from 80 to 88, giving nine values of t, and since both endpoints are used, no smaller n works.

1-42. *Refer to Theorem 1.25. If* $(a, b) = 1$ *and* $c \leq ab$, *does it follow that* $ax + by = c$ *has no positive solutions?*

Solution. No. Consider $8x + 9y = 43$, which has the positive solution $x = 2$, $y = 3$.

1-43. *Solve the following system of equations in positive integers:* $2x+3y+5z = 201$ *and* $3x + 5y + 7z = 315$.

Solution. Eliminate one of the variables, say x; we get $y - z = 27$. The general solution is $y = 27 + t$, $z = t$. Substituting these expressions into the first equation gives $x = 60 - 4t$. For a positive solution, set $x > 0$, $y > 0$, and $z > 0$, which gives $1 \leq t \leq 14$.

1-44. *Find all solutions of the Diophantine equation* $(6x + 15y)(8x + 7y) = 129$.

Solution. Let $6x + 15y = a$ and $8x + 7y = b$, where $ab = 129$; then $39y = 4a - 3b$, and a is divisible by 3. Since a divides 129 and is a multiple of 3, $a = \pm 129$ or $a = \pm 3$. Since y is not an integer when $a = \pm 129$ and $b = \pm 1$, there can be a solution only if $a = \pm 3$ and $b = \pm 43$; in this case, we get $x = 8$, $y = -3$ and $x = -8$, $y = 3$.

1-45. *A customer buys four dozen pieces of fruit – apples and oranges – for $5.68. If an apple costs ten cents more than an orange and more apples than oranges are purchased, how many pieces of each kind of fruit does the customer buy?*

Solution. Let x denote the number of apples and y the number of oranges purchased, and let c be the price of an orange (in cents). Then $x+y = 48$, $x > y$, and $x(c+10)+yc = 568$. Substituting $y = 48-x$ gives $10x+48c = 568$, that is, $5x+24c = 284$. By inspection, one solution is $c = 1$ and $x = 52$; hence all solutions are given by $x = 52+24t$, $c = 1-5t$. Since $x < 48$, we must have $t < -4/24$; since $x > 24$, $t > -28/24$. But t is an integer, so $t = -1$ is the only possibility. Therefore the only solution is $x = 28$, $y = 20$.

1-46. *A farmer buys 120 head of livestock for $8000. Horses cost $100 each, cows $60 each, and sheep $30 each. If the farmer buys at least one animal of* each type and buys more horses than cows, what is the least number of sheep the farmer could buy?

Solution. Let x, y, and z be the number of horses, cows, and sheep, respectively. Then $x + y + z = 120$ and $100x + 60y + 30z = 8000$, i.e., $10x + 6y + 3z = 800$. Eliminating z gives $7x + 3y = 440$. Since $x = 50$ and $y = 30$ is one solution, the general solution is given by $x = 50 + 3t$, $y = 30 - 7t$. Then $z = 120 - x - y = 40 + 4t$. To ensure that $x > y$, let $50 + 3t > 30 - 7t$, i.e., $t \geq -1$. The number of sheep, namely, $40 + 4t$, is minimized by setting $t = -1$; it follows that the least number of sheep that could have been bought is 36.

1-47. Last week, a child purchased a combined total of 60 candy bars and packages of gum. Altogether she spent $19.26 and bought more candy bars than gum. Each package of gum cost over 20 cents, and each candy bar cost 18 cents more than a package of gum. How many candy bars and how many packages of gum did she buy? How much did she pay for each candy bar?

Solution. Let x be the number of candy bars purchased, y the number of packages of gum, and c the cost of a package of gum (in cents). Then $x+y = 60$ and $(c+18)x+cy = 1926$, i.e., $18x+60c = 1926$. Thus $3x+10c = 321$; since $x = 7$ and $c = 30$ is one solution, the general solution is $x = 7 + 10t$, $c = 30 - 3t$. Since $x > y$, we have $30 < x \leq 60$ and therefore $2.3 < t \leq 5.2$. Hence $t = 3$, 4, or 5, and since $c > 20$, the only possible value is $t = 3$. Thus she bought 37 candy bars and 23 packages of gum, and each candy bar cost 39 cents.

1-48. ("Hundred Fowls Problem"; Chang Ch'in Chien, fifth century.) *A cock is worth five ch'ien, a hen three ch'ien, and three chicks one ch'ien. With 100 ch'ien we buy 100 of them. How many cocks, hens, and chicks are there?*

Solution. Let x be the number of cocks, y the number of hens, and z the number of chicks. Then $x + y + z = 100$ and $15x + 9y + z = 300$. We eliminate z and obtain $14x + 8y = 200$, i.e., $7x + 4y = 100$. By inspection, this has the solution $x = 0$, $y = 25$. So the general solution is $x = 4t$, $y = 25 - 7t$, and hence $z = 75 + 3t$. All of these must be nonnegative, so the only possibilities for t are 0, 1, 2, or 3.

1-49. *One egg timer can time an interval of exactly 5 minutes, and a second can time an interval of exactly 11 minutes. How can we boil an egg for exactly 3 minutes?*

Solution. Note that $5 \cdot 5 - 11 \cdot 2 = 3$. Start both timers simultaneously. When either timer expires, reset it. When the 11-minute timer ends its second cycle, put the egg in, and when the 5-minute timer ends its fifth cycle, remove the egg. (The same technique shows that we can time any integral number of minutes by using an a-minute timer and a b-minute timer if a and b are relatively prime.)

1-50. *Let d and e be positive integers. Show that the two arithmetic progressions $a, a + d, a + 2d, \ldots$ and $b, b + e, b + 2e, \ldots$ have a number in common if and only if (d, e) divides $b - a$.*

Solution. The two progressions have an element in common if and only if there exist non-negative integers r and s such that $a + rd = b + se$, i.e., $rd - se = b - a$. This certainly cannot happen unless (d, e) divides $b - a$.

If (d, e) divides $b - a$, then the equation $dx - ey = b - a$ has solutions, by (1.24). The usual formula for the solutions shows that there are solutions with x arbitrarily large; but if $x > (b - a)/d$, then y must be positive. (Thus we have also shown that if the two progressions have a number in common, they have infinitely many numbers in common.)

Miscellaneous Problems

1-51. *Prove that the last nonzero digit of $n!$ is always even if $n \geq 2$.*

Solution. From (1.21), it is clear that if 2^a and 5^b are the largest powers of 2 and 5 that divide $n!$, then $a > b$ since $[n/5^k] \leq [n/2^k]$ for all positive k and $[n/5] < [n/2]$. We can write $n! = 2^a 5^b m$, where $(m, 10) = 1$. Then the greatest power of 10 dividing $n!$ is 10^b, and since $n!/10^b = 2^{a-b} \cdot m$ is even, the result follows.

1-52. *Find the largest power of 15 that divides 60!.*

Solution. Applying (1.21) to the prime factors 3 and 5 of 15 will obviously give a smaller maximum exponent for 5 than for 3. In fact, the largest power of 5 dividing 60! is $[60/5] + [60/25] = 12 + 2 = 14$. Since the largest exponent for 3 is at least 14, it follows that 15^{14} is the largest power of 15 that divides 60!.

1-53. *How many zeros does 169! end in?*

Solution. This is equivalent to finding the largest power of 10 that divides 169!. As in Problem 1-51, it suffices to apply (1.21) to the prime 5, obtaining $[169/5] + [169/25] + [169/125] = 33 + 6 + 1 = 40$. Thus 169! ends in 40 zeros.

1-54. *How many zeros does 500!/200! end in?*

Solution. The largest power of 10 that divides 500! is $[500/5]+[500/25]+[500/125] = 124$ (see Problem 1-51); similarly, the largest power of 10 dividing 200! is $[200/5]+[200/25]+[200/125] = 49$. Thus 500!/200! ends in $124 - 49 = 75$ zeros.

1-55. *Find all positive integers n such that $n!$ ends in exactly 40 zeros.*

Solution. The integer n will end in precisely 40 zeros if and only if the largest power of 5 that divides $n!$ is 5^{40} (see Problem 1-51). Thus, using (1.21), we want n such that $M_n = 40$, where $M_n = [n/5]+[n/25]+[n/125]+\cdots$. If $n = 125$, then $M_n = 31$; if $n = 200$, then $M_n = 49$. Thus if $M_n = 40$, then $125 < n < 200$. Write $n = 125+25s+5u+v$, where $s = 0, 1,$ or 2, $0 \leq u \leq 4$, and $0 \leq v \leq 4$. Then $M_n = 1+(5+s)+(25+5s+u) = 31+6s+u$; thus $M_n = 40$ if and only if $6s + u = 9$. Since $u \leq 4$, we must have $s = 1$ and $u = 3$, and therefore $n = 165 + v$ for $v = 0, 1, \ldots, 4$. Thus $n!$ ends in exactly 40 zeros for $n = 165, 166, 167, 168,$ and 169.

1-56. *Can n! end in exactly 247 or 248 zeros? Explain.*

Solution. As in Problem 1-51, it suffices to consider the largest power of 5 that divides $n!$. For $n = 1000$, $[1000/5] + [1000/25] + [1000/125] + [1000/625] = 249$, but the corresponding sum for $n = 999$ is 246. Thus $n!$ cannot end in either 247 or 248 zeros.

▷ **1-57.** *Suppose that n! ends in exactly M_n zeros. Show that M_n is approximately n/4 for large values of n.*

Solution. The argument used in the solution of Problem 1-51 shows that M_n is equal to the largest power of 5 that divides $n!$. By (1.21), $M_n = [n/5] + [n/5^2] + [n/5^3] + \cdots$; then there are no more than $\log_5 n$ nonzero terms in the sum. Let $S_n = n/5 + n/5^2 + n/5^3 + \cdots$. Comparing the two sums term by term, we find that $0 < S_n - M_n < 1 + \log_5 n$. By the usual formula for the sum of a geometric series, $S_n = n/4$. Dividing through by n, we find that $1/4 - (1 + \log_5 n)/n < M_n/(n/4) < 1/4$. Since $\log_5 n$ grows much more slowly than n, the ratio $M_n/(n/4)$ can be made arbitrarily close to $1/4$ by taking n large enough, and in this sense, M_n is approximately $n/4$.

▷ **1-58.** *(a) Show that $[x] + [y] \le [x + y]$ for any real numbers x and y.*

(b) Use part (a) and (1.21) to show that n! divides the product of any n consecutive positive integers. (Hint. Let the integers be $m+1, m+2, \ldots, m+n$. Apply (a) with $x = m/p^k$ and $y = n/p^k$.)

Solution. (a) Let $x = a + s$ and $y = b + t$, where a and b are integers and $0 \le s, t < 1$. Then $[x] + [y] = a + b$. Since $x + y = a + b + s + t$, we have $[x + y] \ge a + b$, and the result follows.

(b) The product of the n consecutive integers starting with $m+1$ is just $(m+n)!/m!$. For any prime p, let p^d be the largest power of p that divides $n!$ and p^e the largest power of p that divides $(m+n)!/m!$. To prove (b), it suffices to show that $d \le e$. By (1.21), $d = [n/p] + [n/p^2] + \cdots$, while

$$e = \left([(m+n)/p] + [(m+n)/p^2] + \cdots\right) - \left([m/p] + [m/p^2] + \cdots\right).$$

If we use $x = m/p^k$ and $y = n/p^k$ in part (a), it follows that $[n/p^k] \le [(m+n)/p^k] - [m/p^k]$ for all k, and therefore $d \le e$.

Note. An easy proof of part (b) can be given by using properties of binomial coefficients: $\binom{m+n}{m}$ is an integer and is equal to $(m+n)!/m!n!$, so $n!$ divides $(m+n)!/m!$. But the idea used in the solution of (b) is powerful. Employing a similar idea, Chebyshev, in 1852, found the first reasonably good estimate for the number of primes less than a given number x.

1-59. *Suppose that a, r, and s are integers, with s positive. If $a < x < a + 1$, prove that $[(x + r)/s] = [(a + r)/s]$.*

Solution. Let $a + r = qs + t$, with $0 \le t \le s - 1$. Then $qs + t < x + r < qs + t + 1$; hence $q \le q + t/s < (x + r)/s < q + (t+1)/s \le q + 1$. Therefore $[(x + r)/s] = q = [(a + r)/s]$.

CHAPTER 1: DIVISIBILITY AND PRIMES

▷ **1-60.** *How many integers strictly between* 500 *and* 2000 *are divisible by both* 3 *and* 7? *How many are divisible by* 3 *or* 7?

Solution. The number of positive integers not exceeding 500 that are divisible by 21 is just $[500/21] = 23$; similarly, there are $[2000/21] = 95$ positive integers not exceeding 2000 that are multiples of 21. (Note that 2000 is not one of them.) Thus there are $95 - 23 = 72$ integers strictly between 500 and 2000 that are divisible by both 3 and 7.

Now let N be the number of integers strictly between 500 and 2000 that are divisible by 3 or 7 (or both). If M_k denotes the number of positive integers not exceeding k that are divisible by 3 or 7, then $M_k = [k/3] + [k/7] - [k/21]$. (We subtract the last term because multiples of 21 have been counted twice, once as a multiple of 3 and once as a multiple of 7.) Thus $N = M_{1999} - M_{500} = 856 - 214 = 642$.

1-61. *Determine the number of integers strictly between* 500 *and* 2000 *that are not divisible by* 3 *or* 7. *(See the preceding problem.)*

Solution. There are 1499 integers strictly between 500 and 2000. By the preceding problem, 642 of these are divisible by 3 or 7 (or both). Hence the given interval contains $1499 - 642 = 857$ integers that are divisible by neither 3 nor 7.

1-62. *Prove or disprove: No square can be of the form* $3k + 2$.

Solution. This is true. By the Division Algorithm, every integer n can be written as $3k$, $3k+1$, or $3k+2$. If $n = 3k$, then n^2 is also of this form. If $n = 3k+1$, then $n^2 = 3K+1$, with $K = 3k^2 + 2k$. And if $n = 3k+2$, then $n^2 = 3K + 1$, where $K = 3k^2 + 4k + 1$.

1-63. *Let N be a number whose decimal expansion consists of 3^n identical digits. Show by induction that $3^n \mid N$.*

Solution. The result is certainly true for $n = 0$. We show that if it is true for $n = k$, then it is true for $n = k+1$. It is enough to deal with the case that all the digits are 1's. So suppose that N has decimal expansion consisting of 3^{k+1} 1's. Then $N = Ma$, where M has 3^k 1's, and $a = 1 + 10^{3^k} + 10^{2 \cdot 3^k}$. By the induction hypothesis, M is divisible by 3^k, while a, which has three 1's in its decimal expansion, is divisible by 3, so N is divisible by 3^{k+1}.

1-64. *When is the product of four consecutive integers a square? (Hint. Note that $x(x+1)(x+2)(x+3) + 1 = (x(x+3) + 1)^2$.)*

Solution. The identity of the hint, which is easily verified by direct computation, makes the problem very simple. For if P is the product of four consecutive integers, then $P + 1$ is a square, so P cannot be a square unless $P = 0$.

Note. The problem of whether a product of two or more consecutive integers can be a square, or more generally a perfect power, has a long history. In 1939, Paul Erdös proved that it cannot be a square. In 1975, Erdös and Selfridge proved that it cannot be an mth power for any $m \geq 2$. (For a proof, see *Classical Problems in Number Theory* by Narkiewicz.)

1-65. *Prove that every integer of the form $8^n + 1$ is composite. (Hint. If k is odd, then $x^k + 1 = (x+1)(x^{k-1} - x^{k-2} + \cdots - x + 1)$.)*

Solution. Apply the formula to conclude that $2^n + 1$ divides $(2^n)^3 + 1 = 8^n + 1$. Since $2^n + 1 \geq 3$, $8^n + 1$ is composite.

▷ **1-66.** *Show that $e = \sum_0^\infty 1/n!$ is irrational. (Hint. Suppose $e = p/q$ with p and q positive integers. Show that $q!e$ and $q! \sum_0^q 1/n!$ are both integers.)*

Solution. Let $S_q = \sum_0^q 1/n!$, and let $R_q = \sum_{q+1}^\infty 1/n!$. Then $q!e = q!S_q + q!R_q$. Clearly, $q!S_q$ is an integer. If $q!e$ were an integer, then $q!R_q$ would also be. We show that $q!R_q < 1$, and hence $q!R_q$ cannot be an integer.

Note that $q!R_q = 1/(q+1) + 1/(q+1)(q+2) + 1/(q+1)(q+2)(q+3) + \cdots$. Thus $q!R_q < 1/2 + 1/4 + 1/8 + \cdots = 1$, and the result follows.

▷ **1-67.** *Prove that $1 + 1/2 + 1/3 + \cdots + 1/n$ is not an integer for any $n > 1$.*

Solution. Let $S = 1 + 1/2 + 1/3 + \cdots + 1/n$, let m be the largest integer such that $2^m \leq n$, and let P be the product of all the odd numbers not exceeding n. Then each term in $2^{m-1}PS$ is an integer except for $2^{m-1}P(1/2^m)$. Hence S cannot be an integer.

The Number and Sum of Divisors

1-68. *Evaluate $\tau(5112)$ and $\sigma(5112)$.*

Solution. Note that $5112 = 2^3 \cdot 3^2 \cdot 71$. By (1.20), $\tau(5112) = 4 \cdot 3 \cdot 2 = 24$ and $\sigma(5112) = (15/1)(26/2)(72) = 14040$.

1-69. *Find (a) $\tau(509)$; (b) $\tau(9!)$; and (c) $\tau(1128)$.*

Solution. Use (1.20). (a) $\tau(509) = 2$, since 509 is prime. (b) $\tau(9!) = \tau(2^7 \cdot 3^4 \cdot 5 \cdot 7) = 8 \cdot 5 \cdot 2 \cdot 2 = 160$. (c) $\tau(1128) = \tau(2^3 \cdot 3 \cdot 47) = 4 \cdot 2 \cdot 2 = 16$.

1-70. *Find (a) $\sigma(509)$; (b) $\sigma(9!)$; and (c) $\sigma(1128)$.*

Solution. (a) $\sigma(509) = 510$, since 509 is prime. (b) By (1.20), $\sigma(9!) = \sigma(2^7 \cdot 3^4 \cdot 5 \cdot 7) = (255/1)(242/2)(24/4)(48/6) = 1481040$. (c) $\sigma(1128) = \sigma(2^3 \cdot 3 \cdot 47) = (15/1)(8/2)(48) = 2880$. (Since 47 is prime, $\sigma(47)$ is clearly $1 + 47 = 48$; there is no need to use the expression $(47^2 - 1)/46$.)

1-71. *Suppose N is the product of the first seven primes. Find $\tau(N)$ and $\sigma(N)$.*

Solution. Since N has seven prime factors, each occurring to the first power, (1.20) implies that $\tau(N) = 2^7 = 128$. Also, by (1.20), $\sigma(N) = \sigma(2 \cdot 3 \cdot 5 \cdot 7 \cdot 11 \cdot 13 \cdot 17) = (3/1)(8/2)(24/4)(48/6)(120/10)(168/12)(288/16) = 1741824$.

CHAPTER 1: DIVISIBILITY AND PRIMES

1-72. *For which integers n is $\tau(n)$ odd?*

Solution. If $n = p_1^{n_1} p_2^{n_2} \cdots p_r^{n_r}$, (1.20) implies that $\tau(n) = (n_1 + 1)(n_2 + 1) \cdots (n_r + 1)$. Thus $\tau(n)$ is odd if and only if each factor $n_i + 1$ is odd, i.e., if and only if each n_i is even. Therefore $\tau(n)$ will be odd if and only if n is a perfect square.

Another proof: For any d, d divides n if and only if n/d divides n. If $d < \sqrt{n}$ is a divisor of n, pair it with n/d. If n is not a perfect square, all positive divisors of n are members of a pair, so $\tau(n)$ is even. If n is a perfect square, then all but $\sqrt{n}$ are members of a pair, so $\tau(n)$ is odd.

1-73. *For which integers n is $\sigma(n)$ odd?*

Solution. Use (1.20). If $n = p_1^{n_1} p_2^{n_2} \cdots p_r^{n_r}$, then $\sigma(n) = P_1 P_2 \cdots P_r$, where it is convenient to write $P_i = 1 + p_i + \cdots + p_i^{n_i}$ (rather than $P_i = (p_i^{n_i+1} - 1)/(p_i - 1)$). If $p_i = 2$, then P_i is odd. If p_i is an odd prime, then P_i is odd if and only if there is an odd number of terms in the above expression for P_i. Thus for P_i to be odd, n_i must be even. Since $\sigma(n)$ is odd if and only if each P_i is odd, it follows that n must be the product of 2^k ($k \geq 0$) and a perfect square.

Another proof: Note that even divisors do not change the evenness or oddness of the sum, so only odd divisors of n matter. If we write $n = 2^k m$, where m is odd, n has the same odd divisors as m. If $d < \sqrt{m}$ is a positive divisor (necessarily odd) of m, pair d with $m/d > \sqrt{m}$. The sum $d + m/d$ is even. If m is not a perfect square, we have accounted for all positive divisors of m, and hence $\sigma(m)$ is even. If m is a perfect square, we have accounted for all but $\sqrt{m}$, and so $\sigma(m)$ is odd. Thus $\sigma(n)$ is odd if and only if n is of the form $2^k N^2$.

1-74. *Classify the positive integers that have precisely (a) two positive divisors; (b) three positive divisors; (c) four positive divisors.*

Solution. (a) Let $n = p_1^{n_1} p_2^{n_2} \cdots p_r^{n_r}$. It is clear from (1.20) that $\tau(n) = 2$ if and only if $r = 1$ and $n_1 = 1$, that is, if and only if n is a prime.

(b) Similarly, if $\tau(n) = 3$, then $(n_1 + 1) \cdots (n_r + 1) = 3$, and so we must have $r = 1$ and $n_1 = 2$. Thus n must be the square of a prime.

(c) Finally, $\tau(n) = 4$ implies that either $n_1 + 1 = 1$, $n_2 + 1 = 4$, or $n_1 + 1 = n_2 + 1 = 2$. In other words, n must have the form p^3 or pq, where p and q are distinct primes.

1-75. *Let $n > 1$. Prove that the product of the positive divisors of n is $n^{\tau(n)/2}$. (Hint. Pair a given divisor d with the divisor n/d.)*

Solution. With each divisor d of n such that $n/d \neq d$, we associate the divisor n/d. This pair has a product equal to n. If n is not a perfect square, then all positive divisors of n are accounted for, and there are $\tau(n)/2$ pairs; hence the product of the positive divisors of n is $n^{\tau(n)/2}$. If n is a perfect square, say $n = m^2$, there are $(\tau(n) - 1)/2$ pairs, with the factor m left unpaired. Thus the product of the positive divisors of n is $n^{(\tau(n)-1)/2} n^{1/2} = n^{\tau(n)/2}$.

1-76. *Prove or disprove: n is prime if and only if $\sigma(n) = n+1$.*

Solution. If n is prime, then n has only two positive divisors, namely, 1 and n, and hence $\sigma(n) = n+1$. Conversely, if n is not prime, then it has a proper divisor d. Thus $\sigma(n) \geq n+d+1 > n+1$.

1-77. *Prove or disprove: For each $k > 1$, there are infinitely many integers that have precisely k positive divisors.*

Solution. This is true. Let $n = p^{k-1}$, where p is a prime. Then $\tau(n) = k$, by (1.20).

1-78. *Is there an integer k such that the equation $\sigma(n) = k$ has infinitely many solutions n?*

Solution. No. If $n > 1$, then clearly $\sigma(n) \geq n+1$. Thus if $\sigma(n) = k$, we must have $n \leq k-1$, and so the equation $\sigma(n) = k$ has at most $k-1$ solutions.

1-79. *Find all integers $n < 100$ such that $\tau(n) = 12$.*

Solution. If $n = p_1^{n_1} \cdots p_r^{n_r}$, we want $(n_1+1)\cdots(n_r+1) = 12$. Factoring 12 as $4 \cdot 3$ gives $n_1 = 3$, $n_2 = 2$; thus $n = p_1^3 p_2^2$. Since $n < 100$, the only value is $2^3 3^2 = 72$. Using $12 = 2 \cdot 2 \cdot 3$, we get $n_1 = n_2 = 1$ and $n_3 = 2$; hence $n = p_1 p_2 p_3^2$. The only values for n in this case are $3 \cdot 5 \cdot 2^2 = 60$, $7 \cdot 3 \cdot 2^2 = 84$, and $2 \cdot 5 \cdot 3^2 = 90$. Factoring 12 as $6 \cdot 2$ yields $n = p_1^5 p_2$, and $2^5 \cdot 3 = 96$ is the only number of this form less than 100. Finally, writing 12 as $12 \cdot 1$ gives no value of n under 100. Thus the only $n < 100$ such that $\tau(n) = 12$ are 60, 72, 84, 90, and 96.

1-80. *Find an integer n such that $\sigma(n) = 36$.*

Solution. If $n = p_1^{n_1} p_2^{n_2} \cdots p_r^{n_r}$, then $\sigma(n) = P_1 P_2 \cdots P_r$, where $P_i = 1 + p_i + p_i^2 + \cdots + p_i^{n_i}$ (see the proof of (1.20)). If $\sigma(n) = 36$, then each P_i must divide 36. If we look at divisors of 36, it is clear that 1, 2, 9 and 36 cannot be expressed in the form P_i, but $3 = 1+2$, $4 = 1+3$, $6 = 1+5$, $12 = 1+11$, and $18 = 1+17$, and each of these has only one representation as a P_i. Now examine all possible ways of expressing 36 as a product. The only factorization that yields a solution is $36 = 3 \cdot 12$, in which case $p_1 = 2$, $n_1 = 1$ and $p_2 = 11$, $n_2 = 1$. Thus the only solution is $n = 2 \cdot 11 = 22$.

1-81. *Find all integers n such that $\sigma(n) = 72$.*

Solution. Argue as in the previous problem, noting that the only divisors of 72 that do not have a representation as $1 + p + p^2 + \cdots + p^k$ (p prime) are 1, 2, 9, and 36. Thus the factorizations of 72 which yield solutions are $4 \cdot 18$, $3 \cdot 24$, $6 \cdot 12$, and $1 \cdot 72$, which give, respectively, the solutions $3 \cdot 17 = 51$, $2 \cdot 23 = 46$, $5 \cdot 11 = 55$, and 71.

1-82. *Do there exist values of n for which $\sigma(n) = 51$? Explain.*

Solution. No. The numbers 17 and 51 are not of the form $1+p+p^2+\cdots+p^k$ (p prime), and the only proper factorization of 51 is $3 \cdot 17$. Therefore there are no n such that $\sigma(n) = 51$.

1-83. Find all n such that (a) $\sigma(n) = 42$; (b) $\sigma(n) = 91$.

Solution. (a) Of the divisors of 42, only 3, 6, 7, and 14 have a representation as $1 + p + p^2 + \cdots + p^k$, where p is prime (see the solution of Problem 1-80). Hence the only factorizations of 42 that yield a solution are $3 \cdot 14$, which gives $n = 2 \cdot 13 = 26$, and $6 \cdot 7$, which gives $n = 5 \cdot 2^2 = 20$ (since $7 = 1 + 2 + 2^2$).

(b) Since 91 cannot be represented as $1 + p + p^2 + \cdots + p^k$, we need only consider the factorization $91 = 7 \cdot 13$. Note that $7 = 1 + 2 + 2^2$ and $13 = 1 + 3 + 3^2$. Thus $n = 2^2 \cdot 3^2 = 36$.

1-84. (Euclid, Proposition IX.36.) *Suppose that both p and $2^p - 1$ are primes, and let $n = 2^{p-1}(2^p - 1)$. Prove that $\sigma(n) = 2n$. (Integers for which $\sigma(n) = 2n$ are called perfect numbers. They will be studied in detail in Chapter 7.)*

Solution. Let $q = 2^p - 1$. Then all positive divisors of n are of the form 2^k or $2^k q$, where $0 \le k \le p - 1$. Thus $\sigma(n) = 1 + 2 + 2^2 + \cdots + 2^{p-1} + q + 2q + 2^2 q + \cdots + 2^{p-1} q = (q+1)(1 + 2 + 2^2 + \cdots + 2^{p-1}) = (q+1)(2^p - 1) = 2^p(2^p - 1) = 2n$.

Another proof: Using the fact that σ is multiplicative and $(2^{p-1}, 2^p - 1) = 1$, we have $\sigma(n) = \sigma(2^{p-1} q) = \sigma(2^{p-1})\sigma(q) = (2^p - 1)(q + 1) = q 2^p = 2n$.

EXERCISES FOR CHAPTER 1

1. Prove or disprove: If $(m, n) = 1$, then $(m + n, mn) = 1$.
2. Suppose that $(m, 6) = (n, 6) = 3$. Prove that $(m + n, 6) = 6$.
▷ 3. Suppose that $(m, n) = 1$. Prove that the greatest common divisor of $m + 2n$ and $n + 2m$ is either 1 or 3.
4. Find all positive integers m and n for which $(m, n) = 8$ and $[m, n] = 200$.
5. Under what conditions will $(m, n) = [m, n]$?
6. Let p be prime. The expression $p^\alpha || a$ means that $p^\alpha | a$ but $p^{\alpha+1} \nmid a$.
   (a) Show that if $p^\alpha || a$ and $p^\beta || b$, then $p^{\alpha + \beta} || ab$.
   (b) Show that if $p^\alpha || a$ and $p^\beta || b$, where $\alpha < \beta$, then $p^\alpha || a + b$. Does the result hold if $\alpha = \beta$?
7. Suppose that $(r, s) = 1$. What values are possible for $(r + s, r^2 + s^2)$?
8. Prove that $7k + 16$ and $3k + 7$ are relatively prime for every $k \ge 1$.
9. Let k be a positive integer. What is the greatest common divisor of $5k + 4$ and $9k - 7$?
10. Prove that if $(a, b) = 1$, then $(a^2 - b^2, 2ab) = 1$ or 2.
11. Show that if $ab' - a'b = \pm 1$, then $(a + a', b + b') = 1$.
12. Is it true that if r divides u and s divides v, then $r + s$ divides $u + v$? Explain.
13. Prove or disprove: If p and q are distinct primes and $pq | k^2$, then $pq | k$.

EXERCISES

14. For which primes p is $7p + 4$ a perfect square?
15. Does there exist a prime p and integers m and n such that $p = m^4 - n^4$?
16. Find seven consecutive positive integers all of which are composite.
17. Calculate (a, b) and $[a, b]$, where $a = 2^3 \cdot 5^2 \cdot 13^3 \cdot 17$, $b = 2 \cdot 7^3 \cdot 13 \cdot 17^2$.
18. Do there exist four positive integers that have no factor in common greater than 1 but such that no two of them are relatively prime?
19. Use induction to prove that $7 | n^7 - n$ for every $n \geq 1$.
20. Prove or disprove: If $r | s + t$ and $(s, t) = 1$, then $(r, s) = (r, t) = 1$.
21. Prove or disprove: If $(r, s) = (u, v) = 1$ and $r/s + u/v$ is an integer, then $s = \pm v$.
22. Find the smallest positive integer n such that $n!$ is divisible by 7^3 but not by 7^4.
23. How many zeros does 83! end in?
24. For which values of n does $n!$ end in 26 zeros?
25. Is it possible for $n!$ to end in precisely 35 zeros?
26. What is the largest power of 11 that divides $(11^7 - 1)!$?
27. How many zeros does $100!/25!$ end in?
28. Find the largest power of 7 that divides $500!$.
29. How many integers strictly between 2000 and 4000 are divisible by neither 5 nor 7?
▷ 30. Is $216^k + 1$ composite for every $k \geq 1$?
31. Use the Euclidean Algorithm to find the greatest common divisor of 4199 and 38437. Express the greatest common divisor as a linear combination of 4199 and 38437.
32. A person buys a total of one hundred 33¢, 39¢, and 47¢ stamps for $39.98. If the number of 39¢ stamps purchased is between 35 and 40, how many stamps of each type were bought?
33. A child has $4.55 in change consisting entirely of dimes and quarters. How many different possibilities are there?
34. Opera tickets sell for either $87, $73, or $57. For a certain performance, 4900 people paid a total of $355,042. Fewer than 2000 of the $87 tickets and fewer than 1000 of the $73 tickets were sold. How many of each type of ticket were purchased?
35. Is there any combination of 50 coins – each being a penny, dime, or quarter – whose total value is $7.50?
36. A person buys $9.90 worth of 20¢ and 50¢ stamps. How many different combinations are possible?

34   CHAPTER 1: DIVISIBILITY AND PRIMES

37. (From Bachet's *Problèmes plaisants et délectables quis se font par les nombres* (1612).) A group of 41 men, women, and children have meals at an inn, and the bill is for 40 sous. If each man pays 4 sous, each woman 3 sous, and children's meals are 3 to a sou, how many men, women, and children are there?

38. (From Euler's *Algebra* (1770).) A farmer lays out the sum of 1770 crowns in purchasing horses and oxen. He pays 31 crowns for each horse and 21 crowns for each ox. How many horses and oxen did the farmer buy?

39. (From Euler's *Algebra*.) I owe my friend a shilling and have about me nothing but guineas, worth 21 shillings each. He has nothing but louis d'ors, valued at 17 shillings each. How must I acquit myself of the debt?

40. (Bhaskara) Two men are equally rich. One has 5 rubies, 5 pearls, and 90 gold coins; the other has 8 rubies, 9 pearls, and 48 gold coins. If rubies cost more than pearls, find the price in gold coins of each kind of gem.

41. Find all solutions of $63x - 37y = 3$. Do positive solutions exist? If so, how many?

42. Find the greatest common divisor of 2^8+1 and $2^{32}+1$. Express the greatest common divisor as a linear combination of these numbers.

43. Do there exist two integers a and b such that $a/29 + b/37 = 39/3219$?

44. Find a linear combination of 29 and 313 that equals 1.

45. Express 1 as a linear combination of the relatively prime numbers 1895 and 1801.

46. Let d be the greatest common divisor of 20785 and 44350. Find integers x and y such that $20785x + 44350y = d$.

47. Can 21 be expressed as a linear combination of 5278 and 4508?

48. Let a and b be positive integers, and let $d = (a,b)$. Show that there exist positive integers u and v such that $au - bv = d$.

49. How many solutions in positive integers are there for the equation $101x + 99y = 30000$?

50. Find all integer solutions of the following system of equations:

$$2x + 5y - 11z = 1$$
$$x - 12y + 7z = 2.$$

51. (a) Find the greatest common divisor of 791 and 1243.
    (b) Decide whether the Diophantine equation $791x + 1243y = 2825$ has a solution. If so, find the general solution.

52. Calculate $\tau(857500)$ and $\sigma(857500)$.

53. Find $\tau(13!)$ and $\sigma(13!)$.

54. Let n be a positive integer. How many ordered pairs (x, y) of positive integers satisfy the equation $1/x + 1/y = 1/1200$? (Hint. Show that the equation is equivalent to $(x - 1200)(y - 1200) = 1200^2$.)

55. Prove or disprove: If $\sigma(n)$ is prime, then n is a power of a prime.

▷ 56. Prove that $\tau(n) \leq 2\sqrt{n}$ for every $n \geq 1$.

57. Calculate $\sigma(330)$, $\sigma(24500)$, and $\sigma(10!)$.

▷ 58. Let m and n be positive integers. Prove that $\tau(mn) \leq \tau(m)\tau(n)$. (Hint. First prove for the case where m and n are powers of the same prime.)

59. Prove or disprove: n is the product of k distinct primes if and only if $\tau(n) = 2^k$.

60. Find a positive integer such that $\sigma(n) = \sigma(n + 1)$.

61. What positive integers are divisible by 12 and have exactly 14 positive divisors?

NOTES FOR CHAPTER 1

1. The Least Absolute Remainder Algorithm. The algorithm described by Euclid in Proposition VII.2 of the *Elements* is very close to the procedure described in this chapter. The only difference is that instead of dividing a by b, Euclid continually subtracts b from a until the result falls below a. There is a minor complication caused by the fact that, for Euclid, 1 is not a number.

Euclid's Algorithm, despite its venerable age, is still one of the most efficient ways known to find the greatest common divisor, but there is a somewhat faster procedure, which we describe next.

Recall that the Euclidean Algorithm works because $(m, n) = (n, m - tn) = (n, s)$, where $s = m - tn$. If, as in (1.23), we let s be the remainder when m is divided by n, then we are successively seeking the gcd of smaller and smaller numbers, until the problem becomes trivial. Another reasonable choice for s is the number of the form $m - tn$ which has *least absolute value*. Divide m by n as usual and let the remainder be r, where $0 \leq r \leq |n|$. If $r \leq |n|/2$, take $s = r$; otherwise, let $s = r - |n|$. This ensures that $|s| \leq |n|/2$.

It is easy to see that this variant of the Euclidean Algorithm will produce the greatest common divisor. Like the Euclidean Algorithm, it can be used to find integers x and y such that $ax + by = (a, b)$. It is intuitively reasonable, and indeed true, that the Least Absolute Remainder Algorithm never requires more division steps than the Euclidean Algorithm, and it can be considerably better. For example, it takes 15 steps to calculate $(1597, 987)$ using the Euclidean Algorithm, but the Least Absolute Remainder Algorithm takes only eight steps.

2. The Binary Greatest Common Divisor Algorithm. This algorithm is based on an entirely different idea. Let m and n be positive.

(i) If m and n are both even, then $(m,n) = 2(m/2, n/2)$.

(ii) If m and n are both odd, with $m > n$, then $(m,n) = (m-n, n)$.

(iii) If one of m or n is even (say m) and the other is odd, then $(m,n) = (m/2, n)$.

(iv) If $m = n$, then $(m,n) = m$.

Since $m - n$ is even if m and n are odd, we are dividing by 2 at least every second step, so the algorithm terminates quite rapidly. The Binary GCD Algorithm is particularly efficient on a *binary* computer. Division is a fairly slow operation, and divisions account for most of the time spent in running the Euclidean Algorithm. On a binary computer, however, division by 2 is fast (simply remove the final 0 in the binary representation of the number).

The Binary GCD Algorithm can be extended in a straightforward way to produce integers x and y such that $ax + by = (a,b)$. (This observation may be new; Knuth and Koblitz, for example, both assert that the algorithm does not extend in this way. See the references at the end of the chapter.)

Without loss of generality, we may assume that a and b are not both even; if they are, apply (i) repeatedly until at least one is odd, obtaining numbers a' and b'. It is clear that if $a'x + b'y = (a', b')$, then $ax + by = (a,b)$.

If we are applying (ii) and have calculated s, t, u, and v such that $as + bt = m$ and $au + bv = n$, then $m - n = a(s-u) + b(t-v)$.

Finally, suppose that we are applying (iii) and have found u and v such that $au + bv = m$. We want to express $m/2$ as a linear combination of a and b. This is trivial if u and v are even, so suppose at least one is odd. A straightforward examination of cases shows that $u+b$ and $v-a$ are even. Thus we can write $m/2$ as a linear combination of a and b, namely, $m/2 = a((u+b)/2) + b((v-a)/2)$.

3. The Fundamental Theorem of Arithmetic. The first *explicit* statement and proof of this theorem is in Gauss's *Disquisitiones Arithmeticae*, but the result is often credited to Euclid, some 2000 years earlier. The key lemma (1.15) is essentially Proposition 30 of Book VII. But the nearest Euclid gets to the Fundamental Theorem is (in modern language) to show that if N is the smallest positive number which is divisible by the primes $p_1, p_2, \ldots, p_k$, then N is not divisible by any other prime.

It is likely that the Fundamental Theorem was not stated explicitly because our experience with factoring makes it too obvious even to notice. That Gauss felt the result needed proof is a tribute to his insight and meticulousness. By the middle of the nineteenth century, mathematicians were exploring integer-like systems in which the analogue of the Fundamental Theorem can fail. Some of these are discussed in Chapter 11.

We now describe an illustrative example, due to David Hilbert (1862–1943), that shows that the Unique Factorization Theorem is less obvious than it seems. Let H consist of all integers of the form $4k+1$. It is easy to see that the product of elements of H also lies in H. If $m > 1$ is an element of H, m is called an H-prime if m has no positive divisors in H other than 1 and itself. Thus, for example, 21 is an element of H that is composite in the ordinary sense but that is an H-prime. (It is true that $21 = 3 \cdot 7$, but these are not elements of H.)

It is not difficult to show that if m is an element of H greater than 1, then m can be expressed as a product of H-primes. But the representation is not necessarily unique; for example, $441 = 9 \cdot 49 = 21 \cdot 21$, and 9, 49, and 21 are all H-primes.

If we investigate further, we can see that a number of our basic results fail. Define the H-gcd of two elements a and b of H as the largest element of H that divides both a and b. If d is the H-gcd of a and b and e is a common divisor of a and b, it is not necessarily true that $e \mid d$. For example, 21 is the H-gcd of $a = 3^2 \cdot 7 \cdot 11$ and $b = 3^3 \cdot 7$, but 9 is also a common divisor of a and b.

BIOGRAPHICAL SKETCHES

Aryabhata was born in 476, probably in what is now the Indian city of Patna. Like most of the early Indian contributors to mathematics, he was primarily an astronomer. Aryabhata and his successors Bhaskara and Brahmagupta developed a very sophisticated mathematical astronomy in which solving linear Diophantine equations played a part. Like his Greek predecessor Ptolemy, Aryabhata gave an accurate value for π (in this case, 3.1416) and computed a table of sines. His most famous work is the *Aryabhatiya*, of which 33 verses are devoted to mathematics, 25 to the reckoning of time and models of planetary motion, and 50 to the study of eclipses. (There was a tradition in India of writing even technical works in verse.) Aryabhata seems to be the first to have solved linear Diophantine equations by a systematic method (essentially the Euclidean Algorithm).

Euclid flourished probably around 300 B.C. He may have studied mathematics in Athens under the successors of Plato, and he is thought to have been the founder of the great school of mathematics in Alexandria. This city, with its enormous library and museum, became the center of scholarship in the classical world. Beside the *Elements*, Euclid wrote books on conic sections (now lost), optics, mathematical astronomy, and music.

A large part of the *Elements* may be a compilation and systematization of work done by earlier mathematicians, in particular Theaetetus and Eudoxus; there are strands that go back to 500 B.C. and the early Pythagoreans. Euclid's

Elements, shorn of the more difficult and interesting parts, was the staple of advanced mathematics instruction up to the eighteenth century. Simplified versions of parts of the *Elements* were used in high schools well into the twentieth century.

REFERENCES

Thomas L. Heath, *The Thirteen Books of Euclid's Elements, Volume II*, Cambridge University Press, Cambridge, England, 1926.

> This is the standard English edition of Euclid's *Elements*. Volume II contains, in particular, the arithmetical books VII–IX. Heath gives extensive technical commentaries on Euclid's text. This text is very uneven, ranging from the classic proof of the infinitude of primes to a pedantic discussion, in 13 propositions, of trivial properties of odd and even numbers.

Donald E. Knuth, *The Art of Computer Programming, Volume 2* (Second Edition), Addison-Wesley, Reading, Massachusetts, 1981.

> This is an indispensable source book for anyone writing number-theoretic computer programs. It contains a beautifully detailed analysis of Euclid's algorithm and a wealth of other information. The material ranges from the elementary to the difficult, all handled in a masterful expository style.

Neal Koblitz, *A Course in Number Theory and Cryptography*, Springer-Verlag, New York, 1987.

> The book focuses on those parts of number theory that are needed in recent work in public key cryptography. Much attention is devoted to number-theoretic algorithms, particularly algorithms for factoring and primality testing. Some of the material is quite advanced, but the book also gives a superb introduction to basic number theory.

CHAPTER TWO
Congruences

In the opening section of *Disquisitiones Arithmeticae*, Gauss introduced his theory of congruences as follows:

> If a number a divides the difference of the numbers b and c, b and c are said to be congruent relative to a; if not, b and c are incongruent. The number a is called the modulus.

In working with congruences, Gauss was concerned only with the remainder obtained when one integer is divided by another. The congruence notation that he introduced makes it much easier to formulate results about divisibility properties and to carry out the necessary calculations.

The notion of congruence is fundamental in modern number theory, but the underlying ideas precede Gauss's work by many centuries. In India, the sixth-century astronomer and mathematician Aryabhata showed how to solve what we now call systems of two linear congruences. The seventh-century mathematician Brahmagupta was concerned with questions about calendars as they related to planetary cycles; this led to complicated problems that can be solved using the methods of this chapter. In the middle of the twelfth century, Bhaskara gave a complete analysis of systems of linear congruences. Congruences are especially useful in calendar problems – for example, in determining the date of Easter or in finding the day of the week for a particular date.

Problems also appear in early Chinese mathematical literature which involve finding numbers that leave specified remainders when divided by a given set of integers. (See Problem 2-43.) The technique used to solve them is known in Chinese as the *Ta-yen* rule. There is a long tradition of such problems, beginning with Sun-Tzu in the third century and culminating in the work of Ch'in Chiu-shao in 1247. The main result is now referred to as the *Chinese Remainder Theorem*. The first statement and proof of this theorem in more or less modern language is due to Leonhard Euler (1707–1783).

RESULTS FOR CHAPTER 2

(2.1) Definition. Let m be a positive integer. If m divides the difference $a - b$ of two integers, we say that *a is congruent to b modulo m* and write $a \equiv b \pmod{m}$. (Otherwise, we say that *a is not congruent to b modulo m* and write $a \not\equiv b \pmod{m}$.) The integer m is called the *modulus*.

If $a \equiv b \pmod{m}$, then b is called a *residue of a modulo m* (and vice versa). When $0 \leq b \leq m - 1$, b is called the *least nonnegative residue of a modulo m*.

Note. It is common now to denote the least nonnegative residue of a modulo m by $a \bmod m$. Thus a is congruent to b modulo m if and only if $a \bmod m = b \bmod m$. Although this notation is certainly helpful, especially in computer programs, we will not use it in what follows.

An equivalent way of defining $a \equiv b \pmod{m}$ is to say that a and b differ by some multiple of m, that is, $a = b + km$ for some integer k. Prior to Gauss, instead of writing "$a \equiv b \pmod{m}$," mathematicians wrote "a is of the form $km + b$." It is still common to say, for example, that a is of the form $4k + 1$ instead of using the congruence notation $a \equiv 1 \pmod{4}$.

In the special case where a is a multiple of m, we have $a \equiv 0 \pmod{m}$. More generally, for a given integer a, let r be the smallest nonnegative integer congruent to a modulo m. Then r is simply the *remainder* when a is divided by m. Thus *two numbers are congruent modulo m if and only if they leave the same remainder when divided by m*.

Since division by m yields as remainder one of $0, 1, 2, \ldots, m - 1$, it follows that *every* integer is congruent modulo m to one of these m numbers. *The remainder for a given a is therefore the least nonnegative residue of a modulo m*. The set $\{0, 1, \ldots, m - 1\}$ is an example of a *complete residue system* modulo m, that is, a collection of m incongruent numbers modulo m such that *every* integer is congruent to exactly one number in the collection. It is clear that any element in a complete residue system can be replaced by any number congruent to it modulo m. (For example, $\{10, -4, 7, 3, 24\}$ is a complete residue system modulo 5.) We will usually work with the complete residue system $\{0, 1, \ldots, m-1\}$. We note here that any m consecutive integers form a complete residue system modulo m, since the remainders of these m integers when divided by m are just the numbers $0, 1, \ldots, m - 1$ in some order.

The following basic facts about congruences are analogous to those that hold for ordinary equations.

(2.2) Theorem. *Let m be a positive integer.*

(i) *If $a \equiv b \pmod{m}$, then $b \equiv a \pmod{m}$.*

(ii) *If $a \equiv b \pmod{m}$ and $b \equiv c \pmod{m}$, then $a \equiv c \pmod{m}$.*

(iii) *If $a \equiv b \pmod{m}$ and $c \equiv d \pmod{m}$, then $a \pm c \equiv b \pm d \pmod{m}$.*

(iv) *If $a \equiv b \pmod{m}$, then $ca \equiv cb \pmod{m}$ for any integer c.*

(v) *For any common divisor c of a, b, and m, $a \equiv b \pmod{m}$ if and only if $a/c \equiv b/c \pmod{m/c}$.*

(vi) *If $ca \equiv cb \pmod{m}$, then $a \equiv b \pmod{m/(c,m)}$. In particular, if c and m are relatively prime, then $ca \equiv cb \pmod{m}$ implies $a \equiv b \pmod{m}$.*

Proof. The proofs of parts (i) to (v) follow directly from the definition of congruence. To prove (vi), suppose that $ca \equiv cb \pmod{m}$; thus $(a-b)c = km$ for some integer k. Let $d = (c,m)$; then $(a-b)c/d = km/d$. Since the integer m/d divides the right side, it must divide the left side as well. But c/d and m/d are relatively prime, by (1.7.ii); hence m/d divides $a-b$, that is, $a \equiv b \pmod{m/d}$.

Note. We *cannot* in general divide each side of a congruence by the same number without also modifying the modulus. For example, the correct congruence $5 \equiv 15 \pmod{10}$ upon division by 5 yields $1 \equiv 3 \pmod{10}$, which is false. The correct congruence is $1 \equiv 3 \pmod{2}$.

(2.3) Theorem. *Let m be positive, and suppose a, b, c, and d are arbitrary integers.*

(i) *If $a \equiv b \pmod{m}$ and $c \equiv d \pmod{m}$, then $ac \equiv bd \pmod{m}$.*

(ii) *If $a \equiv b \pmod{m}$, then $a^n \equiv b^n \pmod{m}$ for any positive integer n.*

(iii) *If $f(x)$ is any polynomial with integer coefficients and $a \equiv b \pmod{m}$, then $f(a) \equiv f(b) \pmod{m}$.*

Proof. To prove (i), note that $ac \equiv bc \pmod{m}$ by (2.2.iv) and $bc \equiv bd \pmod{m}$, again by (2.2.iv). Part (ii) follows from (i) by multiplying $a \equiv b \pmod{m}$ repeatedly by itself. Part (iii) follows from (ii), using (2.2.iii) and (2.2.iv).

(2.4) Theorem. *Let m be a positive integer.*

(i) *Suppose $d \mid m$ and $d > 0$. If $a \equiv b \pmod{m}$, then $a \equiv b \pmod{d}$.*

(ii) *If $a \equiv b \pmod{m_1}$ and $a \equiv b \pmod{m_2}$, then $a \equiv b \pmod{[m_1, m_2]}$.*

(iii) *In general, $a \equiv b \pmod{m_i}$ ($i = 1, 2, \ldots, r$) if and only if $a \equiv b \pmod{m}$, where $m = [m_1, m_2, \ldots, m_r]$.*

Proof. Part (i) is obvious from the definition of congruence; (ii) follows from the fact that if $r \mid k$ and $s \mid k$, then their least common multiple $[r, s]$ also divides k. To prove part (iii), note that if $m|a-b$, then $m_i|a-b$ for each i. Conversely, if $m_i|a-b$ for each i, then $[m_1,m_2,\ldots,m_r]|a-b$, by (1.12).

If $m = p_1^{k_1}p_2^{k_2}\cdots p_r^{k_r}$ is the prime factorization of m, then the congruence $a \equiv b \pmod{m}$ is equivalent to the system of congruences $a \equiv b \pmod{p_i^{k_i}}$, ($i = 1,2,\ldots,r$). Thus, for example, to solve the linear congruence $ax + b \equiv 0 \pmod{m}$, it suffices to find solutions x_i of $ax + b \equiv 0 \pmod{p_i^{k_i}}$ for $i = 1,2,\ldots,r$ and then produce a solution of the original congruence using the Chinese Remainder Theorem, which we will discuss shortly.

Divisibility Tests

We pause briefly to give an application of congruences. At the end of Section I in *Disquisitiones Arithmeticae*, Gauss notes that congruences can be used to check for divisibility by certain integers. Historically, tests of this type can be found in the work of the ninth-century mathematician al-Khwarizmi; these results spread to Europe in the Middle Ages. Leonardo of Pisa (c. 1175–1250), better known as Fibonacci, gives tests for divisibility by 7, 9, and 11 in his *Liber Abaci*.

(2.5) Theorem. Let $a_k 10^k + a_{k-1}10^{k-1} + \cdots + a_1 10 + a_0$ be the decimal expansion of the positive integer n. (Thus a_k is the first, or leading digit of n, a_{k-1} the second, ..., and a_0 the last.)

(i) n is divisible by 2^r if and only if the number consisting of the last r digits of n is divisible by 2^r.

(ii) n is divisible by 3 if and only if the sum of the digits of n is divisible by 3, that is, if 3 divides $a_k + a_{k-1} + \cdots + a_0$.

(iii) n is divisible by 9 if and only if the sum of the digits of n is divisible by 9.

(iv) n is divisible by 11 if and only if the alternating sum $a_0 - a_1 + a_2 - \cdots + (-1)^k a_k$ is divisible by 11.

Proof. (i) It is clear that $10^j \equiv 0 \pmod{4}$ for $j \geq 2$, so $n \equiv a_1 10 + a_0 \pmod{4}$. Similarly, $10^j \equiv 0 \pmod{8}$ if $j \geq 3$, so $n \equiv a_2 100 + a_1 10 + a_0 \pmod{8}$. The proof for higher powers of 2 is entirely similar.

To prove (ii) and (iii), note that $10 \equiv 1 \pmod{9}$. Hence $10^j \equiv 1 \pmod{9}$ for every positive integer j. Substituting in the decimal expansion for n now gives $n \equiv a_k + a_{k-1} + \cdots + a_1 + a_0$ modulo 9 (and hence modulo 3), that is, n and the sum of the a_j leave the same remainder when divided by 3 or 9.

Part (iv) is proved by observing that $10^j \equiv (-1)^j \pmod{11}$ for every positive integer j. Now substitute in the decimal expansion of n to conclude that $n \equiv a_0 - a_1 + a_2 - \cdots + (-1)^k a_k \pmod{11}$.

Notes. 1. There is an obvious test for divisibility by 5, namely, the last digit of the number must be 0 or 5. Likewise, an integer is divisible by 10 if and only if it ends in 0.

2. Part (iii) is the basis of the technique known as *casting out nines*, a method for checking computations by comparing remainders modulo 9. The proof of (iii) shows that the *remainder* when n is divided by 9 can be found by adding the digits of n modulo 9.

3. We can test for divisibility by other integers by combining the above tests. For example, to see if a number is divisible by 6, test for 2 and 3; for 15, test for 3 and 5. This works as long as the various moduli are relatively prime in pairs.

4. Divisibility tests were once a practical technique for checking the results of computations. They may seem to have diminished relevance in this age of calculators and computers. But computing devices, and even compact disc players, use sophisticated variants of the old divisibility tests, from simple *parity checks* to complex *error-correcting codes*. The latter are often based on subtle number-theoretic ideas.

Linear Congruences

We next investigate solutions of the *linear congruence* $ax \equiv b \pmod{m}$. Unlike the linear equation $ax = b$, which always has a unique real solution if $a \neq 0$, the congruence $ax \equiv b \pmod{m}$ can have more than one (incongruent) solution or indeed may have no solutions, even if $a \not\equiv 0 \pmod{m}$.

We begin by defining what is meant by a solution to a linear congruence.

(2.6) Definition. An integer s is called a *solution* of $ax \equiv b \pmod{m}$ if $as \equiv b \pmod{m}$. Clearly, if s is a solution and $s \equiv t \pmod{m}$, then t is also a solution. In this case, s and t are considered to be the same solution, and we say that $x \equiv s \pmod{m}$ is a solution of $ax \equiv b \pmod{m}$.

In view of this, to solve a linear congruence, it is enough to substitute the elements of a complete residue system, for example, $\{0, 1, \ldots, m-1\}$.

The following result gives a characterization of the linear congruences that have solutions, as well as a complete description of the solutions.

(2.7) Theorem. *Let $d = (a, m)$. The congruence $ax \equiv b \pmod{m}$ is solvable if and only if $d \mid b$. If solutions exist, there are precisely d incongruent solutions modulo m, given by*

$$x \equiv x^* + (m/d)t \pmod{m} \qquad (t = 0, 1, \ldots, d-1),$$

where x^* is any solution of the congruence

$$(a/d)x \equiv b/d \pmod{m/d}.$$

Proof. If $ax \equiv b \pmod{m}$, then $ax = b + km$ for some integer k. Since $b = ax - km$ and d divides a and m, b must be a multiple of d. Conversely, suppose that $d \mid b$. By (1.24), some linear combination of a and m equals b, say, $ax + my = b$. This implies that $ax \equiv b \pmod{m}$. Hence we have shown that solutions exist if and only if b is a multiple of d.

Now suppose that $ax \equiv b \pmod{m}$ is solvable; then $d \mid b$, and hence $(a/d)x \equiv b/d \pmod{m/d}$ is also solvable. If x^* is a solution of the second congruence, then $(a/d)x^* = b/d + km/d$ for some integer k; thus $ax^* - mk = b$, and therefore x^*, k is a solution of $ax - my = b$. By (1.24), the solutions of this equation have x-values given by $x = x^* + (m/d)t$, where t is an arbitrary integer. But $x^* + (m/d)t_1 \equiv x^* + (m/d)t_2 \pmod{m}$ if and only if $(m/d)t_1 \equiv (m/d)t_2 \pmod{m}$. Dividing by m/d, we obtain the equivalent condition $t_1 \equiv t_2 \pmod{d}$ (see (2.2.v)).

Thus, *incongruent* solutions modulo m are obtained by choosing $t_1 \not\equiv t_2 \pmod{d}$. Clearly, then, all incongruent solutions are obtained by setting $t = 0, 1, 2, \ldots, d - 1$.

(2.8) Corollary. *If $(a, m) = 1$, then the congruence $ax \equiv b \pmod{m}$ has a unique solution for any value of b.*

Solutions of the congruence $ax \equiv 1 \pmod{m}$ are particularly important in the theory. This is reflected in the following definition.

(2.9) Definition. *If a' is a solution of the congruence $ax \equiv 1 \pmod{m}$, then a' is called a (multiplicative)* inverse *of a modulo m.*

By (2.7), a has a multiplicative inverse modulo m if and only if a and m are relatively prime, and the inverse of a, if it exists, is unique modulo m.

Note that the inverse of a modulo m behaves very much like the ordinary reciprocal. In particular, if a' is an inverse of a modulo m, then the congruence $ax \equiv b \pmod{m}$ has $x = a'b$ as a solution. This is strongly analogous to the fact that in ordinary arithmetic, the solution of the equation $ax = b$ is $(1/a)b$. Both Euler and Gauss used the notation $1/a$ for the solution of the congruence $ax \equiv 1 \pmod{m}$. Because of the danger of confusion with the reciprocal, this notation is no longer used.

Techniques for Solving $ax \equiv b \pmod{m}$

We come now to the problem of how to *find* solutions of the linear congruence $ax \equiv b \pmod{m}$ if solutions exist. Various methods are used in the problems for this chapter, and they can be roughly described as follows.

1. We can apply the Euclidean Algorithm to find integers r and s such that $ar + ms = b$, using (1.24) and the fact that $(a, m) \mid b$. It follows at once from this equation that $ar \equiv b \pmod{m}$.

TECHNIQUES FOR SOLVING $ax \equiv b \pmod{m}$

2. There is also the technique of replacing a or b (or both) by integers that are congruent to them modulo m and obtaining a congruence where each side can then be divided by a common factor. Repeating this process will generally produce a congruence that is much easier to solve and whose solution is a solution to the original congruence. While this technique works best if the modulus is not too large, it often produces the solution very efficiently.

3. When the modulus is a prime p, the congruence of x can be multiplied by the nearest integer to p/a, yielding an equivalent congruence. If we take the coefficient of x to be the residue of least absolute value, say a', then $|a'| \leq |a|/2$. By repeated application, the solution can be obtained in no more than n steps, where $n = \log_2 a$.

More generally, we can use the same approach for a nonprime modulus, but extraneous solutions may be introduced that must be checked individually. However, if the method leads to only one solution, then it will be the unique solution of the original congruence. Likewise, if this technique produces no solution, then the original congruence is not solvable.

Note. Any of these techniques can be used to solve the equation $ax+by = c$. We first find a solution r of $ax \equiv c \pmod{b}$; thus $c - ar$ is divisible by b. If we let $s = (c - ar)/b$, then $ar + bs = c$.

(2.10) Examples. We first use the Euclidean Algorithm to solve the congruence $11x \equiv 28 \pmod{1943}$. The algorithm shows that $(11, 1943) = 1$, and back substitution yields $11 \cdot 530 - 1943 \cdot 3 = 1$. Multiplying by 28, we obtain $11 \cdot 14840 - 1943 \cdot 84 = 28$, and therefore $x \equiv 14840 \equiv 1239 \pmod{1943}$ is the unique solution of $11x \equiv 28 \pmod{1943}$.

To illustrate the second technique, consider the congruence $143x \equiv 4 \pmod{315}$. If we replace 4 by 319 and divide by 11, we get $13x \equiv 29 \pmod{315}$. Since $29 \equiv -286 \pmod{315}$, dividing by 13 yields $x \equiv -22 \equiv 293 \pmod{315}$. (This is the only solution, since $(143, 315) = 1$.)

We next solve $519x \equiv 311 \pmod{1967}$ using the third technique described above. First multiply by 4, the nearest integer to $1967/519$, and reduce modulo 1967 to get $109x \equiv -723 \pmod{1967}$. Since $1967/109 = 18.04\ldots$, we now multiply by 18 and obtain $-5x \equiv 755 \pmod{1967}$. Hence $x \equiv -151 \equiv 1816 \pmod{1967}$ is the unique solution of $519x \equiv 311 \pmod{1967}$.

Note. As mentioned above, any of these techniques can be used to express the greatest common divisor of a and m as a linear combination of these two integers. For example, it is easy to see that $(519, 1967) = 1$. If we apply the third technique to the congruence $519x \equiv 1 \pmod{1967}$, we obtain successively $109x \equiv 4 \pmod{1967}$ and $-5x \equiv 72 \equiv -1895 \pmod{1967}$; hence $x \equiv 379 \pmod{1967}$ is the unique solution. It follows that $519 \cdot 379 = 1 + 1967s$ for some integer s, and clearly, $s = (519 \cdot 379 - 1)/1967 = 100$. Thus $(519, 1967) = 1 = 519 \cdot 379 - 1967 \cdot 100$.

The Chinese Remainder Theorem

We now consider the problem of finding a common solution to a system of linear congruences where the moduli are assumed to be relatively prime in pairs. Both Gauss and the Swiss mathematician Leonhard Euler (1707–1783) used the method we describe next, but the idea, known as the *Chinese Remainder Theorem*, appears as early as the third century in the writings of the Chinese mathematician Sun-Tzu.

(2.11) Chinese Remainder Theorem. *Let $m_1, m_2, \ldots, m_r$ be positive integers that are relatively prime in pairs, that is, $(m_i, m_j) = 1$ if $i \neq j$. Then for any integers $a_1, a_2, \ldots, a_r$, the r congruences*

$$x \equiv a_i \pmod{m_i} \quad (i = 1, 2, \ldots, r)$$

have a common solution, and any two solutions are congruent modulo the product $m_1 m_2 \cdots m_r$.

Proof. Let $m = m_1 m_2 \cdots m_r$; then m/m_i is an integer that is relatively prime to m_i (use (1.9) and (1.7.ii)). Thus by (2.8), there exist integers b_i such that $(m/m_i)b_i \equiv 1 \pmod{m_i}$; clearly, for $j \neq i$, we have $(m/m_i)b_i \equiv 0 \pmod{m_j}$. Define

$$x^* = (m/m_1)b_1 a_1 + (m/m_2)b_2 a_2 + \cdots + (m/m_r)b_r a_r.$$

Then $x^* \equiv (m/m_i)b_i a_i \equiv a_i \pmod{m_i}$ for each i, and therefore x^* is a common solution of the given congruences.

If both x^* and y^* are common solutions to the system of congruences, then $x^* \equiv y^* \pmod{m_i}$ for $i = 1, 2, \ldots, r$. Hence $x^* \equiv y^* \pmod{m}$, by (2.4); in other words, any two common solutions differ by a multiple of m.

Example. We use the Chinese Remainder Theorem to find all positive integers less than 5000 that leave remainders of 2, 4, and 8 when divided by 9, 10, and 11, respectively. Thus we must solve the system $x \equiv 2 \pmod 9$, $x \equiv 4 \pmod{10}$, $x \equiv 8 \pmod{11}$. We first find b_1, b_2, b_3 such that $110 b_1 \equiv 1 \pmod 9$, $99 b_2 \equiv 1 \pmod{10}$, and $90 b_3 \equiv 1 \pmod{11}$, that is, $2b_1 \equiv 1 \pmod 9$, $-b_2 \equiv 1 \pmod{10}$, and $2b_3 \equiv 1 \pmod{11}$. We can therefore take $b_1 = 5$, $b_2 = -1$ (or 9), and $b_3 = 6$. It follows from the proof of (2.11) that $x^* = 110(5)(2) + 99(-1)(4) + 90(6)(8) = 5024$ is a solution of the system. In this example, $m = 9 \cdot 10 \cdot 11 = 990$. Since $5024 \equiv 74 \pmod{990}$, all solutions of the system are given by $74 + 990t$, where t is any integer. Thus the only integers between 1 and 5000 that satisfy the given system of congruences are 74, 1064, 2054, 3044, and 4034.

An Application: Finding the Day of the Week

The Julian calendar, introduced in 46 B.C., was used in Western nations until 1582 A.D. It called for a leap day every four years, but this introduced an error that made the Julian calendar gain a day about every 128 years. In 1582, Pope Gregory XIII revised the calendar by dropping 10 days to correct for the accumulated error. Years divisible by 4 are leap years, except those years divisible by 100 but not by 400. Most of Europe adopted the Gregorian calendar at once, but England and its possessions, including the American colonies, did not change over until 1752. Thus dates in England and America before September 14, 1752 refer to the Julian calendar; for most other countries that adopted the Gregorian calendar, the changeover came the day after October 4, 1582.

(2.12) Day of the Week. We can use congruences modulo 7 to determine the day of the week for a given date. We use the following coding scheme: Saturday $= 0$, Sunday $= 1$, Monday $= 2, \ldots$, Friday $= 6$. We also need month codes; for January to December, these are, respectively, $144 - 025 - 036 - 146$. (The codes have been given in groups of three for convenience; note that the first three groups happen to be perfect squares, and the last group closely resembles the first.) To begin with, assume that the date is in the twentieth century, for example, May 19, 1945. Find the quotient when the last two digits of the year, 45, are divided by 4; here, we get 11, which is congruent to 4 modulo 7. Add 45 to the 4 and reduce modulo 7, obtaining 0. We now add to 0 the day of the month, 19, and the month code for May, which is 2, obtaining 0 modulo 7. Thus May 19, 1945 fell on a Saturday.

For dates in January or February of a *leap year* (that is, a year after 1900 whose last two digits are divisible by 4), we must *subtract* 1 in our calculation.

The algorithm is not difficult to justify. In going from one year to the next, the day of the week for a given date advances by one *unless* we cross the leap day February 29, in which case the day advances by two. (This follows from the fact that $365 \equiv 1 \pmod 7$ and $366 \equiv 2 \pmod 7$.) For example, compared to the day of the week for May 19, 1900, May 19, 1945 has advanced 45 days, plus an additional 11 days for the 11 intervening leap years. (It is easy to check that the number of leap years here is simply the quotient of 45 divided by 4, that is, 11.)

Thus it is enough to work with dates in 1900, then adjust as above for any other year. (Note that 1900 was not a leap year.) We will use January 1, 1900, which fell on a Monday, as our reference point. The previous method applied to 1900 gives 0 plus the quotient of 0 divided by 4, namely 0. To this we add the date, 1, and the month code for January, which we will call x, obtaining a sum of $x + 1$. Since the code for Monday is 2, we must have $x + 1 = 2$, that is, $x = 1$. The other month codes are determined as follows. In going from a date in January – say, the 19th – to the same date in February, 31 days have intervened; since $31 \equiv 3 \pmod 7$, it is clear that February 19 is three days later in the week than January 19. Hence the month code for February must be three more than the month code for January, namely 4. Proceeding in this fashion, we establish all of the month codes given above.

Finally, we note that this algorithm can be applied to other centuries, past or future. For dates in the 2000s, subtract 1; for (Gregorian) dates in the 1800s, add 2; in the 1700s, add 4; in the 1600s, add 6; and in the 1500s, from October 15, 1582 to December 31, 1599, add 0. The only caution is to be sure that the date given is for the *Gregorian* calendar; otherwise, our method gives an incorrect result.

To calculate the day of the week for a date in the Julian calendar, use a correction of 18 minus the first two digits of the year (we assume that every year is written with four digits). For example, a Julian date in the 1500s requires a correction of $18 - 15 = 3$, while a date in the 800s needs a correction of $18 - 8 = 10$, or 3 modulo 7.

For examples that use this algorithm, see Problems 2-49 to 2-55.

PROBLEMS AND SOLUTIONS

Note. To use a calculator to find the remainder when a is divided by m, first divide a by m and then subtract the integer part of the result, leaving a decimal less than 1. Now multiply this number by m. The result is the remainder when a is divided by m. The answer should, of course, be a nonnegative integer, since it represents the remainder, but because of roundoff in the calculator, you might, for example, get 47.9999999 or 48.0000001 instead of 48.

General Congruences

2-1. *Find the remainder when 17^{17} is divided by 7.*

Solution. $17 \equiv 3 \pmod 7$, so $17^{17} \equiv 3^{17} \pmod 7$, by (2.3). (But note that it is *not* true that $17^{17} \equiv 3^3 \pmod 7$. Why?) To find 3^{17} modulo 7: $3^2 = 9 \equiv 2 \pmod 7$, thus $3^4 = (3^2)^2 \equiv 4 \pmod 7$. Hence $3^8 \equiv (3^4)^2 \equiv 16 \equiv 2 \pmod 7$, and $3^{16} \equiv 4 \pmod 7$. Thus $3^{17} = 3 \cdot 3^{16} \equiv 12 \equiv 5 \pmod 7$, and 5 is the remainder when 17^{17} is divided by 7. (It is also true that $17^{17} \equiv 12 \pmod 7$, but finding the remainder involves finding the *least* nonnegative residue of 17^{17} modulo 7.)

2-2. *What is the remainder when 4^{30} is divided by 23?*

Solution. Since $4^3 = 64 \equiv -5 \pmod{23}$, we have $4^6 \equiv (-5)^2 \equiv 2 \pmod{23}$. Hence $4^{30} \equiv 2^5 \equiv 9 \pmod{23}$. Thus the remainder is 9.

2-3. *Show that $2^{37} - 1$ is a multiple of 223.*

Solution. Since $2^8 \equiv 33 \pmod{223}$, we have $2^{16} \equiv 33^2 \equiv -26 \pmod{223}$; thus $2^{32} \equiv (-26)^2 \equiv 7 \pmod{223}$. Hence $2^{37} = 2^{32} \cdot 2^5 \equiv 7 \cdot 32 \equiv 1 \pmod{223}$.

2-4. *Find the least positive residue of (a) 3^{500} modulo 13; (b) 12! modulo 13; (c) 5^{16} modulo 17; (d) 5^{500} modulo 17.*

Solution. (a) Since $3^3 \equiv 1 \pmod{13}$, we have $3^{498} = (3^3)^{166} \equiv 1 \pmod{13}$. Thus $3^{500} = 3^{498} \cdot 3^2 \equiv 1 \cdot 9 \equiv 9 \pmod{13}$.

(b) $12! = (2 \cdot 3 \cdot 4)(5 \cdot 6)(7 \cdot 8)(9 \cdot 10)(11 \cdot 12) \equiv (-2)(4)(4)(-1)(2) \equiv 12 \pmod{13}$.

(c) $5^2 \equiv 8 \pmod{17}$ implies that $5^4 \equiv 8^2 \equiv -4 \pmod{17}$. Thus $5^8 \equiv 16 \equiv -1 \pmod{17}$ and so $5^{16} \equiv 1 \pmod{17}$.

(d) By (c), $5^{16} \equiv 1 \pmod{17}$, so $5^{496} = (5^{16})^{31} \equiv 1^{31} \pmod{17}$. Hence $5^{500} = 5^{496} \cdot 5^4 \equiv 1 \cdot 5^4 \equiv 13 \pmod{17}$.

2-5. *What are the remainders when 3^{40} and 43^{37} are divided by 11?*

Solution. Since $3^2 \equiv -2 \pmod{11}$, we have $3^4 \equiv 4 \pmod{11}$ and thus $3^8 \equiv 5 \pmod{11}$. Squaring again gives $3^{16} \equiv 3 \pmod{11}$, then $3^{32} \equiv -2 \pmod{11}$, and so $3^{40} = 3^{32} \cdot 3^8 \equiv (-2)(5) \equiv 1 \pmod{11}$. Also, since $43 \equiv -1 \pmod{11}$, we have $43^{37} \equiv (-1)^{37} \equiv -1 \equiv 10 \pmod{11}$. Thus 3^{40} leaves a remainder of 1 and 43^{37} a remainder of 10 when divided by 11.

2-6. *Show that $2^{48} - 1$ is divisible by 97.*

Solution. $2^8 \equiv 62 \equiv -35 \pmod{97}$ implies that $2^{16} \equiv (-35)^2 \equiv 61 \equiv -36 \pmod{97}$; thus $2^{32} \equiv (-36)^2 \equiv 35 \pmod{97}$, and hence $2^{48} = 2^{32} \cdot 2^{16} \equiv 35(-36) \equiv -1260 \equiv -96 \equiv 1 \pmod{97}$. Therefore 97 divides $2^{48} - 1$.

2-7. *Show that 47 divides $5^{23} + 1$.*

Solution. Since $5^4 \equiv 14 \pmod{47}$, it follows that $5^8 \equiv 8 \pmod{47}$ and $5^{16} \equiv 17 \pmod{47}$. Hence $5^{24} = 5^{16} \cdot 5^8 \equiv 17 \cdot 8 \equiv -5 \pmod{47}$, and so 47 divides $5^{24} + 5$. Since $5^{24} + 5 = 5(5^{23} + 1)$ and $(5, 47) = 1$, we conclude that 47 divides $5^{23} + 1$.

2-8. *Does 41 divide $7 \cdot 3^{20} + 6$?*

Solution. $3^4 \equiv -1 \pmod{41}$ implies that $3^{20} \equiv (-1)^5 \equiv -1 \pmod{41}$. Hence $7 \cdot 3^{20} + 6 \equiv 7(-1) + 6 \equiv -1 \pmod{41}$. Thus 41 does *not* divide $7 \cdot 3^{20} + 6$. (In fact, $7 \cdot 3^{20} + 6$ leaves a remainder of 40 when divided by 41.)

2-9. *Prove that 229 divides $13^{2k} + 17^{2k}$ if k is odd. What if k is even?*

Solution. Let $n = 13^{2k} + 17^{2k} = 169^k + 289^k$; then $n \equiv (-60)^k + 60^k \pmod{229}$. So if k is odd, $n \equiv 0 \pmod{229}$. The result does not hold for *any* even k, for then $n \equiv 2 \cdot 60^k \pmod{229}$, so n can never be congruent to 0 modulo 229. (If it were, then 229 would divide 60^k; since 229 is prime, 229 would then divide 60.)

CHAPTER 2: CONGRUENCES

2-10. *Find the least nonnegative residue of* $1! + 2! + \cdots + 100!$ *modulo 45.*

Solution. If $n \geq 6$, then $6! \mid n!$ and hence $45 \mid n!$ (since $45 \mid 6!$). Thus $1! + 2! + \cdots + 100! \equiv 1! + 2! + \cdots + 5! \equiv 18 \pmod{45}$.

2-11. *Prove that if $p \geq 5$ is prime, then $p^2 + 2$ is composite.*

Solution. If $p > 3$ is prime, then $p \equiv \pm 1 \pmod 3$, and therefore $p^2 + 2 \equiv 0 \pmod 3$. Since $p^2 + 2$ is divisible by 3 and greater than 3, it cannot be prime.

2-12. *Show that $2^{2^n} + 5$ is composite for every positive integer n.*

Solution. Let $N = 2^{2^n} + 5$; then $n = 1$ implies $N = 9$ and $n = 2$ implies $N = 21$. So we might conjecture that N is divisible by 3 for every positive integer n. To prove this, note that $2 \equiv -1 \pmod 3$, and the exponent 2^n is even, so $2^{2^n} \equiv 1 \pmod 3$. Hence $N \equiv 1 + 5 \equiv 0 \pmod 3$.

2-13. *Let p_i denote the ith prime. Show that $p_1 p_2 \cdots p_n + 1$ is never a square.* (*Hint. Show that this sum is of the form $4k + 3$.*)

Solution. The product $p_1 p_2 \cdots p_n$ is twice an odd number, so it is congruent to 2 modulo 4. Therefore $p_1 p_2 \cdots p_n + 1 \equiv 3 \pmod 4$ and cannot be a square, since all squares are congruent to 0 or 1 modulo 4. (It cannot even be a sum of two squares.)

2-14. *Let $q_1, q_2, \ldots, q_n$ be odd primes. Can $N = (q_1 q_2 \cdots q_n)^2 + 1$ ever be a perfect cube? Explain.*

Solution. An odd prime must be of the form $4k + 1$ or $4k + 3$, so $q_i \equiv \pm 1 \pmod 4$; it follows that $N \equiv 2 \pmod 4$. Thus 2^1 is the highest power of 2 that divides N, and so N cannot be a perfect kth power for any $k > 1$.

2-15. *Show that any integer x satisfies at least one of the following congruences: $x \equiv 0 \pmod 2$, $x \equiv 0 \pmod 3$, $x \equiv 1 \pmod 4$, $x \equiv 3 \pmod 8$, $x \equiv 7 \pmod{12}$, $x \equiv 23 \pmod{24}$.*

Solution. Every modulus mentioned divides 24, so it is enough to check that if $0 \leq x \leq 23$, then x satisfies at least one of the congruences. The first three congruences together take care of all x except 7, 11, 19, and 23. The fourth congruence takes care of 11 and 19, the fifth takes care of 7, and the last takes care of 23.

Note. Let $m_1 < m_2 < \cdots < m_k$, and consider the system of congruences $x \equiv a_i \pmod{m_i}$ ($i = 1, 2, \ldots, k$). If any integer x satisfies *at least one* of the congruences in the system, then the system is called a *covering system*. Paul Erdös has offered a substantial prize for a proof that there are covering systems with m_1 arbitrarily large (and a smaller prize for a proof that this is not true).

2-16. *Prove that $1 + 2 + \cdots + n$ is divisible by n if n is odd and is divisible by $n + 1$ if n is even.*

Solution. Suppose first that n is odd. Modulo n, the sum is congruent to $1 + 2 + \cdots + (n - 1)$. Note that the first and last terms add to n, as do the second and next-to-last terms, and so on. There are $(n-1)/2$ such pairs adding to n, and hence the sum is congruent to 0 modulo n. If n is even, apply the previous argument to the odd integer $n+1$ to conclude that $n+1$ divides the sum $1+2+\cdots+n$.

2-17. *Show that the product of any three consecutive integers is a multiple of 6.*

Solution. Let $N = (n-1)(n)(n+1)$ be the product of three consecutive integers. At least one of the three integers is even, so 2 divides N. Likewise, 3 divides N, since (exactly) one of $n-1$, n, and $n+1$ is a multiple of 3. (To prove this, consider the three cases $n \equiv 0, 1,$ or 2 (mod 3).) Since 2 and 3 are relatively prime, it follows from Theorem 1.10 that $2 \cdot 3 = 6$ must divide N.

2-18. *Prove that the sum of any three consecutive cubes is a multiple of 9.*

Solution. Let $N = (n-1)^3 + n^3 + (n+1)^3$; use the Binomial Theorem to conclude that $N = 3n^3 + 6n = 3n(n^2 + 2)$. There are three cases to consider. If $n \equiv 0$ (mod 3), then N contains two factors of 3 and so is divisible by 9. If $n \equiv \pm 1$ (mod 3), then $n^2 + 2 \equiv (\pm 1)^2 + 2 \equiv 0$ (mod 3), and again N has a second factor of 3.

2-19. *Show that that no integer of the form $4k+3$ is the sum of two squares.*

Solution. Every integer is congruent to 0, 1, 2, or 3 modulo 4, and hence the square of any integer is congruent modulo 4 to 0^2, 1^2, 2^2, or 3^2, i.e., to 0 or 1. Thus the sum of two squares must be congruent to 0, 1, or 2 modulo 4. But clearly, an integer of the form $4k+3$ is congruent to 3 modulo 4.

▷ **2-20.** *Prove that no integer of the form $8k+7$ is a sum of three squares. Use this to show that no integer of the form $4^m(8k+7)$ is a sum of three squares.*

Solution. Every integer is congruent to 0, ± 1, ± 2, ± 3, or 4 modulo 8, since the collection $\{-3, -2, -1, 0, 1, 2, 3, 4\}$ is a complete residue system modulo 8. Thus every square is congruent modulo 8 to the square of one of these numbers, that is, to 0, 1, or 4. No combination of any three numbers chosen from 0, 1, or 4 can add to 7 modulo 8, and therefore no integer of the form $8k+7$ is a sum of three squares.

Suppose $N = 4^m(8k+7) = x^2 + y^2 + z^2$ is a sum of three squares for some $m \geq 1$. Since $N \equiv 0$ (mod 4) and since any square is congruent to 0 or 1 modulo 4, it follows that x^2, y^2, and z^2 must each be congruent to 0 modulo 4, and hence x, y, and z are all even. If $x = 2r$, $y = 2s$, $z = 2t$, then $N/4 = r^2 + s^2 + t^2$. Repeating this argument eventually shows that $8k+7$ is a sum of three squares, contradicting the first part of the argument.

2-21. *Use congruences to show that the equation $x^2 - 2y^2 = 10$ does not have integer solutions.*

Solution. We calculate modulo 5. It is easy to verify that if $u \not\equiv 0$ (mod 5), then $u^2 \equiv \pm 1$ (mod 5). If neither x nor y is divisible by 5, examination of cases shows that we cannot have $x^2 - 2y^2 \equiv 0$ (mod 5). Thus if $x^2 - 2y^2 \equiv 0$ (mod 5), then at least one of x and y is divisible by 5. It follows that both x and y are divisible by 5, and therefore $x^2 - 2y^2$ is divisible by 25. In particular, we cannot have $x^2 - 2y^2 = 10$.

2-22. *Show that $n^3 + 11n + 1$ is not divisible by the first four primes for any integer n.*

Solution. Let $N = n^3 + 11n + 1$. Since $n \equiv 0$ or $1 \pmod 2$, $n^3 \equiv 0$ or $1 \pmod 2$ and hence $N \equiv 1 \pmod 2$ in each case, i.e., N is not divisible by 2. Similarly, $n \equiv 0, 1$, or $2 \pmod 3$ implies $n^3 \equiv 0, 1$, or $2 \pmod 3$, and it is easy to check that $N \equiv 1 \pmod 3$ in each case, so 3 does not divide N. If $n \equiv 0, 1, 2, 3$, or $4 \pmod 5$, then $n^3 \equiv 0, 1, 3, 2$, or $4 \pmod 5$; hence $N \equiv 1, 3$, or $4 \pmod 5$, i.e., N is not divisible by 5. Finally, $n \equiv 0, 1, \ldots, 6 \pmod 7$ implies $n^3 \equiv 0, 1$, or $6 \pmod 7$; thus $N \equiv 1, 3, 4, 5$, or $6 \pmod 7$ and so 7 does not divide N.

2-23. *Use the fact that $640 = 5 \cdot 2^7$ to prove that the Fermat number $2^{32} + 1$ is divisible by 641.*

Solution. Since $5 \cdot 2^7 \equiv -1 \pmod{641}$, raising each side to the fourth power gives $5^4 \cdot 2^{28} \equiv 1 \pmod{641}$. Note that $5^4 = 625 \equiv -16 \pmod{641}$, and hence $(-2^4)2^{28} \equiv 1 \pmod{641}$. It follows that $2^{32} \equiv -1 \pmod{641}$, i.e., 641 divides $2^{32} + 1$.

2-24. *Show that the sum of the (decimal) digits of a square is congruent to 0, 1, 4, or 7 modulo 9.*

Solution. The sum of the decimal digits of n is congruent to n modulo 9, so it is enough to show that any square is congruent to 0, 1, 4, or 7 (mod 9). This can be done by simple examination of cases. It is only necessary to square 0, 1, 2, 3, and 4 modulo 9, since $(9-x)^2 \equiv x^2 \pmod 9$.

2-25. *Show that $n(n-1)(2n-1)$ is divisible by 6 for every positive integer n.*

Solution. There is an easy "combinatorial" solution if we recall the fact that $1^2 + 2^2 + \cdots + n^2 = (n)(n-1)(2n-1)/6$. We can also find an easy congruential argument. It is clear that $n(n-1)$ is divisible by 2. To show that $n(n-1)(2n-1)$ is divisible by 3, we can either look separately at the cases $n \equiv 0, 1$, and $2 \pmod 3$ or observe that modulo 3, $2 \equiv -1$, so modulo 3 we are looking at $-(n-1)(n)(n+1)$, and that exactly one of any three consecutive integers is divisible by 3.

We can also prove the result by induction. Let $f(n) = (n)(n-1)(2n-1)$. Since $f(1) = 0$, $f(1)$ is a multiple of 6. We show now that for any integer k, if $f(k)$ is divisible by 6, then $f(k+1)$ is divisible by 6. Consider $f(k+1) - f(k)$. An easy calculation shows that this is $6k^2$. So since $f(k+1) = f(k) + 6k^2$, if $f(k)$ is divisible by 6, so is $f(k+1)$.

Linear Congruences

2-26. *Solve $42x \equiv 90 \pmod{156}$.*

Solution. We apply (2.7). Since $d = (42, 156) = 6$ and 6 divides 90, there are exactly 6 incongruent solutions modulo 156. Reduce the given congruence to $7x \equiv 15 \pmod{26}$. Replace 7 by 33 and divide by 3 to get $11x \equiv 5 \pmod{26}$, i.e. $-15x \equiv 5 \pmod{26}$.

Divide by 5 to get $-3x \equiv 1 \equiv 27 \pmod{26}$, and divide by 3 to get $x \equiv -9 \equiv 17 \pmod{26}$. Thus $7x \equiv 15 \pmod{26}$ has the unique solution $x \equiv 17 \pmod{26}$. Therefore by (2.7), all solutions of $42x \equiv 90 \pmod{156}$ are given by $17 + 156t/(42, 156)$, i.e., $17 + 26t$, for $t = 0, 1, \ldots, 5$. Thus all solutions are given by $x \equiv 17, 43, 69, 95, 121, 147 \pmod{156}$.

2-27. *Find all solutions of* $87x \equiv 57 \pmod{105}$.

Solution. Since $(87, 105) = 3$ and 3 divides 57, the congruence has three solutions. Reduce to $29x \equiv 19 \pmod{35}$. Replacing 29 by -6 and 19 by -16 gives $6x \equiv 16 \pmod{35}$, and hence $3x \equiv 8 \pmod{35}$. Replace 8 by -27, then divide by 3 to get $x \equiv -9 \equiv 26 \pmod{35}$. Thus by (2.7), all solutions to the original congruence are given by $x = 26 + 35t$ ($t = 0, 1, 2$), i.e., $x \equiv 26, 61, 96 \pmod{105}$.

2-28. *Solve* $64x \equiv 897 \pmod{1001}$.

Solution. Note that since $897 \equiv -104 \pmod{1001}$, we are solving the congruence $64x \equiv -104 \pmod{1001}$. Divide each side by 8. This gives the equivalent congruence $8x \equiv -13 \pmod{1001}$. Now replace -13 by 988 and divide each side by 4. We get the equivalent congruence $2x \equiv 247 \pmod{1001}$. Replace 247 by 1248 and divide by 2; the solution of the congruence is $x \equiv 624 \pmod{1001}$. (This technique efficiently solves $ax \equiv b \pmod{m}$ whenever a is a power of 2.)

2-29. *Adapt the idea used in the preceding problem to solve the congruence* $3^6 x \equiv 1 \pmod{8180}$.

Solution. Since $1 \equiv 8181 \pmod{8180}$, replace 1 by 8181, and divide both sides of the congruence by 3^4; this produces the equivalent congruence $3^2 x \equiv 101 \pmod{8180}$. Now replace 101 by $101 + 2 \cdot 8180$ and divide by 9. The solution of the congruence is therefore $x \equiv 1829 \pmod{8180}$.

2-30. *Which positive integers less than 15 have inverses modulo 15? Find the inverses.*

Solution. By definition, a has an inverse modulo 15 if and only if the congruence $ax \equiv 1 \pmod{15}$ is solvable. It follows from (2.7) that this is true if and only if $(a, 15)$ divides 1, and hence if and only if $(a, 15) = 1$. Thus a will have an inverse modulo 15 if and only if a is relatively prime to 15, so a must be one of 1, 2, 4, 7, 8, 11, 13, or 14. Calculate: $1 \cdot 1 \equiv 1$, $2 \cdot 8 \equiv 1$, $4 \cdot 4 \equiv 1$, $7 \cdot 13 \equiv 1$, $11 \cdot 11 \equiv 1$, and $14 \cdot 14 \equiv 1$, all modulo 15. So 1, 4, 11, and 14 are their own inverses. Also, 2 and 8 are inverses of each other, as are 7 and 13.

2-31. *What possibilities are there for the number of solutions of a linear congruence modulo 20?*

Solution. According to (2.7), if solutions to $ax \equiv b \pmod{m}$ exist, then there are (a, m) incongruent solutions. If $m = 20$, the only possible values for $(a, 20)$ are 1, 2, 4, 5, 10, and 20. Now the congruences $2x \equiv 1$, $x \equiv 1$, $2x \equiv 2$, $4x \equiv 4, \ldots, 20x \equiv 20$ (all modulo 20) have 0, 1, 2, 4, 5, 10, and 20 solutions, respectively. So these are all the possibilities.

2-32. *(a) Solve* $179x \equiv 283 \pmod{313}$. *(Note that 313 is a prime.)*

(b) Express 283 as a linear combination of 179 and 313. (See the Note before (2.10).)

Solution. (a) We will use the multiplication procedure described in Technique 3 before (2.10). The integer closest to $313/179$ is 2, so multiply the congruence by 2 and reduce so that the absolute value of the coefficient of x is as small as possible. We get $45x \equiv 253 \pmod{313}$. Since the integer closest to $313/45$ is 7, we now multiply by 7 and reduce modulo 313 to obtain $2x \equiv 206 \pmod{313}$. Thus the (unique) solution is $x \equiv 103 \pmod{313}$.

(b) By part (a), we have $179 \cdot 103 = 283 + 313s$ for some integer s, and clearly, $s = (179 \cdot 103 - 283)/313 = 58$. Thus $283 = 179 \cdot 103 - 313 \cdot 58$.

2-33. *Find the unique solution of* $251x \equiv 125 \pmod{521}$. *(521 is a prime.)*

Solution. We again use the multiplication technique. The integer closest to $521/251$ is 2; multiplying by 2 and reducing modulo 521 then gives $-19x \equiv 250 \pmod{521}$. (We use -19 instead of 502 because -19 has a *much* smaller absolute value.) Similarly, mutiplying by 27, the nearest integer to $521/19$, yields $8x \equiv 498 \equiv -23 \pmod{521}$. Since $521/8 = 65.125$, multiply by 65 to get $520x \equiv -453 \pmod{521}$, i.e., $-x \equiv -453 \pmod{521}$. Thus the unique solution to the original congruence is $x \equiv 453 \pmod{521}$.

The next problem provides a technique for reducing a given congruence to a congruence with a smaller modulus. By repeated application of this process if necessary, a congruence is obtained whose solution is easily determined. We then work backward from this solution to produce a solution of the original congruence.

2-34. *Let* y^* *be a solution of the congruence* $my \equiv -b \pmod{a}$. *Then* $(my^* + b)/a$ *is a solution of* $ax \equiv b \pmod{m}$.

Solution. If $my^* \equiv -b \pmod{a}$, then $my^* + b = ka$ for some integer k. Thus $ak \equiv b \pmod{m}$, and hence $k = (my^* + b)/a$ is a solution of $ax \equiv b \pmod{m}$.

Note. The above reduction process can be repeated, but since the modulus, the coefficient of the unknown, and the constant on the right side all change in successive applications, it is important to remember to substitute the appropriate values of a, b, and m at each stage. This technique is illustrated in the solution of the following problem.

2-35. *Find all solutions of* $108x \equiv 171 \pmod{529}$.

Solution. Since $(108, 529) = 1$, there is a unique solution (modulo 529). Use the Euclidean Algorithm to write 1 as a linear combination of 108 and 529, namely, $529 \cdot 49 + 108(-240) = 1$. (Check this!) Hence $108(-240) \equiv 1 \pmod{529}$, and so $108(-240 \cdot 171) \equiv 171 \pmod{529}$. Since $-240 \cdot 171 \equiv 222 \pmod{529}$, we conclude that $x \equiv 222 \pmod{529}$ is the only solution to this congruence.

Alternatively, we can use the reduction method described in the preceding problem. Given $ax \equiv b \pmod{m}$, we first solve $my \equiv -b \pmod{a}$; here, we get $529y \equiv -171 \pmod{108}$, i.e., $-11y \equiv -171 \pmod{108}$ or, equivalently, $11y \equiv 63 \pmod{108}$. (We replace 529 by -11 because 11 is much smaller than the least nonnegative residue 97.) Reduce again to get $108z \equiv -63 \pmod{11}$, i.e., $-2z \equiv -8 \pmod{11}$. This gives $z_0 = 4$ as a solution. Thus $y_0 = (mz_0 + b)/a = (108 \cdot 4 + 63)/11 = 45$. (Note that in the second step of the reduction, writing $11y \equiv 63 \pmod{108}$ in the form $az \equiv b \pmod{m}$ gives $m = 108$, $b = 63$, and $a = 11$.) Finally, $x_0 = (my_0 + b)/a = (529 \cdot 45 + 171)/108 = 222$, since our original congruence has $m = 529$, $b = 171$, and $a = 108$.

2-36. *Find all solutions to the pair of congruences* $3x - 7y \equiv 4 \pmod{19}$, $7x - 3y \equiv 1 \pmod{19}$.

Solution. We need to make only a minor adaptation of the usual method of solving two linear equations in two variables. Since $(7, 19) = 1$, the first congruence is equivalent to the congruence $7(3x - 7y) \equiv 7 \cdot 4 \pmod{19}$, i.e., $21x - 49y \equiv 28 \pmod{19}$. Similarly, the congruence $7x - 3y \equiv 1 \pmod{19}$ is equivalent to $21x - 9y \equiv 3 \pmod{19}$. Subtracting, we obtain $-40y \equiv 25 \pmod{19}$ or, equivalently, $-2y \equiv 6 \pmod{19}$. This has solution $y \equiv -3 \pmod{19}$. Substitute this in the first congruence. We obtain $3x \equiv 2 \pmod{19}$, giving $x \equiv 7 \pmod{19}$. So the solution to the system is $x \equiv 7 \pmod{19}$, $y \equiv 16 \pmod{19}$.

2-37. *Find all solutions to the pair of congruences* $3x - 7y \equiv 4 \pmod{15}$, $7x - 3y \equiv 1 \pmod{15}$.

Solution. As in the previous problem, the first congruence is equivalent to the congruence $21x - 49y \equiv 28 \pmod{15}$. The second congruence *implies* that $21x - 9y \equiv 3 \pmod{15}$ (we do not have *equivalence*, since 3 and 15 are not relatively prime). But as before, if the two given congruences hold, then $-40y \equiv 25 \pmod{15}$ or, equivalently, $5y \equiv 10 \pmod{15}$, and hence $y \equiv 2 \pmod{3}$. Thus modulo 15 the only possibilities for y are 2, 5, 8, 11, and 14. Substitute these values in the congruence $7x - 3y \equiv 1 \pmod{15}$, and solve for x. We get $x \equiv 1, 13, 10, 7$, and $4 \pmod{15}$, respectively.

The Chinese Remainder Theorem

2-38. *Find all integers between 3000 and 5000 that leave remainders of 1, 3, and 5 when divided by 7, 11, and 13, respectively.*

Solution. Apply the Chinese Remainder Theorem to the system $x \equiv 1 \pmod{7}$, $x \equiv 3 \pmod{11}$, $x \equiv 5 \pmod{13}$. Find b_1, b_2, b_3 such that $143b_1 \equiv 1 \pmod{7}$, $91b_2 \equiv 1 \pmod{11}$, and $77b_3 \equiv 1 \pmod{13}$, i.e., $3b_1 \equiv 1 \pmod{7}$, $3b_2 \equiv 1 \pmod{11}$, and $-b_3 \equiv 1 \pmod{13}$. Thus we can take $b_1 = 5$, $b_2 = 4$, and $b_3 = -1$. This gives the solution $x^* = 143(5)(1) + 91(4)(3) + 77(-1)(5) = 1422$. Since $7 \cdot 11 \cdot 13 = 1001$, all solutions are of the form $1422 + 1001t$ (t an integer). It is clear that the only solutions between 3000 and 5000 are $1422 + 2 \cdot 1001 = 3424$ and $3424 + 1001 = 4425$.

2-39. *Find an integer x, with $0 < x < 140$, that satisfies the congruences* $x \equiv 1 \pmod{4}$, $2x \equiv 3 \pmod{5}$, $4x \equiv 5 \pmod{7}$.

Solution. First put the congruences in the form $x \equiv a_i \pmod{m_i}$, then apply the Chinese Remainder Theorem. The first congruence is already in this form; for $2x \equiv 3$

(mod 5), multiply each side by 3 and reduce modulo 5 to get $x \equiv 4$ (mod 5); for $4x \equiv 5$ (mod 7), multiply each side by 2 and reduce modulo 7 to get $x \equiv 3$ (mod 7). Now find b_1, b_2, b_3 so that $5 \cdot 7 b_1 \equiv 1$ (mod 4), $4 \cdot 7 b_2 \equiv 1$ (mod 5), and $4 \cdot 5 b_3 \equiv 1$ (mod 7), i.e., $-b_1 \equiv 1$ (mod 4), $3b_2 \equiv 1$ (mod 5), $-b_3 \equiv 1$ (mod 7). Thus we can take $b_1 = -1$, $b_2 = 2, b_3 = -1$. Hence one solution is $x^* = 35(-1)(1) + 28(2)(4) + 20(-1)(3) = 129$. Since $4 \cdot 5 \cdot 7 = 140$, 129 is the only positive solution to this system that is less than 140.

The next three problems deal with the system $x \equiv a_i$ (mod m_i) ($i = 1, 2, \ldots, r$), where the moduli m_i are not necessarily relatively prime in pairs. It can be shown that the system has a solution if and only if (m_i, m_j) divides $a_j - a_i$ whenever $i \neq j$. The argument for general r is a little delicate (there have been a number of incorrect proofs), so we treat fully only the case $r = 2$.

2-40. Show that the conclusion of (2.11) does not necessarily hold if the moduli m_i are not relatively prime in pairs.

Solution. For example, take $m_1 = 2$, $m_2 = 4$, $a_1 = 1$, and $a_2 = 2$. It is obvious that the system of congruences $x \equiv a_i$ (mod m_i) ($i = 1, 2$) does not have a solution.

2-41. Consider the system $x \equiv a$ (mod m), $x \equiv b$ (mod n), where m and n are not necessarily relatively prime. Show that if (m, n) divides $b - a$, then the system has a solution.

Solution. Let $d = (m, n)$, and suppose $d \mid b - a$. By (1.24), there exist integers u and v such that $mu + nv = b - a$. Let $x = a + mu$; then clearly, $x \equiv a$ (mod m). But $x = a + mu = a + (b - a) - nv = b - nv$, and so $x \equiv b$ (mod n).

2-42. Suppose that the system $x \equiv a_i$ (mod m_i) ($i = 1, 2, \ldots, r$) has a solution. Show that (m_i, m_j) divides $a_j - a_i$ whenever $i \neq j$. Show also that if s is a solution of the system, then the solutions are all the integers congruent to s modulo $[m_1, m_2, \ldots, m_r]$.

Solution. Suppose that $i \neq j$, and let $d = (m_i, m_j)$. If s is a solution of the system, then $s \equiv a_i$ (mod m_i). Thus $m_i \mid s - a_i$, and hence $d \mid s - a_i$; similarly, $d \mid s - a_j$. It follows that d divides $(s - a_i) - (s - a_j) = a_j - a_i$. Therefore the system cannot have a solution unless (m_i, m_j) divides $a_j - a_i$ whenever $i \neq j$.

The number x is a solution of the system if and only if $x \equiv a_i$ (mod m_i) for all i, i.e., if and only if $x \equiv s$ (mod m_i) for all i. But by (2.4.iii), this is true precisely when $x \equiv s$ (mod $[m_1, m_2, \ldots, m_r]$).

2-43. (Ch'in Chiu-shao, thirteenth century.) *Three farmers equally divide the rice that they have grown. One goes to a market where an 83-pound weight is used, another to a market that uses a 110-pound weight, and the third to a market using a 135-pound weight. Each farmer sells as many full measures as possible, and when the three return home, the first has 32 pounds of rice left, the second 70 pounds, and the third 30 pounds. Find the total amount of rice they took to market.*

Solution. Let x be the amount each farmer took to market; then $x \equiv 32 \pmod{83}$, $x \equiv 70 \pmod{110}$, and $x \equiv 30 \pmod{135}$. The problem here is that 83, 110, and 135 are not relatively prime in pairs, since $(110, 135) = 5$.

Since $110 = 2 \cdot 5 \cdot 11$ and $135 = 5 \cdot 27$, the last two congruences are equivalent to $x \equiv 0 \pmod{2}$, $x \equiv 0 \pmod{5}$, $x \equiv 4 \pmod{11}$, and $x \equiv 3 \pmod{27}$. We apply the Chinese Remainder Theorem to these four congruences together with $x \equiv 32 \pmod{83}$, with $m_1 = 2$, $m_2 = 5$, $m_3 = 11$, $m_4 = 27$, $m_5 = 83$ and $a_1 = 0$, $a_2 = 0$, $a_3 = 4$, $a_4 = 3$, $a_5 = 32$. Since $a_1 = a_2 = 0$, we need only find the integers b_3, b_4, b_5 described in the proof of the Chinese Remainder Theorem. Thus we must solve $2 \cdot 5 \cdot 27 \cdot 83 b_3 \equiv 1 \pmod{11}$, $2 \cdot 5 \cdot 11 \cdot 83 b_4 \equiv 1 \pmod{27}$, and $2 \cdot 5 \cdot 11 \cdot 27 b_5 \equiv 1 \pmod{83}$ or, equivalently, $3b_3 \equiv 1 \pmod{11}$, $4b_4 \equiv 1 \pmod{27}$, and $65 b_5 \equiv 1 \pmod{83}$. In the first congruence, replace 1 by 12 and divide each side by 3 to get $b_3 = 4$ (or we could use $b_3 = -7$); in the second, replace 1 by 28 and divide by 4 to get $b_4 = 7$ (or $b_4 = -20$). In the third, replace 65 by -18 and 1 by 84, then divide by 6 to get $-3b_5 \equiv 14 \pmod{83}$; now replace 14 by -69 and divide by 3 to get $b_5 = 23$. (Note that in all of these divisions, the modulus does not change, since the number that we divide by is relatively prime to the modulus.)

Now substitute the appropriate values in the expression for x^* given in the proof of (2.11). Using the values $b_3 = -7$, $b_4 = -20$ (to keep the overall sum smaller), and $b_5 = 23$, we get $x^* = 1010640$. Here, $m = 2 \cdot 5 \cdot 11 \cdot 27 \cdot 83 = 246510$, so the least nonnegative residue of x^* modulo m is 24600. Since the next smallest solution is $x^* + m = 24600 + 246510 = 271110$, which is presumably unreasonably large, we conclude that each farmer takes 24600 pounds of rice to market, and therefore the total amount grown is $3 \cdot 24600 = 73800$.

2-44. (Bhaskara I, sixth century; also al-Haitham, eleventh century; Fibonacci, early thirteenth century.) *If eggs in a basket are taken out 2, 3, 4, 5, and 6 at a time, there are 1, 2, 3, 4, and 5 eggs left over, respectively. If they are taken out 7 at a time, there are no eggs left over. What is the least number of eggs that can be in the basket?*

Solution. We require a positive integer x such that $x \equiv 1 \pmod{2}$, $x \equiv 2 \pmod{3}$, $x \equiv 3 \pmod{4}$, $x \equiv 4 \pmod{5}$, $x \equiv 5 \pmod{6}$, and $x \equiv 0 \pmod{7}$. However, since the moduli are not relatively prime in pairs (for example, 2 and 4 or 3 and 6), the computational procedure described in the proof of the Chinese Remainder Theorem cannot be used directly. But because of the special nature of the congruences, there is an easy solution.

Note that the first five congruences can be written as $x \equiv -1$ modulo 2, 3, 4, 5, and 6. By (2.4.iii), the solution of this system is immediate: $x \equiv -1 \pmod{60}$ (60 is the least common multiple of these moduli). So we want to solve the system $x \equiv -1 \pmod{60}$, $x \equiv 0 \pmod{7}$ or, equivalently, letting $x = 7y$, the congruence $7y \equiv -1 \pmod{60}$. By the Euclidean Algorithm (or by inspection), $y = 17$ is a solution, so $x = 119$ is a solution of the original system. Since solutions differ by $7 \cdot 60 = 420$, $x = 119$ is the smallest solution.

2-45. *Find the smallest positive integer x such that $x \equiv 5 \pmod{12}$, $x \equiv 17 \pmod{20}$, and $x \equiv 23 \pmod{42}$.*

58   CHAPTER 2: CONGRUENCES

Solution. Since the moduli are not relatively prime in pairs, the Chinese Remainder Theorem does not apply directly. We first reduce the given system to one with pairwise relatively prime moduli as follows. By (2.4.iii), $x \equiv 5$ (mod 12) is equivalent to $x \equiv 5$ (mod 3) and $x \equiv 5$ (mod 4), i.e., $x \equiv 2$ (mod 3) and $x \equiv 1$ (mod 4). Similarly, $x \equiv 17$ (mod 20) is equivalent to $x \equiv 1$ (mod 4) and $x \equiv 2$ (mod 5). (Note. If we had gotten, say, $x \equiv 2$ (mod 4) here, this would be inconsistent with the congruence $x \equiv 1$ (mod 4) previously obtained, so the original system would have no solution.) Likewise, $x \equiv 23$ (mod 42) is equivalent to $x \equiv 1$ (mod 2), $x \equiv 2$ (mod 3), and $x \equiv 2$ (mod 7). Since $x \equiv 1$ (mod 4) implies $x \equiv 1$ (mod 2), our reduced system is $x \equiv 1$ (mod 4), $x \equiv 2$ (mod 3), $x \equiv 2$ (mod 5), and $x \equiv 2$ (mod 7). Now we could use the machinery described in the proof of (2.11). But it is simpler to note that the last three congruences are equivalent to $x \equiv 2$ (mod 105) and that $2 - 105 \equiv 1$ (mod 4). So -103 is a solution of the congruence, and $-103 + 4 \cdot 105 = 317$ is the smallest positive solution.

2-46. *Find the smallest positive integer that leaves remainders of $9, 8, \ldots, 2, 1$ when divided by $10, 9, \ldots, 3, 2$, respectively.*

Solution. We want $x \equiv -1$ (mod m) for $m = 10, 9, \ldots, 2$. At first glance, it may be tempting to use the Chinese Remainder Theorem, but there is an easier way to find the answer. One solution is $x = -1$, which unfortunately is not positive. However, by (2.4.iii), this system of congruences is equivalent to the congruence $x \equiv -1$ (mod m), where m is the least common multiple of $2, 3, \ldots, 10$. Thus every solution has the form $-1 + tm$ for some integer t. The smallest positive solution is thus $m - 1$, where $m = 2^3 \cdot 3^2 \cdot 5 \cdot 7$.

Note. The above argument works with -1 replaced by *any* integer; what is important is that the right side of each congruence is the same.

2-47. *Solve the following system of congruences: $x^2 \equiv 2$ (mod 7), $x^2 \equiv 3$ (mod 11), $x^2 \equiv 4$ (mod 13). (Hint. First solve each congruence for x.)*

Solution. In the first congruence, replace 2 by 9 to conclude that $x \equiv \pm 3$ (mod 7); in the second, replace 3 by 25 to get $x \equiv \pm 5$ (mod 11); in the third, clearly, $x \equiv \pm 2$ (mod 13). Now apply the Chinese Remainder Theorem to the system $x \equiv a$ (mod 7), $x \equiv b$ (mod 11), $x \equiv c$ (mod 13), where $a = \pm 3$, $b = \pm 5$, and $c = \pm 2$. Thus there are $2 \cdot 2 \cdot 2 = 8$ different systems to consider. It is easiest to set up the form of the solution in terms of a, b, and c, then substitute the different values. We need b_1, b_2, b_3 such that $143 b_1 \equiv 1$ (mod 7), $91 b_2 \equiv 1$ (mod 11), and $77 b_3 \equiv 1$ (mod 13), i.e., $3 b_1 \equiv 1$ (mod 7), $3 b_2 \equiv 1$ (mod 11), and $-b_3 \equiv 1$ (mod 13). Take $b_1 = -2$, $b_2 = 4$, $b_3 = -1$. Then the general solution is $x^* \equiv 143(-2)a + 91(4)b + 77(-1)c$ (mod $7 \cdot 11 \cdot 13$). Substitute the different values of a, b, and c, taking advantage of the fact that the triples (a, b, c) come in four opposite sign pairs. We obtain that the original system of congruences has the solutions ± 115, ± 171, ± 193, and ± 479 (mod 1001).

▷ **2-48.** *Find the smallest positive integer n such that $n/3$ is a perfect cube, $n/5$ a perfect fifth power, and $n/7$ a perfect seventh power.*

Solution. Since n is divisible by 3, 5, and 7, we may take n to have form $3^a 5^b 7^c$. Because $n/3 = 3^{a-1} 5^b 7^c$ is a cube, $a - 1$, b, and c must be divisible by 3, i.e., $a \equiv 1$,

$b \equiv 0$, $c \equiv 0$ (mod 3). Similarly, $n/5 = 3^a 5^{b-1} 7^c$ a fifth power implies $a \equiv 0$, $b \equiv 1$, $c \equiv 0$ (mod 5); and $n/7 = 3^a 5^b 7^{c-1}$ a seventh power implies $a \equiv 0$, $b \equiv 0$, $c \equiv 1$ (mod 7). The smallest positive solution of the three congruences for a is 70 (since a must be a multiple of 35 congruent to 1 modulo 3). The smallest positive solution of the three congruences for b is 21 (since b must be a multiple of 21 congruent to 1 modulo 5); and the smallest c is 15. Thus $n = 3^{70} \cdot 5^{21} \cdot 7^{15}$.

Day of the Week

The solutions of the following problems use an algorithm for determining the day of the week for a given date. This algorithm is described in detail in (2.12).

2-49. *In the algorithm described in (2.12), the year code is given by $y + [y/4]$, where y is the integer consisting of the last two digits of the year. This code can also be calculated as follows. If $y = 12k + r$, with $0 \le r \le 11$, then the year code is given by $k + r + [r/4]$. (For example, if the year is 1945, the code is equal to the number of 12's in 45 (namely, 3), plus the remainder 9, plus the number of 4's in 9, that is, $3 + 9 + 2$, or 0 modulo 7.) Prove that the two expressions always give the same result; that is, prove that $y + [y/4] \equiv k + r + [r/4]$ (mod 7).*

Solution. If $y = 12k + r$, then $y + [y/4] = 12k + r + [3k + r/4] = 12k + r + 3k + [r/4] = 15k + r + [r/4] \equiv k + r + [r/4]$ (mod 7). We have used the fact that if n is an integer, then $[n + x] = n + [x]$. (This follows immediately from the definition of $[x]$.)

Note. The above method for finding the year code is easier to use if you want to calculate the day of the week in your head. However, if the algorithm is used for something more complicated than finding the day of the week for a given date, then the original way of calculating the year code (namely, $y + [y/4]$) is often more convenient. (See Problem 2-54.)

2-50. *Prove that any date in the twentieth century, beginning with March 1, 1900, falls on the same day of the week 28 years later.*

Solution. It is enough to show that the year code is the same every 28 years, that is, $y' + [y'/4] \equiv y + [y/4]$ (mod 7) when $y' = y + 28k$. This follows easily from the fact that $[y'/4] = [y/4 + 7k] = [y/4] + 7k \equiv [y/4]$ (mod 7). (Note that dates in January or February of 1900 are excluded, since 1900 was *not* a leap year.)

▷ **2-51.** (Friday the 13th.) *Explain why every year has at least one Friday the 13th and no more than three. What day of the week must New Year's Day fall on if the year has three Friday the 13th's?*

Solution. Let x denote the day of the week for January 13 in a given year; thus $x = 0, 1, 2, \ldots, 6$. (Note that Saturday, the seventh day of the week, corresponds to 0.) There are two cases to consider. First assume that we are in a non-leap year. Since January has 31 days and $31 \equiv 3$ (mod 7), the day of the week for February 13 is $x + 3$. Since February has 28 days and $28 \equiv 0$ (mod 7), the day of the week for

March 13 is the same as for February 13, namely, $x+3$. Proceeding this way, we obtain the following days of the week for the 13th of each month: $x, x+3, x+3, x+6, x+1, x+4, x+6, x+2, x+5, x, x+3, x+5$. Regardless of the value of x, the above list will always contain a complete residue system modulo 7, so for any x, at least one of these numbers will be congruent to 6, which represents Friday. This proves that there is at least one Friday the 13th in any non-leap year.

Since the list contains $x+3$ three times, the year has three Friday the 13th's precisely when $x+3 = 6$, i.e., $x = 3$. By our definition of x, this means that January 13 must be the third day of the week, that is, Tuesday. Hence we conclude that in a non-leap year, New Year's Day must fall on Thursday if the year has three Friday the 13th's. It is easy to check that Friday the 13th will then occur in February, March, and November (the months corresponding to $x+3$).

Finally, suppose we are in a leap year. The analysis is exactly the same as above, except that now February has 29 days. Our list for the 13th of each month is $x, x+3, x+4, x, x+2, x+5, x, x+3, x+6, x+1, x+4, x+6$, which again contains a complete residue system modulo 7. But now the year will have three Friday the 13th's precisely when $x = 6$ (since x is the day that appears three times). Thus if there are three Friday the 13th's in a leap year, January 13 ($= x$) must be Friday, and hence New Year's Day will fall on Sunday. In this case, the three Friday the 13th's occur in January, April, and July.

2-52. *The Battle of Hastings was fought on October 14, 1066. What day of the week was this?*

Solution. The year code is $66 + [66/4] \equiv 5 \pmod 7$, the month code is 1, and the correction for the eleventh century is $18 - 10 = 8 \equiv 1 \pmod 7$. Since $5 + 1 + 14 + 1 \equiv 0 \pmod 7$, the battle occurred on Saturday.

▷ **2-53.** *In 1991, a person's birthday fell on American Thanksgiving (the fourth Thursday of November). The first time that happened, he was three years old; the next time it occurred, he was a teenager. Assuming that the person was over 20 and less than 50 years old in 1991, when was the person born?*

Solution. Let d be the day of birth in November. The month code for November is 4, and the year code for 1991 is $91 + [91/4] = 91 + 22 \equiv 1 \pmod 7$, where $[x]$ denotes the greatest integer less than or equal to x. Since the birthday fell on Thursday (the fifth day of the week) in 1991, we have $4 + 1 + d \equiv 5 \pmod 7$, i.e., $d \equiv 0 \pmod 7$. Thus $d = 28$, since the birthday was the fourth Thursday in November. It is clear that if the birthday falls on Thursday, then that Thursday must be Thanksgiving.

We next determine in which years of this century November 28 falls on a Thursday. Denote the year by $1900 + y$. Write $y = 4k + r$ with $r = 0, 1, 2,$ or 3; then the year code is $y + [y/4] = 4k + r + k = 5k + r$. Since November 28 is to fall on a Thursday, we need $5k + r \equiv 1 \pmod 7$. For $r = 0$, we get $k \equiv 3 \pmod 7$, i.e., $k = 7t + 3$ for some nonnegative integer t. Thus $y = 4(7t + 3) + r = 28t + 12$, and this yields the solutions 1912, 1940, 1968, and 1996. Similarly, if $r = 1$, then we get $k \equiv 0 \pmod 7$, giving 1900, 1928, 1956, and 1984; if $r = 2$, then $k \equiv 4 \pmod 7$, which gives 1918, 1946, and 1974; and $r = 3$ implies $k \equiv 1 \pmod 7$, giving 1907, 1935, 1963, and 1991.

Since the person's age in 1991 was over 20 and less than 50, the only possible birth years are 1943, 1953, 1960, and 1965. In the last three cases, the person would not be a teenager the second time his birthday fell on Thanksgiving. Thus we conclude that the person was born November 28, 1943.

2-54. *Find all of the years in the twentieth century such that Christmas falls on a Sunday.*

Solution. The month and day codes for December 25 give $6 + 25 \equiv 3$ (mod 7). Thus if Christmas is to fall on a Sunday, the year code must be congruent to 5 modulo 7. Denote the year by $1900 + y$, and write $y = 4k + r$, where $r = 0, 1, 2,$ or 3. Then, as in the preceding problem, the year code is $y + [y/4] = 5k + r$. Hence we need $5k + r \equiv 5$ (mod 7). For $r = 0$, we have $k \equiv 1$ (mod 7), which yields the solutions 1904, 1932, 1960, and 1988. For $r = 1$, we get $5k \equiv 4$ (mod 7), i.e., $k \equiv 5$ (mod 7), and this gives 1921, 1949, and 1977. If $r = 2$, then $k \equiv 2$ (mod 7), giving 1910, 1938, 1966, and 1994; and $r = 3$ implies that $k \equiv 6$ (mod 7), which gives 1927, 1955, and 1983.

2-55. *Henry VIII married Anne Boleyn in a secret ceremony on January 25, 1533. On what day of the week were they married?*

Solution. The year code is $33 + [33/4] = 41 \equiv 6$ (mod 7), the month code for January is 1, and the correction for the 1500s is $18 - 15 = 3$. Since $6 + 1 + 25 + 3 \equiv 0$ (mod 7), they were married on a Saturday.

Miscellaneous Problems

The divisibility tests in Theorem 2.5 are used in the solutions of the next three problems.

2-56. *An old, partly unreadable receipt shows that 36 turkeys were purchased for \$x73.9y. Assuming that the turkeys cost less than \$10 each, find the missing digits.*

Solution. Since 36 divides $x739y$ cents, $4 \mid x739y$ and $9 \mid x739y$. Thus, by (2.5.i), $4 \mid 9y$ and hence y must be 2 or 6. Also, (2.5.iii) implies that $x + 7 + 3 + 9 + y \equiv 0$ (mod 9), i.e., $x + y \equiv 8$ (mod 9). Since x and y are digits, we have $x + y = 8$. If $y = 2$, then $x = 6$, and each turkey would cost $673.92/36 = 18.72$, which is more than \$10. Thus the only solution is $y = 6$ and $x = 2$; each turkey then costs $273.96/36 = 7.61$ dollars.

2-57. *The seven-digit number $n = 72x20y2$, where x and y are digits, is divisible by 72. What are the possibilities for x and y?*

Solution. Since n is divisible by 8 and 1000 is divisible by 8, $10y + 2 \equiv 0$ (mod 8), so $y \equiv -1$ (mod 4), and therefore $y = 3$ or $y = 7$. Since n is divisible by 9, the sum of the digits is divisible by 9, so $x + y + 4 \equiv 0$ (mod 9). If $y = 3$, then $x = 2$, and if $y = 7$, then $x = 7$.

2-58. *Suppose that 792 divides the integer with decimal representation $13xy45z$. Find the digits x, y, and z.*

Solution. Note that $792 = 8 \cdot 9 \cdot 11$. Since $8 \mid 13xy45z$, (2.5) implies $8 \mid 45z$, and hence $z = 6$. Similarly, $9 \mid 13xy456$ implies $x + y \equiv 8$ (mod 9), and $11 \mid 13xy456$ implies

$6 - 5 + 4 - y + x - 3 + 1 \equiv 0 \pmod{11}$, i.e., $x - y \equiv 8 \pmod{11}$. Thus $x + y = 8$ and $x - y = 8$, so $x = 8$, $y = 0$.

The following problem gives a reasonably easy test for divisibility by 7. The polynomial evaluation procedure described in the solution is usually called *Horner's Method*. It is far more efficient than the naive procedure of computing the powers of x, multiplying by the coefficients, and adding.

▷ **2-59.** *Let $N \geq 10$ be a positive integer. Take the first (leftmost) digit of N, multiply by 3, reduce modulo 7, add the second digit. Multiply the result by 3, reduce modulo 7, add the third digit. Go on in this way until you have added the rightmost digit. Show that the number you get is congruent to N modulo 7.*

Solution. Let $P(x) = b_0 x^n + b_1 x^{n-1} + \cdots + b_n$. Let $y_0 = b_0$, $y_1 = y_0 x + b_1$, $y_2 = y_1 x + b_2$, and so on. It is not difficult to see that $y_n = P(x)$.

Now suppose N has decimal expansion $a_n 10^n + \cdots + a_0$. Let $P(x) = a_n x^n + \cdots + a_0$. Then $N = P(10)$, and since $10 \equiv 3 \pmod 7$, $N \equiv P(3) \pmod 7$ by (2.3.iii). Finally, note that the procedure described in the statement of the problem simply evaluates $P(3)$ modulo 7.

2-60. *Prove or disprove: The set $\{1^2, 2^2, \ldots, m^2\}$ is a complete residue system modulo m.*

Solution. It is easy to see that we get a complete residue system only if m is 1 or 2, for if $m > 2$, then 1 is not congruent to $m - 1$ modulo m, but $1^2 \equiv (m-1)^2 \pmod m$. Thus at least two of the numbers $1^2, 2^2, \ldots, m^2$ are congruent to each other, and hence the set cannot be a complete residue system.

2-61. *Let $\{r_1, r_2, \ldots, r_m\}$ be a complete residue system modulo m. If $(a, m) = 1$, prove that $\{ar_1, ar_2, \ldots, ar_m\}$ is also a complete residue system modulo m.*

Solution. Since any set of m incongruent integers is a complete residue system modulo m, it suffices to show that any two elements of the set $\{ar_1, ar_2, \ldots, ar_m\}$ are incongruent modulo m. But if $ar_i \equiv ar_j \pmod m$, (2.2.vi) implies that $r_i \equiv r_j \pmod m$, which is possible only if $r_i = r_j$ (since the r_i form a complete residue system).

2-62. *Suppose that $(a, m) = 1$. Use the previous problem to show that the linear congruence $ax \equiv b \pmod m$ has a unique solution.*

Solution. By the preceding problem, $\{0, a, 2a, \ldots, (m-1)a\}$ is a complete residue system. Thus, given any integer b, b must be congruent to a unique element of this set, that is, $ar \equiv b \pmod m$ for some unique r between 0 and $m - 1$.

2-63. *Show that if $a \equiv b \pmod m$, then $(a, m) = (b, m)$. Is the converse of this result true?*

Solution. If $a \equiv b \pmod m$, then $a = b + km$ for some integer k, and so $(a, m) = (b + km, m)$. Now apply (1.22) to conclude that $(b + km, m) = (b, m)$. The converse is not true; for example, $(2, 5) = (3, 5) = 1$, but $2 \not\equiv 3 \pmod 5$.

PROBLEMS AND SOLUTIONS   63

The next two problems involve *binomial coefficients*. Recall that $\binom{n}{k} = n!/k!(n-k)! = (n)(n-1)\cdots(n-k+1)/k!$.

2-64. *Let p be prime, and let $0 \le n < p$. Show that the binomial coefficient $\binom{n+p}{p}$ is congruent to 1 modulo p.*

Solution. Let $N = \binom{n+p}{p}$. By the note above, $n!N = (n+p)(n+p-1)\cdots(p+1)$. But $n+p \equiv n \pmod{p}$, $n+p-1 \equiv n-1 \pmod{p}$, ..., and therefore $(n+p)(n+p-1)\cdots(p+1) \equiv n! \pmod{p}$. It follows that $n!N \equiv n! \pmod{p}$. But since $n < p$ and p is prime, $n! \not\equiv 0 \pmod{p}$. Therefore each side of the congruence can be divided by $n!$, giving $N \equiv 1 \pmod{p}$.

2-65. *Let p be prime. Show that $\binom{2p}{p} \equiv 2 \pmod{p}$.*

Solution. Let $N = \binom{2p}{p}$. Then $p!N = (2p)(2p-1)\cdots(p+1)$. Now cancel a p from both sides of this equation, and observe that $2p-1 \equiv p-1 \pmod{p}$, $2p-2 \equiv p-2$, ... It follows that $(p-1)!N \equiv 2(p-1)! \pmod{p}$; dividing both sides of the congruence by $(p-1)!$ yields the result.

▷ **2-66.** *Cup A can hold exactly a ounces of liquid, and cup B can hold exactly b ounces of liquid, where a and b are relatively prime integers and $a < b$. Next to the cups is a large open barrel full of wine. Show that with the help of cup A, we can measure out in cup B any integer number $x \le b$ of ounces of wine.*

Solution. We first show that for any $r < a$, we can measure out r ounces. Since $x = qa + r$ for some integers q, r with $0 \le r \le a-1$, it will follow that x can be measured out.

The construction proceeds in stages. Suppose that at a certain time, we have x_k ounces of wine in cup A. Fill cup B from the barrel, top up cup A from cup B, pour the contents of A into the barrel. Keep on filling A from B and emptying A when it becomes full. After a while, the amount that remains in B is insufficient to fill A, but we still put it in A. Call this amount x_{k+1}. It is clear then that $b = (a - x_k) + an + x_{k+1}$ for some integer n, so $x_{k+1} \equiv x_k + b \pmod{a}$. Thus if we start with A empty, the amounts in A after stages 1, 2, ... will be congruent to $1 \cdot b$, $2 \cdot b$, ... modulo a. Since a and b are relatively prime, the set $\{0 \cdot b, 1 \cdot b, \ldots, (a-1) \cdot b\}$ is a complete residue system modulo a. Thus the sequence $x_0, x_1, \ldots, x_{a-1}$ runs through the numbers $0, 1, \ldots, a-1$ in some order.

▷ **2-67.** *Let n be a positive integer, with $n \ne 1, 2, 3, 4,$ or 6. Prove that there exist integers a and b, with $1 < a < n-1$ and $1 < b < n-1$, such that $ab \equiv -1 \pmod{n}$.*

Solution. We need to find a with $1 < a < n-1$ such that $(a, n) = 1$, for then by (2.7), the congruence $ax \equiv -1 \pmod{n}$ has a solution b, and b cannot be congruent to 1 or $n-1$ modulo n. If $n > 3$ is odd, we can choose $a = 2$. If n is even and ≥ 6, consider the two numbers $n-2$ and $n-4$. Any common divisor of n and $n-2$ must divide 2. Thus, if p is an odd prime that divides $n-2$, then $(p, n) = 1$. It follows that we have found a suitable a unless $n - 2$ is a power of 2. Similarly, any common divisor of n and $n - 4$ must divide 4. Thus if p is an odd prime that divides $n - 4$, then $(p, n) = 1$. It follows that we have found a suitable a unless both $n - 2$ and $n - 4$ are powers of 2. This can only happen with $n = 6$. (A much simpler proof can be given once basic properties of the Euler ϕ-function are developed in Chapter 3.)

EXERCISES FOR CHAPTER 2

1. Find a complete residue system modulo 11 consisting (a) entirely of even integers; (b) entirely of odd integers.
2. Is $\{-3, 34, 8, 12, -1, -11\}$ a complete residue system modulo 6?
3. Determine the least nonnegative residue of $1! + 2! + \cdots + 500!$ modulo 189.
4. Find the remainder when $36!/26!$ is divided by 13.
5. Find the least positive residue of 26^{1000} modulo 29.
6. What are the last two digits in the decimal expansion of 9^{99}? (Calculate modulo 10^2.)
7. Determine the last three digits in the decimal expansion of 7^{493}. (Hint. Work modulo 10^3 and show that $7^{20} \equiv 1 \pmod{1000}$.)
8. Show that $(3^{999} - 1)/2 \equiv 13 \pmod{26}$.
9. Determine if (a) 227 divides $3^{32} + 8$; (b) 117 divides $5^{53} - 1$. (For (b), consider the least positive residue modulo 13.)
10. Prove that $169^{323} + 323^{169}$ is a multiple of 12.
11. What is the remainder when $15^{22} + 22^{15}$ is divided by 330? (Hint. Work modulo 2, 3, 5, and 11, then use the Chinese Remainder Theorem.)
12. Prove that $5^{2n+1} + 2^{8n+9}$ is a multiple of 11 for every $n \geq 1$.
13. Prove or disprove: $3^{2n+5} + 2^{4n+1}$ is divisible by 7 for every $n \geq 1$.
14. Prove that $4^{2n+1} + 3^{n+2} \equiv 0 \pmod{13}$ for every $n \geq 0$.
15. Prove that $n(13n^2 - 1)$ is divisible by 6 for every $n \geq 1$.
16. Does there exist a positive integer n such that $7n^3 - 1$ is a perfect square?
17. Prove or disprove: There exists a prime $p > 5$ such that neither $p^2 - 1$ nor $p^2 + 1$ is divisible by 10.
18. Show that the product of any four consecutive integers is divisible by 24.
19. Show that if $3a^2 - 2b^2 = 1$, then $a^2 - b^2$ is divisible by 40.
20. Find the missing digit: $1751922 \cdot 11012 = 192921x5064$.
21. If $53x0y74z$ is divisible by 264, what are the digits x, y, and z?

22. Find the inverse of (a) 7 modulo 26; (b) 13 modulo 37; (c) 5 modulo 31.
23. Determine all solutions of $51x \equiv 66 \pmod{105}$.
24. Find all solutions of $44x \equiv 76 \pmod{104}$.
25. Use the multiplication procedure described in Technique 3 before (2.10) to solve $263x \equiv 3175 \pmod{9901}$. (9901 is a prime.)
26. For which positive integers a less than 108 is the congruence $30x \equiv a \pmod{108}$ solvable?
27. Solve the following congruences: (a) $37x \equiv 20 \pmod{73}$; (b) $19x \equiv 2 \pmod{97}$; (c) $24x \equiv 30 \pmod{54}$.
28. For which positive integers m is $97 \equiv 25 \pmod{m}$?
29. Determine all solutions, if any exist, of the congruence $28x \equiv 6 \pmod{70}$.
30. Find the least positive residue of each solution of
    (a) $11x \equiv 3 \pmod{32}$;
    (b) $7x \equiv 19 \pmod{37}$;
    (c) $42x \equiv 12 \pmod{90}$.
31. Find the two smallest positive integers that leave a remainder of 2, 3, and 4 when divided by 7, 11, and 13, respectively.
32. Use the Chinese Remainder Theorem to find a solution of $x \equiv 2 \pmod{6}$, $x \equiv 6 \pmod{11}$, $x \equiv 4 \pmod{17}$.
33. What are the two smallest positive integers that leave remainders of 2, 5, and 6 when divided by 4, 7, and 9, respectively?
34. Find all solutions of the following system: $x \equiv 34 \pmod{105}$, $x \equiv 79 \pmod{330}$.
35. Find the four smallest positive integers that leave remainders of 3, 5, and 7 when divided by 9, 10, and 11, respectively.
36. Solve the following system of congruences: $5x \equiv 2 \pmod{9}$, $2x \equiv 5 \pmod{13}$, $3x \equiv 7 \pmod{17}$.
37. Use the Chinese Remainder Theorem to solve $29x \equiv 7 \pmod{1430}$.
38. (China, 1372.) A certain number of coins can be made into 78 equal-sized strings (groups), but we need to add 50 coins to make 77 equal-sized strings. What is the smallest possible number of coins needed?
39. Let r be the number of distinct prime factors of m. Show that that there are exactly 2^r integers x such that $0 \leq x < m$ and $x^2 \equiv x \pmod{m}$.
40. Let n be the product of two consecutive odd numbers. What are the possibilities for the last "digit" in the hexadecimal (base 16) representation of n? The digits are usually called 0, 1, ... , 9, A, B, C, D, E, F. (Hint. Let the integers be $2k-1$ and $2k+1$, and work modulo 16.)
41. Show that the equation $x^2 + xy - y^2 = 3$ does not have integer solutions. (Hint. Solve the equation for x, and show that the expression under the square root sign cannot be a square.)

42. If a and b are relatively prime positive integers, then the equation $ax+by = 1$ has a solution. Prove this by filling in the details of the following idea (Euler, 1760): $1 + ax$ $(x = 0, 1, 2, \ldots, b-1)$ gives distinct remainders when divided by b, so the remainders are $0, 1, \ldots, b-1$ in some order. Therefore there exists x, with $0 \leq x \leq b-1$, such that b divides $1 + ax$.
43. Christopher Columbus landed in the New World on October 12, 1492. What day of the week was this? (Note that the given date is in the Julian calendar.)
44. Carl Friedrich Gauss was born on April 30, 1777, and Pierre Fermat on August 17, 1601. What days of the week were these dates?
45. Abraham Lincoln was shot at Ford's Theatre on the evening of April 14, 1865 and died the following morning. Verify that Lincoln died on a Saturday.
46. Magna Carta was signed on June 15, 1215. Determine on which day of the week this occurred.

NOTES FOR CHAPTER 2

An Algebraic Approach to Congruences. Let m be a fixed positive integer. The symbol $\equiv$ that Gauss chose as an abbreviation for "is congruent to" encourages us to think of congruence modulo m as a kind of equality. Since m will be fixed for this discussion, we will write $a \equiv b$ rather than $a \equiv b$ (mod m) to bring out more clearly the close connection between congruence and equality.

If we now look back at (2.2.i)–(2.2.iv) and (2.3), we can see that they assert that congruence modulo m has many of the familiar properties of equality. But while there are infinitely many integers, there are only m really "different" integers modulo m. In calculations that involve addition and multiplication modulo m, the integer a can be replaced by b as long as $a \equiv b$ (mod m).

Since any integer is congruent modulo m to one of $0, 1, \ldots, m-1$, a reasonable set of names for the m different objects is $[0]_m, [1]_m, \ldots, [m-1]_m$. This is too unwieldy; a better set of names is $0, 1, \ldots, m-1$. There are dangers in this simple notation. For example, $[3]_4$ has properties that are very different from those of $[3]_5$, so calling them both 3 can cause serious problems. In particular, if we are working simultaneously with several moduli, the notation can produce hopeless confusion. But if we are working with a single modulus m, the notation has many advantages.

On this collection $\{0, 1, \ldots, m-1\}$ of abstract objects, define two operations, called "addition" and "multiplication," by letting the "sum" of i and j be the remainder when the ordinary number $i + j$ is divided by m and letting the "product" of i and j be the remainder when the ordinary product ij is divided by m. These two operations are often called *addition modulo m* and *multiplication modulo m*.

Consider, for example, the case $m = 12$; thus $7+8 = 3$, $4+8 = 0$, $5 \cdot 10 = 2$, $3 \cdot 4 = 0$, and so on. The arithmetic modulo 12 is sometimes called "clock arithmetic." If the hour hand is advanced by 7 hours and then by 8 hours, the net effect is a 3-hour advance. If it is advanced by 5 hours 10 times, the net effect is a 2-hour advance.

Addition and multiplication modulo m have a number of properties in common with ordinary addition and multiplication. In what follows, + denotes addition modulo m, and · denotes multiplication modulo m. (The dot is normally omitted.) Variables x, y, z range over the set $\{0, 1, \ldots, m-1\}$.

(i) For any x, y, and z, $x + (y + z) = (x + y) + z$.

(ii) There is an object n, called the *additive identity*, such that for any x, $x + n = n + x = x$. (Clearly, 0 is that object.)

(iii) For any x, there exists a y such that $x + y = y + x = 0$. This y is called the *additive inverse* of x. It is obvious that the additive inverse of 0 is 0, and if $0 < x < m$, the additive inverse of x is $m - x$.

A set G with an addition operation + that satisfies properties (i), (ii), and (iii) is called a *group*. For example, the set $\{0, 1, \ldots, m-1\}$ is a group under addition modulo m. Another number-theoretic example is $\{r_1, r_2, \ldots, r_k\}$, where the r_i are the numbers in the interval $0 \leq x < m$ that are relatively prime to m and the operation is multiplication modulo m. Properties (i) and (ii) are easy to verify, and property (iii) follows from (2.8). There are important examples of groups in nearly every area of mathematics.

The objects $0, 1, 2, \ldots, m-1$ under the operation of addition modulo m also satisfy the following property:

(iv) For any x and y, $x + y = y + x$.

Groups that satisfy property (iv) are called *Abelian* groups, after the nineteenth-century Norwegian mathematician Niels Abel (1802–1829).

Multiplication modulo m has the following properties in common with ordinary multiplication:

(v) for any x, y, and z, $(xy)z = x(yz)$;

(vi) for any x, y, and z, $x(y + z) = xy + xz$, and $(y + z)x = yx + zx$.

A set R with an addition and multiplication that satisfy properties (i)–(vi) is called a *ring*. If in addition

(vii) there is an object e in R, called the *multiplicative identity*, such that for all x, $ex = xe = x$ and

(viii) for all x and y, $xy = yx$,

then R is called a *commutative ring with unit*.

The set $\{0, 1, \ldots, m-1\}$ under addition and multiplication modulo m is a *commutative ring for any* $m > 1$. (The e referred to in property (vii) is the object 1.)

Let R be a commutative ring with unit. If for every element $x \neq 0$ there is an object y such that $xy = e$, then R is called a *field*. The (unique) y such that $xy = e$ is called the *multiplicative inverse* of x. One important and familiar example of a field is the set of real numbers, with the usual addition and multiplication; here, the inverse of x is the number $1/x$. Other examples include the rational numbers and the complex numbers.

The objects $0, 1, \ldots, m-1$ under addition and multiplication modulo m do not in general form a field, since if $0 < x < m$, there does not necessarily exist a y such that $xy = 1$ (i.e., $xy \equiv 1 \pmod{m}$). In fact, such a y is precisely what we have defined as the inverse of x modulo m, and if $(x, m) > 1$, x does not have an inverse modulo m. *The set $\{0, 1, \ldots, m-1\}$ under addition and multiplication modulo m is a field if and only if m is prime.*

Biographical Sketches

Ch'in Chiu-shao was born in 1202 in the province of Szechwan. After studies at the Board of Astronomy, he was appointed a military official. After that came a series of administrative appointments, despite repeated charges of corruption. Ch'in had interest in many things – astronomy, mathematics, poetry, archery, sword play.

In 1247, he published the *Shu-shu chiu-chang* ("Mathematical Treatise in Nine Sections"). The book consists of a series of solved problems, many of considerable complexity. Ch'in dealt easily with systems of linear equations and knew how to compute good approximations to zeros of polynomials. Ch'in set and solved ten problems that lead to systems of linear congruences in one variable. There is a tradition of such problems in the Chinese literature, dating back to Sun-Tzu (third century). But Ch'in's collection goes well beyond problems posed by his predecessors. His solutions make it clear that he was in possession of a general method.

Ch'in Chiu-shao died in Kuangtung province, probably in 1261.

Carl Friedrich Gauss was born in 1777 in the German city of Brunswick. Though he grew up in relative poverty, his enormous intellectual gifts were soon noticed. By 1795, he had conjectured the Prime Number Theorem and the Law of Quadratic Reciprocity and had devised the method of least squares. In 1796, he settled a 2000-year-old problem by characterizing those regular polygons that can be constructed by ruler and compass. In 1798, he gave the first proof of the Fundamental Theorem of Algebra (that every nonconstant polynomial with complex coefficients has a zero in the complex numbers). In 1801, his *Disquisitiones Arithmeticae* appeared. In addition to introducing the notion of congruence and showing its usefulness in elementary number theory, the book gave the first proof of the law of quadratic reciprocity and made fundamental advances in the analysis of quadratic forms.

In 1801, Gauss computed the orbit of the asteroid Ceres, which had been briefly observed and then lost. Ceres was found again in 1802 using Gauss's calculations, and this achievement brought Gauss world fame. In 1807, Gauss became professor of astronomy and director of the observatory at Göttingen. There he continued to make fundamental contributions in number theory, analysis, probability theory, and many other branches of mathematics. Concurrently, he was doing important work in observational astronomy, celestial mechanics, electromagnetism, optics, mechanics, and geodesy.

Gauss died in Göttingen on February 23, 1855. He is universally acknowledged to have been the greatest mathematician of his time, and perhaps of all time.

Leonardo of Pisa (Fibonacci) was born in 1175 in the city-state of Pisa. Around 1192, his father was sent to Algeria on city business. Leonardo joined him and was taught there how to calculate with the Indian-Arabic notation. On later business trips to Egypt, Syria, Sicily, and elsewhere, he had extensive contact with Muslim scholars. In 1202, he published *Liber Abaci*, an exposition of the Indian-Arabic notation that included also a large number of puzzles, including the famous rabbit problem that gives rise to the sequence now called the Fibonacci sequence. Leonardo wrote a number of other books. The deepest one mathematically is *Liber Quadratorum* (1225), which has significant results on quadratic Diophantine equations.

In 1240, Pisa recognized her famous son and awarded a yearly stipend to the "serious and learned Master Leonardo Bigollo." Nothing is known about Leonardo after this date.

REFERENCES

Carl Friedrich Gauss, *Disquisitiones Arithmeticae*, translated by Arthur A. Clarke, Yale University Press, New Haven, Connecticut, 1966.

> The most influential book in number theory ever written was published in 1801, when the author was 24 years old. In *Disquisitiones*, Gauss introduces the modern definition of congruence and residues, as well as the notation $\equiv$. The book contains the first statement and proof of the Fundamental Theorem of Arithmetic, a detailed treatment of linear congruences, the first complete proof of the Law of Quadratic Reciprocity (which we cover in Chapter 5), and a comprehensive discussion of primitive roots (see Chapter 6). All of this is done in the first quarter of *Disquisitiones*. Much of the rest is devoted to a deep and detailed study of quadratic forms.

Gauss's treatment of the basic topics is concise and elegant, and the first part of the book is surprisingly easy to read.

Donald E. Knuth, *The Art of Computer Programming, Volume 2*. (See Chapter 1.)

Ulrich Libbrecht, *Chinese Mathematics in the Thirteenth Century*, The MIT Press, Cambridge, Massachusetts, 1973.

This book deals primarily with the very influential *Shu-shu chiu-chang* of Ch'in Chiu-shao. Libbrecht devotes about 200 pages to a history of the Chinese Remainder Theorem, paying particular attention to Chinese contributions.

CHAPTER THREE

The Theorems of Fermat, Euler, and Wilson

The first mention of Fermat's Theorem in the European literature was in June, 1640, in a letter from Pierre Fermat to the Franciscan friar Marin Mersenne. In it, he asserts that if p is prime, then $2^p - 2$ is a multiple of $2p$, and that if q is a prime divisor of $2^p - 1$, then $q - 1$ is a multiple of p. In his letter of October 18, 1640 to the Parisian number-hobbyist Frenicle de Bessy (1605–1675), Fermat claims that if p is prime and a a positive integer, then p divides $a^n - 1$ for some n, and that the smallest n for which this holds divides $p - 1$.

In the letter to Frenicle, Fermat writes that he has a proof and that he would send it if he did not fear its being too long. Unfortunately, Fermat systematically withheld proofs of his results. There was a long tradition for this kind of behavior; mathematicians challenged each other with problems and so were loath to make public any special techniques they might have found. By Fermat's time, this tradition was dying, and not many years later, scientific journals began to appear.

Leibniz proved Fermat's Theorem around 1680, but the proof was left among his manuscripts and came to light only in 1863. After Fermat, number theory entered a long period of dormancy. In 1730, the subject was revived at the hands of Euler, who rediscovered Fermat's Theorem, published a proof based on the Binomial Theorem in 1736, and published a more algebraic proof that he himself preferred in 1758, which led quickly to the generalization to composite moduli that we call Euler's Theorem.

Wilson's Theorem states that if p is prime, then $(p - 1)! + 1$ is divisible by p. The first mention in print of Wilson's Theorem was in 1770, by the English mathematician Edward Waring (1734–1798). He gave credit to his former student John Wilson, though in fact the result appeared a hundred years earlier in a manuscript of Leibniz. Leibniz's version states that if p is prime, then p divides $(p-2)! - 1$, but the two versions can be easily shown to be equivalent.

Waring wrote that he could not prove the result and that a proof must be very difficult, since there is no notation for primes. What he meant is that in the absence of a "formula" for primes, he could not produce a proof; the only sort of proof he could imagine was by a symbolic manipulation. (Gauss acerbically wrote that truths of this kind should be drawn from notions, not notations.) The first published proof was given by Lagrange in 1771. Lagrange's proof uses a fairly complicated manipulation of the polynomial $(x+1)(x+2)\cdots(x+p-1)$. In 1773, Euler gave a proof using primitive roots. A simpler conceptual proof (essentially the same as the unpublished proof of Leibniz) was given by Gauss in the pivotal *Disquisitiones Arithmeticae*.

RESULTS FOR CHAPTER 3

Fermat's Theorem and Wilson's Theorem

The proof of Fermat's Theorem that we give in (3.6) is due to Dirichlet. It is a mild variant of Gauss's proof of Wilson's Theorem and has the advantage of giving simultaneously proofs of Fermat's Theorem, Wilson's Theorem, and information about the congruence $x^2 \equiv a \pmod{p}$, which will be taken up systematically in Chapter 5. We require the following lemma.

(3.1) Lemma. *Let p be an odd prime, and suppose $p \nmid a$. If there exists a number b such that $b^2 \equiv a \pmod{p}$, then the congruence $x^2 \equiv a \pmod{p}$ has precisely two incongruent solutions modulo p.*

Proof. There are at least two incongruent solutions, since $(-b)^2 \equiv a \pmod{p}$ and $b \not\equiv -b \pmod{p}$ because $p \neq 2$. To show that there are only two incongruent solutions, suppose $x^2 \equiv a \pmod{p}$. Then $x^2 \equiv b^2 \pmod{p}$, so $p \mid x^2 - b^2$, and therefore $p \mid (x-b)(x+b)$. Hence $p \mid x-b$ or $p \mid x+b$. In the first case, $x \equiv b \pmod{p}$, and in the second, $x \equiv -b \pmod{p}$.

Informally, (3.1) says that if p is an odd prime and the number a has a "square root" modulo p, then a has precisely two square roots modulo p.

(3.2) Theorem (Dirichlet, 1828). *Let p be prime, and suppose $1 \leq a \leq p-1$. If the congruence $x^2 \equiv a \pmod{p}$ does not have a solution, then $(p-1)! \equiv a^{(p-1)/2} \pmod{p}$. If the congruence has a solution, then $(p-1)! \equiv -a^{(p-1)/2} \pmod{p}$.*

Proof. The result is obvious if $p = 2$, so assume that p is odd. By (2.8), if $1 \leq m \leq p-1$, there exists a unique n such that $1 \leq n \leq p-1$ and $mn \equiv a$

(mod p). Call m and n *corresponding numbers*. If $x^2 \equiv a$ (mod p) does not have a solution, then $m \ne n$ for corresponding numbers m, n. Therefore the numbers from 1 to $p-1$ can be paired off into $(p-1)/2$ corresponding pairs. The product of all the corresponding pairs is congruent to $a^{(p-1)/2}$ (mod p) but is clearly also equal to $(p-1)!$. Therefore $(p-1)! \equiv a^{(p-1)/2}$ (mod p).

If $x^2 \equiv a$ (mod p) is solvable, it has a solution b such that $1 \le b \le p-1$. The only other solution in this range is $p-b$. Arrange the remaining $p-3$ numbers from 1 to $p-1$ into corresponding pairs; their product is then congruent to $a^{(p-3)/2}$ (mod p). The product of the two remaining numbers b and $p-b$ is congruent to $-a$ (mod p). Hence $(p-1)! \equiv (-a)a^{(p-3)/2} = -a^{(p-1)/2}$ (mod p).

One significant consequence of Dirichlet's result is Wilson's Theorem. Lagrange gave a somewhat complicated proof in 1771, but Wilson's Theorem follows very simply from (3.2).

(3.3) Wilson's Theorem. *If p is prime, then $(p-1)! \equiv -1$ (mod p).*

Proof. In (3.2), choose $a = 1$. Certainly, $x^2 \equiv 1$ (mod p) has a solution, so $(p-1)! \equiv -1^{(p-1)/2} = -1$ (mod p).

Note. The converse of Wilson's Theorem is also true (see Problem 3-22). Thus Wilson's Theorem provides a *primality test*, although a highly impractical one. If n is a large integer, say roughly of size 10^{100}, there seems to be no feasible procedure for directly computing the remainder when $(n-1)!$ is divided by n.

As an easy consequence of (3.2) and Wilson's Theorem, we have the following useful characterization of when the congruence $x^2 \equiv a$ (mod p) has a solution.

(3.4) Euler's Criterion. *Let p be an odd prime, and suppose $p \nmid a$. Then $x^2 \equiv a$ (mod p) is solvable or not solvable according as $a^{(p-1)/2} \equiv 1$ (mod p) or $a^{(p-1)/2} \equiv -1$ (mod p).*

Proof. Suppose the congruence has a solution. Then by (3.2) and Wilson's Theorem, $-1 \equiv -a^{(p-1)/2}$ (mod p), and multiplication by -1 gives the result. The argument when the congruence has no solution is essentially the same.

The following important result is an immediate consequence of Euler's Criterion.

(3.5) Theorem. *Let p be an odd prime. Then the congruence $x^2 \equiv -1$ (mod p) has a solution if p is of the form $4k+1$ and does not have a solution if p is of the form $4k+3$.*

74   CHAPTER 3: THE THEOREMS OF FERMAT, EULER, AND WILSON

We now state Fermat's Theorem. It is one of the basic results of elementary number theory and will be used often in subsequent proofs and calculations.

(3.6) Fermat's Theorem. *Let p be prime, and suppose $p \nmid a$. Then $a^{p-1} \equiv 1 \pmod{p}$. Equivalently, if a is any integer, then $a^p \equiv a \pmod{p}$.*

Proof. By Euler's Criterion, $a^{(p-1)/2} \equiv \pm 1 \pmod{p}$. Square both sides of the congruence. To get the alternative form of Fermat's Theorem, multiply by a, and observe that the modified form is also true when $a \equiv 0 \pmod{p}$.

Notes. 1. A more direct route to Fermat's Theorem is to use the argument in the proof of Euler's Theorem (3.13). For Euler's 1758 proof of Fermat's Theorem, see Note 2 at the end of the chapter.

2. The converse of Fermat's Theorem is false: m need not be prime even if $a^{m-1} \equiv 1 \pmod{m}$ holds for *every* a relatively prime to m (see Problems 3-39 and 3-40). A correct converse of Fermat's Theorem is given in Problem 3-55 and also in Chapter 6 (see Theorem 6.9).

We can make a conjectural reconstruction of Fermat's reasoning for the assertion that $2p$ divides $2^p - 2$. At some earlier time, probably by 1636, Fermat had discovered an identity for binomial coefficients that in modern language can be written as $n\binom{n+m-1}{m-1} = m\binom{n+m-1}{m}$. This identity is an easy consequence of the usual formula for binomial coefficients in terms of factorials, but Pascal did not systematically explore binomial coefficients until 1654. From Fermat's identity, one can prove by induction that if p is prime and $1 \le k \le p-1$, then p divides $\binom{p}{k}$. The Binomial Theorem gives

$$2^p = (1+1)^p = 1 + \binom{p}{1} + \binom{p}{2} + \cdots + \binom{p}{p-1} + 1,$$

and it follows that $2^p - 2$ is divisible by p. Since $2^p - 2$ is also divisible by 2, Fermat's result follows for odd primes p.

Similar reasoning can be used to prove that if p is prime, then $a^p - a$ is divisible by p (see Problem 3-14), but the wording in Fermat's original statement of his theorem suggests that he may have used a more conceptual approach, such as the one used by Euler in his 1758 proof of Fermat's Theorem. (See the Notes at the end of this chapter for a sketch.)

We prove next, in Theorems 3.8 and 3.9, the rest of Fermat's 1640 assertion. We need a preliminary lemma that will also be useful elsewhere.

(3.7) Lemma. *Let a, u, v be integers, with u and v positive, let $d = (u, v)$, and let $m > 0$. If $a^u \equiv 1 \pmod{m}$ and $a^v \equiv 1 \pmod{m}$, then $a^d \equiv 1 \pmod{m}$.*

Proof. By (1.5), d can be expressed as an integer linear combination of u and v, say, $su + tv = d$. One of s and t will not be positive. Without loss of generality we may assume that it is t, so $su = d + |t|v$. Then $(a^u)^s = a^d(a^v)^{|t|}$. Since $a^u \equiv a^v \equiv 1 \pmod{m}$, it follows that $a^d \equiv 1 \pmod{m}$.

(3.8) Theorem. *Let q be a prime divisor of $2^p - 1$, where p is an odd prime. Then q is of the form $2kp + 1$.*

Proof. By Fermat's Theorem, $2^{q-1} \equiv 1 \pmod{q}$, and by assumption, $2^p \equiv 1 \pmod{q}$. Therefore (3.7) implies that $2^d \equiv 1 \pmod{q}$, where $d = (p, q-1)$. But since p is prime, the greatest common divisor of p and $q - 1$ is either 1 or p. Now $(p, q-1)$ cannot be 1, since if it were, we would have $2^1 \equiv 1 \pmod{q}$, a contradiction. Thus $(p, q-1) = p$, so p divides $q - 1$. But q is odd since $2^p - 1$ is odd, and therefore $q - 1$ is even. It follows that $2p$ divides $q - 1$, and hence $q \equiv 1 \pmod{2p}$, that is, q is of the form $2kp + 1$.

The preceding result can be used to check $2^p - 1$ for primality (see Chapter 7, Problems 7-15 to 7-19).

(3.9) Theorem. *Let p be prime, and suppose $p \nmid a$. Let n be the smallest positive integer such that p divides $a^n - 1$. (Such an n exists by Fermat's Theorem.) Then n divides $p - 1$.*

Proof. Fermat's Theorem implies $a^{p-1} \equiv 1 \pmod{p}$. Let $d = (n, p-1)$. By (3.7), $a^d \equiv 1 \pmod{p}$. Since n is the smallest positive integer such that $p | a^n - 1$, it follows that $d = n$ and therefore that $n | p - 1$.

Euler's Theorem and the Euler ϕ-function

It is natural to ask whether a result similar to Fermat's Theorem holds when the modulus is not prime. The answer is yes. Euler found an appropriate generalization and published a proof in 1760. In the same paper, Euler studied basic properties of the ϕ-*function*, which is key to formulating the generalization.

(3.10) Definition. *If $m > 1$, let $\phi(m)$ be the number of positive integers less than m that are relatively prime to m. Define $\phi(1)$ to be 1. The function ϕ is usually called the* Euler ϕ-*function*.

It is clear that $\phi(m) \leq m - 1$ for every $m > 1$. Also, $\phi(m) = m - 1$ if and only if m is prime. (See Problem 3-59.)

(3.11) Definition. *Let m be positive. A* reduced residue system modulo m *is a set of integers such that every number relatively prime to m is congruent modulo m to a unique element of the set.*

Since any two reduced residue systems modulo m have the same number of elements, they all have $\phi(m)$ elements. We will generally (but not always) take the elements of a reduced residue system modulo m to be the $\phi(m)$ positive integers less than m and relatively prime to m. In particular, if p is a prime, note that $\{1, 2, \ldots, p-1\}$ is a reduced residue system modulo p.

(3.12) Lemma. *Let $r_1, r_2, \ldots, r_k$ be a reduced residue system modulo m, and suppose $(a, m) = 1$. Then $ar_1, ar_2, \ldots, ar_k$ is a reduced residue system modulo m.*

Proof. We must show that no two elements of the sequence $ar_1, ar_2, \ldots, ar_k$ are congruent to each other modulo m and that $(ar_i, m) = 1$ for $1 \leq i \leq k$. Suppose $ar_i \equiv ar_j \pmod{m}$ for some i and j with $i \neq j$. Since $(a, m) = 1$, by (2.2.vi) we may divide both sides of the congruence by a to get $r_i \equiv r_j \pmod{m}$, which contradicts the fact that $r_1, r_2, \ldots, r_k$ is a reduced residue system. Also, since $(r_i, m) = 1$ for all i and $(a, m) = 1$, it follows from (1.10) that $(ar_i, m) = 1$. Thus $ar_1, ar_2, \ldots, ar_k$ is a collection of k incongruent integers, each of which is relatively prime to m, and so forms a reduced residue system modulo m.

We are now ready to prove Euler's Theorem. The proof given above for Fermat's Theorem could be modified to deal with a composite modulus, but the following argument is much tidier. It was originally used by Ivory in 1804 to prove Fermat's Theorem and modified by Horner in 1826 to yield Euler's Theorem.

(3.13) Euler's Theorem. *Let m be a positive integer and suppose $(a, m) = 1$. Then $a^{\phi(m)} \equiv 1 \pmod{m}$.*

Proof. Let $r_1, r_2, \ldots, r_{\phi(m)}$ be a reduced residue system modulo m. By the preceding lemma, $ar_1, ar_2, \ldots, ar_{\phi(m)}$ is also a reduced residue system modulo m. Thus we have

$$(ar_1)(ar_2) \cdots (ar_{\phi(m)}) \equiv r_1 r_2 \cdots r_{\phi(m)} \pmod{m},$$

since $ar_1, ar_2, \ldots, ar_{\phi(m)}$, being a reduced residue system, must be congruent in some order to $r_1, r_2, \ldots, r_{\phi(m)}$. Hence $a^{\phi(m)} r_1 r_2 \cdots r_{\phi(m)} \equiv r_1 r_2 \cdots r_{\phi(m)} \pmod{m}$. Since each r_i is relatively prime to m, repeated application of (2.2.vi) yields $a^{\phi(m)} \equiv 1 \pmod{m}$.

Note. It is a simple consequence of Euler's Theorem that if $(a, m) = 1$, then $a^{\phi(m)-1}$ is a multiplicative inverse of a modulo m.

In applying Euler's Theorem, we must first determine $\phi(m)$. If we know the prime factorization of m, an efficient way of calculating $\phi(m)$ is to use the fact that ϕ is *multiplicative*. It is then a simple matter to derive a formula for $\phi(m)$ in terms of the prime factorization of m.

We first show that ϕ is a multiplicative function. The proof is a nice application of the Chinese Remainder Theorem.

EULER'S THEOREM AND THE EULER ϕ-FUNCTION 77

(3.14) Theorem. *If m and n are relatively prime, then $\phi(mn) = \phi(m)\phi(n)$.*

Proof. For $0 \leq x < mn$, let x_m be the remainder when x is divided by m and x_n the remainder when x is divided by n. Distinct numbers in the interval from 0 to $mn - 1$ give rise to distinct remainder pairs. For if $x_m = y_m$ and $x_n = y_n$, then $x \equiv y \pmod{m}$ and $x \equiv y \pmod{n}$; hence $x \equiv y \pmod{mn}$, since $(m, n) = 1$. But there are mn numbers in the range $0 \leq x < mn$ and mn pairs a, b with $0 \leq a < m$ and $0 \leq b < n$. Hence every such pair a, b is the pair x_m, x_n for some x.

Since $x \equiv x_m \pmod{m}$, we have $x = x_m + qm$ for some q. Thus $(x, m) = (x_m, m)$ by (1.22). Similarly, $(x, n) = (x_n, n)$. Hence, by (1.10), $(x, mn) = 1$ if and only if $(x_m, m) = 1$ and $(x_n, n) = 1$. Since there are $\phi(m)\phi(n)$ ways of choosing pairs a, b with $(a, m) = (b, n) = 1$ and $\phi(mn)$ ways of choosing $x_{a,b}$ relatively prime to mn (by the definition of ϕ), it follows that $\phi(mn) = \phi(m)\phi(n)$.

Note. The first paragraph of the proof above shows that if $(m, n) = 1$, then for any a and b, there exists a unique x in the interval $0 \leq x < mn$ such that $x \equiv a \pmod{m}$ and $x \equiv b \pmod{n}$. This proves the Chinese Remainder Theorem for a pair of congruences. The proof of the Chinese Remainder Theorem for k congruences can be obtained in exactly the same way by considering k-tuples of remainders. The argument given in Chapter 2, however, is more informative, for it provides a practical method of finding the solution of a system of linear congruences.

(3.15) Lemma. *If p is a prime, then $\phi(p^k) = p^k - p^{k-1}$ for every $k \geq 1$.*

Proof. Suppose that a is less than p^k and *not* relatively prime to p^k; then a must be divisible by p and thus $a = mp$, where $1 \leq m \leq p^{k-1}$. Clearly, there are precisely p^{k-1} choices for a. Hence there are exactly $p^k - p^{k-1}$ integers less than p^k that *are* relatively prime to p^k.

(3.16) Theorem. *Let $m = p_1^{m_1} p_2^{m_2} \cdots p_r^{m_r}$. Then*

$$\phi(m) = (p_1^{m_1} - p_1^{m_1-1})(p_2^{m_2} - p_2^{m_2-1}) \cdots (p_r^{m_r} - p_r^{m_r-1})$$
$$= p_1^{m_1-1} p_2^{m_2-1} \cdots p_r^{m_r-1}(p_1 - 1)(p_2 - 1) \cdots (p_r - 1)$$
$$= m(1 - 1/p_1)(1 - 1/p_2) \cdots (1 - 1/p_r).$$

Proof. Since ϕ is multiplicative and the $p_i^{m_i}$ are relatively prime in pairs, the first equation follows from (3.15). The other two equations are simple variants of the first.

PROBLEMS AND SOLUTIONS

Fermat's Theorem and Wilson's Theorem

3-1. *Find an integer n, with $0 \le n \le 16$, such that $3^{100} \equiv n$ (mod 17).*

Solution. By Fermat's Theorem, $3^{16} \equiv 1$ (mod 17), so $3^{96} = (3^{16})^6 \equiv 1$ (mod 17). Therefore $3^{100} = 3^4 3^{96} \equiv 3^4$ (mod 17). But $3^4 \equiv 13$ (mod 17), so $n = 13$.

3-2. *Find the remainders when the following numbers are divided by 37: (a) 2^{52}; (b) 2^{70}.*

Solution. (a) $52 \equiv 16$ (mod 36), so $2^{52} \equiv 2^{16}$ (mod 37), by Fermat's Theorem. We can calculate 2^{16} modulo 37 efficiently by repeated squaring modulo 37: $2^2 \equiv 4$, $2^4 \equiv 16$, $2^8 \equiv (16)^2 \equiv 34 \equiv -3$, all modulo 37, so $2^{16} \equiv 9$ (mod 37), and therefore the remainder is 9.

(b) $2^{70} \equiv 2^{34}$ (mod 37). Since by (a) we already know that $2^{16} \equiv 9$ (mod 37), we find that $2^{32} \equiv 81 \equiv 7$ (mod 37), so $2^{34} \equiv 2^2 \cdot 7 \equiv 28$ (mod 37). But it is perhaps easier to observe that $2^2 \cdot 2^{34} \equiv 1$ (mod 37), so 2^{34} is a solution of the congruence $4x \equiv 1$ (mod 37). Since $4 \cdot 9 \equiv -1$ (mod 37), $x \equiv -9$ (mod 37), so the remainder is 28.

3-3. *What is the remainder when 55^{142} is divided by 143?*

Solution. Note that $143 = 11 \cdot 13$. Clearly, $55^{142} \equiv 0$ (mod 11), since $55 \equiv 0$ (mod 11). Since $142 \equiv 10$ (mod 12), $55^{142} \equiv 55^{10}$ (mod 13) by Fermat's Theorem. But $55 \equiv 3$ (mod 13) and $3^3 \equiv 1$ (mod 13), so $3^{10} \equiv 3$ (mod 13). We want the smallest positive integer that is divisible by 11 and congruent to 3 (mod 13). By the Chinese Remainder Theorem, or by inspection, the answer is 55.

3-4. *Find the remainder when $13 \cdot 12^{45}$ is divided by 47.*

Solution. By Fermat's Theorem, $12 \cdot 12^{45} \equiv 1$ (mod 47). Multiplying each side by 4 gives $12^{45} \equiv 4$ (mod 47), since $12 \cdot 4 \equiv 1$ (mod 47). It follows that $13 \cdot 12^{45} \equiv 5$ (mod 47); thus the remainder is 5.

3-5. *Show that $5555^{2222} + 2222^{5555}$ is divisible by 7.*

Solution. $2222 \equiv 2$ (mod 6) and $5555 \equiv 5$ (mod 6), and 5555 and 2222 are congruent to 4 and 3 (mod 7), respectively. So $5555^{2222} + 2222^{5555} \equiv 4^2 + 3^5$ (mod 7). Direct calculation modulo 7 now gives the result.

3-6. *Show that for any nonnegative integer n, $3^{6n} - 2^{6n}$ is divisible by 35.*

Solution. By Fermat's Theorem or otherwise, $(3^6)^n \equiv 1$ (mod 7) and $(2^6)^n \equiv 1$ (mod 7), so $7 | 3^{6n} - 2^{6n}$. Since $2 \equiv -3$ (mod 5), we have $2^6 \equiv (-3)^6 = 3^6$ (mod 5), and hence $5 | 3^{6n} - 2^{6n}$. Since 5 and 7 are relatively prime, it follows that $3^{6n} - 2^{6n}$ is a multiple of 35.

3-7. *Show that for any positive integer n, the number $1^n+2^n+3^n+4^n$ is divisible by 5 if and only if n is not divisible by 4.*

Solution. Let $n = 4q + r$, where $0 \le r \le 3$. If $a \not\equiv 0 \pmod 5$, then $a^n = (a^4)^q a^r \equiv a^r$ (mod 5) by Fermat's Theorem, and so $1^n + 2^n + 3^n + 4^n \equiv 1^r + 2^r + 3^r + 4^r \pmod 5$.

If n is a multiple of 4, then $r = 0$ and $1^r + 2^r + 3^r + 4^r = 4$, so 5 does not divide $1^n + 2^n + 3^n + 4^n$. If 4 does not divide n, then $r = 1, 2,$ or 3. We can now compute directly, finding easily that 5 divides $1^r + 2^r + 3^r + 4^r$ for these values of r.

3-8. *Let $S_k = 1^k + 2^k + \cdots + (p-1)^k$, where p is an odd prime. If k is divisible by $p - 1$, prove that $S_k \equiv -1 \pmod p$.*

Solution. By Fermat's Theorem, $t^{p-1} \equiv 1 \pmod p$ for $1 \le t \le p-1$, and therefore $t^k \equiv 1 \pmod p$. So modulo p, S_k is a sum of $p - 1$ 1's.

3-9. *Show that if n is an integer, then so is $n^5/5 + n^3/3 + 7n/15$.*

Solution. Since $n^5/5 + n^3/3 + 7n/15 = (3n^5 + 5n^3 + 7n)/15$, we need to show that $3n^5 + 5n^3 + 7n$ is divisible by 15 for all n. It is enough to show that for all n, $3n^5 + 5n^3 + 7n$ is divisible by 3 and by 5. But $3n^5 + 5n^3 + 7n \equiv 2n^3 + n \equiv 2n^3 - 2n \pmod 3$, and $n^3 - n \equiv 0$ (mod 3) for all n by Fermat's Theorem. Also, $3n^5 + 5n^3 + 7n \equiv 3n^5 + 2n \equiv 3n^5 - 3n$ (mod 5), and $n^5 - n \equiv 0 \pmod 5$ for all n by Fermat's Theorem.

3-10. *Use Fermat's Theorem to solve the congruence $x^{35} + 5x^{19} + 11x^3 \equiv 0$ (mod 17).*

Solution. By Fermat's Theorem, $x^{17} \equiv x \pmod{17}$ for every x, and thus $x^{35} = x(x^{17})^2 \equiv x^3 \pmod{17}$ for all x. Similarly, $5x^{19} \equiv 5x^3 \pmod{17}$ for every x. Hence, for each x, $x^{35} + 5x^{19} + 11x^3 \equiv 17x^3 \equiv 0 \pmod{17}$, and therefore the original congruence holds for every x. Another way of handling the problem is to divide $x^{35} + 5x^{19} + 11x^3$ by $x^{17} - x$ using ordinary long division of polynomials. We get that $x^{35} + 5x^{19} + 11x^3 = (x^{17} - x)(x^{18} + 6x^2) + 17x^3$. Since $x^{17} - x \equiv 0 \pmod{17}$ for all x by Fermat's Theorem, and $17x^3 \equiv 0 \pmod{17}$, the congruence hold for all x.

3-11. *Reduce the congruence $304x^{303} + 204x^{202} - 104x^{101} \equiv 0 \pmod{101}$ to one of degree 3, and find all solutions. (Note that 101 is prime.)*

Solution. Since $x^{101} \equiv x \pmod{101}$ for all x, by Fermat's Theorem, we obtain the equivalent congruence $x^3 + 2x^2 - 3x \equiv 0 \pmod{101}$, or $x(x-1)(x+3) \equiv 0 \pmod{101}$. The solutions are then all numbers congruent to one of 0, 1, or -3 modulo 101.

3-12. *Suppose that p is prime and $a^p + b^p = c^p$. Show that p divides $a + b - c$.*

Solution. By Fermat's Theorem, $a^p + b^p - c^p \equiv a + b - c \pmod p$. So if $a^p + b^p = c^p$, then $a + b - c \equiv 0 \pmod p$.

3-13. Let p and q be distinct odd primes such that $p-1$ divides $q-1$. If $(a, pq) = 1$, prove that $a^{q-1} \equiv 1 \pmod{pq}$.

Solution. By Fermat's Theorem, $a^{q-1} \equiv 1 \pmod{q}$. We need to show also that $a^{q-1} \equiv 1 \pmod{p}$. Since p and q are relatively prime, it will follow that $a^{q-1} \equiv 1 \pmod{pq}$. Let $q - 1 = k(p-1)$. Then $a^{q-1} = (a^{p-1})^k \equiv 1^k \pmod{p}$.

▷ **3-14.** Suppose p is prime. By using the binomial expansion of $(a+b)^p$, show that $(a+b)^p \equiv a^p + b^p \pmod{p}$. Do not use Fermat's Theorem.

Solution. By the Binomial Theorem, $(a+b)^p = \sum_{k=0}^{p} \binom{p}{k} a^{p-k} b^k$. Hence $(a+b)^p - (a^p + b^p)$ is a sum of terms of the form $\binom{p}{k} a^{p-k} b^k$, where $1 \le k \le p-1$. But $\binom{p}{k}$ is divisible by p. We can see this, for example, by noting that $\binom{p}{k} k!(p-k)! = p!$, and since p divides $p!$ but does not divide either $k!$ or $(p-k)!$, p must divide $\binom{p}{k}$. Thus $(a+b)^p - (a^p + b^p)$ is a sum of terms each divisible by p, and the result follows.

Note. Let $b = 1$ in the above result. Then $(a+1)^p \equiv a^p + 1 \pmod{p}$. So if we already know that $a^p \equiv a \pmod{p}$, it follows that $(a+1)^p \equiv a+1 \pmod{p}$. This provides the induction step in the first published proof of Fermat's Theorem (Euler, 1736).

3-15. Use Fermat's Theorem to show that any prime $p > 5$ divides infinitely many numbers of the form $999\ldots 99$ (i.e., numbers whose representation to the base 10 has 9's only).

Solution. We are asked to show that for any prime $p > 5$, $10^n \equiv 1 \pmod{p}$ for infinitely many n. Since 10 and p are relatively prime, $10^{p-1} \equiv 1 \pmod{p}$, so any positive multiple n of $p-1$ has the required property.

▷ **3-16.** Show that if $n \mid 2^n - 1$, then $n = 1$. (Hint. If $n > 1$, let p be the smallest prime divisor of n, and use (3.7).)

Solution. Suppose $n > 1$ and $n \mid 2^n - 1$, and let p be the smallest prime divisor of n. By Fermat's Theorem, we have $2^{p-1} \equiv 1 \pmod{p}$; since $p \mid n$ and $n \mid 2^n - 1$, it follows that $2^n \equiv 1 \pmod{p}$. Let $d = (n, p-1)$. If $d > 1$, then n has a divisor greater than 1 but less than p, contradicting the choice of p. Thus $d = 1$. But (3.7) implies that $2^d \equiv 1 \pmod{p}$, that is, $2^1 \equiv 1 \pmod{p}$, which is impossible.

3-17. (a) Arrange the numbers $2, 3, \ldots, 17$ in pairs $\{x, y\}$ such that $xy \equiv 1 \pmod{19}$. (b) Use part (a) to find the least positive residue of 18! modulo 19.

Solution. (a) The pairs are $\{2, 10\}, \{17, 9\}, \{3, 13\}, \{16, 6\}, \{4, 5\}, \{15, 14\}, \{7, 11\}, \{12, 8\}$. (The pairs $\{x, y\}$ and $\{19 - x, 19 - y\}$ have been put next to each other, since the work can be cut in half by noting that $ab \equiv 1$ implies $(-a)(-b) \equiv 1$.) (b) Paired elements have product 1 modulo 19, so $18! \equiv 18 \pmod{19}$, and hence the least positive residue is 18.

3-18. *Show that if p is an odd prime, then $2(p-3)! \equiv -1 \pmod{p}$. Find the remainder when $56!$ is divided by 59.*

Solution. $(p-1)! = (p-1)(p-2)(p-3)! \equiv (-1)(-2)(p-3)! \pmod{p}$. But $(p-1)! \equiv -1 \pmod{p}$ by Wilson's Theorem, and thus $2(p-3)! \equiv -1 \pmod{p}$. In particular, $2(56)! \equiv -1 \equiv 58 \pmod{59}$, and therefore the remainder is 29.

3-19. *Find the remainder when $90!$ is divided by 97.*

Solution. Since Wilson's Theorem implies that $96! \equiv -1 \pmod{97}$, it is better to proceed backward from 96 rather than forward from 1. To keep numbers small, we use the fact that $97 - x \equiv -x \pmod{97}$. So $96 \cdot 95 \cdots 91 \cdot 90! \equiv (-1)(-2) \cdots (-6) \cdot 90! \equiv -1 \pmod{97}$. But $6! \equiv 41 \pmod{97}$, and therefore $56 \cdot 90! \equiv 1 \pmod{97}$. Solving the congruence $56x \equiv 1 \pmod{97}$ by the Euclidean Algorithm, we find that $x \equiv 26 \pmod{97}$, so the remainder is 26.

3-20. *Let $y = 82!/21$. What is the remainder when y is divided by 83?*

Solution. By Wilson's Theorem, $21y \equiv -1 \pmod{83}$. Now $21 \cdot 4 \equiv 1 \pmod{83}$ and therefore $y \equiv -4 \pmod{83}$. Thus the remainder is 79.

3-21. *Find the remainder when $18!$ is divided by 437. (First factor 437.)*

Solution. $437 = 19 \cdot 23$. Since $18! \equiv -1 \pmod{19}$, it remains to calculate modulo 23. Now $22! \equiv -1 \pmod{23}$ by Wilson's Theorem, so $18! \cdot 19 \cdot 20 \cdot 21 \cdot 22 \equiv -1 \pmod{23}$. But $22, 21, 20, 19$ are congruent to $-1, -2, -3$, and -4 modulo 23, respectively, so their product is congruent to 1 modulo 23. Therefore $18! \equiv -1 \pmod{23}$, and hence the remainder is 436.

3-22. *Prove the converse of Wilson's Theorem: If $m > 1$ and m is not prime, then $(m-1)! \not\equiv -1 \pmod{m}$.*

Solution. Since m is not prime, there exists an integer t, with $1 < t < m$, such that $t \mid m$. But then $t \mid (m-1)!$, so if $(m-1)! \equiv -1 \pmod{m}$, it would follow that $t \mid -1$, which is false.

Note. We can prove a stronger result. Note that $(4-1)! \equiv 2 \pmod{4}$. We show that if $m > 4$ is composite, then $(m-1)! \equiv 0 \pmod{m}$. Let p be a prime dividing m, and suppose $m \neq p^2$. Then $p < m$, $m/p < m$, and $p \neq m/p$, so $m \mid (m-1)!$. Now we deal with squares of primes. If $m = p^2$ and $p \neq 2$, then $p < m$ and $2p < m$, so again $m \mid (m-1)!$.

▷ **3-23.** *Find all integers $n > 1$ such that $n(n+1) \mid (n-1)!$. (Hint. See the preceding Note.)*

Solution. If n is prime, then $(n-1)! \equiv -1 \pmod{n}$ by Wilson's Theorem, so in particular n cannot divide $(n-1)!$. If $n+1$ is prime, then $n! \equiv -1 \pmod{n+1}$. But $n! = n(n-1)! \equiv -(n-1)! \pmod{n+1}$, so $(n-1)! \equiv 1 \pmod{n+1}$ and hence $n+1$ cannot divide $(n-1)!$. We have thus ruled out all n that are prime or 1 less than a prime.

82 CHAPTER 3: THE THEOREMS OF FERMAT, EULER, AND WILSON

We now show that for *all* other $n > 1$, $n(n+1)$ divides $(n-1)!$. Since $(n, n+1) = 1$, it is enough to prove that if neither n nor $n+1$ is prime, then each divides $(n-1)!$. The preceding Note showed that except in the case $n = 4$ (which is not relevant here, since 5 is prime), $n \mid (n-1)!$ if n is composite. Essentially the same argument shows that if $n+1$ is composite and not equal to 4, then $n+1$ divides $(n-1)!$.

3-24. *Show that for every prime number p and every integer a, the number $a^p + (p-1)!a$ is divisible by p.*

Solution. Fermat's Theorem implies $a^p \equiv a \pmod{p}$, so $a^p + (p-1)!a \equiv a(1 + (p-1)!) \pmod{p}$. But $1 + (p-1)! \equiv 0 \pmod{p}$ by Wilson's Theorem.

Note. The above result "contains" Wilson's Theorem (take $a = 1$). It also "contains" Fermat's Theorem: Since $(p-1)! + 1 \equiv 0 \pmod{p}$, it follows from $a^p + (p-1)!a \equiv 0 \pmod{p}$ that $a^p - a \equiv 0 \pmod{p}$.

3-25. (a) *Let $r_1, r_2, \ldots, r_{p-1}$ and $s_1, s_2, \ldots, s_{p-1}$ be reduced residue systems modulo the odd prime p. Show that $r_1 s_1, r_2 s_2, \ldots, r_{p-1} s_{p-1}$ cannot be a reduced residue system modulo p. (Hint. Use Wilson's Theorem.)*

(b) *Let $r_1, r_2, \ldots, r_p$ and $s_1, s_2, \ldots, s_p$ be complete residue systems modulo the odd prime p. Show that $r_1 s_1, r_2 s_2, \ldots, r_p s_p$ cannot be a complete residue system modulo p.*

Solution. (a) By Wilson's Theorem, the product of the r_i is congruent to -1 modulo p, as is the product of the s_i. Thus the product of the $r_i s_i$ is congruent to 1 modulo p. If the $r_i s_i$ formed a reduced residue system modulo p, then this product would be congruent to $-1 \pmod{p}$ by Wilson's Theorem. But if $p > 2$, then $1 \not\equiv -1 \pmod{p}$.

(b) Without loss of generality we may assume that $r_p \equiv 0 \pmod{p}$. We must then have $s_p \equiv 0 \pmod{p}$, for if $s_j \equiv 0 \pmod{p}$ for some $j \ne p$, then both $r_j s_j$ and $r_p s_p$ are congruent to 0 $\pmod{p}$, and hence $r_1 s_1, r_2 s_2, \ldots, r_p s_p$ is not a complete residue system modulo p. But if $r_p \equiv s_p \equiv 0 \pmod{p}$, then $r_1, r_2, \ldots, r_{p-1}$ and $s_1, s_2, \ldots, s_{p-1}$ are reduced residue systems, so the result follows from part (a).

▷ **3-26.** (*A proof of Theorem 3.5 via Wilson's Theorem.*) *Let p be a prime of the form $4k + 1$. Show that $((p-1)/2)!$ is a solution of $x^2 \equiv -1 \pmod{p}$. (Hint. For $1 \le x \le (p-1)/2$, $p - x \equiv -x \pmod{p}$.)*

Solution. Let $p = 4k + 1$. As x runs from 1 to $2k$, $p - x$ runs from $4k$ down to $2k + 1$. Thus $4k(4k - 1) \cdots (2k + 1) \equiv (-1)^{2k} (2k)! \pmod{p}$, and therefore $(4k)! \equiv ((2k)!)^2 (-1)^{2k} \pmod{p}$. But $(4k)! \equiv -1 \pmod{p}$ by Wilson's Theorem, and $(-1)^{2k} = 1$, so $((2k)!)^2 \equiv -1 \pmod{p}$.

3-27. *Find solutions of $x^2 \equiv -1 \pmod{37}$ and $x^2 \equiv -1 \pmod{41}$ using a calculation based on the preceding problem.*

Solution. If p is of the form $4k + 1$, the preceding problem shows that $x = ((p-1)/2)!$ is a solution of $x^2 \equiv -1 \pmod{p}$. For $p = 37$, this gives the solution $18!$, and for $p = 41$, it gives $20!$. While these are correct, it may be better to find the least positive residues. With some work, it turns out that $18! \equiv 31 \pmod{37}$ and $20! \equiv 9 \pmod{41}$.

(For large primes p, however, this is not a computationally feasible way to solve the congruence $x^2 \equiv -1 \pmod{p}$.)

▷ **3-28.** Let p be an odd prime. Prove that $[1 \cdot 3 \cdot 5 \cdots (p-2)]^2 \equiv [2 \cdot 4 \cdot 6 \cdots (p-1)]^2 \equiv (-1)^{(p+1)/2} \pmod{p}$.

Solution. As x runs through the $(p-1)/2$ even integers from 2 to $p-1$, $p-x$ runs through the odd integers from $p-2$ down to 1. Therefore $[2 \cdot 4 \cdot 6 \cdots (p-1)] \equiv (-1)^{(p-1)/2}[1 \cdot 3 \cdot 5 \cdots (p-2)] \pmod{p}$. If we square both sides of this congruence, we obtain $[1 \cdot 3 \cdot 5 \cdots (p-2)]^2 \equiv [2 \cdot 4 \cdot 6 \cdots (p-1)]^2 \pmod{p}$.

By Wilson's Theorem, $(p-1)! = [1 \cdot 3 \cdots (p-2)][2 \cdot 4 \cdots (p-1)] \equiv -1 \pmod{p}$. Thus $(-1)^{(p-1)/2}[1 \cdot 3 \cdots (p-2)]^2 \equiv -1 \pmod{p}$, and therefore $[1 \cdot 3 \cdot (p-2)]^2 \equiv (-1)^{(p+1)/2} \pmod{p}$.

3-29. (a) Show that there are infinitely many integers n for which $n! - 1$ is composite. (b) Show that there are infinitely many n for which $n! + 1$ is composite.

Solution. (a) Let $n = p - 2$, where p is a prime greater than 5. By Wilson's Theorem, $(p-1)(p-2)! \equiv -1 \pmod{p}$. Since $p - 1 \equiv -1 \pmod{p}$, it follows that $(p-2)! \equiv 1 \pmod{p}$. Therefore p divides $n! - 1$. Since $p > 5$, we have $(p-2)! - 1 > p$, and thus $(p-2)! - 1$ is composite.

(b) Let $n = p - 1$, where p is prime. By Wilson's Theorem, p divides $n! + 1$, which is greater than p except when $p = 2$.

▷ **3-30.** Show by induction on s that if p is prime and $1 \leq s \leq p - 1$, then $(s-1)!(p-s)! \equiv (-1)^s \pmod{p}$.

Solution. The result says that if s is increased by 1, the expression $(s-1)!(p-s)!$ changes sign modulo p. This should lend itself to proof by induction. Let $A = (s-1)!(p-s)!$ and $B = ((s+1)-1)!(p-(s+1))!$. It is easy to see that $As = B(p-s)$, so $As \equiv -Bs \pmod{p}$ and therefore $B \equiv -A \pmod{p}$. Since the case $s = 1$ is just Wilson's Theorem, the result follows.

The prime numbers p and q are said to be *twin primes* if they differ by 2. It is widely believed that there are infinitely many pairs of twin primes – certainly, they keep appearing fairly regularly in tables of primes – but no proof has ever been given.

3-31. Show that if n and $n + 2$ are both prime, then $4[(n-1)! + 1] + n \equiv 0 \pmod{n(n+2)}$. (The converse also holds.) (*Hint.* $n(n+1) \equiv (-2)(-1) \pmod{n+2}$.)

Solution. By Wilson's Theorem, if n is prime, then $(n-1)! + 1 \equiv 0 \pmod{n}$, and therefore $4[(n-1)! + 1] + n \equiv 0 \pmod{n}$. If $n + 2$ is prime, then $(n+1)! + 1 \equiv 0 \pmod{n+2}$. But since $n + 1 \equiv -1 \pmod{n+2}$ and $n \equiv -2 \pmod{n+2}$, we have $4[(n-1)! + 1] + n \equiv 2(n+1)! + 2 + n + 2 \equiv 0 \pmod{n+2}$. Since n is odd, it follows that $4[(n-1)! + 1] + n \equiv 0 \pmod{n(n+2)}$.

▷ **3-32.** (Liouville, 1856.) *If p is one of the primes 2, 3, or 5, then $(p-1)! + 1$ is a power of a prime. Show that this is false for $p > 5$ by filling in the details of the following argument.*

(a) *If $p > 5$, then $(p-1)^2 | (p-1)!$.*
(b) *If $(p-1)! + 1$ is a prime power, then it is a power of the prime p.*
(c) *If $(p-1)! + 1 = p^k$, then $(p-1) | p^{k-1} + \cdots + 1$. This can happen only if $(p-1) | k$, but then $p^k = (p-1)! + 1$ is impossible.*

Solution. (a) Since p is odd, $p-1$ is divisible by 2 and by $(p-1)/2$. Since $p > 5$, these are distinct and less than $p-1$. Therefore $(p-1)!$ is divisible by $(p-1)^2$.

(b) By Wilson's Theorem, p divides $(p-1)! + 1$; hence if $(p-1)! + 1$ is to be a power of a prime, p must be that prime.

(c) If $(p-1)! + 1 = p^k$, then $(p-1)! = p^k - 1 = (p-1)(p^{k-1} + \cdots + 1)$. From (a), it then follows that $(p-1) | p^{k-1} + \cdots + 1$. But $p^{k-1} + \cdots + 1 \equiv k \pmod{p-1}$, for note that $p \equiv 1 \pmod{p-1}$; thus k must be a multiple of $p-1$. But then $p^k \geq p^{p-1}$, and since p^{p-1} is larger than $(p-1)! + 1$, we cannot have $(p-1)! + 1 = p^k$.

Note. Leibniz, in 1680, gave an incorrect argument that if n is not prime, then n does not divide $2^n - 2$. The first composite n for which $2^n \equiv 2 \pmod n$ is 341, so it is not surprising that it was believed that this congruence gave a *primality test*.

If n is a composite number, but $a^{n-1} \equiv 1 \pmod n$, the number n is called a *pseudoprime* to the base a. The next question gives some examples of pseudoprimes to the base 2. In addition, every Fermat number $2^{2^n} + 1$ and every Mersenne number $2^p - 1$, with p prime, is either prime or pseudoprime to the base 2. (See the following seven problems.)

3-33. *Show that if p and q are distinct primes such that $2^p \equiv 2 \pmod q$ and $2^q \equiv 2 \pmod p$, then $2^{pq} \equiv 2 \pmod{pq}$. Verify that these conditions hold for $p = 11, q = 31$; $p = 19, q = 73$; and $p = 17, q = 257$. (It follows that in each case, pq is a pseudoprime to the base 2.)*

Solution. Using Fermat's Theorem, we have $2^{pq} = (2^p)^q \equiv 2^q \equiv 2 \pmod q$ and, similarly, $2^{pq} \equiv 2 \pmod p$. Thus $2^{pq} \equiv 2 \pmod{pq}$. The numerical computations are straightforward. For example, $2^{17} = 2 \cdot 2^8 \cdot 2^8 \equiv 2(-1)^2 \equiv 2 \pmod{257}$. Similarly, $2^{257} = 2(2^4)^{64} \equiv 2 \pmod{17}$.

3-34. (E. Lucas, 1877.) *Show that if $n = 37 \cdot 73$, then $2^{n-1} \equiv 1 \pmod n$.*

Solution. Here, $n - 1 = 2700$. We want to show that 2^{n-1} is congruent to 1 modulo 37 and 73. By Fermat's Theorem, $2^{36} \equiv 1 \pmod{37}$; since 36 divides 2700, $2^{n-1} \equiv 1 \pmod{37}$. Similarly, $2^{72} \equiv 1 \pmod{73}$. This is not quite good enough, since 72 does not divide 2700. But in fact, $2^{36} \equiv 1 \pmod{73}$. There are various ways of seeing this, but direct calculation is not hard: $2^6 \equiv -9 \pmod{73}$, so $2^{12} \equiv 8 \pmod{73}$, and $2^{18} \equiv (-9)(8) \equiv 1 \pmod{73}$.

▷ **3-35.** *Show that if $n = 161038$, then n divides $2^n - 2$. (The question of whether there exists an even number n such that n divides $2^n - 2$ was open until 1950, when D.H. Lehmer found this example.)*

Solution. It is easy to verify that $n = 2 \cdot 73 \cdot 1103$ and $n - 1 = 3^2 \cdot 29 \cdot 617$. Hence $2^{n-1} - 1$ is divisible by $2^9 - 1 = 7 \cdot 73$ and by $2^{29} - 1$, which in turn is divisible by 1103. (This is done more or less by brute force: $2^{10} \equiv -79 \pmod{1103}$, so $2^{20} \equiv 726 \pmod{1103}$, and $2^{29} \equiv 1 \pmod{1103}$.) Thus $2^n - 2$ is divisible by 2, 73, and 1103, and hence it is divisible by n.

3-36. *Suppose that $2^{n-1} \equiv 1 \pmod n$. If $N = 2^n - 1$, show that $2^{N-1} \equiv 1 \pmod N$. (Hint. Let $2^{n-1} - 1 = nk$.)*

Solution. If $2^{n-1} - 1 = nk$, then $N = 2nk + 1$, and hence $2^{N-1} = (2^n)^{2k} = (1+N)^{2k} \equiv 1 \pmod N$.

Note. When p is prime, we have $2^{p-1} \equiv 1 \pmod p$ by Fermat's Theorem. Thus we have shown that if $N = 2^p - 1$, then N is either a prime or a pseudoprime to the base 2.

▷ **3-37.** *Use the result of the preceding problem to show that there are infinitely many pseudoprimes to the base 2. (Hint. Let $n_1 = 2^{11} - 1$, $n_2 = 2^{n_1} - 1$, and so on.)*

Solution. Let $n_1, n_2, \ldots$ be as in the Hint. Fermat's Theorem implies that $2^{10} \equiv 1 \pmod{11}$, and hence $2^{n_1-1} \equiv 1 \pmod{n_1}$ by the preceding problem. Similarly, since $2^{n_1-1} \equiv 1 \pmod{n_1}$, the preceding problem shows that n_2 satisfies the congruence $2^{n_2-1} \equiv 1 \pmod{n_2}$. Continuing this way, we find that $2^{n_k-1} \equiv 1 \pmod{n_k}$ for all $k \geq 1$.

We complete the proof by showing that n_k is composite for every $k \geq 1$. Note that $n_1 = 2047$, and 2047 is divisible by 23. But in general, if n is composite, then $2^n - 1$ is composite, for if $n = rs$, then $2^r - 1 \mid 2^{rs} - 1$. Thus, since n_1 is composite, it follows that n_2 is composite, and so on.

▷ **3-38.** *Show that if $F_k = 2^{2^k} + 1$, then $2^{F_k} \equiv 2 \pmod{F_k}$. (Hint. Use the fact that $2^k \mid F_k - 1$, and argue as in Problem 3-36.)*

Solution. $F_k - 1 = 2^k \cdot 2^{2^k - k}$, and $2^k > k$ for all k. Thus $F_k = 1 + 2^k m$, where m is even. Therefore $2^{F_k} = 2 \cdot (2^{2^k})^m \equiv 2(-1)^m \pmod{F_k}$. Since m is even, the result follows.

Note. We have shown that any F_k is either prime or pseudoprime to the base 2. Perhaps this result led Fermat to his false conjecture that all of the F_k are prime. The conjecture was refuted by Euler in 1730, when he showed that 641 is a factor of F_5. It was Euler's first number-theoretic result.

3-39. *Show that $a^{560} \equiv 1 \pmod{561}$ for every a relatively prime to 561. (561 is not prime.)*

Solution. Since $561 = 3 \cdot 11 \cdot 17$, it is enough to show that if $(a, 561) = 1$, then $a^{560} \equiv 1$ modulo 3, 11, and 17. If a is relatively prime to 561, then a is not divisible by 3, 11, or 17. Thus, by Fermat's Theorem, $a^2 \equiv 1 \pmod 3$, $a^{10} \equiv 1 \pmod{11}$, and $a^{16} \equiv 1 \pmod{17}$. But 560 is a multiple of 2, 10, and 16, and therefore $a^{560} \equiv 1$ modulo 3, 11, and 17.

Note. A composite number n such that $a^{n-1} \equiv 1 \pmod n$ for every a relatively prime to n is called a *Carmichael number*. So a Carmichael number n is a pseudoprime to every base relatively prime to n. The preceding problem shows that 561 is a Carmichael number. For a long time, it was not known whether there are infinitely many Carmichael numbers. The problem was finally settled in 1993, when W.R. Alford, Andrew Granville, and Carl Pomerance proved a much stronger result. They showed that if x is sufficiently large, there are more than $x^{2/7}$ Carmichael numbers up to x. There are exactly 105212 Carmichael numbers up to 10^{15}.

3-40. *Show that $6601 = 7 \cdot 23 \cdot 41$ is a Carmichael number.*

Solution. We have to show that if a is any number relatively prime to 6601, then $a^{6600} \equiv 1 \pmod{6601}$. By Fermat's Theorem, $a^6 \equiv 1 \pmod 7$, $a^{22} \equiv 1 \pmod{23}$, and $a^{40} \equiv 1 \pmod{41}$. Since 6, 22, and 40 each divide 6600, it follows that $a^{6600} \equiv 1$ modulo 7, 23, and 41, and hence modulo 6601, whenever $(a, 6601) = 1$.

Euler's Theorem

Note. The value of $\phi(n)$ needs to be calculated to solve some of the problems. This can be done using any of the representations of $\phi(n)$ given in Theorem 3.16.

3-41. *True or false: The fourth power of any number that does not have 2 or 5 as a divisor has 1 as its last digit.*

Solution. The question is equivalent to asking whether $(a, 10) = 1$ implies $a^4 \equiv 1 \pmod{10}$. Since $\phi(10) = 4$, this is true by Euler's Theorem. We can also do a direct calculation; we only need to check that the fourth powers of 1, 3, 7, and 9 all have last digit 1.

3-42. *What are the possible remainders when the 100th power of an integer is divided by 125?*

Solution. $\phi(125) = 100$, so if $5 \nmid a$, then $a^{100} \equiv 1 \pmod{125}$. If $5 \mid a$, then a^{100} is divisible by 125. So the possible remainders are 1 and 0.

3-43. *Find the last two digits in the decimal representation of 9^{9^9}. (Hint. Show that $9^{9^9} \equiv 9^9 \pmod{100}$.)*

Solution. Since $\phi(100) = 40$, we first find the remainder when the exponent 9^9 is divided by 40. Since $9 \equiv 1 \pmod 8$ and $9 \equiv -1 \pmod 5$, the same congruences hold for 9^9, so $9^9 \equiv 9 \pmod{40}$. Now calculate the remainder when 9^9 is divided by 100. For example, $9^2 \equiv 81 \pmod{100}$, $9^4 \equiv 61 \pmod{100}$, and $9^8 \equiv 21 \pmod{100}$; thus $9^9 \equiv 89 \pmod{100}$, and so the remainder is 89. (Many calculators will display 9^9 correctly, so in fact the answer can be read off very simply. We could also save some calculation by noting that $9^{10} = (10-1)^{10} \equiv 1 \pmod{100}$, since a glance at the binomial expansion shows that all terms except for the last one are divisible by 100. Thus $9 \cdot 9^9 \equiv 1 \equiv -99 \pmod{100}$, and hence $9^9 \equiv -11 \pmod{100}$, so the remainder is 89.)

3-44. *Show that if a is not divisible by 2 or by 5, then a^{101} ends in the same three decimal digits as does a. (Here we use the convention that 21, for example, "ends" with 021.)*

Solution. We need to show that $a^{101} \equiv a \pmod{1000}$. Note that $(a, 125) = (a, 8) = 1$. Now $a^{100} \equiv 1 \pmod{125}$ by Euler's Theorem, since $\phi(125) = 100$; also, $a^{100} \equiv 1 \pmod{8}$, since $\phi(8) | 100$. Therefore $a^{100} \equiv 1 \pmod{1000}$, and it follows that $a^{101} \equiv a \pmod{1000}$.

3-45. *Use Euler's Theorem to show that $n^{12} \equiv 1 \pmod{72}$ if $(n, 72) = 1$.*

Solution. Since $\phi(8) = 4$, $\phi(9) = 6$, and 12 is a multiple of 4 and of 6, Euler's Theorem implies that $n^{12} \equiv 1$ modulo 8 and modulo 9, and hence modulo 72, whenever $(n, 72) = 1$.

3-46. *Does there exist an integer $n > 1$ such that 1729 divides $n^{36} - 1$? Do there exist infinitely many?*

Solution. Since $1729 | n^{36} - 1$ when $n = 1$, we also have $1729 | n^{36} - 1$ whenever $n \equiv 1 \pmod{1729}$, giving infinitely many solutions $n = 1 + 1729k$, where $k \geq 1$. In fact, there are many more solutions: since $1729 = 7 \cdot 13 \cdot 19$ and $(p-1) | 36$ for $p = 7, 13$, and 19, Fermat's Theorem implies that $n^{36} \equiv 1 \pmod{1729}$ for *every* n relatively prime to 1729.

3-47. *Use Euler's Theorem to show that $n^{20} - n^4$ is divisible by 4080 for all n.*

Solution. $4080 = 2^4 \cdot 3 \cdot 5 \cdot 17$. Work separately modulo 3, 5, 17, and 2^4. Note that $n^{20} - n^4 = n^4(n^{16} - 1)$ and $\phi(p) | 16$ for $p = 3, 5$, and 17. If $p \nmid n$, then $p | n^{16} - 1$, and if $p | n$, then $p | n^4$; thus $p | n^{20} - n^4$ for all n. Now work modulo 2^4. If n is odd, then $2^4 | n^{16} - 1$, since $\phi(2^4) = 8$; if n is even, then $2^4 | n^4$. Thus $n^{20} - n^4$ is divisible by 3, 5, 17, and 2^4 and hence by 4080.

3-48. *Let $(m, n) = 1$. Prove that $m^{\phi(n)} + n^{\phi(m)} \equiv 1 \pmod{mn}$.*

Solution. By Euler's Theorem, $m^{\phi(n)} \equiv 1 \pmod{n}$, and clearly $n^{\phi(m)} \equiv 0 \pmod{n}$, so the sum is congruent to 1 modulo n. By symmetry, the sum is also congruent to 1 modulo m, and therefore, by (2.4), the sum is congruent to 1 modulo mn.

3-49. Find all integers between 0 and 44 that satisfy the congruence $5x^{13} + 3x^3 + 2 \equiv 0 \pmod{45}$. *(Use Fermat's Theorem and Euler's Theorem to simplify the calculations.)*

Solution. First work modulo 5. We want $3x^3 + 2 \equiv 0 \pmod 5$, or, equivalently, $3x^3 \equiv 3 \pmod 5$. By inspection, $x \equiv 1 \pmod 5$ is the only solution. Now work modulo 9. A solution of the congruence cannot be divisible by 3 and so is relatively prime to 9. Since $\phi(9) = 6$, Euler's Theorem implies that $5x^{13} = 5x(x^6)^2 \equiv 5x \pmod 9$ for any solution x. Thus $5x^{13} + 3x^3 + 2 \equiv 0 \pmod 9$ has the same solutions as $3x^3 + 5x + 2 \equiv 0 \pmod 9$. Note that if x is a solution of the last congruence, then $x \equiv -1 \pmod 3$ and hence $x^3 \equiv -1 \pmod 9$, by the Binomial Theorem. This reduces the original congruence to $5x - 1 \equiv 0 \pmod 9$, giving $x \equiv 2 \pmod 9$. Finally, solve the system $x \equiv 1 \pmod 5$ and $x \equiv 2 \pmod 9$, using, for example, the Chinese Remainder Theorem. The solution is $x \equiv 11 \pmod{45}$, so 11 is the only integer between 0 and 44 that satisfies the original congruence.

3-50. (a) Show that if p is prime and $p \nmid a$, then ba^{p-2} is a solution of the congruence $ax \equiv b \pmod p$. Use this technique to solve the congruence $5x \equiv 4 \pmod{17}$.

(b) Adapt the idea of (a) to find a solution of the congruence $ax \equiv b \pmod m$, where $(a, m) = 1$ and m is not necessarily prime. Use the resulting formula to solve the congruence $5x \equiv 4 \pmod{42}$.

Solution. (a) $a(ba^{p-2}) = ba^{p-1} \equiv b \pmod p$ by Fermat's Theorem. So in the numerical example, $x \equiv 4 \cdot 5^{15} \pmod{17}$. This is the answer, but if we want the least positive residue, some calculation is needed. Now $5^2 \equiv 8 \pmod{17}$, so $5^4 \equiv -4 \pmod{17}$, $5^8 \equiv -1 \pmod{17}$, and hence $5^{15} = 5^1 5^2 5^4 5^8 \equiv 5(8)(-4)(-1) \equiv 7 \pmod{17}$. Thus $x \equiv 11 \pmod{17}$.

(b) Using Euler's Theorem, we find in exactly the same way as in (a) that $x = ba^{\phi(m)-1}$ is a solution of the congruence. So in the numerical example, $x \equiv 4 \cdot 5^{11} \pmod{42}$. Fortunately, $5^3 \equiv -1 \pmod{42}$, so $4 \cdot 5^{11} \equiv 4(-1)^3(5^2) \equiv 26 \pmod{42}$.

Note. For large primes p and large numbers a, the technique described in part (a) can be roughly as fast as the Euclidean Algorithm if we use an efficient way of finding powers modulo p, such as the repeated squaring method. For large p but small a, the Euclidean Algorithm is more efficient, since after one step we are dealing with small numbers. The technique of part (b) is almost always an *inefficient* way of solving linear congruences, since to calculate $\phi(m)$, we need to factor m, a computationally very difficult problem.

3-51. (Bunyakovskii, 1831.) Let a, b be relatively prime positive integers. Show that the equation $ax + by = c$ has the solution $x = ca^{\phi(b)-1}$, $y = (-c/b)(a^{\phi(b)} - 1)$.

Solution. For the given values of x and y, $ax + by = ca^{\phi(b)} - ca^{\phi(b)} + c = c$. We need to check that the solution is indeed an *integer* solution, that is, b divides $a^{\phi(b)} - 1$. But this is precisely the content of Euler's Theorem.

3-52. *Suppose $(a,m) = 1$ and $n \mid t\phi(m) + 1$ for some integer t. Prove that $x^n \equiv a \pmod{m}$ has the unique solution a^k, where $k = (t\phi(m) + 1)/n$.*

Solution. By Euler's Theorem, $(a^k)^n = a^{t\phi(m)+1} = (a^{\phi(m)})^t a \equiv a \pmod{m}$. To show that the solution is unique, suppose $s^n \equiv a \pmod{m}$. Then Euler's Theorem implies that $s \equiv (s^{\phi(m)})^t s = s^{t\phi(m)+1} = (s^n)^k \equiv a^k \pmod{m}$.

3-53. *Use the preceding problem to solve (a) $x^{11} \equiv 3 \pmod{68}$; (b) $x^{13} \equiv 7 \pmod{68}$; (c) $x^{23} \equiv 5 \pmod{68}$.*

Solution. (a) $\phi(68) = \phi(4)\phi(17) = 2 \cdot 16 = 32$; thus $11 \mid \phi(68) + 1$. Hence the solution is given by $x \equiv 3^{33/11} \equiv 27 \pmod{68}$.

(b) Since $13 \mid 2\phi(68) + 1$, the solution is given by $x \equiv 7^{65/13} \equiv 11 \pmod{68}$.

(c) $23 \mid 5\phi(68) + 1$, so $x \equiv 5^{161/23} \equiv 61 \pmod{68}$ is the only solution of $x^{23} \equiv 5 \pmod{68}$.

▷ **3-54.** (Chinese Remainder Theorem via Euler's Theorem.) *Let $m_1, m_2, \ldots, m_k$ be pairwise relatively prime positive integers, and let $a_1, a_2, \ldots, a_k$ be integers. Let $M = m_1 m_2 \cdots m_k$. Show that the system of congruences $x \equiv a_1 \pmod{m_1}, \ldots, x \equiv a_k \pmod{m_k}$ has a solution x given by $x = a_1(M/m_1)^{\phi(m_1)} + a_2(M/m_2)^{\phi(m_2)} + \cdots + a_k(M/m_k)^{\phi(m_k)}$.*

Solution. By symmetry, it is enough to verify that the given x is congruent to $a_1 \pmod{m_1}$. If $i > 1$, then m_1 divides M/m_i, so $x \equiv a_1(M/m_1)^{\phi(m_1)} \pmod{m_1}$. But since M/m_1 is relatively prime to m_1, we have $(M/m_1)^{\phi(m_1)} \equiv 1 \pmod{m_1}$ by Euler's Theorem, and hence $x \equiv a_1 \pmod{m_1}$.

Lucas, in 1878, gave a partial converse of Fermat's Theorem, which is the object of the next problem.

3-55. *(a) Suppose that m is composite but $a^{m-1} \equiv 1 \pmod{m}$, where $a \not\equiv 1 \pmod{m}$. Use Euler's Theorem and (3.7) to show that $a^d \equiv 1 \pmod{m}$ for some proper divisor d of $m - 1$.*

(b) Use part (a) to show that if there exists a such that $a^{m-1} \equiv 1 \pmod{m}$ but $a^{(m-1)/p} \not\equiv 1 \pmod{m}$ for every prime divisor p of $m - 1$, then m is prime.

Solution. (a) Let $d = (\phi(m), m - 1)$. Since $a^{\phi(m)} \equiv 1 \pmod{m}$ by Euler's Theorem and $a^{m-1} \equiv 1 \pmod{m}$ by assumption, (3.7) shows that $a^d \equiv 1 \pmod{m}$. Because $a \not\equiv 1 \pmod{m}$, it follows that $d \neq 1$; also, $\phi(m) < m - 1$ since m is composite, and hence $d \neq m - 1$. Thus d is a proper divisor of $m - 1$.

(b) Suppose to the contrary that m is composite. Since the d produced in part (a) is a proper divisor of $m - 1$, it follows that $(m - 1)/d$ is divisible by some prime p. Let $(m-1)/d = kp$; then $a^{(m-1)/p} = (a^d)^k \equiv 1 \pmod{m}$.

▷ **3-56.** (Crelle, 1829.) *Let m be a positive integer not divisible by 2 or by 5. Show that m divides N for infinitely many numbers N whose decimal* expansion has the form 147147147...147. *(Any string of digits, of any length, may be substituted for 147).*

Solution. Let N be a number that has k repetitions of the block 147, where k will be chosen later. Then $N = 147(1 + 10^3 + \cdots + 10^{3(k-1)}) = 147((a^k - 1)/(a - 1))$, where $a = 10^3$. Since $(a, m) = 1$, we can ensure that m divides $a^k - 1$ by taking k to be a multiple of $\phi(m)$. This is not enough, however, because of the $a - 1$ in the denominator. So let k be any multiple of $\phi(m(a-1))$; then $a^k \equiv 1 \pmod{m(a-1)}$ by Euler's Theorem. Thus $m(a-1)$ divides $a^k - 1$, and hence m divides $(a^k - 1)/(a - 1)$. The argument for any string of digits is essentially the same.

The Euler ϕ-function

Note. The remaining problems for this chapter deal with properties of the Euler ϕ-function. Most of them can be solved by using one of the formulas for $\phi(n)$ given in (3.16).

3-57. Find $\phi(5040)$ and $\phi(496125)$.

Solution. Note that $5040 = 2^4 \cdot 3^2 \cdot 5 \cdot 7$. Hence by (3.16), $\phi(5040) = \phi(16)\phi(9)\phi(5)\phi(7) = (16 - 8)(9 - 3) \cdot 4 \cdot 6 = 1152$. Similarly, $496125 = 3^4 \cdot 5^3 \cdot 7^2$, and thus $\phi(496125) = (81 - 27)(125 - 25)(49 - 7) = 226800$.

3-58. Prove that $\phi(n)$ is even if $n \geq 3$.

Solution. Use the second formula for $\phi(m)$ given in (3.16). If n has at least one odd prime factor p_i, then $\phi(n)$ is even since $(p_i - 1) | \phi(n)$. Otherwise, $n = 2^k$ with $k \geq 2$, and therefore $\phi(n) = 2^{k-1}$, so $\phi(n)$ is even.

3-59. Suppose $m > 1$. Show that $\phi(m) = m - 1$ if and only if m is prime.

Solution. If m is prime, then clearly, $1, 2, \ldots, m-1$ are all relatively prime to m, and so it follows from the definition that $\phi(m) = m - 1$. Conversely, if m is not prime, then m has a proper divisor d, which cannot be relatively prime to m. Thus there is at least one positive integer less than m that is not relatively prime to m, and hence $\phi(m) \leq m - 2$.

3-60. For which n is it true that $\phi(n) = n - 2$?

Solution. Obviously, all but one of the numbers $1, 2, \ldots, n-1$ must be relatively prime to n. Thus n must have exactly one proper divisor, and it follows that $n = p^2$ for some prime p. Hence $\phi(n) = p(p - 1)$. Setting this equal to $p^2 - 2$, we find that $p = 2$, so $n = 4$.

3-61. Prove or disprove: If $d | n$, then $\phi(d) | \phi(n)$.

Solution. Let p_i be any prime divisor of d, let $p_i^{n_i}$ be the largest power of p_i that divides n, and let $p_i^{d_i}$ be the largest power of p_i that divides d. By (3.16), $p_i^{n_i}$ contributes a factor of $p_i^{n_i-1}(p_i - 1)$ to the expression for $\phi(n)$, while $p_i^{d_i}$ contributes a factor of $p_i^{d_i-1}(p_i - 1)$ to the expression for $\phi(d)$. Since $d_i \leq n_i$, the result follows.

3-62. *Suppose that $a = 2^k b$, where b is odd. If $\phi(x) = a$, prove that x has at most k distinct odd prime factors.*

Solution. It is clear from the second formula for ϕ in (3.16) that if p is an odd prime that divides x, then the term $p - 1$ in the expression for $\phi(x)$ contributes at least one factor of 2. Since a has precisely k factors of 2, it is clear that x cannot have more than k distinct odd prime divisors. (Of course, x can have fewer than k odd prime factors; for example, if $b = 1$, take $x = 2^{k+1}$, which has no odd prime factor.)

3-63. *Use Problem 3-62 to characterize those integers n for which $\phi(n)$ is not divisible by 4.*

Solution. First suppose $n = 2^a$ for $a \geq 1$. Since $\phi(n) = 2^{a-1}$, a must be 1 or 2 if $4 \nmid \phi(n)$, that is, $n = 2$ or 4. Now suppose $n = 2^a m$, where $m > 1$ is odd. If $\phi(n)$ is not divisible by 4, the preceding problem implies that n can have at most one odd prime factor. Thus let $n = 2^a p^k$, where $k \geq 1$ and p is an odd prime. Since $4 \mid p - 1$ if $p \equiv 1 \pmod 4$, p must be of the form $4t + 3$. In this case, $\phi(p^k)$ is divisible by 2 but not by 4. Thus a must be chosen so that $\phi(2^a)$ is odd; hence a must be 0 or 1. It follows that the only n for which $\phi(n)$ is not divisible by 4 are 1, 2, 4 and numbers of the form p^k or $2p^k$, where p is a prime of the form $4t + 3$.

3-64. *Prove that $\phi(2n) = \phi(n)$ if and only if n is odd.*

Solution. If n is odd, then $(2, n) = 1$ and so $\phi(2n) = \phi(2)\phi(n) = \phi(n)$. Conversely, suppose that $n = 2^k m$, where m is odd. If $k \geq 1$, then $\phi(n) = \phi(2^k)\phi(m) = 2^{k-1}\phi(m)$ and $\phi(2n) = 2^k \phi(m)$; hence $\phi(2n) = 2\phi(n) \neq \phi(n)$. Thus $k = 0$ and therefore n is odd.

3-65. *Suppose that n is even. Prove that $\phi(n) = n/2$ if and only if $n = 2^k$ for some $k \geq 1$.*

Solution. If $n = 2^k$ with $k \geq 1$, then $\phi(n) = 2^{k-1} = n/2$. Now suppose $n = 2^k m$, where m is odd. Since $(2^k, m) = 1$, we have $\phi(n) = 2^{k-1}\phi(m)$. Thus if $\phi(n) = n/2$, then $\phi(m) = 1$, that is, $m = 1$. Hence $n = 2^k$.

3-66. *For a fixed positive integer n, prove that there are only a finite number of x such that $\phi(x) = n$. (There may in fact be no solutions.)*

Solution. If $x = \prod p^a$, then $\phi(x) = n$ implies that $\prod_{p \mid x}(p^a - p^{a-1}) = n$. Since there are only a finite number of ways to factor n as a product of integers, there are at most a finite number of x that solve the equation $\phi(x) = n$.

3-67. *Find all integers n such that (a) $\phi(n) = 18$; (b) $\phi(n) = 80$. (Hint. For part (a), use Problem 3-63.)*

Solution. (a) Since 18 has precisely one factor of 2, Problem 3-63 implies that $n = p^k$ or $n = 2p^k$, where p is prime and $p \equiv 3 \pmod 4$. If $n = p^k$ or $n = 2p^k$, then $\phi(n) = p^{k-1}(p - 1)$. If $k > 1$, then $p = 3$, and n is one of 27 or 54. If $k = 1$, then $p = 19$, and n is one of 19 or 38.

(b) Let $n = \prod p^a$. Since n is not a power of 2, some odd prime p divides n. If $p^2 \mid n$, then $p \mid \phi(n)$ (see (3.16)). Thus if $p^2 \mid n$ for an odd prime p, then $p = 5$ and $n = 25m$, where $(m, 5) = 1$ and hence $\phi(m) = 80/20 = 4$. It is easy to see that m must be 2^3 or $2^2 \cdot 3$, and thus $n = 200$ or $n = 300$. If n is not divisible by p^2 for any odd prime p, then the factor of 5 in $\phi(n)$ must come from $p - 1$, where p is an odd prime dividing n. Thus $p - 1 = 2^j \cdot 5$, where $1 \le j \le 4$. It is easy to check that p is prime only for $j = 1$ and $j = 3$. If $j = 1$, then $p = 11$ and $n = 11m$, where $\phi(m) = 8$. It is easy to see that m is 2^4, $2^3 \cdot 3$, $2^2 \cdot 5$, $2 \cdot 3 \cdot 5$, or $3 \cdot 5$. This implies that n is 176, 264, 220, 330, or 165. If $j = 3$, then $p = 41$ and $n = 41m$, where $\phi(m) = 2$; thus n is 164, 123, or 246. Hence the only n for which $\phi(n) = 80$ are 123, 164, 165, 176, 200, 220, 246, 264, 300, and 330.

3-68. (a) Show that there is no integer n such that $\phi(n) = 14$.
(b) Prove that there is no integer n such that $\phi(n) = 2 \cdot 7^e$, where $e \ge 1$. (Hint. Show that $2 \cdot 7^e + 1$ is never a prime.)
(c) Find other cases where twice an odd number is not $\phi(n)$ for any n.

Solution. Note that (a) is a special case of (b). Since $2 \cdot 7^e$ is divisible by 2 but not by 4, $\phi(n) = 2 \cdot 7^e$ implies that n must be of the form p^k or $2p^k$, where p is a prime of the form $4t + 3$ (see Problem 3-63). In either case, $\phi(n) = p^{k-1}(p - 1)$, and $\phi(n) = 2 \cdot 7^e$ implies that $k = 1$ and $p - 1 = 2 \cdot 7^e$, i.e., $p = 2 \cdot 7^e + 1$. However, since $7^e \equiv 1 \pmod 3$, $2 \cdot 7^e + 1$ is divisible by 3 and therefore cannot be prime if $e \ge 1$.

(c) Precisely the same argument works if 7 is replaced by any prime of the form $3t + 1$ or if 7 is replaced by p^2, where p is a prime of the form $3t + 2$. Thus, for example, there is no n for which $\phi(n) = 2 \cdot 13^e$ or $2 \cdot 11^{2e}$. Similarly, instead of 7^e, we can use 3^{4e+3}, since $2 \cdot 3^{4e+3} + 1$ is always divisible by 5. With little change in the proof, we can also use $7 \cdot 13$ instead of 7.

3-69. If $\phi(n)$ divides $n - 1$, prove that n is square-free (that is, n is divisible by no square greater than 1).

Solution. If n is not square-free, then p^2 divides n for some prime p. It is clear from (3.16) that p divides $\phi(n)$, and thus if $\phi(n)$ divides $n - 1$, then p divides $n - 1$. Hence p divides n and $n - 1$, which is impossible.

Note. There is a long-standing conjecture that if $n > 1$ and $\phi(n) \mid n - 1$, then n is prime.

▷ **3-70.** Prove that $\phi(mn) = m\phi(n)$ if and only if every prime that divides m also divides n. In particular, $\phi(n^e) = n^{e-1}\phi(n)$ for any $e \ge 1$.

Solution. Let P_k denote the product $\prod_{p \mid k}(1 - 1/p)$. First suppose that every prime that divides m also divides n. Then clearly $P_{mn} = P_n$, and so it follows from (3.16) that $\phi(mn) = mnP_{mn} = m(nP_n) = m\phi(n)$.

Conversely, suppose $\phi(mn) = m\phi(n)$. Then it follows from (3.16) that $P_{mn} = P_n$. If there exists a prime p that divides m but not n, then the term $(1 - 1/p)$ occurs in the product P_{mn} but not in the product P_n, while for every prime q that divides n, $1 - 1/q$ occurs in both products. Thus $P_{mn} \le (1 - 1/p)P_n$, contradicting the fact that $P_{mn} = P_n$.

▷ **3-71.** *Prove that $\phi(m)/m = \phi(n)/n$ if and only if m and n have exactly the same prime divisors (possibly to different powers).*

Solution. Let $P_k = \prod_{p|k}(1 - 1/p)$; then $\phi(k) = kP_k$, by (3.16). If m and n have the same prime divisors, then clearly $P_m = P_n$, and hence $\phi(m)/m = \phi(n)/n$. (This also follows from the preceding problem, since the hypothesis that m and n have the same prime divisors implies that $\phi(mn) = m\phi(n)$ and also $\phi(mn) = n\phi(m)$, whence $\phi(m)/m = \phi(n)/n$.)

Conversely, suppose that $\phi(m)/m = \phi(n)/n$; thus $P_m = P_n$. Let $p_1, p_2, \ldots, p_s$ be the primes that divide m, listed in *increasing order*, and $q_1, q_2, \ldots, q_t$ the primes that divide n, again in increasing order. Since $P_m = P_n$, it follows that $(p_1 - 1) \ldots (p_s - 1)q_1 \ldots q_t = (q_1 - 1) \ldots (q_t - 1)p_1 \ldots p_s$. Suppose $q_t \geq p_s$. Since q_t divides the left side of the preceding equation, and since it is larger than any term on the right side except possibly p_s, it follows that $q_t = p_s$. Now cancel the terms involving q_t and p_s from both sides. (If $p_s > q_t$, argue similarly.) Continuing this way, we find that $s = t$ and $p_i = q_i$ for all i.

3-72. *Suppose that $n \geq 2$. Prove that the sum of all positive integers less than n that are relatively prime to n is $n\phi(n)/2$. (Hint. First show that $(n-a, n) = 1$ if $(a, n) = 1$.)*

Solution. Observe that a is relatively prime to n if and only if $n - a$ is (see (1.22)). Pair each $a \leq n/2$ that is relatively prime to n with $n - a$. Except in the case $n = 2$, where the result is trivial, $(n/2, n) \neq 1$, so a is never paired with itself. Since there are exactly $\phi(n)/2$ such pairs and each pair adds to n, the result follows.

▷ **3-73.** *Let P be the product of the distinct prime divisors of (m, n) (where we define an empty product to be 1). Prove that $\phi(mn)/(\phi(m)\phi(n)) = P/\phi(P)$. In particular, show that if $(m, n) > 1$, then $\phi(mn) > \phi(m)\phi(n)$.*

Solution. We may suppose that exactly the same primes divide m and n. For if p divides m but not n, let p^e be the largest power of p dividing m, and let $m' = m/p^e$. Then $\phi(mn) = \phi(m'n)\phi(p^e)$, and $\phi(m)\phi(n) = \phi(m')\phi(p^e)\phi(n)$; hence the ratio $\phi(mn)/(\phi(m)\phi(n))$ is unchanged if m is replaced by m'.

Thus let $m = \prod p_i^{a_i}$ and $n = \prod p_i^{b_i}$; then (m, n) is divisible by each p_i. By (3.16), $\phi(mn)$ is the product of terms $p^{a+b-1}(p-1)$, where p ranges over the p_i. The corresponding term in $\phi(m)\phi(n)$ is $p^{a-1}(p-1)p^{b-1}(p-1)$, so the ratio of these terms is $p/(p-1)$, which is precisely the contribution that p makes to the ratio $P/\phi(P)$.

Finally, if $(m, n) > 1$, then $P > 1$ and hence $P/\phi(P) > 1$, since $\phi(k) \leq k - 1$ for every $k \geq 2$. Thus by the previous argument, we have $\phi(mn)/(\phi(m)\phi(n)) > 1$, that is, $\phi(mn) > \phi(m)\phi(n)$.

▷ **3-74.** *If $n \geq 1$, prove that $\sum_{d|n} \phi(d) = n$.*

Solution. Let N be the complete residue system $\{0, 1, 2, \ldots, n-1\}$. If d is any divisor of n, let N_d consist of all elements $k \in N$ such that $(k, n) = n/d$. Thus N_d consists of the elements of N of the form $e(n/d)$, where $0 \leq e \leq d$ and $(e, d) = 1$; in particular, there are $\phi(d)$ numbers in the set N_d. It is clear that $N_d \neq N_{d'}$ if $d \neq d'$. As d ranges over the divisors of n, n/d also ranges over the divisors of n, and therefore every element of N belongs to a uniquely determined N_d. Since N has n elements and any N_d has $\phi(d)$ elements, it follows that $n = \sum_{d|n} \phi(d)$.

EXERCISES FOR CHAPTER 3

1. Find the remainder when 24! is divided by 29.
2. What is the remainder when 3(26!) is divided by 29?
3. It is true that $1991! \equiv 1 \pmod{1993}$. Does it follow from this that 1993 is prime?
4. What is the least positive residue of 53! modulo 59?
5. Use Wilson's Theorem to find the remainder when 27! is divided by 899. (Hint. First factor 899.)
6. Use Wilson's Theorem to show that if p is prime, then $(p-1)! \equiv p-1 \pmod{p(p-1)}$.
7. Find the remainder when 15! is divided by 323.
8. Is $16 \cdot 77! + 7!$ a multiple of 79?
▷ 9. Use Wilson's Theorem to find the remainder when 42! is divided by 2021.
10. Use Euler's Theorem to find the last two digits of 7^{209}.
11. Is $54^{103} + 69^{67}$ a multiple of 13?
12. Solve the congruence $x^{200} - 200x \equiv 0 \pmod{199}$. (Note that 199 is prime.)
13. For which primes p is $2^p + 1$ divisible by p?
14. Prove or disprove: If p is an odd prime, then $n^{2p-1} \equiv n \pmod{2p}$.
15. Suppose p and q are odd primes, with $q > p$. If $q-1$ is divisible by $p-1$, prove that $4^{q-1} - 1$ is a multiple of pq.
16. Use Euler's Theorem to calculate the last three digits of 3^{9610}.
17. What is the least positive residue of 3^{725} modulo 675?
18. Find the remainder when 11^{196} is divided by 144.
19. Find the remainder when 3^{10000} is divided by 35. Justify the calculations.
20. What is the least positive residue of 7^{243} modulo 144? Of 11^{484} modulo 288?
21. What is the remainder when 1177^{1177} is divided by 9?
22. Determine the last two digits of 7^{8^9}.
23. Prove that $n^{25} - n$ is a multiple of 5460 for every odd n.
24. Show that $n^{50} - n^2$ is divisible by 12240 for every odd n.

EXERCISES

25. Show that $m^{18} - n^{18}$ is divisible by 133 for all integers m and n that are relatively prime to 133.
26. Show that $mn(m^{60} - n^{60})$ is divisible by the number $56,786,730$ for all integers m and n.
27. Is it true that $n^{37} - n$ is divisible by 54 for every n?
28. Prove that $n^{13} - n$ is divisible by 273 for every n.
▷ 29. Use Euler's Theorem to prove that $x^{89} \equiv 3 \pmod{2200}$ has a unique solution, and find the solution.
30. (a) Use Fermat's Theorem to solve $18x \equiv 23 \pmod{37}$.
    (b) Use Euler's Theorem to solve $7x \equiv 39 \pmod{54}$.
    (c) Solve the congruences in parts (a) and (b) using the Euclidean Algorithm.
31. Let m be a positive integer that is relatively prime to $a(a-1)$. Show that $1 + a + a^2 + \cdots + a^{\phi(m)-1} \equiv 0 \pmod{m}$.
32. Does there exist a positive integer m such that 2^m leaves a remainder of 1 when divided by m?
33. Find $\phi(330)$ and $\phi(857500)$.
34. Calculate $\phi(12!)$ and $\phi(17!)$.
35. Prove that $\phi(415800)$ is a multiple of 16.
36. Find the number of positive rational numbers r/s, in lowest terms, such that $r/s < 1$ and $1 \leq s \leq 10$.
37. Find all positive integers n less than 100 such that $\phi(n) | n$.
38. Is it true that $\phi(n) \geq \sqrt{n}/2$?
39. Prove or disprove: $\phi(12^k) = 12^{k-1}\phi(12)$.
40. Do there exist infinitely many positive integers n such that $n = 3\phi(n)$?
41. Are there infinitely many n such that $\phi(n) = n/4$?
42. Suppose p is an odd prime such that $2p + 1$ is composite. Prove that there are no positive integers n for which $\phi(n) = 2p$.
43. For which n is $\phi(2n) > \phi(n)$?
44. Prove or disprove: $\phi(n)$ is a perfect square for only finitely many odd values of n.
45. If $n > 1$ and $\phi(n)$ divides $n - 1$, prove that n is the product of distinct primes.
46. Find six values of n for which $\phi(n) = \phi(n+2)$.
▷ 47. Let a, b be relatively prime positive integers, and let p be an odd prime. Show that the greatest common divisor of $a + b$ and $(a^p + b^p)/(a + b)$ is either 1 or p. (Hint. Let $c = a + b$; then $a^p + b^p = (c - b)^p + b^p$. Expand $(c - b)^p$ using the Binomial Theorem.)

NOTES FOR CHAPTER 3

1. Fermat's Theorem is a fundamental result of elementary number theory, so it is interesting to look at the motivation that might have led Fermat to it. In the 1630s, Parisian mathematicians, including Frenicle, Mersenne, and even the aloof Descartes, were looking at problems connected with "perfect numbers" and the primality of what are now known as *Mersenne numbers* (see Chapter 7).

In Fermat's time, the Mersenne number $M_p = 2^p - 1$ was known to be prime for $p = 2, 3, 5, 7, 13, 17, 19$ and composite for $p = 11, 23$: M_{11} is the product of 23 and 89, while M_{23} is divisible by 47 (a fact discovered by Fermat). The form of these divisors for M_{11} and M_{23} may have led Fermat to conjecture that every prime divisor of M_p is of the form $2kp + 1$ (see Theorem 3.8).

Fermat's original result – namely, that $2^p - 2$ is a multiple of the prime p – can be obtained as an easy consequence of (3.8), because (3.8) implies that *all* divisors of $2^p - 1$ are of the form $2kp + 1$ (since every prime divisor is); in particular, $2^p - 1$ is itself of this form, and hence $2^p - 2$ is a multiple of p.

2. We sketch Euler's 1758 proof of Fermat's Theorem that was mentioned in the introduction. The argument is historically important, since it presages a basic result of the branch of modern mathematics called group theory.

Suppose that a is greater than 1 and not divisible by p, and consider the remainders when the p numbers $1, a, a^2, \ldots, a^{p-1}$ are divided by p. There are at most $p - 1$ possible remainders, but the list has p members, so at least two of the remainders are equal. Thus there exist i, j with $0 \leq i < j \leq p - 1$ such that $a^i \equiv a^j \pmod{p}$. Since a is relatively prime to p, we can divide i times by a and obtain $a^{j-i} \equiv 1 \pmod{p}$. Hence there is a number n with $1 \leq n \leq p - 1$ such that $a^n \equiv 1 \pmod{p}$.

Now we prove that if n is the *smallest* positive integer such that $a^n \equiv 1 \pmod{p}$, then n divides $p - 1$. This quickly yields $a^{p-1} \equiv 1 \pmod{p}$.

Let S be the set $\{1, a, a^2, \ldots, a^{n-1}\}$. These powers of a are all incongruent modulo p; otherwise, we could, as above, obtain a number m with $1 \leq m < n$ such that $a^m \equiv 1 \pmod{p}$. Now divide the numbers $1, 2, \ldots, p-1$ into *families* by saying that x and y belong to the same family if $y \equiv sx \pmod{p}$ for some $s \in S$. It can be shown that the family of any x has exactly n members, namely, the remainders when $x, ax, \ldots, a^{n-1}x$ are divided by p. Thus if f is the number of families, then $nf = p - 1$; it follows that n divides $p - 1$.

3. Fermat's Theorem and Euler's Theorem greatly simplify the task of finding the least nonnegative residue (that is, the remainder) when a number raised to a large power is divided by a positive integer (prime or otherwise).

The computational methods in the solutions of Problems 3-1, 3-2, 3-5, and 3-43 should be studied carefully.

The important points are these: First reduce the exponent by removing as many multiples of $p-1$ as possible when the modulus is a prime p or, in the case of a general modulus m, as many multiples of $\phi(m)$ if $\phi(m)$ is easy to compute. (This effectively means finding the least nonnegative residue of the exponent modulo $p-1$ or $\phi(m)$, respectively.) This leaves an expression of the form a^k, with $0 \leq k < \phi(m)$, which must be reduced modulo m. In many cases, it may be advantageous to work separately with the prime powers in the factorization of m and then put things together using the Chinese Remainder Theorem.

For large values of k, it is often most expedient first to write the exponent k as a sum of distinct powers of 2. (For example, if $k = 29$, then $k = 16+8+4+1$; thus $a^{29} = a^{16}a^8a^4a^1$.) The residues modulo m of the powers $a, a^2, a^4, \ldots$ can be found on a calculator by repeatedly squaring and reducing modulo m (see Problem 3-2). Finally, the least nonnegative residue of a_k itself is found by multiplying together the least nonnegative residues of the appropriate powers of a and (if necessary) reducing at each step modulo m.

4. Pseudoprimes to the base 2 are much rarer than primes. For example, there are 455,052,512 primes less than 10^{10} but only 14,884 pseudoprimes to the base 2. If a large number n (for example, one with 100 digits) satisfies the congruence $2^{n-1} \equiv 1 \pmod{n}$, it is overwhelmingly likely that n is actually prime. Henri Cohen has called a number $n > 1$ an "industrial grade prime" if $2^{n-1} \equiv 1 \pmod{n}$. They can be used in applications where very large primes are needed, such as cryptography or the generation of secure pseudorandom numbers.

If $a^{n-1} \equiv 1 \pmod{n}$ for several values of a, for example, 2, 3, and 5, and if n has about 100 digits, then the likelihood that n is not prime is much less, for instance, than the likelihood of an asteroid obliterating the computer doing the calculations.

5. The RSA Encryption Method. This procedure, first described by R.L. Rivest, A. Shamir, and L.M. Adleman, is the first commercially important application of number theory. Using the RSA method, you can *reveal publicly* how secret messages intended for you should be encoded. Despite this, it is extremely difficult for anyone but you to decode these messages.

We can encode keyboard symbols using two-digit integers. By breaking up the message appropriately, we can assume that messages are made up of k-digit numbers, where, for example, we can take $k = 150$. Now choose two primes p and q, where p and q each have at least 100 digits, and let $n = pq$. We also choose an *encryption index* e which is relatively prime to $\phi(n)$ and reasonably large; for example, e could be a prime larger than p or q. You *reveal publicly* the ordered pair (e, n), while keeping p and q secret.

Someone who wishes to send you the 150-digit message x calculates the remainder y when x^e is divided by n, and transmits y to you. The computation of y can be done quite quickly by using the method of repeated squaring. The probability that p or q divides x is negligibly small, so we may assume that x is relatively prime to n. We now show how to recover x from y.

Using your knowledge of p and q, calculate $\phi(n) = (p-1)(q-1)$ and then find the unique integer d, with $0 < d < \phi(n)$, such that $de \equiv 1 \pmod{\phi(n)}$. This integer d is called the *decoding index*. Let $de = 1 + t\phi(n)$. Then

$$y^d \equiv (x^e)^d = x \cdot (x^{\phi(n)})^t \equiv x \pmod{n}$$

by Euler's Theorem, and hence x is the remainder when y^d is divided by n. Thus, knowing the decoding index d and the encrypted message y, we can readily recover x.

It is believed that the RSA encryption method is very secure. There does not appear to be a way of decoding RSA encrypted messages without finding the factorization of n, and factorization of 200-digit integers seems to be beyond the reach of today's algorithms. Variants of the RSA method are in widespread use and have sparked renewed interest in finding efficient algorithms for primality testing and factoring. For details, see the books by D. Bressoud, P. Giblin, N. Koblitz, and H. Riesel listed in the Bibliography.

BIOGRAPHICAL SKETCHES

Pierre Simon de Fermat was born in France in 1601. A magistrate by profession, he came to mathematics fairly late in life – after the age of 30 – and pursued mathematics as a hobby. Perhaps the last great "amateur" mathematician, Fermat corresponded with many of the leading mathematical figures of his time, challenging them (as was the custom) to solve problems he had posed.

Fermat's mathematical notes were not organized, perhaps because they were never intended to be published. (Indeed, in his lifetime, Fermat published almost none of his number-theoretic results.) His notes were often written in the margins of his books, most notably his edition of Diophantus's *Arithmetica*, whose margin was "unfortunately too narrow" to contain the proof of his famous Last Theorem.

Fermat laid the foundations of analytic geometry some ten years before Descartes published his own work, and in his correspondence with Pascal, Fermat helped to establish the mathematical concepts of probability theory.

As well, his method of finding tangents at points of a curve inspired Newton in his development of differential calculus. But perhaps most of all, Fermat is remembered as the founder of modern number theory, with his investigations into primes, divisibility, sums of squares, and Diophantine equations, including the method of infinite descent.

Fermat died on January 12, 1665.

Leonhard Euler was born in 1707, in Basel, Switzerland. At university, Euler decided not to pursue a career in theology but instead to study mathematics under the tutelage of Johann Bernoulli. Most of Euler's life was spent in Berlin and St. Petersburg. Unlike Fermat, Euler was very open in explaining how he arrived at his results. Although he was blind for the last 17 years of his life, Euler was nevertheless the most prolific mathematician in history. His collected works – nearly 900 books and papers – are expected to fill 75 volumes.

Much of our modern mathematical notation is due to Euler (for example, the functional notation $f(x)$ and the summation symbol $\sum$). He founded analytic number theory and was the first to study power residues systematically. Euler also worked on Diophantine equations, provided proofs for many of Fermat's results, and gave a systematic treatment of continued fractions. Euler contributed in many other areas as well – including mechanics, the calculus of variations, hydrodynamics, differential equations, and the theory of functions – and published four volumes giving a unified presentation of the differential and integral calculus.

Euler died on September 18, 1783, at the age of 76.

REFERENCES

David M. Bressoud, *Factorization and Primality Testing*, Springer-Verlag, New York, 1989.

> The book deals with issues that have become very important in recent years, since number-theoretic ideas are used extensively in modern cryptography. There is a thorough discussion of pseudoprimes and Carmichael numbers, and the theorems of Fermat and Euler play a leading role. Bressoud pays a lot of attention to computational matters and gives detailed computer algorithms.

Leonard Eugene Dickson, *History of the Theory of Numbers* (3 volumes), Chelsea, New York, 1952 (originally published in 1919).

> The material relevant to this chapter can be found in Volume I. Dickson's treatment is encyclopedic but very cryptic and not at all analytical. It gives the bare sketch of a proof, or no proof at all, and does not clearly distinguish between important results and puzzles. These are wonderful books for browsing, in small doses.

André Weil, *Number Theory: An approach through history from Hammurapi to Legendre*, Birkhäuser, Boston, 1984.

This is an analytical treatment by one of the masters of modern number theory, concentrating on the most important themes in the work of Fermat, Euler, Lagrange, and Legendre. In parts, it requires a fairly sophisticated knowledge of number theory, for it approaches the history through very modern eyes, but much of the material on Fermat and Euler is accessible.

CHAPTER FOUR
Polynomial Congruences

In this chapter, we investigate the general polynomial congruence $f(x) \equiv 0$ (mod m), where $f(x)$ is a polynomial with integer coefficients. Almost all the material of this chapter can be found in some form in the writings of Lagrange, although Lagrange's work was done well before Gauss defined the notion of congruence. Gauss analyzed two special cases in his *Disquisitiones Arithmeticae*, and he was the first to consider the problem of finding solutions to polynomial congruences with nonprime modulus. The simplest case, studied in detail in Chapter 2, is the linear congruence $ax \equiv b$ (mod m), where $f(x) = ax - b$ is a polynomial of degree 1. Polynomial congruences of the second degree, or *quadratic congruences*, will be covered extensively in Chapter 5, and some special congruences of higher degree will be treated in Chapter 6.

RESULTS FOR CHAPTER 4

The main results of this chapter deal with the number of solutions of a general polynomial congruence $f(x) \equiv 0$ (mod p), where p is a prime. In a later section, we consider the problem of generating a solution modulo p^k from a solution modulo p.

General Polynomial Congruences

Henceforth, the polynomials that we consider will be assumed to have integer coefficients. We begin with a definition.

(4.1) Definition. A *solution* of the polynomial congruence $f(x) \equiv 0$ (mod m) is an integer c such that $f(c) \equiv 0$ (mod m). In this case, c will also be called a *root* of $f(x)$ modulo m (by analogy with the root of an ordinary polynomial equation).

Note. Suppose that $a \equiv b \pmod{m}$. By (2.3), $f(a) \equiv 0 \pmod{m}$ if and only if $f(b) \equiv 0 \pmod{m}$. Hence we do not consider a and b to be different solutions if $a \equiv b \pmod{m}$.

Since every integer is congruent to exactly one element in the complete residue system $0, 1, \ldots, m-1$, any solution of $f(x) \equiv 0 \pmod{m}$ must be congruent to one of these m numbers. In particular, a *polynomial congruence modulo m can have at most m incongruent solutions*, which may be found, for example, by checking each of the integers $0, 1, \ldots, m-1$ separately. (It is often easier to check instead all integers x in the interval $-m/2 < x \le m/2$.) Neither procedure is efficient if m is large.

To find solutions in general, let us first write $m = p_1^{a_1} p_2^{a_2} \cdots p_r^{a_r}$ as a product of prime powers. Since $p_i^{a_i}$ divides m for each i, it is clear that any root of $f(x)$ modulo m is also a root of $f(x)$ modulo $p_i^{a_i}$ for $i = 1, 2, \ldots, r$. Conversely, suppose that $f(c_i) \equiv 0 \pmod{p_i^{a_i}}$ for each i. Since the $p_i^{a_i}$ are relatively prime in pairs, we can use the Chinese Remainder Theorem to find an integer c (which is unique modulo m) such that $c \equiv c_i \pmod{p_i^{a_i}}$ for each i; thus $f(c) \equiv 0 \pmod{m}$. We have therefore proved the following result.

(4.2) Theorem. *Let $m = p_1^{a_1} p_2^{a_2} \cdots p_r^{a_r}$. If c is a solution of $f(x) \equiv 0 \pmod{m}$, then c is a solution of $f(x) \equiv 0 \pmod{p_i^{a_i}}$ for $i = 1, 2, \ldots, r$. Conversely, if c_i is a solution of $f(x) \equiv 0 \pmod{p_i^{a_i}}$ for each i, then there is exactly one solution c of $f(x) \equiv 0 \pmod{m}$ such that $c \equiv c_i \pmod{p_i^{a_i}}$ for $i = 1, 2, \ldots, r$.*

It follows from (4.2) that every distinct set of solutions of the polynomial congruences $f(x) \equiv 0 \pmod{p_i^{a_i}}$ corresponds to a single solution of $f(x) \equiv 0 \pmod{m}$. Suppose that, for each i, there are t_i incongruent solutions of $f(x) \equiv 0 \pmod{p_i^{a_i}}$; in the notation of (4.2), there are then t_1 choices for c_1, t_2 choices for c_2, and so forth. Hence there will be exactly $t_1 t_2 \cdots t_r$ roots of the polynomial congruence $f(x) \equiv 0 \pmod{m}$. Clearly, if even one of the congruences $f(x) \equiv 0 \pmod{p_i^{a_i}}$ has no solution (whence $t_i = 0$), then there cannot be any roots of $f(x)$ modulo m. Thus we have the following result.

(4.3) Theorem. *Suppose that $m = p_1^{a_1} p_2^{a_2} \cdots p_r^{a_r}$. If t_i denotes the number of incongruent solutions of $f(x) \equiv 0 \pmod{p_i^{a_i}}$, then the number of solutions of $f(x) \equiv 0 \pmod{m}$ is precisely $t_1 t_2 \cdots t_r$.*

In view of (4.2), to find solutions of a polynomial congruence, it suffices to consider the case in which the modulus is a prime power p^a. Since a solution of $f(x) \equiv 0 \pmod{p^a}$ must also be a solution of $f(x) \equiv 0 \pmod{p}$, all roots of $f(x)$ modulo p^a can be found among the integers x such that $f(x) \equiv 0$

(mod p). In Theorem 4.10, we will give a method for generating solutions of $f(x) \equiv 0$ (mod p^a) from solutions of $f(x) \equiv 0$ (mod p).

Recall that the polynomial $f(x) = a_n x^n + a_{n-1} x^{n-1} + \cdots + a_1 x + a_0$ is said to have *degree* n if $a_n \neq 0$. The zero polynomial is not assigned a degree. It is a familiar fact that the degree of the product $f(x)g(x)$ of two polynomials is the sum of the degrees of $f(x)$ and $g(x)$. (One reason that the zero polynomial is not assigned a degree is that this result would no longer always hold.)

Often, in dealing with congruences modulo m, the usual definition of degree is changed somewhat. The polynomial $f(x) = a_n x^n + a_{n-1} x^{n-1} + \cdots + a_1 x + a_0$ is said to have *degree k modulo m* if k is the largest integer such that $m \nmid a_k$. If all coefficients of $f(x)$ are divisible by m, the degree modulo m is undefined. This definition reflects the fact that when we are studying polynomial congruences modulo m, coefficients that are divisible by m can be treated as if they were zero. *The notion of degree modulo m is not needed in this book, and henceforth, degree will mean ordinary degree.*

The next result deals with the familiar process of dividing one polynomial by another. If $f(x)$ is divided by $g(x)$ in the usual way, the quotient and remainder need not have integer coefficients even if $f(x)$ and $g(x)$ do, so we assume that the leading coefficient of $g(x)$ is equal to 1. Now suppose that $f(x) = a_n x^n + a_{n-1} x^{n-1} + \cdots + a_0$, where $a_n \neq 0$, and that $g(x)$ has degree $m \leq n$. Let $f_1(x) = f(x) - a_n x^{n-m} g(x)$; then $f_1(x)$ either is the zero polynomial or has degree less than n. If we continue to subtract terms of the form $bx^k g(x)$ for appropriate b and k, we will eventually be left with either the zero polynomial or a polynomial of degree less than m. This procedure is exactly the same as the ordinary "long division" algorithm for polynomials. We therefore have the following result.

(4.4) Theorem (A Division Algorithm for Polynomials). *Let $f(x)$ and $g(x)$ be polynomials with integer coefficients, and suppose that $g(x)$ has leading coefficient equal to 1. Then there exist polynomials $q(x)$ and $r(x)$, each with integer coefficients, such that $f(x) = q(x)g(x) + r(x)$, where either the degree of $r(x)$ is less than the degree of $g(x)$ or $r(x)$ is the zero polynomial.*

Since the study of polynomial congruences for a general modulus can be reduced to the case where the modulus is prime, we next investigate the nature of solutions of $f(x) \equiv 0$ (mod p), where p is prime. As a consequence of Fermat's Theorem, we may suppose the degree of $f(x)$ to be less than p. For if $f(x)$ is any polynomial, we can use the Division Algorithm to write $f(x) = q(x)(x^p - x) + r(x)$, where $q(x)$ and $r(x)$ are polynomials and either $r(x)$ has degree less than p or $r(x)$ is the zero polynomial. Since $x^p - x \equiv 0$ (mod p) for all x, $r(x)$ clearly has the same roots as $f(x)$.

Alternatively, consider x^n where n is positive, and write $n - 1 = k(p-1) + r$, where $0 \leq r < p - 1$. If $x \not\equiv 0$ (mod p), then $x^{p-1} \equiv 1$ (mod p) by Fermat's

Theorem, and hence $x^{n-1} = (x^{p-1})^k x^r \equiv x^r \pmod{p}$. Thus, $x^n \equiv x^{r+1}$ (mod p) if $x \not\equiv 0 \pmod{p}$. Clearly, $x^n \equiv x^{r+1} \pmod{p}$ also holds for $x \equiv 0 \pmod{p}$. Since $r + 1 \le p - 1$, we can replace $f(x)$ by a polynomial $g(x)$ of degree not exceeding $p - 1$ such that $f(x) \equiv g(x) \pmod{p}$ for all x. In particular, the polynomials $f(x)$ and $g(x)$ have precisely the same roots modulo p. (See Problem 4-19.)

We are therefore led to consider solutions of $f(x) \equiv 0 \pmod{p}$, where p is a prime and $f(x)$ has degree less than or equal to $p - 1$. The principal result, proved in 1768 by Lagrange, states that for a prime modulus, a polynomial congruence of degree n has at most n solutions. We require the following lemma.

(4.5) Lemma. *Suppose p is prime, and let $f(x)$ be a polynomial of degree n. If a is a root of $f(x)$ modulo p, then $f(x) = (x - a)q(x) + c$, where $p \mid c$ and $q(x)$ is a polynomial of degree $n - 1$ with integer coefficients.*

Proof. By (4.4), $f(x) = q(x)(x - a) + c$ for some polynomial $q(x)$ and a constant c. Let $x = a$ in this identity. It follows that $c = f(a)$, and hence p divides c. Clearly, $q(x)$ has degree $n - 1$.

Note. We do not need to assume that the modulus is prime; exactly the same proof works for an arbitrary modulus m. A similar argument can be used to prove the familiar fact that if $f(x)$ is a polynomial with real coefficients and $f(a) = 0$, then the polynomial $x - a$ divides $f(x)$.

(4.6) Lagrange's Theorem. *Let p be prime, and suppose $f(x)$ is a polynomial of degree n not all of whose coefficients are divisible by p. Then the congruence $f(x) \equiv 0 \pmod{p}$ cannot have more than n incongruent solutions.*

Proof. The proof is by induction on the degree of $f(x)$. If $f(x)$ has degree 0, then $f(x)$ is the constant polynomial a_0. Since $p \nmid a_0$, the congruence $f(x) \equiv 0 \pmod{p}$ has no solutions.

Now suppose the result is true for polynomials of degree $n - 1$, and let $f(x)$ be a polynomial of degree n. Either $f(x) \equiv 0 \pmod{p}$ has no solutions (in which case the theorem holds) or it has at least one solution, say a; thus $f(a) \equiv 0 \pmod{p}$. Let $q(x)$ be as in the preceding lemma; then it is easy to verify that not every coefficient of $q(x)$ is divisible by p.

If b is another solution of $f(x) \equiv 0 \pmod{p}$ that is incongruent to a modulo p, (4.5) implies that $0 \equiv f(b) \equiv (b - a)q(b) \pmod{p}$; since $a \not\equiv b \pmod{p}$, we must have $q(b) \equiv 0 \pmod{p}$. Thus any solution of $f(x) \equiv 0 \pmod{p}$ that is different from a is a solution of $q(x) \equiv 0 \pmod{p}$, which by the induction hypothesis has no more than $n - 1$ incongruent solutions. Thus $f(x) \equiv 0 \pmod{p}$ has at most n incongruent solutions.

Notes. **1.** It follows from Lagrange's Theorem that if k is the largest integer such that $p \nmid a_k$, then the congruence $f(x) \equiv 0 \pmod{p}$ cannot have more than k incongruent solutions.

2. If the modulus is *not* prime, then a polynomial congruence of degree n can have more than n solutions; for example, the congruence $x^2 - 1 \equiv 0 \pmod{8}$ has the four solutions 1, 3, 5, 7. The reason that the proof of Lagrange's Theorem does not go through in this case is that if $m > 1$ is not prime, then we cannot conclude from $uv \equiv 0 \pmod{m}$ that either u or v is congruent to zero modulo m.

3. The proof of Lagrange's Theorem given above is essentially the same as the usual proof that a polynomial of degree n with real coefficients has no more than n real roots.

Under certain conditions, we can prove that a polynomial $f(x)$ of degree n has exactly n roots modulo p. If $f(x)$ is a polynomial of degree n whose leading coefficient a_n is not congruent to 0 modulo p, then there exists an integer c such that $a_n c \equiv 1 \pmod{p}$. Let $f_1(x)$ be the polynomial obtained by replacing the leading coefficient of $cf(x)$ by 1. Then $f_1(x) \equiv cf(x) \pmod{p}$ for all x, and thus $f_1(x)$ has the same roots modulo p as $f(x)$.

(4.7) Theorem (Chebyshev, 1849). *Let p be prime, and suppose that the polynomial $f(x)$ has degree n, with $n \leq p$, and leading coefficient 1. Use the division algorithm to write $x^p - x = q(x)f(x) + r(x)$, where $r(x)$ is the zero polynomial or $r(x)$ has degree less than n. Then $f(x)$ has exactly n roots modulo p if and only if every coefficient of $r(x)$ is divisible by p.*

Proof. Suppose that every coefficient of $r(x)$ is divisible by p. Then $q(x)f(x)$ has the same roots modulo p as $x^p - x$. By Fermat's Theorem, $x^p - x$ has p roots modulo p. Thus $q(x)f(x)$ also has p roots, and since p is a prime, each of these p roots must be a root of either $q(x)$ or $f(x)$ (or both). But since $q(x)$ has degree $p - n$ and leading coefficient 1, it has no more than $p - n$ roots, by Lagrange's Theorem. Hence $f(x)$ has at least n roots and therefore exactly n roots.

Now suppose that $f(x) \equiv 0 \pmod{p}$ has precisely n solutions. By Fermat's Theorem, $x^p - x \equiv 0 \pmod{p}$ for every x. Thus any root of $f(x)$ modulo p will also be a root of $r(x)$ modulo p, and therefore $r(x)$ has at least n roots. Either $r(x)$ is the zero polynomial (and there is nothing to prove) or the degree of $r(x)$ is less than n, in which case Lagrange's Theorem implies that every coefficient of $r(x)$ is divisible by p.

(4.8) Corollary. *Suppose p is prime and d divides $p - 1$. Then the polynomial congruence $x^d - 1 \equiv 0 \pmod{p}$ has exactly d incongruent solutions.*

Proof. If $p - 1 = kd$, then
$$x^p - x = (x^{p-1} - 1)x = (x^d - 1)(x^{d(k-1)} + x^{d(k-2)} + \cdots + x^d + 1)x.$$

Thus the remainder is 0 when $x^p - x$ is divided by $x^d - 1$, and the result follows from (4.7).

Solutions of $f(x) \equiv 0 \pmod{p^k}$

We now look at the problem of determining which solutions of $f(x) \equiv 0 \pmod{p}$ are also solutions of $f(x) \equiv 0 \pmod{p^k}$. The general procedure, as detailed in the summary and example following (4.12), is to start with a root modulo p and use it to generate a root (or roots) modulo p^2. Using the same technique, we produce roots modulo p^3, p^4, and so on, until we finally obtain a root (or roots) for the original modulus p^k.

We require the following lemma. In this section, $f'(x)$ denotes the *derivative* of the polynomial $f(x)$.

(4.9) Lemma. *Let p be a prime and k a positive integer. Then for every choice of x and t,*

$$f(x + p^k t) \equiv f(x) + f'(x) p^k t \pmod{p^{k+1}}.$$

Proof. The proof is by induction on the degree of $f(x)$. The result is trivial if $f(x)$ has degree 0. Suppose the result is true for polynomials of degree n, and let $f(x)$ have degree $n+1$. Then $f(x) = a + xg(x)$, where a is a constant and $g(x)$ has degree n. By the induction assumption, $g(x + p^k t) \equiv g(x) + g'(x) p^k t \pmod{p^{k+1}}$. Thus

$$f(x + p^k t) = a + (x + p^k t) g(x + p^k t) \equiv a + (x + p^k t)(g(x) + g'(x) p^k t)$$
$$\equiv a + xg(x) + (xg'(x) + g(x)) p^k t \pmod{p^{k+1}}.$$

Since $a + xg(x) = f(x)$ and $xg'(x) + g(x) = f'(x)$, the result follows.

Note that any root of $f(x)$ modulo p^{k+1} is clearly a root of $f(x)$ modulo p^k. Suppose that the roots of $f(x)$ modulo p^k are given by $r_1, r_2, \ldots, r_n$, and let S be any root modulo p^{k+1}; then $S \equiv r_i \pmod{p^k}$ for some i. Thus *all solutions of $f(x) \equiv 0 \pmod{p^{k+1}}$ are generated from solutions of $f(x) \equiv 0 \pmod{p^k}$*. We now show how to produce roots of $f(x)$ modulo p^{k+1} from roots modulo p^k.

(4.10) Theorem. *Let p be a prime and k an arbitrary positive integer, and suppose that s is a solution of $f(x) \equiv 0 \pmod{p^k}$.*

(i) If $p \nmid f'(s)$, then there is precisely one solution s_{k+1} of $f(x) \equiv 0 \pmod{p^{k+1}}$ such that $s_{k+1} \equiv s \pmod{p^k}$. The solution s_{k+1} is given by $s_{k+1} = s + p^k t$, where t is the unique solution of $f'(s) t \equiv -f(s)/p^k \pmod{p}$.

(ii) If $p \mid f'(s)$ and $p^{k+1} \mid f(s)$, then there are p solutions of $f(x) \equiv 0$ (mod p^{k+1}) that are congruent to s modulo p, given by $s + p^k j$ for $j = 0, 1, \ldots, p - 1$.

(iii) If $p \mid f'(s)$ and $p^{k+1} \nmid f(s)$, then there are no solutions of $f(x) \equiv 0$ (mod p^{k+1}) that are congruent to s modulo p^k.

Proof. Let S be a solution of $f(x) \equiv 0$ (mod p^{k+1}) such that $S \equiv s$ (mod p^k); then $S = s + p^k t$ for some integer t. Thus the problem is to find values of t such that that $s + p^k t$ is a root of $f(x) \equiv 0$ (mod p^{k+1}), that is, integers t for which $f(s + p^k t) \equiv 0$ (mod p^{k+1}). In view of the preceding lemma, we therefore want an integer t such that

$$f(s) + f'(s)p^k t \equiv 0 \pmod{p^{k+1}}.$$

Since $f(s) \equiv 0$ (mod p^k), it follows that $f(s)/p^k$ is an integer, and hence we can divide by p^k to get

$$f'(s)t \equiv -f(s)/p^k \pmod{p}.$$

By (2.8), the latter congruence has a unique solution if $(f'(s), p) = 1$ or, equivalently, if p does not divide $f'(s)$. If p does divide $f'(s)$, then the left side is congruent to 0 modulo p, and hence the right side must also be congruent to 0. Thus we must have $f(s)/p^k \equiv 0$ (mod p), that is, $p^{k+1} \mid f(s)$, in which case any value of t in a complete residue system will be a solution (for example, $t = 0, 1, \ldots, p - 1$). Finally, if $p \mid f'(s)$ but $p^{k+1} \nmid f(s)$, then the right side is not congruent to 0 modulo p, and so there will be no t that solves the congruence. This completes the proof of the theorem.

Since any root of $f(x)$ modulo p^k is a root modulo p, it follows that *if $f(x) \equiv 0$ (mod p) has no solutions, then there are no solutions of $f(x) \equiv 0$ (mod p^k) for any $k \geq 1$.*

(4.11) Corollary. *Let p be a prime and k an arbitrary positive integer. If s_1 is a solution of $f(x) \equiv 0$ (mod p) and $p \nmid f'(s_1)$, then there exists precisely one solution s_k of $f(x) \equiv 0$ (mod p^k) such that $s_k \equiv s_1$ (mod p).*

Proof. Since $p \nmid f'(s_1)$, we can use (4.10.i) to find a unique solution s_2 of $f(x) \equiv 0$ (mod p^2) such that $s_2 \equiv s_1$ (mod p). Since $s_2 \equiv s_1$ (mod p) and $f'(s_1) \not\equiv 0$ (mod p), it follows from (2.3) that $f'(s_2) \not\equiv 0$ (mod p). Thus we can apply (4.10.i) to s_2 to find the unique root s_3 of $f(x) \equiv 0$ (mod p^3) such that $s_3 \equiv s_2$ (mod p^2). Clearly, $s_3 \equiv s_2$ (mod p^2) implies $s_3 \equiv s_2$ (mod p); since $s_2 \equiv s_1$ (mod p), we have $s_3 \equiv s_1$ (mod p). We therefore proceed in this way until a root s_k of $f(x) \equiv 0$ (mod p^k) has been found such that $s_k \equiv s_1$ (mod p).

Summary. The general procedure for finding all solutions of $f(x) \equiv 0$ (mod p^k) can be summarized as follows.

1. First find all solutions of $f(x) \equiv 0$ (mod p).

2. Select one, say s_1; then by (4.10), there are either 0, 1, or p solutions of $f(x) \equiv 0$ (mod p^2) congruent to s_1 modulo p; if solutions exist, they are found by solving the linear congruence $f'(s_1)t \equiv -f(s_1)/p$ (mod p). If there are no solutions, start again with a different s_1.

3. If there are solutions of $f(x) \equiv 0$ (mod p^2), select one, say s_2, and find the corresponding roots of $f(x)$ modulo p^3 by solving the congruence $f'(s_2)t \equiv -f(s_2)/p^2$ (mod p). Do this for each root of $f(x)$ modulo p^2. Note that since $s_2 \equiv s_1$ (mod p), $f'(s_2) \equiv f'(s_1)$ (mod p), so we do not need to calculate $f'(s_2)$.

4. Proceeding in this fashion, we will eventually determine all solutions of $f(x) \equiv 0$ (mod p^k).

Note. It is worth emphasizing that if at any step in this procedure we get multiple solutions (that is, if $p \mid f'(s)$ and $p^{k+1} \mid f(a)$, case (ii) of Theorem 4.10), then we must apply the above process to *each* solution.

(4.12) Example. We will go through the details of this technique and find all of the solutions of the polynomial congruence $13x^7 - 42x + 674 \equiv 0$ (mod 1323).

Let $f(x) = 13x^7 - 42x + 674$. Since $1323 = 3^3 \cdot 7^2$, we first find all solutions of $f(x) \equiv 0$ (mod 3^3) and $f(x) \equiv 0$ (mod 7^2), then use the Chinese Remainder Theorem to find all solutions of the original congruence. To solve $f(x) \equiv 0$ (mod 27), first consider $f(x) \equiv 0$ (mod 3). Since 0 is not a solution, we can use Fermat's Theorem to conclude that $x^2 \equiv 1$ (mod 3) for any solution x, and hence $x^7 \equiv x$ (mod 3). Thus $f(x) \equiv 0$ (mod 3) reduces to $-29x + 674 \equiv 0$ (mod 3), that is, $x + 2 \equiv 0$ (mod 3). This has the unique solution $s_1 = 1$. (In fact, the solution is obvious here, but the same technique is useful for larger primes.)

Note that $f'(x) = 91x^6 - 42 \equiv x^6 = (x^2)^3 \equiv 1$ (mod 3) for any solution x, and hence $f'(1) \not\equiv 0$ (mod 3). Thus (4.11) guarantees that $f(x) \equiv 0$ (mod 9) and $f(x) \equiv 0$ (mod 27) each have exactly one solution, and these solutions must be congruent to 1 modulo 3. We look for a solution of $f(x) \equiv 0$ (mod 9) of the form $s_1 + 3t = 1 + 3t$; hence, by (4.10.i), we want t such that $f'(1)t \equiv -f(1)/3$ (mod 3), that is, $t \equiv 1$ (mod 3), since $f(1) \equiv 6$ (mod 9). Thus $t = 1$ and therefore $s_2 = 1 + 3t = 4$ is the unique solution of $f(x) \equiv 0$ (mod 9).

We next look for the unique root of $f(x)$ modulo 27, which must be of the form $4 + 9t$ since a root of $f(x)$ modulo 27 will also be a root modulo 9. By (4.10.i), t must satisfy $f'(4)t \equiv -f(4)/9$ (mod 3). Since $f'(x) \equiv 1$ (mod 3) for any solution x, we have $f'(4) \equiv 1$ (mod 3). To simplify the calculation of $f(4)/9$, observe that $f(x) \equiv 13x^7 + 12x - 1$ (mod 27); thus $f(4) \equiv$

9 (mod 27). Hence $f(4)/9 \equiv 1$ (mod 3), by (2.2.v), and so $f'(4)t \equiv -f(4)/9$ (mod 3) reduces to $t \equiv -1$ (mod 3), which has the unique solution $t = 2$. Hence $s_3 = 4 + 9t = 22$ is the unique solution of $f(x) \equiv 0$ (mod 27).

Similarly, use Fermat's Theorem to reduce $f(x) \equiv 0$ (mod 7) to $-29x + 674 \equiv 0$ (mod 7), that is, $-x + 2 \equiv 0$ (mod 7), which has the unique solution $s_1 = 2$. Thus we look for a root of $f(x)$ modulo 49 of the form $2 + 7t$, where $f'(2)t \equiv -f(2)/7$ (mod 7). Since $f'(x) = 91x^6 - 42 \equiv 0$ (mod 7) for all x, 7 divides $f'(2)$; also, 49 divides $f(2) = 2254$, so (4.10.ii) implies that *any* value of $t = 0, 1, \ldots, 6$ yields a solution of $f(x) \equiv 0$ (mod 49). Hence there are seven roots of $f(x)$ modulo 49: 2, 9, 16, 23, 30, 37, and 44.

Now use the Chinese Remainder Theorem to solve $x \equiv 22$ (mod 27) and $x \equiv a$ (mod 49), where a is any of the seven roots of $f(x)$ modulo 49. In applying the Chinese Remainder Theorem, we need only calculate the b_i once and then substitute the various values of a in the expression for x^*. (See the proof of (2.11).) Thus all solutions of $f(x) \equiv 0$ (mod 1323) are given by $x^* \equiv 49 + 540a$ (mod 1323), that is, 184, 373, 562, 751, 940, 1129, and 1318.

The Congruence $x^2 \equiv a$ (mod p^k)

Finally, we consider a special type of polynomial congruence, namely, the quadratic congruence $x^2 \equiv a$ (mod p^k), where p is a prime. These congruences play an important role in the general theory of quadratic congruences, which will be presented in the next chapter. (As its name implies, a quadratic polynomial congruence is one in which the polynomial is of degree 2.)

(4.13) Theorem. *Let p be an odd prime and suppose $k \geq 1$. If $(a, p) = 1$, then $x^2 \equiv a$ (mod p^k) has either no solutions or exactly two solutions, according as $x^2 \equiv a$ (mod p) is or is not solvable.*

Proof. If the congruence $x^2 \equiv a$ (mod p) has no solutions, then there are no solutions of $x^2 \equiv a$ (mod p^k). Now suppose there is a solution of $x^2 \equiv a$ (mod p), say s; then $-s$ is also a solution. Since s and $-s$ are incongruent modulo p, they are the only roots of $x^2 - a$ modulo p, by (4.6). Clearly, s is not divisible by p, since $(a, p) = 1$. Thus if $f(x) = x^2 - a$, then $f'(s) = 2s$ is not divisible by p, and so the result follows from (4.10.i). (In particular, the roots s and $-s$ modulo p each produce exactly one root modulo p^k for any $k \geq 1$.)

(4.14) Theorem. *Suppose that a is an odd integer. Then*

(i) *$x^2 \equiv a$ (mod 2) is always solvable and has exactly one solution;*

(ii) *$x^2 \equiv a$ (mod 4) is solvable if and only if $a \equiv 1$ (mod 4), in which case there are precisely two solutions;*

(iii) $x^2 \equiv a \pmod{2^k}$, with $k \geq 3$, is solvable if and only if $a \equiv 1 \pmod 8$, in which case there are exactly four solutions. In particular, if s is any solution, then all of the solutions are given by $\pm s$ and $\pm s + 2^{k-1}$.

Proof. Parts (i) and (ii) are obvious. Now suppose $k \geq 3$. If we square the 2^{k-3} odd numbers from 1 to 2^{k-2}, no two of the squares are congruent modulo 2^k. For if $a^2 \equiv b^2 \pmod{2^k}$, with $a > b$ and a and b odd, then $2^k \mid (a-b)(a+b)$. But exactly one of $a-b$ and $a+b$ is congruent to 2 modulo 4 and hence has only one factor of 2. Thus the other must be divisible by 2^{k-1}, which is impossible since $a - b$ and $a + b$ are both less than 2^{k-1}.

The square of an odd number is congruent to 1 modulo 8, and there are exactly 2^{k-3} positive integers less than 2^k that are congruent to 1 modulo 8. It follows that the squares of the 2^{k-3} odd numbers from 1 to 2^{k-2} are congruent modulo 2^k, in some order, to the positive integers less than 2^k that are congruent to 1 modulo 8. Thus if $a \equiv 1 \pmod 8$, the congruence $x^2 \equiv a \pmod{2^k}$ clearly has a solution s, with $1 \leq s < 2^{k-2}$. It is obvious that there cannot be a solution if a is odd and $a \not\equiv 1 \pmod 8$.

If s is a solution of $x^2 \equiv a \pmod{2^k}$, then squaring $-s$ and $\pm s + 2^{k-1}$ modulo 2^k shows that these are also solutions; by taking least positive residues, we may suppose that all of the solutions are positive and less than 2^k. It is easily checked that no two of these numbers are congruent modulo 2^k. Thus the congruence $x^2 \equiv a \pmod{2^k}$ has at least four solutions whenever $a \equiv 1 \pmod 8$. There are 2^{k-3} such a and at least four solutions for each a; this accounts for $4 \cdot 2^{k-3} = 2^{k-1}$ odd numbers less than 2^k, namely, all of them. It follows that if $a \equiv 1 \pmod 8$, the congruence $x^2 \equiv a \pmod{2^k}$ has exactly four solutions.

Note. Additional results concerning the congruences $x^2 \equiv a \pmod{p^k}$ and $x^2 \equiv a \pmod{2^k}$ are given in Problems 4-37 to 4-47.

PROBLEMS AND SOLUTIONS

4-1. Find all solutions of $x^4 + 2x + 36 \equiv 0 \pmod{875}$.

Solution. Let $f(x) = x^4 + 2x + 36$. Note that $875 = 5^3 \cdot 7$; thus we first find all solutions of $f(x) \equiv 0 \pmod{5^3}$ and $f(x) \equiv 0 \pmod 7$, then apply the Chinese Remainder Theorem to find all of the solutions to the original congruence. It is easy to check that the only solutions of $f(x) \equiv 0 \pmod 7$ are $x \equiv -1, 2 \pmod 7$; also, $f(x) \equiv 0 \pmod 5$ has the unique solution $x \equiv -1 \pmod 5$.

Since $f'(x) = 4x^3 + 2$ and hence $f'(-1) \not\equiv 0 \pmod 5$, (4.11) guarantees that $f(x) \equiv 0 \pmod{5^3}$ has exactly one solution (and this solution is necessarily congruent to -1

modulo 5). Let $s_1 = -1$; then we look for a solution of $f(x) \equiv 0$ (mod 5^2) of the form $-1 + 5t$; thus, by (4.10.i), we want t such that $f'(-1)t \equiv -f(-1)/5$ (mod 5). Since $f(-1) = 35$ and $f'(-1) = -2$, this congruence reduces to $-2t \equiv -7$ (mod 5), which has the unique solution 1. Hence $s_2 = -1 + 5t = 4$ is the only solution of $f(x) \equiv 0$ (mod 25).

We now look for a solution of $f(x) \equiv 0$ (mod 125) of the form $4 + 25t$; by (4.10.i), t must satisfy $f'(4)t \equiv -f(4)/25$ (mod 5). Since $f(4) \equiv 50$ (mod 125) and $f'(4) \equiv f'(-1) \equiv -2$ (mod 5), this congruence reduces to $-2t \equiv -2$ (mod 5), which has the unique solution $t = 1$. Thus $s_3 = 4 + 25t = 29$ is the unique solution of $f(x) \equiv 0$ (mod 125).

Finally, apply the Chinese Remainder Theorem to solve $x \equiv 29$ (mod 125) and $x \equiv -1$ or 2 (mod 7), obtaining 279 and 779 as the only solutions of the original congruence $f(x) \equiv 0$ (mod 875).

4-2. *Find all solutions of the congruence $x^6 - 2x^5 - 35 \equiv 0$ (mod 6125).*

Solution. Let $f(x) = x^6 - 2x^5 - 35$; then $f'(x) = 6x^5 - 10x^4$. Since $6125 = 5^3 \cdot 7^2$, we must find the roots of $f(x)$ modulo 125 and 49. The congruence $f(x) \equiv 0$ (mod 5) reduces, by Fermat's Theorem, to $x^2 - 2x \equiv 0$ (mod 5), which has the two solutions 0 and 2. Since 5 divides $f'(0) = 0$ and 25 does not divide $f(0) = -35$, the root 0 does not generate any root modulo 25 and hence none modulo 125. Since 5 does not divide $f'(2) = 32 \equiv 2$ (mod 5), there is a unique root modulo 25 of the form $2 + 5t$, where t satisfies $32t \equiv -f(2)/5$ (mod 5), i.e., $2t \equiv 2$ (mod 5). Thus $t = 1$ and so $2 + 5 \cdot 1 = 7$ is a root modulo 25. Now look for a root modulo 125 of the form $7 + 25t$. Since $f(7) \equiv 0$ (mod 125), 7 is already a root modulo 125. Since $f'(2) \not\equiv 0$ (mod 5), $x \equiv 7$ (mod 125) is the only root modulo 125. (See (4.11).)

For any x, $f(x) \equiv x^5(x - 2)$ (mod 7); hence the roots of $f(x)$ modulo 7 are 0 and 2. Since 7 divides $f'(0) = 0$ but 49 does not divide $f(0) = -35$, 0 does not generate a root modulo 49. Since 7 does not divide $f'(2) = 32$, 2 generates a unique root modulo 49, of the form $2 + 7t$ where $32t \equiv -f(2)/7$ (mod 7), that is, $4t \equiv 5$ (mod 7). Hence $t = 3$ and therefore $2 + 7 \cdot 3 = 23$ is the only root of $f(x)$ modulo 49.

Finally, the Chinese Remainder Theorem applied to the system $x \equiv 7$ (mod 125) and $x \equiv 23$ (mod 49) yields 3257 as the only solution of $f(x) \equiv 0$ (mod 6125).

4-3. *Solve the congruence $x^2 - 31x - 12 \equiv 0$ (mod 36).*

Solution. Let $f(x) = x^2 - 31x - 12$; then $f'(x) = 2x - 31$. Clearly, the only solutions modulo 4 are 0 and 3. Also, $f(x)$ has the roots 0 and 1 modulo 3. Look for a solution modulo 9 of the form $0 + 3t = 3t$, where $f'(0)t \equiv -f(0)/3$ (mod 3), i.e., $2t \equiv 1$ (mod 3). Thus $t = 2$, and so 0 generates the root $0 + 3 \cdot 2 = 6$ modulo 9. Similarly, the root 1 modulo 3 generates the root $1 + 3t$ modulo 9, where $f'(1)t \equiv -f(1)/3$ (mod 3), i.e., $-29t \equiv 14$ (mod 3). Hence $t = 2$, and so the only other root modulo 9 is $1 + 3 \cdot 2 = 7$.

Now use the Chinese Remainder Theorem to solve $x \equiv a$ (mod 4) and $x \equiv b$ (mod 9), where $a = 0$ or 3 and $b = 6$ or 7. Thus, modulo 36, $f(x)$ has the four roots 7, 15, 16, and 24.

Note. In Chapter 5, we describe a much better way of solving quadratic congruences, using a suitable modification of the familiar process of completing the square.

4-4. *Find all solutions of $x^7 - 14x - 2 \equiv 0 \pmod{27}$.*

Solution. Let $f(x) = x^7 - 14x - 2$. We first solve $f(x) \equiv 0 \pmod 3$. This is easy to do by inspection; we need only test 0, 1, and -1. It turns out that 1 is the only solution.

Note that $f'(1) \not\equiv 0 \pmod 3$ since $f'(x) = 7x^6 - 14$. Thus, by (4.11), $f(x) \equiv 0 \pmod 9$ and $f(x) \equiv 0 \pmod{27}$ each have exactly one solution (and these solutions are congruent to 1 modulo 3). Now look for a solution of $f(x) \equiv 0 \pmod 9$ of the form $s_1 + 3t = 1 + 3t$; in view of (4.10.i), we want t such that $f'(1)t \equiv -f(1)/3 \pmod 3$. Since $f(1) = -15$ and $f'(1) = -7$, we solve $-7t \equiv 5 \pmod 3$, i.e., $2t \equiv 2 \pmod 3$, which has the unique solution $t = 1$. Hence $s_2 = 1 + 3t = 4$ is the unique solution of $f(x) \equiv 0 \pmod 9$.

Finally, we look for the unique solution of $f(x) \equiv 0 \pmod{27}$, which must be of the form $4 + 9t$ since a root of $f(x)$ modulo 27 will also be a root modulo 9. By (4.10.i), t must satisfy $f'(4)t \equiv -f(4)/9 \pmod 3$. Since $f(4) \equiv 18 \pmod{27}$ and $f'(4) \equiv f'(1) \equiv 2 \pmod 3$, this congruence reduces to $2t \equiv -2 \pmod 3$, which has the unique solution $t = 2$. Hence $s_3 = 4 + 9t = 22$ is the unique solution of $f(x) \equiv 0 \pmod{27}$.

4-5. *Find all solutions of $x^7 - 14x - 2 \equiv 0 \pmod{49}$.*

Solution. Let $f(x) = x^7 - 14x - 2$. Use Fermat's Theorem to reduce $f(x) \equiv 0 \pmod 7$ to $x - 2 \equiv 0 \pmod 7$, which has the unique solution $s_1 = 2$. Now look for roots of $f(x)$ modulo 49 of the form $2 + 7t$, where $f'(2)t \equiv -f(2)/7 \pmod 7$. Since 7 divides $f'(2) = 434$ and 49 divides $f(2) = 98$, (4.10.ii) implies that any value of $t = 0, 1, \ldots, 6$ provides a solution of $f(x) \equiv 0 \pmod{49}$. Hence there are seven roots of $f(x)$ modulo 49: 2, 9, 16, 23, 30, 37, and 44.

4-6. *Let p be prime, and suppose $f(x) \equiv 0 \pmod p$ has k solutions $a_1, a_2, \ldots, a_k$. Prove that there is a polynomial $q(x)$ of degree $n-k$ with the same leading coefficient as $f(x)$ such that for all x,*

$$f(x) \equiv (x - a_1)(x - a_2) \cdots (x - a_k) q(x) \pmod p.$$

(Hint. Divide $f(x)$ by $(x - a_1)(x - a_2) \cdots (x - a_k)$.)

Solution. Let $g(x) = (x - a_1)(x - a_2) \cdots (x - a_k)$. Then $g(x)$ has leading coefficient 1 and degree k. Let $f(x) = q(x)g(x) + r(x)$, where $q(x), r(x)$ are as in the Division Algorithm. Clearly, $q(x)$ has degree $n - k$ and the same leading coefficient as $f(x)$. Since $a_1, a_2, \ldots, a_k$ are roots of both $f(x)$ and $g(x)$ modulo p, they are also roots of $r(x)$. But since $r(x)$ either is the zero polynomial or has degree less than k, it follows from Lagrange's Theorem that all the coefficients of $r(x)$ are divisible by p, and hence $f(x) \equiv q(x)g(x) \pmod p$ for all x.

4-7. *Suppose that $f(x)$ and $g(x)$ have the same degree and have exactly the same roots modulo p. Does it follow that $f(x) \equiv g(x) \pmod p$ for every x? What if $f(x)$ and $g(x)$ also have the same leading coefficient?*

Solution. No. Let $p = 5$ and consider $f(x) = 3(x - 1)$ and $g(x) = 4(x - 1)$. Then f and g each have the unique root 1, but $f(x) \equiv g(x) \pmod 5$ holds only when $x \equiv 1$

(mod 5). We can also arrange for $f(x)$ and $g(x)$ to have the same leading coefficient. For example, let $p = 5$, $f(x) = x(x^2 - 2)$, and $g(x) = x(x^2 - 3)$. Since a square can never be congruent to 2 or 3 modulo 5, $f(x)$ and $g(x)$ each have the unique root 0. But it is clear that $f(x) \equiv g(x) \pmod 5$ only when $x \equiv 0 \pmod 5$.

4-8. Refer to the preceding problem. Suppose that $f(x)$ and $g(x)$ are each of degree n, have n distinct roots, and have the same leading coefficient. Is it true now that $f(x) \equiv g(x) \pmod p$ for every integer x?

Solution. Yes. Consider the polynomial $D(x) = f(x) - g(x)$. We show that for all x, $D(x) \equiv 0 \pmod p$, and hence $f(x) \equiv g(x) \pmod p$, by showing that every coefficient of $D(x)$ is divisible by p.

Any common root of $f(x)$ and $g(x)$ modulo p is a root of $D(x)$ modulo p. Hence $D(x)$ has at least n roots. But the coefficient of x^n in $D(x)$ is 0, so $D(x)$ either is the zero polynomial or has degree less than n. Thus by Lagrange's Theorem, every coefficient of $D(x)$ is divisible by p, and the result follows.

4-9. Use Fermat's Theorem to prove that

$$x^{p-1} - 1 \equiv (x-1)(x-2)\cdots(x-(p-1)) \pmod p$$

holds for every x. (Hint. Use Problem 4-8.)

Solution. Let $f(x) = x^{p-1} - 1$ and let $g(x) = (x-1)(x-2)\cdots(x-(p-1))$. Without loss of generality, we can assume that all roots are in the complete residue system $\{0, 1, \ldots, p-1\}$. Fermat's Theorem then implies that the congruence $f(x) \equiv 0 \pmod p$ has precisely $p-1$ solutions, namely, $1, 2, \ldots, p-1$. It is clear that the congruence $g(x) \equiv 0 \pmod p$ has exactly the same solutions. But $f(x)$ and $g(x)$ each have leading coefficient equal to 1, so by Problem 4-8, $f(x) \equiv g(x)$ for all x. (Indeed, we can conclude that corresponding coefficients of $f(x)$ and $g(x)$ match modulo p.)

4-10. Prove Wilson's Theorem by comparing the constant terms on each side of the congruence in the preceding problem.

Solution. If p is prime, it follows from Problem 4-9 that

$$x^{p-1} - 1 \equiv (x-1)(x-2)\cdots(x-(p-1)) \pmod p$$

for all x. Taking $x = 0$, we obtain -1 on the left side and $(-1)^{p-1}(p-1)!$ on the right. If $p \neq 2$, then $(-1)^{p-1} = 1$, while if $p = 2$, then $-1 \equiv 1 \pmod p$. Thus we conclude that $-1 \equiv (p-1)! \pmod p$, which is Wilson's Theorem.

Note. Chebyshev seems to be the first to have proved Wilson's Theorem using the above argument. Lagrange's original proof also used polynomials, but in a much more complicated way.

4-11. *Refer to Problem 4-9. If p is a prime greater than 2, what can be deduced by comparing the coefficients of x^{p-2} on each side of the congruence?*

Solution. The coefficient of x^{p-2} in $x^{p-1} - 1$ is 0, while the coefficient of x^{p-2} in $(x-1)(x-2)\cdots(x-(p-1))$ is $-S$, where $S = 1 + 2 + \cdots + (p-1)$. By the remark at the end of the solution of Problem 4-9, corresponding coefficients match modulo p, and hence $S \equiv 0 \pmod{p}$.

The result of the preceding problem can be generalized. Let p be an odd prime. Multiply out the product $(x-1)(x-2)\cdots(x-(p-1))$, and define $a_1, a_2, \ldots, a_{p-1}$ by

$$(x-1)(x-2)\cdots(x-(p-1)) = x^{p-1} + a_1 x^{p-2} + \cdots + a_{p-1}. \quad (1)$$

Then $a_1 = -(1 + 2 + \cdots + (p-1))$ and $a_{p-1} = (-1)^{p-1}(p-1)! = (p-1)!$.

4-12. *Let p be an odd prime, and suppose $a_1, a_2, \ldots, a_{p-1}$ are defined as in (1). Show that if $1 \le k \le p-2$, then $a_k \equiv 0 \pmod{p}$. (Hint. See Problem 4-10.)*

Solution. The coefficient of x^k in $x^{p-1} - 1$ is 0, while the coefficient of x^k in $(x-1)(x-2)\cdots(x-(p-1))$ is a_k. Thus we conclude exactly as in Problem 4-10 that $a_k \equiv 0 \pmod{p}$.

▷ **4-13.** *(Wolstenholme, 1862.) Let p be a prime greater than 3. Show that $a_{p-2} \equiv 0 \pmod{p^2}$, where a_{p-2} is defined as in formula (1) above. (Hint. Take $x = p$ in the formula and use Problem 4-12.)*

Solution. Substituting p for x in (1), we obtain

$$(p-1)! = p^{p-1} + a_1 p^{p-2} + \cdots + a_{p-2} p + a_{p-1}.$$

Note that $a_{p-1} = (p-1)!$. By cancelling and rearranging, we obtain the equation $-a_{p-2} = p^{p-2} + a_1 p^{p-3} + \cdots + a_{p-3} p$. Since $p \ge 5$, it follows from Problem 4-12 that p divides a_{p-3}. Thus every term on the right side of the equation is divisible by p^2, and hence $a_{p-2} \equiv 0 \pmod{p^2}$.

Note. Consider the sum $1 + 1/2 + 1/3 + \cdots + 1/(p-1)$, where p is an odd prime. If this is brought to the common denominator $(p-1)!$, then the numerator is precisely $-a_{p-2}$.

4-14. *Find the number of solutions of $x^3 + x^2 + 2 \equiv 0 \pmod{3^7 \cdot 7^3}$.*

Solution. The congruence clearly has no solution modulo 3 and hence no solution modulo 3^7. Thus by (4.3), the original congruence has no solutions.

4-15. *Find the number of solutions of $x^2 - 3 \equiv 0 \pmod{11^4 \cdot 23^3}$.*

Solution. Modulo 11, the congruence has two solutions, 5 and -5. Since neither $f'(5)$ nor $f'(-5)$ is divisible by 11, (4.11) guarantees that 5 and -5 each generate a unique solution modulo 11^4. Similarly, there are two solutions, 7 and -7, modulo 23, and 23 does not divide $f'(7)$ or $f'(-7)$. Thus there are two solutions modulo 23^3. It follows from (4.3) that the original congruence has $2 \cdot 2 = 4$ solutions.

4-16. Find the number of solutions of $x^3 - 2x^2 - 4x - 17 \equiv 0 \pmod{25}$.

Solution. There are two solutions, 2 and 3, modulo 5. Since $f'(x) = 3x^2 - 4x - 4$, we have $f'(2) = 0$. Also, $f(2) = -25$ is divisible by 25, so (4.10.ii) implies that 2 generates five distinct roots modulo 25. Similarly, $f'(3)$ is not divisible by 5, and hence 3 generates a unique root modulo 25, by (4.10.i). Thus there are precisely six roots of $f(x)$ modulo 25.

4-17. Let p be prime. Suppose that $f(x)$ has r roots $x_1, x_2, \ldots, x_r$ modulo p and that $f'(x_i)$ is not divisible by p for any i. Prove that $f(x)$ has precisely r roots modulo p^k for any positive integer k.

Solution. For each i, it follows from (4.11) that there is exactly one solution of $f(x) \equiv 0 \pmod{p^k}$ that is congruent to x_i modulo p. Thus $f(x) \equiv 0 \pmod{p^k}$ has precisely r solutions.

4-18. Find the number of solutions of $x^3 - 18x^2 + 72 \equiv 0 \pmod{1125}$.

Solution. Let $f(x) = x^3 - 18x^2 + 72$; then $f'(x) = 3x^2 - 36x$. Note that $1125 = 9 \cdot 125$. It is easily checked that 0 is the only root of $f(x)$ modulo 3. Since $3 \mid f'(0)$ and $9 \mid f(0)$, (4.10.i) implies that $f(x)$ has three roots modulo 9. Similarly, $f(x)$ has the unique root 1 modulo 5. Since 5 does not divide $f'(1)$, it follows from (4.11) that $f(x)$ has a unique root modulo 5^k for any $k \geq 1$. Thus by (4.3), $f(x)$ has exactly three roots modulo 1125. (The same argument shows that there are precisely three roots modulo $9 \cdot 5^k$ for any $k \geq 1$.)

4-19. For each polynomial $f(x)$ and modulus p, find a polynomial $g(x)$ of degree less than p such that $f(x) \equiv g(x) \pmod{p}$ for all x:
 (a) $p = 7$ and $f(x) = x^{16} + 5x^4 - 3x^2 + 1$;
 (b) $p = 11$ and $f(x) = x^{40} + x^{39} + \cdots + x + 1$.

Solution. (a) By Fermat's Theorem, $x^7 \equiv x \pmod 7$ for all x and hence $x^{16} = (x^7)^2 x^2 \equiv x^4 \pmod 7$ for all x. Therefore take $g(x) = 6x^4 - 3x^2 + 1$.
 (b) By Fermat's Theorem, for all x we have $x^{11} \equiv x \pmod{11}$, $x^{12} \equiv x^2 \pmod{11}$, $\ldots$, $x^{20} \equiv x^{10} \pmod{11}$. The pattern repeats four times, and hence for all x, $f(x) \equiv 4(x^{10} + x^9 + \cdots + x) + 1 \pmod{11}$.

4-20. Find the number of solutions of $x^{361} - 1 \equiv 0 \pmod{3^j \cdot 5^k}$, where j and k are positive integers.

Solution. Clearly, the only root modulo 3 is 1, since $(-1)^{361} \equiv -1 \pmod 3$. Since 3 does not divide $f'(1) = 361$, (4.11) implies that there is exactly one root modulo 3^j for any $j \geq 1$. Similarly, note that 0 is not a root modulo 5, and hence Fermat's Theorem implies that $s^4 \equiv 1 \pmod 5$ for any root s modulo 5. Thus $s^{360} \equiv 1 \pmod 5$, and so the original congruence reduces to $x - 1 \equiv 0 \pmod 5$. Therefore 1 is the only root modulo 5. Since $5 \nmid f'(1)$, it follows from (4.11) that $x^{361} - 1 \equiv 0 \pmod{5^k}$ has only one solution. Now apply (4.3) to conclude that the original congruence has exactly one solution for any choice of j and k.

116   CHAPTER 4: POLYNOMIAL CONGRUENCES

4-21. *If the odd number m has exactly r distinct prime factors, show that the congruence $x^2 \equiv 1 \pmod{m}$ has exactly 2^r solutions.*

Solution. If $m = 1$, then $r = 0$ and the number of solutions is 1, namely, 2^0. Now let $m = \prod p_i^{a_i}$. Then for any i, the congruence $x^2 \equiv 1 \pmod{p_i^{a_i}}$ has exactly two solutions. (It obviously has a solution; hence by (4.13), it has exactly two solutions, ± 1.) In producing a solution of the original congruence via the Chinese Remainder Theorem, we have two choices for every i, and hence there are 2^r solutions.

4-22. *(a) Use the preceding problem to find the smallest odd number m such that the congruence $x^2 \equiv 1 \pmod{m}$ has 16 solutions. (b) What is the smallest such even number m?*

Solution. (a) By the previous problem, we want m to have four distinct prime factors. The smallest such odd m is $3 \cdot 5 \cdot 7 \cdot 11 = 1155$.
 (b) Let $m = 2^k n$, where n is odd. The congruence $x^2 \equiv 1 \pmod{2^k}$ has one solution if $k = 1$, two if $k = 2$, and four if $k \geq 3$ (see (4.14)). It is easy to see that the smallest choice for m is $8 \cdot 3 \cdot 5 = 120$.

4-23. *Find the number of solutions of $10x^4 + 4x + 1 \equiv 0 \pmod{27}$.*

Solution. Since $x^4 \equiv x^2 \pmod{3}$ by Fermat's Theorem, any solution of the original congruence satisfies $x^2 + x + 1 \equiv 0 \pmod{3}$, which has 1 as its only solution. Note that $f'(x) = 40x^3 + 4$. Since $f'(1) \equiv 2 \pmod{3}$, it follows from (4.11) that 1 generates a unique solution modulo 27. Therefore the original congruence has exactly one solution.

4-24. *Find the number of solutions of $7x^2 - 17x - 2 \equiv 0 \pmod{128}$.*

Solution. Both 0 and 1 are roots modulo 2. Since $f'(x) = 14x - 17$, neither $f'(0)$ nor $f'(1)$ is divisible by 2. So (4.11) implies that there are precisely two solutions modulo 128.

4-25. *Find the number of solutions of $7x^5 - 3x^3 + 2x - 5 \equiv 0 \pmod{27 \cdot 25 \cdot 49}$.*

Solution. Fermat's Theorem implies $x^3 \equiv x \pmod{3}$, so modulo 3 the congruence reduces to $x - 3x + 2x - 5 \equiv 0 \pmod{3}$, which clearly has no solutions. Thus the original congruence has no solutions.

4-26. *Find the number of solutions of $3x^3 + x + 1 \equiv 0 \pmod{125}$.*

Solution. The only roots modulo 5 are 1 and 3. Since $f'(x) = 9x^2 + 1$, $f'(3) = 82$ is not divisible by 5, and so 3 generates a unique solution modulo 125, by (4.11). Since 5 divides $f'(1) = 10$ but $f(1) = 5$ is not divisible by 25, 5 generates no solution modulo 25 and hence none modulo 125, by (4.10.iii). Thus there is exactly one solution to the original congruence.

4-27. *Find the number of solutions of (a) $x^2 \equiv 49$ (mod $53^3 \cdot 61^4$); (b) $x^2 \equiv 851$ (mod $5^2 \cdot 7^3 \cdot 11^4$); (c) $x^2 \equiv -1$ (mod $5^3 \cdot 7^2$).*

Solution. (a) Solutions obviously exist modulo 53 and 61 (namely, 7 and -7). Thus by (4.11), there are exactly two solutions modulo each of 53^3 and 61^4. Now apply (4.3) to conclude that the original congruence has $2 \cdot 2 = 4$ solutions.

(b) We first consider $x^2 \equiv 851 \equiv 1$ (mod 5), $x^2 \equiv 851 \equiv 4$ (mod 7), and $x^2 \equiv 851 \equiv 4$ (mod 11). Since each right side is a square, apply (4.11) and (4.3) to conclude that the original congruence has $2 \cdot 2 \cdot 2 = 8$ solutions.

(c) Any solution x of the congruence must satisfy $x^2 \equiv -1$ (mod 7). It is easy to check that this has no solutions. Thus the original congruence has no solutions.

4-28. *Suppose that p is prime and p divides neither a nor n. Show that that for any positive integer k, the congruence $x^n \equiv a$ (mod p^k) has a solution if and only if the congruence $x^n \equiv a$ (mod p) has a solution.*

Solution. Any solution of the congruence $x^n \equiv a$ (mod p^k) is a solution of the congruence $x^n \equiv a$ (mod p), so if the first congruence has a solution, so does the second. Conversely, suppose that the congruence $x^n \equiv a$ (mod p) has a solution s. Since p does not divide a, it cannot divide s. If we let $f(x) = x^n - a$, then $f'(x) = nx^{n-1}$. Since p divides neither n nor s, p cannot divide $f'(s)$. Thus by (4.11), s generates a solution of $x^n \equiv a$ (mod p^k), and it follows that the congruence has a solution. (In fact, the two congruences have the same number of solutions.)

4-29. *Show that for any prime p, there is a polynomial $f(x)$ of degree p with leading coefficient 1 such that the congruence $f(x) \equiv 0$ (mod p) has no solutions.*

Solution. Let $f(x) = x^p - x + 1$. By Fermat's Theorem, $x^p - x \equiv 0$ (mod p) for all x, and hence $f(x) \equiv 1$ (mod p) for all x.

4-30. *Let k be a positive integer. Show that the congruence $x^2 + x + a \equiv 0$ (mod 2^k) has no solutions if a is odd and two solutions if a is even.*

Solution. If a is odd, then the congruence does not have any solutions modulo 2, for it is clear that neither 0 nor 1 is a solution; thus there are no solutions modulo 2^k for any positive k. If a is even, 0 and 1 are solutions modulo 2. Let $f(x) = x^2 + x - a$. Then $f'(x) = 2x + 1$, so $f'(x)$ is never congruent to 0 modulo 2. It follows from (4.11) that each of the two solutions modulo 2 extends to a unique solution modulo 2^k for any k.

4-31. *Which five-digit numbers x have the property that the last five digits of x^2 are the same as the corresponding digits of x?*

Solution. Such a number must satisfy the congruence $x^2 \equiv x$ (mod 10^5) or, equivalently, $x(x - 1) \equiv 0$ (mod 10^5). Since x and $x - 1$ are always relatively prime and $10^5 = 2^5 \cdot 5^5$, x must be congruent to either 0 or 1 modulo 32 and also modulo 3125.

118   CHAPTER 4: POLYNOMIAL CONGRUENCES

There are four nonnegative solutions less than 10^5, which can be obtained by the Chinese Remainder Theorem. These are 0, 1, 9376 and 90625. Thus the only five-digit solution is 90625.

4-32. Let $f(x) = x^{99} + x^{98} + \cdots + x + 1$. How many solutions are there to the congruence $f(x) \equiv 0 \pmod{101}$?

Solution. By the usual formula for the sum of a geometric progression, or by direct multiplication, $(x-1)f(x) = x^{100} - 1$. Since 101 is prime, the congruence $x^{100} - 1 \equiv 0 \pmod{101}$ has 100 solutions by Fermat's Theorem, namely, $1, 2, \ldots, 100$. If $x \not\equiv 1 \pmod{101}$ and $x^{100} - 1 \equiv 0 \pmod{101}$, then we must have $f(x) \equiv 0 \pmod{101}$. Thus $f(x) \equiv 0 \pmod{101}$ has at least 99 solutions. But $f(x)$ has degree 99, so the congruence has exactly 99 solutions.

Another proof: Note that $x(x-1)f(x) = x^{101} - x$. Since $f(x)$ has degree 99, it follows from (4.7) that $f(x) \equiv 0 \pmod{101}$ has exactly 99 solutions.

4-33. Suppose p is an odd prime and $(a, p) = 1$. Prove that for any positive integer k, $x^2 \equiv a \pmod{p^k}$ has a solution if and only if $x^2 \equiv a \pmod{p^{k+1}}$ has a solution.

Solution. Let $f(x) = x^2 - a$. It is clear that any root of $f(x)$ modulo p^{k+1} is also a root modulo p^k. Conversely, suppose s is a root modulo p^k; then p does not divide s since $(a, p) = 1$. Thus $f'(s) = 2s$ is not divisible by p, and hence (4.10.i) implies that $f(x)$ has a root modulo p^{k+1}.

4-34. Suppose p is an odd prime and $(a, p) = 1$. If $x^2 \equiv a \pmod{p^k}$ is solvable for some $k \geq 1$, prove directly (without using (4.10)) that $x^2 \equiv a \pmod{p^{k+1}}$ is solvable and has exactly two solutions. (Hint. If s is a solution modulo p^k, look for a solution modulo p^{k+1} of the form $s + tp^k$.)

Solution. Suppose $s^2 \equiv a \pmod{p^k}$; then $s^2 = a + mp^k$ for some integer m. We look for solutions of $x^2 \equiv a \pmod{p^{k+1}}$ of the form $s + tp^k$. Thus we want $(s + tp^k)^2 \equiv a \pmod{p^{k+1}}$, i.e., $s^2 + 2stp^k \equiv 0 \pmod{p^{k+1}}$. Substituting $s^2 = a + mp^k$, we have $mp^k + 2stp^k \equiv 0 \pmod{p^{k+1}}$, i.e., $2st \equiv -m \pmod{p}$. By (2.8), this congruence has a unique solution. It follows that every solution modulo p^k generates a unique solution modulo p^{k+1}.

Finally, we show that the congruence $x^2 \equiv a \pmod{p^{k+1}}$ has exactly two solutions. Let u be a solution; then clearly $-u$ is also a solution. Note that since $(a, p) = 1$, we must have $(u, p) = 1$. If v is any solution of the congruence, then $v^2 \equiv u^2 \equiv a \pmod{p^{k+1}}$, and so $p^{k+1} \mid (v-u)(v+u)$. But p cannot divide both $v-u$ and $v+u$, since otherwise we would have $p \mid u$, contradicting the fact that $(u, p) = 1$. Thus p^{k+1} divides exactly one of $v - u$ and $v + u$, and hence $v \equiv u \pmod{p^{k+1}}$ or $v \equiv -u \pmod{p^{k+1}}$.

4-35. Let p be an odd prime, and suppose $k \geq 1$. Prove that $x^2 \equiv 0 \pmod{p^k}$ has exactly p^m solutions, where $m = k/2$ if k is even and $m = (k-1)/2$ if k is odd.

Solution. Let s be a solution and write $s = p^n t$, where $(t, p) = 1$. We may suppose that $0 \le s < p^k$; hence $n \le k - 1$. Since $p^k | s^2$, we must have $2n \ge k$. Thus all solutions between 0 and p^k have the form $p^r t$, where $r = k/2$ if k is even, $r = (k+1)/2$ if k is odd, and $t = 0, 1, 2, \ldots, p^{k-r} - 1$. Hence there are p^{k-r} solutions, and since $k - r = m$ in each case, the result follows.

▷ **4-36.** *Let m be a positive integer, $f(x)$ a polynomial with integer coefficients, and a a root of $f(x)$ modulo m. Use the Division Algorithm to express $f(x)$ as $q(x)(x - a)^2 + r(x)$, where $r(x)$ is the zero polynomial or a polynomial of degree less than or equal to 1. Show that $f'(a) \equiv 0 \pmod{m}$ if and only if every coefficient of $r(x)$ is divisible by m.*

Solution. Note that $f'(x) = 2(x - a)q(x) + q'(x)(x - a)^2 + r'(x)$. If every coefficient of $r(x)$ is divisible by m, then $r'(x) \equiv 0 \pmod{m}$ for all x. Substituting a in the expression for $f'(x)$, we find that $f'(a) \equiv 0 \pmod{m}$.

Conversely, suppose that $f'(a) \equiv 0 \pmod{m}$. If we divide $r(x)$ by $x - a$, we get $r(x) = b(x - a) + c$, where b and c are constants. It is clear that $r(a) \equiv 0 \pmod{m}$, and hence $c \equiv 0 \pmod{m}$. Since $f'(a) \equiv 0 \pmod{m}$, substituting a in the expression for $f'(x)$ shows that $r'(a) \equiv 0 \pmod{m}$, and hence $b \equiv 0 \pmod{m}$. Thus every coefficient of $r(x)$ is divisible by m.

Note. In the ordinary algebra of polynomials, the real number a is called a *multiple root* of $f(x)$ if $(x - a)^2$ divides $f(x)$, and it is easy to show that this is the case if and only if $f'(a) = 0$. This problem shows that an analogous result holds for congruences.

The Congruence $x^2 \equiv a \pmod{2^k}$

4-37. *Prove that the congruence $x^2 \equiv 0 \pmod{2^k}$ has precisely 2^m solutions, where $m = k/2$ if k is even and $m = (k-1)/2$ if k is odd. The solutions are given, respectively, by $2^m t$ and $2^{m+1} t$, where $0 \le t \le 2^m - 1$.*

Solution. Suppose s is a solution of $x^2 \equiv 0 \pmod{2^k}$, with $0 \le s < 2^k$. If $k = 2m$, then $2^{2m} | s^2$ and hence $2^m | s$, that is, $s = 2^m t$, where $0 \le t \le 2^m - 1$. Thus, since any such s is a solution, there are exactly 2^m solutions. Now suppose $k = 2m + 1$, i.e., $m = (k - 1)/2$. Since the largest power of 2 that divides a square is even, we have $2^{2m+2} | s^2$, and hence $2^{m+1} | s$. Thus $s = 2^{m+1} t$, where $0 \le t \le 2^m - 1$. Since any such s is a solution, there are exactly 2^m solutions.

4-38. *Suppose a is even, with $a \not\equiv 0 \pmod{2^k}$, and let 2^b be the largest power of 2 that divides a. If b is odd, prove that the congruence $x^2 \equiv a \pmod{2^k}$ is not solvable for any value of k.*

Solution. Note that $b \le k - 1$. If $2^k | s^2 - a$ for some s, then $2^b | s^2 - a$ and hence $2^b | s^2$. Thus $2^{b+1} | s^2$ (since b is odd) and therefore $2^{b+1} | a$, which contradicts the definition of b.

▷ **4-39.** Suppose $a \not\equiv 0 \pmod{2^k}$, and let $a = 2^b c$, where c is odd. Suppose that $b = 2t$ is even, and prove the following.

(i) If $b \leq k - 3$, $x^2 \equiv a \pmod{2^k}$ is solvable if and only if $c \equiv 1 \pmod{8}$, in which case there are exactly 2^{t+2} solutions, given by $2^t s + 2^{k-t} j$ for $j = 0, 1, \ldots, 2^t - 1$, where s ranges over the four solutions of $y^2 \equiv c \pmod{2^{k-b}}$.

(ii) If $b = k - 2$, $x^2 \equiv a \pmod{2^k}$ is solvable if and only if $c \equiv 1 \pmod 4$, in which case there are precisely 2^{t+1} solutions, given by $\pm 2^t + 2^{t+2} j$ for $j = 0, 1, \ldots, 2^t - 1$.

(iii) If $b = k - 1$, there are exactly 2^t solutions of $x^2 \equiv a \pmod{2^k}$ for any odd c, given by $2^t + 2^{t+1} j$ for $j = 0, 1, \ldots, 2^t - 1$.

Solution. Note that if $b = 0$, this is simply Theorem 4.14. Let s be a solution of $x^2 \equiv a \pmod{2^k}$; we may assume that $0 < s < 2^k - 1$. Since $2^k \mid s^2 - a$ and $2^b \mid a$, it follows that $2^b \mid s^2$ and hence $2^t \mid s$. Divide each side of $s^2 \equiv a \pmod{2^k}$ by 2^b to get the equivalent congruence $(s/2^t)^2 \equiv c \pmod{2^{k-b}}$.

(i) Since c is odd, (4.14) implies that for $k - b \geq 3$, this congruence is solvable if and only if $c \equiv 1 \pmod 8$, in which case there are exactly four solutions, say, s_1, s_2, s_3, s_4. Thus $s/2^t = s_i + 2^{k-b} j$ and hence $s = 2^t s_i + 2^{k-t} j$, where $j = 0, 1, 2, \ldots, 2^t - 1$. Thus the original congruence has $4 \cdot 2^t = 2^{t+2}$ incongruent solutions.

(ii) If $b = k - 2$, we obtain the equivalent congruence $(s/2^t)^2 \equiv c \pmod 4$; thus solutions exist if and only if $c \equiv 1 \pmod 4$. In this case, we have $s/2^t = \pm 1 + 4j$ and hence $s = \pm 2^t + 2^{t+2} j$, where $j = 0, 1, \ldots, 2^t - 1$. Hence there are exactly $2 \cdot 2^t = 2^{t+1}$ solutions.

(iii) Finally, if $b = k - 1$, we get $(s/2^t)^2 \equiv c \equiv 1 \pmod 2$, since c is odd. Hence $s/2^t = 1 + 2j$ and so $s = 2^t + 2^{t+1} j$, where $j = 0, 1, \ldots, 2^t - 1$. Thus there are precisely 2^t solutions in this case.

4-40. Use Problem 4-39 to find the number of solutions of (a) $x^2 \equiv 0 \pmod{512}$; (b) $x^2 \equiv 0 \pmod{1024}$; (c) $x^2 \equiv 0 \pmod{2^{15}}$.

Solution. (a) Since $512 = 2^8$, there are 16 solutions. (b) Since $1024 = 2^{10}$, there are 32 solutions. (c) Since $(15 - 1)/2 = 7$, there are $2^7 = 128$ solutions.

4-41. Find the solutions of (a) $x^2 \equiv 17 \pmod{512}$; (b) $x^2 \equiv 7 \pmod{32}$; (c) $x^2 \equiv -1 \pmod{128}$; (d) $x^2 \equiv 9 \pmod{256}$.

Solution. By (4.14.iii), each congruence has either 0 or 4 solutions.

(a) Replacing 17 by $17 + 512 = 529 = 23^2$, it is clear that there are solutions, and hence there will be exactly four solutions, namely, ± 23 and $\pm 23 + 128$, i.e., 23, 105, 151, and 233.

(b) If $s^2 \equiv 7 \pmod{32}$, then $s^2 \equiv 7 \equiv 3 \pmod 4$, which has no solutions. Thus the original congruence has no solutions.

(c) If this congruence has a solution, then $x^2 \equiv -1 \pmod 4$ has a solution, which it clearly does not. So there are no solutions of $x^2 \equiv -1 \pmod{128}$.

(d) The congruence obviously has solutions (3 and -3, for example). Thus $x^2 \equiv 9 \pmod{256}$ has exactly four solutions, namely, ± 3 and $\pm 3 + 128$, i.e., 3, 125, 131, 253.

4-42. *Use Problems 4-38 and 4-39 to find the number of solutions of (a)* $x^2 \equiv 2$ (mod 128); *(b)* $x^2 \equiv 48$ (mod 256); *(c)* $x^2 \equiv 164$ (mod 512).

Solution. (a) In the notation of Problem 4-38, b is odd and hence there are no solutions.

(b) Write $48 = 2^4 \cdot 3$. Thus, in the notation of Problem 4-39, $b = 4$, and hence there are either 0 or 4 solutions. Dividing each side by 16 yields a congruence of the form $y^2 \equiv 3$ (mod 16), which has no solution, since $y^2 \equiv 3$ (mod 4) has no solution. Thus $x^2 \equiv 48$ (mod 256) has no solutions.

(c) Write $164 = 2^2 \cdot 41$; in the notation of Problem 4-39, $b = 2$, $t = 1$, $c = 1$, and $k = 9$. Hence there are $2^3 = 8$ solutions.

4-43. *Find all solutions of (a)* $x^2 \equiv 0$ (mod 64); *(b)* $x^2 \equiv 0$ (mod 128). *(Hint. Refer to Problem 4-37.)*

Solution. (a) Since $64 = 2^6$ and 6 is even, there are $2^3 = 8$ solutions, given by all multiples of 8, namely, 0, 8, 16, 24, 32, 40, 48, and 56.

(b) $128 = 2^7$; therefore (in the notation of Problem 4-37) $m = 3$ and so there are $2^3 = 8$ solutions. Since $(k + 1)/2 = 4$, all solutions are given by multiples of $2^4 = 16$, namely, 0, 16, 32, 48, 64, 80, 96, and 112.

4-44. *Find all solutions of (a)* $x^2 \equiv 25$ (mod 256); *(b)* $x^2 \equiv 21$ (mod 32); *(c)* $x^2 \equiv 41$ (mod 128).

Solution. Apply (4.14). (a) Clearly, 5 is a solution; thus all solutions are given by ± 5 and $\pm 5 + 128$. Hence the only solutions are 5, 123, 133, and 251.

(b) The given congruence implies that $x^2 \equiv 21 \equiv 5$ (mod 8), which has no solutions since the squares of 1, 3, 5, and 7 are all congruent to 1 modulo 8. Thus the original congruence has no solutions.

(c) Replace 41 by $41 + 128 = 13^2$. Thus 13 is a solution. Then (4.14.iii) implies that solutions are ± 13 and $\pm 13 + 64$. Thus the only solutions are 13, 51, 77, and 115.

4-45. *Use Problems 4-38 and 4-39 to find all solutions of (a)* $x^2 \equiv 24$ (mod 512); *(b)* $x^2 \equiv 144$ (mod 256).

Solution. (a) Write $24 = 2^3 \cdot 3$; since the exponent 3 is odd, Problem 4-38 implies that there are no solutions. Alternatively, if $x^2 \equiv 24$ (mod 512), then x is even, say, $x = 2y$. Thus $y^2 \equiv 6$ (mod 128), which is impossible since, in particular, this implies that $y^2 \equiv 2$ (mod 4).

(b) Write $144 = 2^4 \cdot 9$; in the notation of Problem 4-39, we have $b = 4$, $t = 2$, $c = 9$, and $k = 8$. Hence the congruence has $2^{2+2} = 16$ solutions. Dividing each side by 2^4, we get the equivalent congruence $(x/4)^2 \equiv 9$ (mod 16), which has the solutions ± 3 and ± 5. It follows from the solution of Problem 4-39 that all solutions of the original congruence are given by $\pm 4 \cdot 3 + 64j$ and $\pm 4 \cdot 5 + 64j$, where $j = 0, 1, 2, 3$. Thus the 16 solutions are (check!) ± 12, ± 20, ± 44, ± 52, ± 76, ± 84, ± 108, and ± 11.

We can also find the solutions without appealing to Problem 4-39. If $s^2 \equiv 144$ (mod 256), then $(s/4)^2 \equiv 9$ (mod 16) and so $s/4 \equiv \pm 3, \pm 5$ (mod 16). Thus $s \equiv \pm 12, \pm 20$ (mod 64), and hence all solutions of $x^2 \equiv 144$ (mod 256) are given by α, $\alpha + 64$, $\alpha + 128$, and $\alpha + 192$, where $\alpha = \pm 12, \pm 20$.

4-46. Suppose that s is a solution of the congruence $x^2 \equiv a \pmod{2^k}$, where a is odd and $k \geq 3$. Show that exactly one of s and $s + 2^{k-1}$ is a solution of $x^2 \equiv a \pmod{2^{k+1}}$. *(Hint. Consider $(s + 2^{k-1})^2 - s^2$.)*

Solution. We have $(s+2^{k-1})^2 - s^2 = s2^k + 2^{2k-2}$. Because s is odd, $s2^k \equiv 2^k \pmod{2^{k+1}}$. Also, $2k-2 \geq k+1$ since $k \geq 3$; thus $(s+2^{k-1})^2 - s^2 \equiv 2^k \pmod{2^{k+1}}$. Let $s^2 = a + t2^k$. If t is even, then $s^2 \equiv a \pmod{2^{k+1}}$, while $(s + 2^{k-1})^2 \equiv a + 2^k \pmod{2^{k+1}}$; therefore s is a solution of $x^2 \equiv a \pmod{2^{k+1}}$ and $s + 2^{k-1}$ is not. If t is even, $s + 2^{k-1}$ is a solution and s is not.

Note. The proof of Theorem 4.14 shows that a solution exists but does not provide a computationally feasible algorithm for finding a solution when the modulus is large. This problem provides such an algorithm. Note the resemblance to the procedure given in the proof of (4.10).

4-47. It is easy to check that 23 is one of the solutions of $x^2 \equiv 17 \pmod{512}$. Find a solution of $x^2 \equiv 17 \pmod{2048}$. *(Hint. See the preceding problem.)*

Solution. In the preceding problem, it is shown that if $k \geq 3$ and s is a solution of $x^2 \equiv a \pmod{2^k}$, then one of s or $s + 2^{k-1}$ is a solution of $x^2 \equiv a \pmod{2^{k+1}}$. Here, $a = 17$, $k = 9$, and $s = 23$. Thus one of 23 or $23 + 256$ is a solution of $x^2 \equiv 17 \pmod{1024}$. It is easy to see that 23 is not a solution; thus 279 is a solution. Now let $a = 17$, $s = 279$, and $k = 10$. Then one of 279 or $279 + 512$ is a solution of $x^2 \equiv 17 \pmod{2048}$. Calculation shows that 279 is a solution. (A similar calculation shows that 279 is also a solution if the congruence is taken modulo 4096. If the congruence is modulo 8192, then $279 + 2048$ is a solution.)

EXERCISES FOR CHAPTER 4

1. For each polynomial $f(x)$ and modulus p, find a polynomial $g(x)$ of degree less than p such that $f(x) \equiv g(x) \pmod{p}$ for all x:
   (a) $p = 13$ and $f(x) = 2x^{29} - x^{17} + 3x^{13} - 4$;
   (b) $p = 5$ and $f(x) = x^{16} + x^{15} + \cdots + x + 1$.
2. Determine the number of solutions of $6x^3 + 13x^2 + x - 2 \equiv 0 \pmod{25}$.
3. Find the number of solutions of $64x^3 + 26x^2 + 108 \equiv 0 \pmod{1125}$.
4. How many solutions does the congruence $4x^4 + 7x + 11 \equiv 0 \pmod{27}$ have?
5. Find the number of solutions of $10x^5 - 9x^3 + 11x + 1 \equiv 0 \pmod{3^3 \cdot 5^2 \cdot 11^4}$.
6. Determine the number of solutions of $x^2 - 39x - 46 \equiv 0 \pmod{128}$.
7. Find all solutions of the congruence $7x^2 - x + 24 \equiv 0 \pmod{36}$.
8. Determine the solutions of $14x^3 + 11x - 13 \equiv 0 \pmod{27}$.
9. Solve the congruence $10x^7 - 21x - 13 \equiv 0 \pmod{1323}$.

EXERCISES

10. Find the solutions of $x^4 + 177x - 139 \equiv 0 \pmod{875}$.
11. Solve $7x^7 + 10x + 13 \equiv 0 \pmod{27}$.
12. Find the number of solutions of $x^{71} - 1 \equiv 0 \pmod{7^j \cdot 11^k}$, where j and k are positive integers.
13. How many solutions does the congruence $x^2 \equiv 4 \pmod{4725}$ have?
14. Find the number of solutions of the congruence $4x^3 + 43x - 82 \equiv 0 \pmod{125}$.
15. Determine the number of solutions of the following congruences:
    (a) $x^2 \equiv 25 \pmod{37^2 \cdot 59^3}$;
    (b) $x^2 \equiv 764 \pmod{5^3 \cdot 11^2 \cdot 13^5}$;
    (c) $x^2 \equiv 3 \pmod{427}$.
16. How many solutions does the congruence $x^{35} + x^{34} + \cdots + x + 1 \equiv 0 \pmod{37}$ have?
17. Determine the number of solutions of (a) $x^2 \equiv 0 \pmod{7^3}$; (b) $x^2 \equiv 0 \pmod{7^6}$.
18. Find the number of solutions of (a) $x^2 \equiv 11 \pmod{32}$; (b) $x^2 \equiv 33 \pmod{256}$; (c) $x^2 \equiv 25 \pmod{512}$.
19. Use (4.14.iii) to find all solutions of (a) $x^2 \equiv 49 \pmod{128}$; (b) $x^2 \equiv 139 \pmod{256}$; (c) $x^2 \equiv 113 \pmod{512}$.
20. Using Problem 4-35, determine the solutions of (a) $x^2 \equiv 0 \pmod{3^4}$; (b) $x^2 \equiv 0 \pmod{3^5}$.
21. Apply Problem 4-37 to find the solutions of (a) $x^2 \equiv 0 \pmod{64}$; (b) $x^2 \equiv 0 \pmod{128}$.
22. Use Problems 4-38 and 4-39 to calculate all solutions of (a) $x^2 \equiv 224 \pmod{512}$; (b) $x^2 \equiv 64 \pmod{128}$; (c) $x^2 \equiv 64 \pmod{256}$; (d) $x^2 \equiv 272 \pmod{1024}$.
23. Suppose $x^2 \equiv a \pmod{p^k}$ is solvable. Is $x^2 \equiv a \pmod{p^{k+1}}$ solvable?
24. Find all solutions of $7x^4 - 5x + 1 \equiv 0 \pmod{27}$.
25. Solve the congruence $6x^4 - 23x^3 + 13x - 16 \equiv 0 \pmod{35}$.
26. Find the solutions of $3x^7 + 7x - 6 \equiv 0 \pmod{49}$.
27. Determine the solutions of the congruence $64x^4 - 51x^3 - 3x - 13 \equiv 0 \pmod{225}$.
28. Solve the congruence $x^7 - 14x - 2 \equiv 0 \pmod{1323}$.
29. Suppose p is an odd prime and a and k are integers, with k positive. Determine the solutions of $x^{p^k} \equiv a \pmod{p}$.

NOTES FOR CHAPTER 4

1. It is customary to use Taylor's Theorem or the Binomial Theorem to prove Lemma 4.9 and hence Theorem 4.10. In the case of a polynomial $f(x)$ of degree n, Taylor's Theorem reduces to

$$f(x+y) = f(x) + y^2 f'(x) + y^3 f''(x)/2! + \cdots + y^n f^{(n)}(x)/n!,$$

where $f', f'', \ldots, f^{(n)}$ denote the successive derivatives of f. We have chosen a somewhat different approach to make the induction run more smoothly.

2. The usual notion of congruence modulo m can be extended from integers to polynomials. Let $f(x)$ and $g(x)$ be polynomials with integer coefficients, and let m be a positive integer. We say that the polynomial $f(x)$ is *congruent* to $g(x)$ modulo m if all the coefficients of the difference $f(x) - g(x)$ are divisible by m. If $f(x)$ is congruent to $g(x)$ modulo m, it is usual to write $f(x) \equiv g(x)$ (mod m). There is some risk in this notation, since we have used the notation $f(x) \equiv 0$ (mod m) to refer to a polynomial congruence, to be solved for x. If we now think of 0 as the zero polynomial, then $f(x) \equiv 0$ (mod m) could also be viewed as asserting that the *polynomials* $f(x)$ and 0 are congruent modulo m, that is, all the coefficients of $f(x)$ are divisible by m. These are two entirely different notions, and it is in principle dangerous to use the same notation for both. In practice, however, confusion seldom arises.

If $f(x) \equiv g(x)$ (mod m), then $f(a) \equiv g(a)$ (mod m) for any integer a. It is important to be aware that the converse does not hold. If $f(a) \equiv g(a)$ (mod m) for all a, it does not necessarily follow that $f(x) \equiv g(x)$ (mod m). For example, let p be a prime and let $f(x) = x^p$, $g(x) = x$. Then $f(x) \not\equiv g(x)$ (mod p), since not all the coefficients of the polynomial $x^p - x$ are divisible by p. But by Fermat's Theorem, $f(a) \equiv g(a)$ (mod p) for every integer a. So fundamentally *different-looking* polynomials such as x^p and x can determine, modulo p, the same functions. This cannot happen, for example, when we are calculating over the real numbers. If $f(a) = g(a)$ for all real numbers a, then $f(x)$ and $g(x)$ are the *same* polynomial.

REFERENCES

Trygve Nagell, *Introduction to Number Theory*, Wiley, New York, 1951.

> Nagell's book is an excellent treatment of basic number theory. In particular, it gives a much more thorough analysis of polynomial congruences than usual. The book also contains Selberg's "elementary" (but difficult) proof of the Prime Number Theorem.

CHAPTER FIVE

Quadratic Congruences and the Law of Quadratic Reciprocity

While no efficient procedure is known for solving polynomial congruences in general, or even for deciding if a solution exists, a great deal more can be said in the special case of *quadratic congruences*, that is, congruences of degree 2. These will be studied in detail in the present chapter. In particular, we will present a technique, using Gauss's Law of Quadratic Reciprocity, for deciding when a quadratic congruence is solvable. However, the problem of actually determining the solutions when they exist is still difficult, although there are methods that can be given in certain cases.

The Law of Quadratic Reciprocity is one of the most famous results in number theory. It first appeared in a paper by Euler in 1783, but he was not able to prove it. (Euler had in fact conjectured an equivalent result as early as 1746.) In 1785, Adrien-Marie Legendre (1752–1833) stated the result in the form given in (5.18), but his proof had many gaps. (Legendre assumed that there are infinitely many primes in any arithmetic progression of the form $ak + b$, where $(a, b) = 1$. This is Dirichlet's Theorem, which was not proved until 1837. But even with this result, Legendre's argument works only in certain cases.) Legendre was the first to refer to the result as a "law of reciprocity," and in 1798, he offered another proof in his *Essai sur la théorie des nombres*, but it also contained an error. The first complete demonstration of the Law of Quadratic Reciprocity was given by Gauss in 1796 and appeared in his *Disquisitiones Arithmeticae* five years later. Gauss eventually gave six proofs of this result, and since then, more than 100 have appeared. The proof we give in this chapter relies on a result (known as Gauss's Lemma) that Gauss discovered in 1808 and that leads to a fairly simple proof of the reciprocity law.

RESULTS FOR CHAPTER 5

As indicated in Chapter 4, the study of polynomial congruences can be reduced to the case where the modulus is the power of a prime p. The cases $p = 2$ and p odd will be considered separately. (This is necessary because we are considering *quadratic* congruences. The prime 5, for example, must be treated differently for polynomial congruences of degree 5.)

General Quadratic Congruences

If p is odd, the study of quadratic congruences modulo p^m reduces to the case where the modulus is simply p. (See (4.10) and (4.11).) We therefore consider the general quadratic congruence

$$ax^2 + bx + c \equiv 0 \pmod{p},$$

where p is an odd prime and $p \nmid a$. As in the case of ordinary quadratic equations, we begin by completing the square on the left side of the congruence. (This is, incidentally, the usual method for deriving the well-known quadratic formula.) Since $(a, p) = 1$ implies $(4a, p) = 1$, we multiply the congruence by $4a$ to get the equivalent congruence

$$(2ax)^2 + 4abx + 4ac \equiv 0 \pmod{p},$$

that is,

$$(2ax + b)^2 \equiv b^2 - 4ac \pmod{p}.$$

Since this last congruence has exactly the same solutions as the original, we have proved the following result.

(5.1) Theorem. *Let p be an odd prime and suppose $(a, p) = 1$. Then all solutions of the congruence $ax^2 + bx + c \equiv 0 \pmod{p}$ can be found by solving the chain of congruences*

$$y^2 \equiv b^2 - 4ac \pmod{p}, \quad 2ax \equiv y - b \pmod{p}.$$

Thus, to solve a general quadratic congruence modulo p when p is an odd prime, it suffices to solve a congruence of the form $x^2 \equiv a \pmod{p}$. The following example illustrates this technique.

(5.2) Example. We will find the solutions of $11x^2 + 5x + 18 \equiv 0 \pmod{29}$. Complete the square to get $(22x+5)^2 \equiv b^2 - 4ac \equiv 16 \pmod{29}$; thus $22x + 5 \equiv \pm 4 \pmod{29}$. Solving $22x + 5 \equiv 4 \pmod{29}$ gives $x = 25$, and $22x + 5 \equiv -4$

(mod 29) yields $x = 22$. Hence 22 and 25 are the only solutions of the original congruence.

We now consider $ax^2 + bx + c \equiv 0 \pmod{2^m}$. Since 4 is not relatively prime to 2, the preceding argument must be modified somewhat. We can still multiply by $4a$, but to obtain a congruence with the same solutions, the modulus now must be multiplied by an appropriate power of 2.

(5.3) Theorem. *Let $a = 2^r s$, with s odd. Then all solutions of the congruence $ax^2 + bx + c \equiv 0 \pmod{2^m}$ can be found by solving the chain of congruences*

$$y^2 \equiv b^2 - 4ac \pmod{2^{m+r+2}}, \quad 2ax \equiv y - b \pmod{2^{m+r+2}}.$$

Proof. Multiply the original congruence by s to get the equivalent congruence $s(ax^2 + bx + c) \equiv 0 \pmod{2^m}$. The modulus need not be changed, since $(s, 2^m) = 1$. Now multiply by $4 \cdot 2^r$; this time, to get an equivalent congruence, we must also multiply the modulus by $4 \cdot 2^r$. The net effect is to multiply by $4a$, and we obtain the equivalent congruence

$$(2ax + b)^2 \equiv b^2 - 4ac \pmod{2^{m+r+2}}.$$

This is obviously equivalent to the chain of congruences given in the statement of the theorem.

The Congruence $x^2 \equiv a \pmod{m}$

We have already noted that if s is odd, the congruence $x^2 \equiv a \pmod{s}$ can be reduced to the study of $x^2 \equiv a \pmod{p}$, where p is an odd prime. Thus, in view of (5.1) and (5.3), the analysis of a general quadratic congruence $ax^2 + bx + c \equiv 0 \pmod{m}$ reduces to an investigation of

$$x^2 \equiv a \pmod{2^k} \quad \text{and} \quad x^2 \equiv a \pmod{p} \quad (p \text{ an odd prime}).$$

Roughly speaking, we are then trying to determine which integers are "perfect squares" modulo 2^k and modulo p.

There are two problems to consider. First, when do solutions *exist* for these congruences? Second, if these congruences are solvable, *how many* solutions are there? The first question is very difficult to answer for an odd prime p; for $p = 2$, the answer appears in (4.14). The second question, which is considerably easier than the first, was covered in (4.13) and (4.14). For completeness, we state the relevant results here.

(5.4) Theorem. (i) *If $(a,p) = 1$, then $x^2 \equiv a \pmod{p^k}$ has no solutions if $x^2 \equiv a \pmod{p}$ is not solvable and exactly two solutions if $x^2 \equiv a \pmod{p}$ is solvable.*

(ii) *Suppose a is odd. If the congruence $x^2 \equiv a \pmod{2^k}$ is solvable, then it has 1, 2, or 4 solutions according as $k = 1$, $k = 2$, or $k \geq 3$.*

These results can be combined with (4.3) to give the number of solutions of $x^2 \equiv a \pmod{m}$, where m is an arbitrary positive integer.

(5.5) Theorem. *Let $m = 2^k p_1^{k_1} \cdots p_r^{k_r}$, and suppose $(a,m) = 1$. Then the congruence $x^2 \equiv a \pmod{m}$ is solvable if and only if $x^2 \equiv a \pmod{2^k}$ and $x^2 \equiv a \pmod{p_i^{k_i}}$ $(i = 1, 2, \ldots, r)$ are solvable. If $x^2 \equiv a \pmod{m}$ is solvable, there are 2^r solutions if $k = 0$ or $k = 1$, 2^{r+1} solutions if $k = 2$, and 2^{r+2} solutions if $k \geq 3$.*

Quadratic Residues

The preceding discussion largely focused on the *number* of solutions of $x^2 \equiv a \pmod{p^k}$ when p is prime. We turn our attention now to the question of the *existence* of solutions. When $p = 2$, (4.14) provides a complete answer. If p is odd and $p \mid a$, the problem can be reduced in a straightforward way to the case where $(a,p) = 1$.

We will therefore assume from now on that p is an odd prime and $(a,p) = 1$. Let $f(x) = x^2 - a$; then $f'(x) = 2x$. If s is a root of $f(x)$ modulo p, then p does not divide s, since $(a,p) = 1$. Hence $f'(s) = 2s$ is not divisible by p, and so it follows from (4.11) that $f(x)$ has a root modulo p^k for any $k \geq 1$.

Thus we may restrict our attention to the existence of solutions of $x^2 \equiv a \pmod{p}$, where p is an odd prime. This requires a much more sophisticated approach than for $p = 2$, and we will eventually use the *Law of Quadratic Reciprocity*, one of the most important results in number theory.

We begin with the following important definition.

(5.6) Definition. *Let m be an integer greater than 1, and suppose $(a,m) = 1$. Then a is called a* quadratic residue of m *if $x^2 \equiv a \pmod{m}$ has a solution. If there is no solution, then a is called a* quadratic nonresidue of m.

Notes. 1. If $a \equiv b \pmod{m}$, then clearly, a is a quadratic residue of m if and only if b is a quadratic residue of m.

2. Since any solution of $x^2 \equiv a \pmod{m}$ must be relatively prime to m if a is relatively prime to m, all of the quadratic residues of m can be found by squaring the elements of a reduced residue system modulo m. In particular,

in the case of a prime modulus p, it is enough to square $\pm 1, \pm 2, \ldots, \pm(p-1)/2$. It is easily checked that the squares of any two of $1, 2, \ldots, (p-1)/2$ are incongruent modulo p.

Since there are precisely $p-1$ elements in any reduced residue system modulo p when p is prime, we have the following result.

(5.7) Theorem. *Let p be an odd prime. Then there are exactly $(p-1)/2$ incongruent quadratic residues of p and exactly $(p-1)/2$ quadratic nonresidues of p.*

The Legendre symbol, defined next, was introduced by Legendre in 1798, in his *Essai sur la théorie des nombres*, which was the first significant work (apart from translations of Diophantus and Fibonacci's *Liber Quadratorum*) devoted entirely to the theory of numbers.

(5.8) Definition. If p is an odd prime and $(a, p) = 1$, define the *Legendre symbol* (a/p) to be 1 if a is a quadratic residue of p and -1 if a is a quadratic nonresidue of p.

It is worth emphasizing that *the Legendre symbol (a/p) is defined only when p is an odd prime and p does not divide a.*

The next result is a restatement of Euler's Criterion (Theorem 3.4), and (5.10) follows as a simple consequence.

(5.9) Euler's Criterion. *Let p be an odd prime, and suppose $(a, p) = 1$. Then $(a/p) \equiv a^{(p-1)/2} \pmod{p}$.*

(5.10) Theorem. *Suppose that p is an odd prime. Then*

(i) $a \equiv b \pmod{p}$ *implies* $(a/p) = (b/p)$;

(ii) $(ab/p) = (a/p)(b/p)$;

(iii) $\left(a^2/p\right) = 1$;

(iv) $\left(a^2 b/p\right) = (b/p)$.

Note. Part (ii) of (5.10) can be rephrased in the following way: *The product of two quadratic residues (or two nonresidues) is again a quadratic residue, whereas the product of a quadratic residue and a nonresidue is a nonresidue of p.*

If we take $a = -1$ in (5.9) and note that $(p-1)/2$ is even if and only if $p \equiv 1 \pmod{4}$, we obtain a characterization of the odd primes for which -1 is a quadratic residue.

(5.11) Theorem. *Let p be an odd prime. Then $(-1/p) = 1$ if and only if $p \equiv 1 \pmod{4}$.*

Fermat was aware of the fact that $x^2 \equiv -1 \pmod{p}$ is solvable if and only if p is of the form $4k+1$, a result that was first proved by Euler around 1750. (Euler's Criterion was proved some five years later.) When the congruence is solvable, the solutions are given by $\pm(2k)!$, where $p = 4k+1$. (See Problem 3-26.) This is not a computationally feasible way of solving the congruence for large primes p. But a solution of $x^2 \equiv -1 \pmod{p}$ can be found by raising *any* quadratic nonresidue of p to the power $(p-1)/4$. (This follows at once from Euler's Criterion.)

In applying the Law of Quadratic Reciprocity, we will also need a classification of the primes which have 2 as a quadratic residue. Instead of the usual method of employing Gauss's Lemma to obtain this characterization, the following proof uses Euler's Criterion.

(5.12) Theorem. *Let p be an odd prime. Then 2 is a quadratic residue of p if $p \equiv \pm 1 \pmod{8}$ and a quadratic nonresidue of p if $p \equiv \pm 3 \pmod{8}$.*

Proof. If $p \equiv 1$ or $5 \pmod 8$, it is straightforward to check that

$$2^{(p-1)/2} \left(\frac{p-1}{2}\right)! = 2 \cdot 4 \cdot 6 \cdots (p-1)$$

$$\equiv 2 \cdot 4 \cdot 6 \cdots \frac{p-1}{2} \cdot (-\frac{p-3}{2}) \cdots (-5)(-3)(-1)$$

$$\equiv (-1)^{(p-1)/4} \left(\frac{p-1}{2}\right)! \pmod{p}.$$

Dividing by $((p-1)/2)!$ then gives $2^{(p-1)/2} \equiv (-1)^{(p-1)/4} \pmod{p}$. Hence, by Euler's Criterion, $(2/p) = (-1)^{(p-1)/4}$. Thus $(2/p)$ is 1 or -1 according as $p \equiv 1$ or $5 \pmod 8$.

Similarly, if $p \equiv 3$ or $7 \pmod 8$, it is easily checked that

$$2^{(p-1)/2} \left(\frac{p-1}{2}\right)! = 2 \cdot 4 \cdot 6 \cdots \frac{p-3}{2}(-\frac{p-1}{2}) \cdots (-5)(-3)(-1)$$

$$= (-1)^{(p+1)/4} \left(\frac{p-1}{2}\right)! \pmod{p}.$$

Dividing each side by $((p-1)/2)!$ then gives $2^{(p-1)/2} \equiv (-1)^{(p+1)/4} \pmod{p}$. Hence, by Euler's Criterion, $(2/p)$ is -1 or 1 according as $p \equiv 3$ or $7 \pmod 8$.

The next result is useful for many of the problems in this chapter. The proofs can be found in Problems 5-49, 5-51, 5-52, and 5-53.

(5.13) Theorem. *Let p be an odd prime. Then*
(i) *-2 is a quadratic residue of p if and only if $p \equiv 1, 3 \pmod{8}$;*
(ii) *3 is a quadratic residue of p if and only if $p \equiv \pm 1 \pmod{12}$;*
(iii) *-3 is a quadratic residue of p if and only if $p \equiv 1 \pmod{6}$;*
(iv) *5 is a quadratic residue of p if and only if $p \equiv \pm 1 \pmod{5}$.*

The Law of Quadratic Reciprocity

Although there are many proofs of the Law of Quadratic Reciprocity, the one that we will give is perhaps the most straightforward. The following two results play a key role in the proof. The first, proved by Gauss in 1808, gives a criterion for an integer to be a quadratic residue of the prime p. Note the similarity between its proof and the proof of Euler's Theorem.

(5.14) Gauss's Lemma. *Let p be an odd prime, and suppose $(a, p) = 1$. Consider the least positive residues modulo p of the numbers $a, 2a, \ldots, \frac{p-1}{2}a$. If N is the number of these residues that are greater than $p/2$, then $(a/p) = (-1)^N$.*

Proof. The integers $a, 2a, \ldots, \frac{p-1}{2}a$ are relatively prime to p and incongruent modulo p. Let $u_1, u_2, \ldots, u_N$ represent the least positive residues of these numbers that exceed $p/2$, and let $v_1, v_2, \ldots, v_M$ be the least positive residues that are less than $p/2$; then $N + M = (p-1)/2$.

The numbers $p - u_1, p - u_2, \ldots, p - u_N$ are positive and less than $p/2$, relatively prime to p, and no two are congruent modulo p. Also, no $p - u_i$ is a v_j. For suppose $p - u_i = v_j$; let $u_i \equiv ra \pmod{p}$ and $v_j \equiv sa \pmod{p}$, where r and s are distinct integers between 1 and $(p-1)/2$. Then $p \equiv a(r+s) \pmod{p}$, and since $(a, p) = 1$, we must have $p \mid r + s$, a contradiction since $0 < r + s < p$.

Thus $p - u_1, p - u_2, \ldots, p - u_N, v_1, v_2, \ldots, v_M$ is a collection of $(p-1)/2$ incongruent integers that are positive and less than $p/2$. Hence they are equal, in some order, to $1, 2, \ldots, (p-1)/2$, and therefore

$$(p - u_1)(p - u_2) \cdots (p - u_N) v_1 v_2 \cdots v_M = ((p-1)/2)!,$$

that is,

$$(-1)^N u_1 u_2 \ldots u_N v_1 v_2 \ldots v_M \equiv ((p-1)/2)! \pmod{p}.$$

But it is also true that $u_1, u_2, \ldots, u_N, v_1, v_2, \ldots, v_M$ are congruent in some order to $a, 2a, \ldots, \frac{p-1}{2}a$, and hence

$$\left(\frac{p-1}{2}\right)! \equiv (-1)^N a \cdot 2a \cdots \frac{p-1}{2}a = (-1)^N a^{(p-1)/2} \left(\frac{p-1}{2}\right)! \pmod{p}.$$

Since each factor in $((p-1)/2)!$ is relatively prime to p, we can divide each side of the last congruence by $((p-1)/2)!$ to get $a^{(p-1)/2} \equiv (-1)^N \pmod{p}$. The conclusion of the lemma now follows from Euler's Criterion.

(5.15) Example. Since Gauss's Lemma is crucial to the proof of (5.16), it is instructive to look at an example in detail. We will use Gauss's Lemma to determine the primes p that have 5 as a quadratic residue. If $P = (p-1)/2$, then all of the integers $5, 10, 15, \ldots, 5P$ are less than $5p/2$. Thus the only ones whose least positive residues are greater than $p/2$ are the integers $5j$ that lie in the interval $(p/2, p)$ or $(3p/2, 2p)$. For reasons that will become clear shortly, write $p = 20k + r$, where $r = 1, 3, 7, 9, 11, 13, 17,$ or 19. (For other values of r, p is not prime.) Then we want to find the number of j for which $p/2 < 5j < p$ or $3p/2 < 5j < 2p$. The first inequality becomes $p/10 < j < p/5$, and hence $2k + r/10 < j < 4k + r/5$; similarly, the second inequality becomes $3p/10 < j < 2p/5$, and so $6k + 3r/10 < j < 8k + 2r/5$.

Note that the parity (that is, the evenness or oddness) of the number of integers that satisfy each of these inequalities is not affected by adding even integers (which need not be the same) to each side of the inequality. This follows from the fact that the number of integers in the new interval differs from the number of integers in the original interval by an even number. More precisely, if $n(a, b)$ denotes the number of integers in the interval (a, b), then $n(a, b) = n(a + 2k, b + 2k)$, since we have just translated the interval by $2k$; also, $n(a + 2k, b + 4k) = n(a + 2k, b + 2k) + 2k$, since we have expanded the interval $(a + 2k, b + 2k)$ by $2k$ to get $(a + 2k, b + 4k)$. Thus we can remove $2k$ and $4k$ from the first inequality above and $6k$ and $8k$ from the second. (The point of writing p in the form $20k + r$ is to get *even* integers such as $2k$, $4k$, $6k$, and $8k$ in these inequalities.) We are then left with $r/10 < j < r/5$ and $3r/10 < j < 2r/5$. It is now an easy matter to check that if $r = 1$, for example, there are no such j; if $r = 3$, there is one such j (none satisfying the first inequality and one satisfying the second); if $r = 17$, there are three such j (two integers satisfy the first inequality and one satisfies the second); and so on.

Thus if N is the number of least positive residues of $5, 10, 15, \ldots, 5P$ that exceed $p/2$, then N is even if and only if $r = 1, 9, 11,$ or 19. Hence it follows from Gauss's Lemma that 5 is a quadratic residue of p if and only if $p \equiv \pm 1, \pm 9 \pmod{20}$.

Note. It is clear from this example that the value of $(5/p)$ depends only on the *remainder* when p is divided by $20 = 4 \cdot 5$; thus if odd primes p and q leave the same remainder when divided by 20, then $(5/p) = (5/q)$. The same type of argument shows that the value of $(3/p)$ depends only on the remainder when p is divided by $12 = 4 \cdot 3$, the value of $(6/p)$ only on the remainder when p is divided by $24 = 4 \cdot 6$, the value of $(7/p)$ only on the remainder when p is divided by $28 = 4 \cdot 7$, and so forth. A similar statement also holds for $(2/p)$; see (5.12).

On the basis of numerical examples such as these, it is reasonable to ask whether this is true in general. The next result, which is equivalent to the Law of Quadratic Reciprocity, was originally formulated (but not proved) by Euler. It shows in fact that the quadratic nature of a modulo the prime p is determined completely by the remainder when p is divided by $4a$. In particular, if two primes leave the same remainder upon division by $4a$, then a is either a quadratic residue of both primes or a quadratic nonresidue of both.

(5.16) Theorem. *Let p and q be distinct odd primes, and suppose that $(a,p) = (a,q) = 1$, where a is a positive integer. If $q \equiv p$ (mod $4a$) or $q \equiv -p$ (mod $4a$), then $(a/q) = (a/p)$.*

Proof. Let $S = \{a, 2a, 3a, \ldots, Pa\}$, where $P = (p-1)/2$. Denote by N the number of elements of S whose least positive residues are greater than $p/2$; then $(a/p) = (-1)^N$, by Gauss's Lemma. To find N, we have to determine, as in the preceding example, the number of elements of S that lie in the intervals $(p/2, p)$, $(3p/2, 2p)$, $(5p/2, 3p)$, ..., since the least positive residue of ja ($1 \leq j \leq P$) will exceed $p/2$ if and only if ja lies in one of these intervals. Clearly, there are only a finite number of intervals to examine; denote the last interval by $((c-1/2)p, cp)$. (Since $Pa < (a/2)p$, it is easy to check that $c = a/2$ if a is even and $c = (a-1)/2$ if a is odd.) Note that the elements of S cannot be the endpoints of any of these intervals, for if $ja = mp$ (or $ja = (m/2)p$), then $p | ja$ and hence $p | j$, since $(a,p) = 1$; this is impossible since $0 < j < p$.

If an element ja of S lies in the first interval, then $p/2 < ja < p$, that is, $p/2a < j < p/a$. Similarly, if ja is in the second interval, then $3p/2a < j < 2p/a$, and so on. Therefore N is the total number of integers in the intervals

$$\left(\frac{p}{2a}, \frac{p}{a}\right), \quad \left(\frac{3p}{2a}, \frac{2p}{a}\right), \ldots, \left(\frac{(2c-1)p}{2a}, \frac{cp}{a}\right). \tag{1}$$

Now suppose that $q \equiv p$ (mod $4a$); then $q = 4am + p$ for some integer m. If M is the number of elements of $a, 2a, 3a, \ldots, ((q-1)/2)a$ whose least positive residues exceed $q/2$, then $(a/q) = (-1)^M$. If we argue as above, it follows that M is just the number of integers contained in the intervals

$$\left(\frac{q}{2a}, \frac{q}{a}\right), \quad \left(\frac{3q}{2a}, \frac{2q}{a}\right), \ldots, \left(\frac{(2c-1)q}{2a}, \frac{cq}{a}\right). \tag{2}$$

Since $q = 4am + p$, this collection can be written as

$$\left(2m + \frac{p}{2a}, 4m + \frac{p}{a}\right), \left(6m + \frac{3p}{2a}, 8m + \frac{2p}{a}\right), \ldots, \left(2(2c-1)m + \frac{(2c-1)p}{2a}, 4cm + \frac{cp}{a}\right). \tag{3}$$

As we pointed out in Example 5.15, the parity of M is not changed if even integers, which can be negative, are added to the endpoints. Different even integers can be used for the left endpoint and the right endpoint. Subtracting $2m$ and $4m$ from the endpoints of the first interval in (3), $6m$ and $8m$ from the endpoints of the second interval, and so on, gives

$$\left(\frac{p}{2a}, \frac{p}{a}\right), \left(\frac{3p}{2a}, \frac{2p}{a}\right), \ldots, \left(\frac{(2c-1)p}{2a}, \frac{cp}{a}\right).$$

Since this is the same collection of intervals as in (1), M and N have the same parity, and hence it follows from Gauss's Lemma that $(a/p) = (a/q)$.

Suppose now that $q \equiv -p \pmod{4a}$; then $q = 4am - p$ for some m. Substituting this expression for q in (2), we get the following collection of disjoint intervals:

$$\left(2m - \frac{p}{2a}, 4m - \frac{p}{a}\right), \left(6m - \frac{3p}{2a}, 8m - \frac{2p}{a}\right), \ldots, \left(2(2c-1)m - \frac{(2c-1)p}{2a}, 4cm - \frac{cp}{a}\right).$$
(4)

If M is the total number of integers in these intervals, then $(a/q) = (-1)^M$. Let $n(u,v)$ denote the number of integers in the interval (u,v). If we consider the first interval in (4), we can subtract $2m$ from each endpoint to get $n(2m - p/2a, 4m - p/a) = n(-p/2a, 2m - p/a)$. Reflecting this last interval about the point 0, we have $n(2m - p/2a, 4m - p/a) = n(p/a - 2m, p/2a)$. The union of the intervals $(p/a - 2m, p/2a)$ and $(p/2a, p/a)$ is the interval $(p/a - 2m, p/a)$ minus the point $p/2a$; note that the omitted point $p/2a$ cannot be an integer, since $(a,p) = 1$. Thus $n(p/a - 2m, p/2a) + n(p/2a, p/a) = n(p/a - 2m, p/a)$. Since the interval $(p/a - 2m, p/a)$ contains $2m$ integers and the endpoints are not integers, $n(p/a - 2m, p/2a)$ and $n(p/2a, p/a)$ have the same parity, and therefore $n(2m - p/2a, 4m - p/a)$ and $n(p/2a, p/a)$ have the same parity.

The same argument works for any interval in (1) and the corresponding interval in (4). It follows that M and N have the same parity, and therefore $(a/p) = (a/q)$ by Gauss's Lemma. This completes the proof.

We are now ready to prove the Law of Quadratic Reciprocity. The formulation given in (5.17) was originally stated by Legendre in 1785 as a conjecture. The first complete proof of this result was given by Gauss in 1801 in *Disquisitiones Arithmeticae*.

(5.17) Gauss's Law of Quadratic Reciprocity. *Suppose that p and q are distinct odd primes. Then*

$(p/q) = (q/p)$ *if at least one of p and q is of the form $4k+1$, and*
$(p/q) = -(q/p)$ *if both p and q are of the form $4k+3$.*

Proof. First suppose that $q \equiv p \pmod{4}$; then $q = p + 4m$ for some m. It follows from (5.10) that $(q/p) = ((p+4m)/p) = (4m/p) = (m/p)$; similarly, $(p/q) = ((q-4m)/q) = (-4m/q) = (-1/q)(m/q)$. The preceding theorem implies that $(m/q) = (m/p)$, and so $(q/p)(p/q) = (-1/q)$. Therefore, by (5.11), this product is 1 if q and hence p are of the form $4k+1$, and -1 if both are of the form $4k+3$.

Now suppose that $q \equiv -p \pmod{4}$; then one of p and q is of the form $4k+1$ and the other is of the form $4k+3$. Let $q = -p + 4m$. Arguing as above, we have $(q/p) = (m/p)$ and $(p/q) = (m/q)$. It follows from (5.16) that $(m/q) = (m/p)$, and hence $(q/p) = (p/q)$.

The following equivalent formulation of the reciprocity law is a simple consequence of (5.17), noting that the exponent $\frac{p-1}{2} \cdot \frac{q-1}{2}$ is odd if and only if both primes are of the form $4k+3$.

(5.18) Law of Quadratic Reciprocity. *Let p and q be distinct odd primes. Then*

$$(p/q)(q/p) = (-1)^{\frac{p-1}{2}\frac{q-1}{2}}.$$

The significance of the Law of Quadratic Reciprocity lies in the fact that a given Legendre symbol can be replaced (using (5.10)) by a product of Legendre symbols with smaller "denominators" (necessarily prime). However, this reduction process requires the prime factorization of the numerator in the Legendre symbol, and hence it is not practical when the given integers are large. A much more efficient method utilizes the Jacobi symbol, whose definition and properties are given before Problems 5-67 to 5-71.

Notes. 1. Suppose a is an integer greater than 1. Since a is a product of prime powers, we can use (5.10.ii) to write (a/p) as a product of the Legendre symbols $\left(q^k/p\right)$, where q is a prime (possibly 2). It follows from (5.10.iv) that if q is prime and $(q,p) = 1$, then

$$\left(q^k/p\right) = 1 \text{ when } k \text{ is even and } \left(q^k/p\right) = (q/p) \text{ when } k \text{ is odd}.$$

(If a is negative, observe that $(a/p) = (-1/p)(|a|/p)$.) Thus *to evaluate any Legendre symbol (a/p), it suffices to know how to evaluate $(-1/p)$, $(2/p)$, and (q/p), where q is an odd prime.* We use (5.11) and (5.12) to calculate $(-1/p)$ and $(2/p)$, and apply the Law of Quadratic Reciprocity to invert (q/p). By (5.10.i), we can write (p/q) as (b/q), where $b \equiv p \pmod{q}$ and $b < p$. Now express b as a product of prime powers and repeat this process. The reciprocity law can be applied as often as necessary until we get Legendre symbols whose values are obvious. For examples of this technique, see Problem 5-46.

2. Euler's Criterion is in fact a very efficient way to evaluate (a/p) and *much faster* than using the Law of Quadratic Reciprocity when p is large.

(To use the reciprocity law, we must factor the numerator of the Legendre symbol into primes, and there is no known way of factoring large numbers quickly.) However, the Law of Quadratic Reciprocity can produce an answer fairly quickly when p is not large.

One important use of the Law of Quadratic Reciprocity is to determine the solvability of the congruence $x^2 \equiv a \pmod{p}$, where p is a given odd prime, but its main theoretical use is to characterize the primes for which a given integer is a quadratic residue. (See (5.20).) One rather surprising application of the reciprocity law is to prove the existence of infinitely many primes in certain arithmetic progressions (for example, infinitely many primes of the form $5k - 1$, $6k + 1$, $8k + 3$, $8k - 1$, and so on – see Problems 5-23, 5-48, 5-50, and 5-54).

All of these applications will be treated in detail in the problems. For now, we give an example of how the reciprocity law can be used to decide if a given polynomial congruence has solutions.

(5.19) Example. Determine if the congruence $18x^2 - 74x + 67 \equiv 0 \pmod{311}$ is solvable. (311 is a prime.)

Solution. By (5.1), it suffices to determine the solvability of $y^2 \equiv b^2 - 4ac \equiv 30 \pmod{311}$. (Here, $y = 2ax + b = 36x - 74$, but since we only want to determine if solutions exist, we simply need to see if $b^2 - 4ac$ is a quadratic residue of 311.) Since $30 = 2 \cdot 3 \cdot 5$, we have $(30/311) = (2/311)(3/311)(5/311)$. Note that 311 is of the form $8k + 7$, so (5.12) implies that $(2/311) = 1$. Since both 3 and 311 are of the form $4k + 3$, use (5.17), (5.10.i), and (5.12) to get $(3/311) = -(311/3) = -(2/3) = 1$. (Note that $(2/3) = -1$ since 3 is of the form $8k + 3$.) Similarly, use (5.17) to conclude that $(5/311) = (311/5) = (1/5) = 1$. Thus $(30/311) = 1 \cdot 1 \cdot 1 = 1$, and so the given congruence is solvable. In fact, since $b^2 - 4ac \not\equiv 0 \pmod{311}$, it has two solutions (see Problem 5-2).

Finally, if q is an odd prime, the following general result can be used to determine all primes that have q as a quadratic residue. (See Problems 5-45 and 5-53.)

(5.20) Theorem. *Suppose q is an odd prime. Then the odd primes p such that q is a quadratic residue of p are given by*

$$p = 4qk \pm a, \quad \text{where } 0 < a < 4q, \quad a \equiv 1 \pmod{4}, \quad \text{and } (a/q) = 1.$$

The integers a satisfying these conditions are simply the least positive residues modulo $4q$ of $1^2, 3^2, 5^2, \ldots, (q-2)^2$.

Proof. Write $p = 4qm + r$, where r is odd and $0 < r < 4q$. If $r \equiv 1 \pmod{4}$, let $k = m$ and $a = r$. If $r \equiv 3 \pmod{4}$, let $k = m + 1$ and $a = 4q - r$.

In either case, then, we have $p = 4qk \pm a$, where $0 < a < 4q$ and $a \equiv 1$ (mod 4).

If $p \equiv a$ (mod $4q$), then $p \equiv 1$ (mod 4). Hence (5.17) and (5.10.i) imply that $(q/p) = (p/q) = (a/q)$. If $p \equiv -a$ (mod $4q$), then $p \equiv 3$ (mod 4) and hence $(q/p) = -(p/q) = -(-a/q) = (a/q)$, using (5.10.ii) and (5.11). Thus in both cases, we have $(q/p) = (a/q)$.

Now suppose $(a/q) = 1$; then there is an integer s such that $s^2 \equiv a$ (mod q). We may assume that s is odd (otherwise, use the solution $q - s$). Since $a \equiv 1$ (mod 4), it follows that $s^2 \equiv a$ (mod 4), and hence $s^2 \equiv a$ (mod $4q$), by (2.4.ii).

PROBLEMS AND SOLUTIONS

General Quadratic Congruences and $x^2 \equiv a$ (mod m)

5-1. *Consider $ax^2 + bx + c \equiv 0$ (mod m), where $(2a, m) = 1$. Prove that finding solutions of this congruence reduces to finding solutions of a congruence of the form $y^2 \equiv d$ (mod m).*

Solution. The argument is essentially the same as the argument for (5.1). The given congruence is equivalent to the congruence $4a(ax^2 + bx + c) \equiv 0$ (mod m), since $(2a, m) = 1$ and therefore $(4a, m) = 1$. But this second congruence can be rewritten as $(2ax + b)^2 \equiv b^2 - 4ac$ (mod m). Thus to find all solutions, we solve the congruence $y^2 \equiv b^2 - 4ac$ (mod m), and for every solution y, we solve the linear congruence $2ax \equiv y - b$ (mod m).

For the quadratic equation $ax^2 + bx + c = 0$ (where a, b, and c are real numbers), the well-known quadratic formula gives $x = (-b \pm \sqrt{b^2 - 4ac})/2a$. Thus the number of solutions is 0, 1, or 2 according as the quantity $b^2 - 4ac$, called the *discriminant*, is negative, zero, or positive. An analogous result holds for quadratic congruences.

5-2. *Consider $ax^2 + bx + c \equiv 0$ (mod p), where p is an odd prime and $p \nmid a$. Let $D = b^2 - 4ac$. Prove that the congruence has (a) no solutions if D is a quadratic nonresidue of p; (b) a unique solution if $p \mid D$; and (c) exactly two solutions if D is a quadratic residue of p.*

Solution. This follows from (5.1). (a) If D is not a quadratic residue of p, then $y^2 \equiv D$ (mod p) has no solution, and hence the system of congruences of (5.1) has no solution. (b) Note that $y^2 \equiv 0$ (mod p) has the unique solution $y = 0$, and the congruence $2ax \equiv -b$ (mod p) has a unique solution. (c) The congruence $y^2 \equiv D$ (mod p) has two solutions. Solving $2ax \equiv y - b$ (mod p) then yields two solutions of the original congruence.

5-3. Suppose that $(a/p) = 1$, where p is a prime of the form $4k + 3$. Prove that the solutions of $x^2 \equiv a \pmod{p}$ are given by $\pm a^{k+1}$.

Solution. Since $(p - 1)/2 = 2k + 1$, Euler's Criterion (5.9) implies that $a^{2k+1} \equiv 1 \pmod{p}$. Thus $(\pm a^{k+1})^2 = a^{2k+1} a \equiv a \pmod{p}$.

Note. This result can also be used to determine if $x^2 \equiv a \pmod{p}$ is solvable. If solutions exist, they must be $\pm a^{k+1}$; thus it is enough to check if these values *are* solutions. In fact, the same argument can be used to show that if $\pm a^{k+1}$ are not solutions of $x^2 \equiv a \pmod{p}$, then they are solutions of $x^2 \equiv -a \pmod{p}$.

5-4. Use the preceding problem to find the solutions of (a) $x^2 \equiv 46 \pmod{59}$; (b) $9x^2 - 24x + 13 \equiv 0 \pmod{59}$. *(You may assume that these congruences are solvable.)*

Solution. (a) According to the previous problem, the solutions are given by $\pm 46^{15}$ or, equivalently, $\mp 13^{15}$. Modulo 59, we have $13^2 \equiv -8$, $13^4 \equiv 5$, $13^8 \equiv 25$, $13^{12} \equiv 5 \cdot 25 \equiv 7$, $13^{14} \equiv 7(-8) \equiv 3$, and $13^{15} \equiv 39$. Thus the solutions are ± 39, that is, 20 and 39 modulo 59.

(b) Complete the square as in (5.1) to get $(18x - 24)^2 \equiv b^2 - 4ac = 108 \equiv -10 \pmod{59}$. This congruence is solvable since the original congruence is. Thus the solutions are given by $18x - 24 \equiv \pm 10^{15} \pmod{59}$. Check that $\pm 10^{15} \equiv \pm 7 \pmod{59}$. Solving $18x - 24 \equiv \pm 7 \pmod{59}$ now yields 5 and 37 as the solutions of the original congruence.

5-5. Determine if the following congruences are solvable:
 (a) $42x^2 - 51x + 91 \equiv 0 \pmod{311}$;
 (b) $42x^2 - 51x + 91 \equiv 0 \pmod{622}$.

Solution. (a) In view of (5.1), it is enough to decide if $b^2 - 4ac = 51^2 - 4(42)(91) \equiv 64 \pmod{311}$ is a quadratic residue. Obviously, $(64/311) = 1$ since $64 = 8^2$, and hence the given congruence is solvable. (In fact, it has exactly two solutions, by Problem 5-2.)

(b) For this congruence to have a solution, there must be a solution modulo 2 and 311. There is clearly a solution modulo 2, since the congruence reduces to $x + 1 \equiv 0 \pmod{2}$. By (a), there is also a solution modulo 311. Thus the congruence in (b) has a solution.

5-6. Prove or disprove: If $x^2 \equiv a \pmod{p}$ and $x^2 \equiv a \pmod{q}$ are not solvable, then $x^2 \equiv a \pmod{pq}$ is solvable.

Solution. This is false. If $x^2 \equiv a \pmod{pq}$ were solvable, then $x^2 \equiv a \pmod{p}$ would be as well, since $x^2 \equiv a \pmod{pq}$ implies $x^2 \equiv a \pmod{p}$.

5-7. Use (5.1) to reduce each of the following congruences to the form $y^2 \equiv d \pmod{p}$: (a) $11x^2 - 7x + 12 \equiv 0 \pmod{29}$; (b) $5x^2 + 9x - 21 \equiv 0 \pmod{23}$.

Solution. In (a), $(2ax + b)^2 \equiv b^2 - 4ac \pmod{p}$ becomes $(22x - 7)^2 \equiv 14 \pmod{29}$. In (b), we get $(10x + 9)^2 \equiv 18 \pmod{23}$.

5-8. *Find all solutions of $x^2 \equiv 8 \pmod{287}$.*

Solution. Note that $287 = 7 \cdot 41$. The congruence $x^2 \equiv 8 \pmod 7$ has the solutions ± 1, and the congruence $x^2 \equiv 8 \equiv 49 \pmod{41}$ has the solutions ± 7. Thus by the Chinese Remainder Theorem, $x^2 \equiv 8 \pmod{287}$ has the four solutions ± 34 and ± 48.

5-9. *Prove that the congruence $x^2 \equiv 1 \pmod m$ has only the solutions ± 1 if and only if $m = 2, 4, p^n,$ or $2p^n$, where p is an odd prime.*

Solution. If $m = 2$ or 4, it is clear that the only solutions of $x^2 \equiv 1 \pmod m$ are ± 1. (Note that for $m = 2$, the solutions 1 and -1 are congruent.) If $m = p^n$ or $2p^n$, where p is an odd prime, the result follows from (5.5).

Conversely, suppose that $x^2 \equiv 1 \pmod m$ has only the two solutions ± 1. If $m > 2$, let $m = 2^k p_1^{k_1} \cdots p_r^{k_r}$ be the prime factorization of m. It follows from (5.5) that $k = 0$ or 1 and $r = 1$, in which case $m = p^n$ or $2p^n$, or $k = 2$ and $r = 0$, in which case $m = 4$.

▷ **5-10.** *Prove Gauss's generalization of Wilson's Theorem: Suppose that $m > 2$, and let P be the product of the positive integers less than m that are relatively prime to m. Then $P \equiv -1 \pmod m$ if $m = 4, p^n,$ or $2p^n$, where p is an odd prime, and $P \equiv 1 \pmod m$ otherwise. (Hint. Imitate the proof of (3.2), with $a = 1$, and use (5.5).)*

Solution. Let $1 \le x \le m - 1$, where $(x, m) = 1$ and $x^2 \not\equiv 1 \pmod m$. Pair x with the unique number y such that $1 \le y \le m - 1$ and $xy \equiv 1 \pmod m$ (the inverse of x). The product of all the numbers that occur in some pair is clearly congruent to 1 modulo m. Thus $P \equiv Q \pmod m$, where Q is the product of all numbers x such that $1 \le x \le m - 1$ and $x^2 \equiv 1 \pmod m$.

We now consider another kind of pairing. If $x^2 \equiv 1 \pmod m$, pair x with $m - x$. Clearly, $(m - x)^2 \equiv 1 \pmod x$, $m - x \not\equiv x \pmod m$, and $x(m - x) \equiv -1 \pmod m$. Thus $Q \equiv (-1)^{N/2} \pmod m$, where N is the number of solutions of the congruence $x^2 \equiv 1 \pmod m$. By (5.5), $N = 2$ if $m = 4, p^n,$ or $2p^n$, where p is an odd prime, in which case $Q \equiv -1 \pmod m$. For any other m, we have $Q \equiv 1 \pmod m$, since N is divisible by 4.

Quadratic Residues

5-11. *If p is an odd prime, prove that $(2/p) = (-1)^{(p^2-1)/8}$.*

Solution. Apply (5.12), noting that $(p^2 - 1)/8$ is even if $p \equiv \pm 1 \pmod 8$ and is odd if $p \equiv \pm 3 \pmod 8$.

5-12. *Suppose a is a quadratic residue of the odd prime p. Prove that $-a$ is also a quadratic residue of p if and only if $p \equiv 1 \pmod 4$.*

Solution. $(-a/p) = (-1/p)(a/p) = (-1/p)$, since $(a/p) = 1$. Thus $-a$ is a quadratic residue of p if and only if -1 is a quadratic residue of p, i.e., if and only if $p \equiv 1 \pmod 4$, by (5.11).

5-13. If p is an odd prime, prove that $(1/p) + (2/p) + \cdots + ((p-1)/p) = 0$.

Solution. By (5.7), there are $(p-1)/2$ quadratic residues and $(p-1)/2$ quadratic nonresidues of p. Hence half of the terms in this sum are 1 and half are -1.

5-14. Let p be an odd prime, and let q be the smallest positive quadratic nonresidue of p. Show that q is prime.

Solution. By (5.10.ii), a product of quadratic residues is a quadratic residue. Thus if $q = ab$ with $1 < a < q$, then one of a or b must be a quadratic nonresidue of p. This contradicts the assumption that q is the *least* quadratic nonresidue of p.

▷ **5-15.** Let p be an odd prime, and let q be the smallest positive quadratic nonresidue of p. Show that $q < \sqrt{p} + 1$. (Hint. Show that $q, 2q, \ldots, (q-1)q$ are all less than p.)

Solution. Let k be the smallest positive integer such that $kq > p$, and let r be the remainder when kq is divided by p. Since $(k-1)q < p$, it follows that $r < q$, and thus r (and hence kq) is a quadratic residue of p. Therefore $k \geq q$, for if $1 \leq k < q$, then k is a quadratic residue of p, and so, by (5.10.ii), kq is a quadratic nonresidue of p. Thus $(q-1)q < p$, and it follows that $q < \sqrt{p}+1$, since otherwise, $(q-1)q \geq \sqrt{p}(\sqrt{p}+1) > p$.

▷ **5-16.** Let p be an odd prime. Prove that

$$((1 \cdot 2)/p) + ((2 \cdot 3)/p) + \cdots + (((p-2)(p-1))/p) = -1.$$

(Hint. First show that $(a(a+1)/p) = ((a^* + 1)/p)$, where $aa^* \equiv 1 \pmod{p}$.)

Solution. Let $(a, p) = 1$, and let a^* be such that $aa^* \equiv 1 \pmod{p}$. (The existence of a^* follows from (2.7).) Then by (5.10),

$$(a(a+1)/p) = (a(a+aa^*)/p) = \left(a^2(1+a^*)/p\right) = ((1+a^*)/p).$$

Note that as a ranges from 1 to $p-2$, the least positive residue of a^*+1 ranges through the integers from 2 to $p-1$. Thus the above sum is the same as $(2/p) + (3/p) + \cdots + ((p-1)/p)$, which equals $-(1/p) = -1$ by the preceding problem.

▷ **5-17.** Let $p > 5$ be prime. Use the preceding problem to prove that there are always consecutive integers that are quadratic residues of p and consecutive integers that are quadratic nonresidues of p.

Solution. Suppose there are no two consecutive integers that are quadratic residues of p; then $(a/p)((a+1)/p) = -1$ for every a. Hence $(a(a+1)/p) = (a/p)((a+1)/p) = -1$ for every a, which cannot happen in view of the preceding problem. A similar argument works if we assume that there are no consecutive quadratic nonresidues of p, since in this case as well, $(a/p)((a+1)/p) = -1$.

5-18. *Let $p > 5$ be prime. Show that at least one of 2, 5, or 10 is a quadratic residue of p. Use this to conclude that there are always consecutive integers that are quadratic residues of p. Show then that there are always consecutive integers that are quadratic nonresidues of p.*

Solution. If 2 is a quadratic residue of p, then 1 and 2 are consecutive quadratic residues. Likewise, if 5 is a residue, then 4 and 5 are consecutive residues. If 2 and 5 are both quadratic nonresidues of p, then their product 10 must be a quadratic residue of p, and therefore 9 and 10 are consecutive residues.

We now consider quadratic nonresidues. If 2 and 3 are both nonresidues, we are finished. Otherwise, at least three of 1, 2, 3, and 4 are residues. If, in the interval $1 \leq x \leq p-1$, we never had two consecutive quadratic nonresidues, then for any such x, the number of residues in the interval from 1 to x would always be greater than the number of nonresidues. But this is impossible, since there are exactly as many residues as nonresidues in the interval from 1 to $p-1$.

5-19. *What are the least positive residues of the quadratic residues of 29?*

Solution. Find the least positive residues of $1^2, 2^2, 3^2, \ldots, ((29-1)/2)^2$ modulo 29. We get 1, 4, 9, 16, 25, 7, 20, 6, 23, 13, 5, 28, 24, and 22.

5-20. *Is $x^2 \equiv -2 \pmod{263}$ solvable? (263 is a prime.)*

Solution. $(-2/263) = (-1/263)(2/263) = (-1)(+1) = -1$, since 263 is of the form $4k+3$ and $8k+7$.

▷ **5-21.** *Let p be an odd prime. Prove that the product of the quadratic residues of p is congruent to -1 or 1 modulo p, according as $p \equiv 1 \pmod 4$ or $p \equiv 3 \pmod 4$. (Hint. Note that if $ab \equiv 1 \pmod p$, then $(a/p) = (b/p)$.)*

Solution. Pair each quadratic residue a different from 1 and -1 with its multiplicative inverse a^*, where $aa^* \equiv 1 \pmod p$. If $a \not\equiv \pm 1 \pmod p$, then $a^* \not\equiv a \pmod p$. The product of each pair is congruent to 1. Thus the product of the quadratic residues of p is congruent to -1 modulo p if -1 is a quadratic residue of p (i.e., if $p \equiv 1 \pmod 4$) and to 1 otherwise.

5-22. *Let $p > 3$ be an odd prime. Prove that the sum S of the quadratic residues of p in the interval $1 \leq x \leq p-1$ is divisible by p. (Hint. Use the formula $1^2 + 2^2 + \cdots + n^2 = n(n+1)(2n+1)/6$.)*

Solution. The quadratic residues of p are congruent to $1^2, 2^2, \ldots, ((p-1)/2)^2$; thus $6S = ((p-1)/2)((p+1)/2)p \equiv 0 \pmod p$, and hence $p \mid 6S$. Since p is not 2 or 3, it follows that p divides S.

5-23. *Prove that there are infinitely many primes of the form $4k+1$. (Hint. Let $p_1, p_2, \ldots, p_n$ be primes of this form, and consider $N = (2p_1 p_2 \cdots p_n)^2 + 1$.)*

Solution. Suppose p is a (necessarily odd) prime divisor of N; then $(2p_1 p_2 \cdots p_n)^2 \equiv -1 \pmod p$, and hence p must be of the form $4k+1$, by (5.11). Clearly, p is not one of the p_i. Thus given any finite collection of primes of the form $4k+1$, we can find another prime of this form. Therefore there exist infinitely many primes of the form $4k+1$.

▷ **5-24.** *Suppose $p > 3$ is a prime of the form $4k+3$, and let N be the number of quadratic nonresidues of p between 1 and $p/2$. Prove that $((p-1)/2)! \equiv (-1)^N \pmod{p}$. (Hint. Let $P = (p-1)/2$ and show that $P! \equiv \pm 1 \pmod{p}$.)*

Solution. If $P = (p-1)/2$, it is clear that $(P!/p) = (1/p)(2/p)\cdots(P/p) = (-1)^N$. By Wilson's Theorem, we have

$$-1 \equiv (p-1)! = 1 \cdot 2 \cdots P(P+1) \cdots (p-2)(p-1)$$
$$\equiv 1 \cdot 2 \cdots P(-P) \cdots (-2)(-1)$$
$$= (-1)^P (P!)^2 = -(P!)^2 \pmod{p}.$$

Thus $(P!)^2 \equiv 1 \pmod{p}$, and so $P! \equiv \pm 1 \pmod{p}$. If $P! \equiv 1 \pmod{p}$, then $(P!/p) = (1/p) = 1$, and if $P! \equiv -1 \pmod{p}$, then $(P!/p) = (-1/p) = -1$, by (5.11). Hence in either case, we have $(P!/p) \equiv P! \pmod{p}$, and the result follows.

5.25. *Find the number of solutions of the following congruences:*
  (a) $x^2 \equiv 19 \pmod{170}$;
  (b) $x^2 \equiv -73 \pmod{2^4 \cdot 71^3 \cdot 79^2}$;
  (c) $x^2 \equiv 76 \pmod{165}$;
  (d) $x^2 \equiv 38 \pmod{79}$;
  (e) $x^2 \equiv 33 \pmod{2^6 \cdot 37^3 \cdot 83^4}$;
  (f) $x^2 \equiv 4 \pmod{11025}$.

Solution. (a) $170 = 2 \cdot 5 \cdot 17$; clearly, $x^2 \equiv 19 \equiv 1 \pmod{2}$ has one solution. Also, $(19/5) = (4/5) = 1$, so $x^2 \equiv 19 \pmod 5$ has two solutions. And $(19/17) = (2/17) = 1$, by (5.12), so $x^2 \equiv 19 \pmod{17}$ has two solutions. Thus the original congruence has $1 \cdot 2 \cdot 2 = 4$ solutions, by (4.3) (or use (5.5)).

(b) $(-73/71) = (-2/71) = -1$ by (5.12). Thus there are no solutions modulo 71, and hence the given congruence is not solvable.

(c) Note that $165 = 3 \cdot 5 \cdot 11$; then $(76/3) = (1/3) = 1$, $(76/5) = 1$, and $(76/11) = (-1/11) = -1$, by (5.11). So there are no solutions modulo 11 and hence none modulo 165.

(d) $(38/79) = (2/79)(19/79) = -(79/19) = -(3/19) = 1$, using Gauss's Lemma. Hence there are two solutions.

(e) $(33/37) = (-4/37) = (-1/37)(4/37) = (-1/37) = 1$, by (5.11); thus there are two solutions modulo 37 and hence two solutions modulo 37^k for any $k \geq 1$, by (5.3). Also, $(33/83) = (-50/83) = (-2/83)(25/83) = (-2/83) = 1$, by (5.12), so there are two solutions modulo 83^4. Since $x^2 \equiv 33 \pmod{64}$ is solvable by (4.14), (5.5) implies that there are $2^{2+2} = 16$ solutions of the original congruence.

(f) Note that $11025 = 3^2 5^2 7^2$. Since 4 is a perfect square, there is a solution. Thus by (5.5), there are $2^3 = 8$ solutions modulo 11025.

5-26. *Find the number of solutions of $x^4 \equiv 4 \pmod{71^3 \cdot 97^5}$.*

Solution. Reduce the given congruence to $x^2 \equiv \pm 2$ modulo 71^3 and 97^5. Note that $x^2 \equiv 2 \pmod{71}$ has two solutions, by (5.12); thus it follows from (5.3) that $x^2 \equiv 2 \pmod{71^3}$ has two solutions. Also, $x^2 \equiv -2 \pmod{71}$ has no solutions, since $(-2/71) = (-1/71)(2/71) = (-1)(+1) = -1$, by (5.11) and (5.12). Hence $x^4 \equiv 4 \pmod{71^3}$ has exactly two solutions.

Similarly, $(2/97) = 1$ and $(-2/97) = 1$; thus $x^4 \equiv 4 \pmod{97}$ has four solutions, and therefore there are four solutions modulo 97^5. (Apply (5.3) to $x^2 \equiv 2 \pmod{97^5}$ and to $x^2 \equiv -2 \pmod{97^5}$.) Hence there are $2 \cdot 4 = 8$ solutions modulo $71^3 \cdot 97^5$.

5-27. *Show that if p is a prime of the form $4k + 1$, then $(1/p) + (2/p) + \cdots + (P/p) = 0$, where $P = (p - 1)/2$. (Hint. Note that $(a/p) = ((p - a)/p)$ and use Problem 5-13.)*

Solution. By (5.11), $((p - a)/p) = (-a/p) = (-1/p)(a/p) = (a/p)$. It follows from Problem 5-13 that $0 = (1/p) + (2/p) + \cdots + ((p - 1)/p) = 2[(1/p) + (2/p) + \cdots + (P/p)]$, which proves the result.

5-28. *Let n be a positive integer of the form $4k + 3$. If $q = 2n + 1$ is prime, prove that q divides $2^n - 1$.*

Solution. By Euler's Criterion, $(2/q) \equiv 2^{(q-1)/2} = 2^n \pmod q$. Since $q = 2n + 1$ and n is of the form $4k + 3$, q is of the form $8k + 7$, so $(2/q) = 1$ by (5.12). Thus $2^n \equiv 1 \pmod q$, i.e., $q \mid 2^n - 1$.

5-29. *Determine if 83 divides $2^{41} - 1$. (Hint. Use Euler's Criterion.)*

Solution. Note that $41 = (83 - 1)/2$. By Euler's Criterion, $(2/83) \equiv 2^{41} \pmod{83}$. But $(2/83) = -1$, by (5.12), so $2^{41} \equiv -1 \pmod{83}$, i.e., $83 \mid 2^{41} + 1$, and hence 83 does not divide $2^{41} - 1$.

5-30. *Use Euler's Criterion to decide if the prime 1999 divides $2^{999} - 1$.*

Solution. Note that $1999 \equiv 7 \pmod 8$, so $(2/1999) = 1$, by (5.12). Thus by Euler's Criterion, $1 = (2/1999) \equiv 2^{999} \pmod{1999}$, and hence $1999 \mid 2^{999} - 1$.

5-31. *Suppose that q is odd and $p = 4q + 1$ is prime.*
  (a) Prove that 2 is a quadratic nonresidue of p.
  (b) Prove that p divides $4^q + 1$. (Use Euler's Criterion.)

Solution. (a) Let $q = 2k + 1$. Then $p = 4q + 1 = 8k + 5$; now apply (5.12).
  (b) By Euler's Criterion, $(2/p) \equiv 2^{(p-1)/2} = 2^{2q} = 4^q \pmod p$. By (a), $(2/p) = -1$, so $4^q \equiv -1 \pmod p$, i.e., $p \mid 4^q + 1$.

5-32. *Let p be a prime of the form $8k + 5$, and suppose that the congruence $x^2 \equiv a \pmod p$ has a solution. Show that either a^{k+1} or $2^{2k+1}a^{k+1}$ is a solution. (Hint. Use Euler's Criterion and (5.12).)*

Solution. By Euler's Criterion, $a^{4k+2} \equiv 1 \pmod p$, and so $a^{2k+1} \equiv \pm 1 \pmod p$. If $a^{2k+1} \equiv 1 \pmod p$, then $a^{2k+2} \equiv a \pmod p$ and a is congruent modulo p to the square of a^{k+1}. Now suppose $a^{2k+1} \equiv -1 \pmod{p}$. Since p is of the form $8k+5$, 2 is a quadratic nonresidue of p, and therefore $2^{4k+2} \equiv -1 \pmod{p}$. It follows that $2^{4k+2}a^{2k+1} \equiv 1 \pmod{p}$, and hence $2^{4k+2}a^{2k+2} \equiv a \pmod{p}$. So in this case, a is congruent modulo p to the square of $2^{2k+1}a^{k+1}$.

5-33. *Let p be an odd prime. Find the number of quadratic residues of p^n.*

Solution. The quadratic residues of p^n consist of the squares of numbers between 1 and p^n that are relatively prime to p. But if the congruence $x^2 \equiv a \pmod{p^n}$ is solvable, it has precisely two solutions by (5.3), and thus there are $\phi(p^n)/2 = p^{n-1}(p-1)/2$ quadratic residues of p^n.

5-34. *Let $m = 2^k p_1^{a_1} p_2^{a_2} \cdots p_r^{a_r}$. Find the number of quadratic residues of m. (Hint. See the preceding problem.)*

Solution. If we square the $\phi(m)$ numbers from 1 to m that are relatively prime to m, we will obtain the quadratic residues of m. But when $x^2 \equiv a \pmod{m}$ is solvable, then the number of solutions is given by (5.5) and does not depend on a. Thus the number of quadratic residues of m is $\phi(m)/2^r$ if $k = 0$ or $k = 1$, $\phi(m)/2^{r+1}$ if $k = 2$, and $\phi(m)/2^{r+2}$ if $k \geq 3$.

5-35. *Show that 3 is a quadratic nonresidue of all primes of the form $4^n + 1$.*

Solution. Let $p = 4^n + 1$. Since $4 \equiv 1 \pmod{3}$, we have $p \equiv 2 \pmod{3}$; also, $p \equiv 1 \pmod{4}$. Thus, by the Law of Quadratic Reciprocity, $(3/p) = (p/3) = (2/3) = -1$.

5-36. *Does there exist a square of the form $55k - 1$? Explain.*

Solution. If $55k - 1 = n^2$, then n^2 is congruent to -1 modulo 55 and hence modulo 11. This is impossible, since 11 is not of the form $4k + 1$ (see (5.11)).

5-37. *Suppose that a is not a multiple of 71. Show that the congruences $x^{26} \equiv a \pmod{71}$ and $x^{26} \equiv -a \pmod{71}$ cannot both be solvable.*

Solution. If the congruences were solvable, then a and $-a$ would both be quadratic residues of p. But $(-a/71) = (-1/71)(a/71) = -(a/71)$, by (5.11), and thus precisely one of a and $-a$ is a quadratic residue of p.

5-38. *Let p be a prime. Prove that $(n^2 - 3)(n^2 - 5)(n^2 - 15)$ is divisible by p for infinitely many integers n.*

Solution. If $p = 2$, then any odd integer n may be used; if $p = 3$ or $p = 5$, then any multiple of 15 may be used. Thus suppose $p > 5$. If 3 or 5 is a quadratic residue of p, then there exists n such that $p \mid n^2 - 3$ or $p \mid n^2 - 5$. If neither 3 nor 5 is a quadratic residue of p, then $(15/p) = (3/p)(5/p) = (-1)(-1) = 1$, so $p \mid n^2 - 15$ for some n. Thus in either case, p divides the given product for some integer n. To show there are infinitely many such n, note, for example, that if $n^2 \equiv 3 \pmod{p}$, then $(n + kp)^2 \equiv 3 \pmod{p}$ for any $k \geq 1$.

▷ **5-39.** Let $p \geq 7$ be a prime of the form $4k+3$. Show that the sum of the squares of the quadratic residues of p is a multiple of p and also that the sum of the squares of the quadratic nonresidues of p is a multiple of p (Hint. Use Problem 5-22.)

Solution. If a and b are incongruent quadratic residues (or nonresidues) of p, then $a^2 \not\equiv b^2 \pmod{p}$. For if $a^2 \equiv b^2 \pmod{p}$ and $a \not\equiv b \pmod{p}$, then $a \equiv -b \pmod{p}$. But then $(a/p) = (-b/p) = (-1/p)(b/p) = -(b/p)$, and thus a and b cannot be both quadratic residues (or nonresidues) of p.

By (5.7), there are $(p-1)/2$ quadratic residues of p and $(p-1)/2$ nonresidues. Thus the squares of the residues (or nonresidues) form a complete set of $(p-1)/2$ incongruent quadratic residues, and the result now follows from Problem 5-22.

5-40. Suppose that p is a prime of the form $8k+3$. Does p divide $2^{(p-1)/2}-1$?

Solution. No. If $p \mid 2^{(p-1)/2}-1$, then $2^{(p-1)/2} \equiv 1 \pmod{p}$, and hence by Euler's Criterion, we would then have $(2/p) = 1$. But $(2/p) = -1$, by (5.12).

5-41. (a) Suppose that p is an odd prime that divides the sum $r^2 + s^2$, where $(r,p) = (s,p) = 1$. Prove that p is of the form $4k+1$.

(b) Show that if n divides the sum $r^2 + s^2$, where $(r,n) = (s,n) = 1$, then n is the product, or twice the product, of prime powers with each prime of the form $4k+1$.

Solution. (a) If $p \mid r^2 + s^2$, then $r^2 \equiv -s^2 \pmod{p}$ and hence $1 = (r^2/p) = (-s^2/p) = (-1/p)(s^2/p) = (-1/p)$. Thus by (5.11), p must be of the form $4k+1$.

(b) Suppose $n \mid r^2 + s^2$ and let p be an odd prime divisor of n. Then $p \equiv 1 \pmod{4}$, by (a). Also, if n is even, then n can have only one factor of 2, for then r and s must be odd, in which case $r^2 + s^2 \equiv 2 \pmod{4}$ and so $4 \nmid r^2 + s^2$. Thus n must be of the form $\prod p_i^{r_i}$ or $2\prod p_i^{r_i}$, where each p_i is a prime of the form $4k+1$.

5-42. Prove that $1! + 2! + \cdots + n!$ is never a square if $n > 3$.

Solution. Let $N = 1! + 2! + \cdots + n!$. Then $N \equiv 1! + 2! + 3! + 4! = 33 \equiv 3 \pmod{5}$. Thus if $N = m^2$, then $m^2 \equiv 3 \pmod{5}$. But it is easy to see that 3 is a quadratic nonresidue of 5.

The Law of Quadratic Reciprocity

5-43. Let p and q be distinct odd primes. Show that the Law of Quadratic Reciprocity can be stated as follows: If p is of the form $4k+1$, then $(p/q) = (q/p)$. If p is of the form $4k+3$, then $(-p/q) = (q/p)$. (This was essentially Gauss's original formulation.)

Solution. When p is of the form $4k+1$, Gauss's version and (5.17) obviously give the same result. Now suppose that p is of the form $4k+3$. Note that $(-p/q) = (-1/q)(p/q)$. If q is of the form $4k+1$, then $(-1/q) = 1$, and Gauss's version agrees with (5.17). Finally, let q be of the form $4k+3$. Then $(-1/q) = -1$, so Gauss's version implies that $(q/p) = -(p/q)$, the same result as in (5.17).

5-44. *Use Gauss's Lemma directly to show that 2 is a quadratic residue of the prime p if p is of the form $8k+1$ or $8k+7$, and a quadratic nonresidue if p is of the form $8k+3$ or $8k+5$.*

Solution. If $1 \le j \le (p-1)/2$, then $2 \le 2j \le p-1$. Let N be the number of integers in the set $A = \{2, 4, 6, \ldots, p-1\}$ that are larger than $p/2$. Then by Gauss's Lemma, $(2/p) = (-1)^N$. Now $2j < p/2$ if and only if $j < p/4$. If $p = 8k+1$, then $j < p/4$ is equivalent to $j < 2k + 1/4$. There are $2k$ integers satisfying this last inequality; since A contains $(p-1)/2 = 4k$ elements, it follows that $N = 4k - 2k = 2k$. Thus $(2/p) = 1$ if $p = 8k+1$.

Similarly, if p is $8k+3$, $8k+5$, or $8k+7$, then N is, respectively, $(4k+1)-2k = 2k+1$, $(4k+2)-(2k+1) = 2k+1$, or $(4k+3)-(2k+1) = 2k+2$. Hence it follows from Gauss's Lemma that $(2/p) = 1$ or -1 according as $p \equiv 1, 7 \pmod{8}$ or $p \equiv 3, 5 \pmod{8}$.

5-45. *Characterize the odd primes $p \ne 7$ such that $x^2 \equiv 7 \pmod{p}$ is solvable.*

Solution. Use (5.20). Then $p = 28k \pm a$, where a ranges over the least positive residues modulo 28 of $1^2, 3^2, \ldots, (7-2)^2$, i.e., 1, 9, and 25. Thus $(7/p) = 1$ if and only if $a = 28k \pm 1, 28k \pm 3, 28k \pm 9$. (Note that $\pm 25 \equiv \mp 3 \pmod{28}$.)

Another proof: First we deal with primes p of the form $4k+1$. Then by the Law of Quadratic Reciprocity, $(7/p) = (p/7)$. But this is $(r/7)$, where r is the remainder when p is divided by 7, and it is easy to check that $(r/7) = 1$ for $r = 1, 2,$ and 4. Thus p is of the form $28k+1$, $28k+9$, or $28k+25$.

Next we deal with primes p of the form $4k+3$. By the Law of Quadratic Reciprocity, $(7/p) = -(r/7)$, where r is the remainder when p is divided by 7. Thus $(7/p) = 1$ if and only if $(r/7) = -1$, i.e., if and only if $r = 3, 5,$ or 6. Thus p must be of the form $28k+3$, $28k+19$, or $28k+27$.

5-46. *Calculate (a) $(70/97)$; (b) $(-14/83)$; (c) $(263/331)$; (d) $(-219/383)$; (e) $(461/773)$. (263, 331, 383, and 773 are primes.)*

Solution. We use (5.10)–(5.12), together with the Law of Quadratic Reciprocity.

(a) $(70/97) = (2/97)(5/97)(7/97)$. Note that $(2/97) = 1$ since $97 \equiv 1 \pmod{8}$. Also, $(5/97) = (97/5) = (2/5) = -1$ and $(7/97) = (97/7) = (-1/7) = -1$. Thus $(70/97) = 1$.

(b) $(-14/83) = (-1/83)(2/83)(7/83) = (-1)(-1)(7/83) = (7/83) = -(83/7) = -(-1/7) = 1$.

(c) $(263/331) = -(331/263) = -(68/263) = -(4/263)(17/263) = -(17/263) = -(263/17) = -(8/17) = -(2/17) = -1$.

(d) $(-219/383) = (164/383) = (4/383)(41/383) = (41/383) = (383/41) = (14/41) = (2/41)(7/41) = (7/41) = (41/7) = (6/7) = (-1/7) = -1$.

(e) $(461/773) = (773/461) = (312/461) = (4/461)(2/461)(3/461)(13/461) = (+1)(-1)(461/3)(461/13) = (6/13) = (2/13)(3/13) = -(3/13) = -(13/3) = -(1/3) = -1$.

5-47. Prove that 10 is a quadratic residue of the odd prime p if and only if $p \equiv \pm 1, \pm 3, \pm 9, \pm 13 \pmod{40}$.

Solution. $(10/p) = 1$ if and only if $(2/p) = (5/p) = 1$ or $(2/p) = (5/p) = -1$. The first case holds if and only if $p \equiv \pm 1 \pmod 8$ and $p \equiv \pm 1 \pmod 5$, using (5.12) and (5.13.iv); thus $p \equiv \pm 1, \pm 9 \pmod{40}$ by the Chinese Remainder Theorem. The second case holds if and only if $p \equiv \pm 3 \pmod 8$ and $p \equiv \pm 2 \pmod 5$; hence $p \equiv \pm 3, \pm 13 \pmod{40}$.

5-48. Prove that there are infinitely many primes ending in the digit 9. (Hint. First show that there are infinitely many primes of the form $10k - 1$ by considering $N = 5(n!)^2 - 1$, where $n > 1$, and using (5.13).)

Solution. Let p be a prime divisor of N; note that p is odd. Then $5(n!)^2 \equiv 1 \pmod p$, and hence $1 = (5(n!)^2/p) = (5/p)$, by (5.10). By (5.13), p is therefore of the form $5k+1$ or $5k-1$. However, if all of the prime divisors of N were of the form $5k+1$, then N would also be of this form. But N is plainly of the form $5k-1$, so N must have at least one prime divisor p of the form $5k-1$; in fact, p is of the form $10k-1$, since $10k+4$ cannot be prime. Note that $p > n$ (for if $p \le n$, then $p|n!$ and $p|N$, so $p|1$, a contradiction). We have therefore shown that for any positive integer n, there is a prime greater than n of the form $10k-1$. Thus there are infinitely many primes of the form $10k-1$, all of which end in the digit 9.

5-49. Prove part (i) of Theorem 5.13: -2 is a quadratic residue of the odd prime p if and only if p is of the form $8k+1$ or $8k+3$.

Solution. $(-2/p) = 1$ if and only if $(-1/p) = (2/p) = 1$ or $(-1/p) = (2/p) = -1$. The first case holds precisely when $p \equiv 1 \pmod 4$ and $p \equiv \pm 1 \pmod 8$, by (5.11) and (5.12); thus it holds if and only if $p \equiv 1 \pmod 8$. The second case holds if and only if $p \equiv 3 \pmod 4$ and $p \equiv \pm 3 \pmod 8$, i.e., $p \equiv 3 \pmod 8$.

5-50. Prove that there are infinitely many primes of the form (a) $8k+3$; (b) $8k+5$; (c) $8k+7$. (Hint. Let $p_1, p_2, \ldots, p_n$ be primes of the given form. In (a), consider $N = (p_1 p_2 \cdots p_n)^2 + 2$; in (b), $N = (p_1 p_2 \cdots p_n)^2 + 4$; and in (c), $N = (p_1 p_2 \cdots p_n)^2 - 2$.)

Solution. In each case, let p be a prime divisor of N. (a) Since $(p_1 p_2 \cdots p_n)^2 \equiv -2 \pmod p$, we have $(-2/p) = 1$. Thus by (5.13.i), p is of the form $8k+1$ or $8k+3$. If every prime divisor of N were of the form $8k+1$, then N would also be of this form; but clearly, $N \equiv 3 \pmod 8$ since each $p_i^2 \equiv 1 \pmod 8$. Hence N has at least one prime divisor of the form $8k+3$, which cannot be one of the p_i (otherwise $p|2$, a contradiction since N is odd).

(b) Note that $N \equiv 5 \pmod 8$, since $p_i^2 \equiv 1 \pmod 8$ for each i. Also, $(p_1 p_2 \cdots p_n)^2 \equiv -4 \pmod p$; thus $(-4/p) = 1$ and so $(-1/p) = 1$. By (5.11), p is then of the form $4k+1$ and hence $8k+1$ or $8k+5$. If every prime divisor of N were of the form $8k+1$, N would also be of this form. Thus N has at least one prime divisor p of the form $8k+5$ that is different from all of the p_i.

(c) Since $(p_1 p_2 \cdots p_n)^2 \equiv 2 \pmod p$, 2 is a quadratic residue of p and so $p \equiv \pm 1 \pmod 8$, by (5.12). If every prime divisor of N were of the form $8k+1$, then N would

148   CHAPTER 5: QUADRATIC CONGRUENCES

also be of this form, which it is not. (In fact, N is of the form $8k - 1$.) Hence N has at least one prime factor p of the form $8k - 1$, and clearly, p is not one of the p_i.

Note. The case $8k + 1$ requires a different technique and will be considered in the next chapter. (See Problem 6-23.)

5-51. *Use (5.20) to show that 3 is a quadratic residue of the odd prime p if and only if $p \equiv \pm 1$ (mod 12). (This proves part (ii) of (5.13).)*

Solution. By (5.20), p is of the form $(4 \cdot 3k) \pm a$, where a ranges over the least positive residues modulo 12 of $1^2, 3^2, \ldots, (q-2)^2$; here $q = 3$, so $a = 1$. Thus $(3/p) = 1$ if and only if $p = 12k \pm 1$.

5-52. *Prove part (iii) of Theorem 5.13: -3 is a quadratic residue of the odd prime p if and only if p is of the form $6k + 1$.*

Solution. Note that $(-3/p) = 1$ if and only if (i) $(-1/p) = (3/p) = 1$ or (ii) $(-1/p) = (3/p) = -1$. By (5.13.ii), condition (i) holds if and only if p is of the form $4k + 1$ and also of the form $12k \pm 1$, that is, for p of the form $12k + 1$. Also, (ii) holds if and only if p is of the form $4k + 3$ and $12k \pm 5$, that is, for p of the form $12k + 7$. Since p is of the form $6k + 1$ if and only if it is of the form $12k + 1$ or $12k + 7$, the result follows.

5-53. *Prove part (iv) of Theorem 5.13: 5 is a quadratic residue of the odd prime p if and only if $p \equiv \pm 1$ (mod 5).*

Solution. We can apply (5.20) or use the Law of Quadratic Reciprocity to conclude that $(5/p) = (p/5)$. If $p = 5k + r$, then $(p/5) = (r/5)$. Hence $(5/p) = 1$ if and only if $(r/5) = 1$. Since the only quadratic residues of 5 are 1 and 4, p must be of the form $5k + 1$ or $5k + 4$, i.e., $5k \pm 1$.

5-54. *Prove that there are infinitely many primes of the form $6k + 1$ and therefore infinitely many of the form $3k + 1$. (Hint. Let $p_1, p_2, \ldots, p_n$ be primes of the form $6k + 1$, and consider $N = (p_1 p_2 \cdots p_n)^2 + 3$.)*

Solution. Let p be a prime divisor of N; clearly, $p \neq 3$. Then $(p_1 p_2 \cdots p_n)^2 \equiv -3$ (mod p) and so -3 is a quadratic residue of p. By (5.13.iii), p is therefore of the form $6k + 1$ and hence also of the form $3k + 1$. It is clear that p is not one of the p_i. Thus, given any finite collection of primes of the form $6k + 1$, we can always find another prime of this form.

5-55. *Find all solutions of (a) $x^2 \equiv 41$ (mod 43); (b) $x^2 \equiv -6$ (mod 103). (Hint. Use Problem 5-3.)*

Solution. Note that 43 and 103 are both of the form $4k+3$, so we can apply Problem 5-3 to find the solutions if they exist.

(a) To simplify calculations, use -2 instead of 41. Since $(-2/43) = 1$ by (5.13.i), the congruence has solutions that, according to Problem 5-3, are given by $\pm(-2)^{11}$, i.e., 16 and 27 modulo 43.

(b) Check that $(-6/103) = (-1/103)(2/103)(3/103) = (-1)(+1)(-1) = 1$, so the congruence is solvable. The solutions are given by $\pm 6^{26}$, i.e., ± 32 modulo 103.

PROBLEMS AND SOLUTIONS 149

▷ **5-56.** *Let p be prime, with $p = 4k + 1$. If d is odd and $d \mid k$, prove that $x^2 \equiv d$ (mod p) is solvable.*

Solution. By (5.10.ii), or directly, we can see that if the congruences $x^2 \equiv a$ (mod p) and $x^2 \equiv b$ (mod p) are solvable, so is $x^2 \equiv ab$ (mod p). Thus to prove that $x^2 \equiv d$ (mod p) is solvable, we need only show that $x^2 \equiv q$ (mod p) is solvable for any prime divisor q of d. Accordingly, suppose that $q \mid k$ and q is prime. Because $p \equiv 1$ (mod 4), the Law of Quadratic Reciprocity implies that $(q/p) = (p/q) = ((4k+1)/q) = (1/q)$, using the fact that $4k \equiv 0$ (mod q). Therefore $x^2 \equiv q$ (mod p) is solvable.

5-57. *Use Problem 5-3 to find all solutions of the congruence $9x^4 - 19x^2 + 30 \equiv 0$ (mod 59).*

Solution. Complete the square to get $(18x^2 - 19)^2 \equiv 48 \equiv -11$ (mod 59). Since $(48/59) = (3/59) = 1$ by (5.10.iv) and (5.10.ii), the congruence $y^2 \equiv -11$ (mod 59) has solutions. By Problem 5-3, these solutions are given by $\pm 11^{15}$, i.e., ± 15. Now solve $18x^2 - 19 \equiv \pm 15$ (mod 59); this gives $18x^2 \equiv 34$ (mod 59) and $18x^2 \equiv 4$ (mod 59). The congruence $18x^2 \equiv 34$ (mod 59) is equivalent to $9x^2 \equiv 17 \equiv 135$ (mod 59), i.e., $x^2 \equiv 15$ (mod 59). Since $(15/59) = (3/59)(5/59) = 1$, solutions exist; by Problem 5-3, they are given by $\pm 15^{15}$, i.e., ± 29. Now consider $18x^2 \equiv 4 \equiv 63$ (mod 59), i.e., $2x^2 \equiv 7 \equiv 66$ (mod 59). Thus $x^2 \equiv 33$ (mod 59), and since $(33/59) = (3/59)(11/59) = (59/3)(59/11) = (2/3)(4/11) = -1$, there are no solutions.

Thus the only solutions of the original congruence are ± 29, i.e., 29 and 30.

5-58. *Use Gauss's Lemma to evaluate $(14/23)$.*

Solution. As k runs from 1 to $(23-1)/2 = 11$, the least positive residues of $14k$ are 14, 5, 19, 10, 1, 15, 6, 20, 11, 2, and 16. Of these, 5 are greater than $23/2$, so by Gauss's Lemma, $(14/23) = (-1)^5 = -1$.

5-59. *Suppose that $q > 2$ is prime. If $p = 2^q - 1$ is also prime, prove that $x^2 \equiv 3$ (mod p) is not solvable.*

Solution. Since q is odd, $p = 2^q - 1 \equiv (-1)^q - 1 = -2 \equiv 1$ (mod 3); thus $(p/3) = 1$. Note that p is of the form $4k + 3$, so by the Law of Quadratic Reciprocity, $(3/p) = -(p/3) = -1$.

5-60. *Describe the odd prime divisors of $n^2 + 1$; $n^2 + 2$; and $n^2 + 3$.*

Solution. Let p be an odd prime. If $p \mid n^2 + 1$, then $n^2 \equiv -1$ (mod p), so p is of the form $4k + 1$, by (5.11). If $p \mid n^2 + 2$ and $p \neq 2$, then $(-2/p) = 1$, and hence p is of the form $8k + 1$ or $8k + 3$, by (5.13.i). And if $p \mid n^2 + 3$ and $p > 3$, then $(-3/p) = 1$; thus p is of the form $6k + 1$, by (5.13.iii).

5-61. *Calculate $(6/19)$ using (a) Euler's Criterion; (b) Gauss's Lemma; (c) the Law of Quadratic Reciprocity.*

Solution. (a) $(6/19) \equiv 6^9 \equiv 6(6^2)^4 \equiv 6(-2)^4 \equiv 6(-3) \equiv 1$ (mod 19).

(b) The least positive residues of $6, 2\cdot 6, 3\cdot 6, \ldots, 9\cdot 6$ are 6, 12, 18, 5, 11, 17, 4, 10, and 16. Of these, six are greater than 19/2, and hence Gauss's Lemma implies that $(6/19) = (-1)^6 = 1$.

(c) $(6/19) = (2/19)(3/19) = -(3/19) = (19/3) = (1/3) = 1$, using the Law of Quadratic Reciprocity and (5.12).

5-62. *(a) Prove that the odd prime divisors of $9n^2 - 6n + 4$ are of the form $6k + 1$. (Hint. Complete the square and use (5.13.iii).)*

(b) Prove that the odd prime divisors of $n^2 + 4n + 6$ are of the form $8k + 1$ or $8k + 3$.

(c) Prove that the prime divisors >5 of $n^2 - 2n - 4$ are of the form $10k \pm 1$.

Solution. (a) If $p \mid 9n^2 - 6n + 4$, then $9n^2 - 6n + 4 \equiv 0 \pmod{p}$. Complete the square to get $y^2 \equiv b^2 - 4ac \equiv -108 \pmod{p}$. Note that $(-108/p) = (-3/p)$, and apply (5.13.iii).

(b) Argue as in (a). Complete the square to get $y^2 \equiv -8 \pmod{p}$. Since $(-8/p) = (-2/p)$, the result follows from (5.13.i).

(c) Completing the square gives $y^2 \equiv 20 \pmod{p}$. Note that $(20/p) = (5/p)$ and apply (5.13.iv) to conclude that p is of the form $5k \pm 1$. Since p is odd, k must be even, and hence p is of the form $10k \pm 1$.

5-63. *Prove or disprove: If p and q are odd primes such that $p \equiv q \pmod{26}$, then $(13/p) = (13/q)$.*

Solution. The result is true. Since $p \equiv q \pmod{26}$ and 13 is of the form $4k + 1$, it follows from the Law of Quadratic Reciprocity and (5.10.i) that $(13/p) = (p/13) = ((q + 26k)/13) = (q/13) = (13/q)$.

5-64. *Determine if the following congruences are solvable: (a) $x^2 \equiv 1993 \pmod{1997}$; (b) $x^2 \equiv 1993 \pmod{1999}$. (1993, 1997, and 1999 are primes.)*

Solution. Use the Law of Quadratic Reciprocity. (a) $(1993/1997) = (1997/1993) = (4/1993) = 1$, so the congruence is solvable. (b) $(1993/1999) = (1999/1993) = (6/1993) = (2/1993)(3/1993) = (3/1993) = (1993/3) = (1/3) = 1$, and hence the congruence has solutions.

5-65. *For which primes p does $13x^2 + 7x + 1 \equiv 0 \pmod{p}$ have a solution?*

Solution. If $p = 13$, then the congruence reduces to $7x \equiv 1 \pmod{13}$, which has a solution. If $p \neq 13$, use (5.1) to reduce the congruence to $y^2 \equiv b^2 - 4ac = -3 \pmod{p}$. Thus by (5.13.iii), the given congruence is solvable if and only if p is of the form $6k + 1$.

5-66. *Use the Law of Quadratic Reciprocity to determine if $x^4 - 6x^2 + 35 \equiv 0 \pmod{37}$ is solvable.*

Solution. Complete the square to get $(2x^2 - 6)^2 \equiv 44 \pmod{37}$. (Simplify the calculation by replacing 35 with -2 modulo 37.) Since $44 \equiv 81 \pmod{37}$, we have $2x^2 - 6 \equiv \pm 9 \pmod{37}$, and hence $2x^2 \equiv 15 \pmod{37}$ or $2x^2 \equiv -3 \pmod{37}$. These are equivalent to $x^2 \equiv 26 \pmod{37}$ and $x^2 \equiv 17 \pmod{37}$. Note that $(17/37) = (37/17) =$

$(3/17) = (17/3) = (2/3) = -1$; also, $(26/37) = (2/37)(13/37) = -(37/13) = -(11/13) = -(13/11) = -(2/11) = 1$, using (5.12). Thus the original congruence has two solutions.

The Jacobi Symbol

To facilitate calculation, the Legendre symbol can be extended to the case where the number at the bottom is not prime. The *Jacobi symbol* (a/m), introduced in 1846 by Carl Gustav Jacobi (1804–1851), assumes only the values 1 and -1 and coincides with the Legendre symbol when m is prime. Unlike the Legendre symbol, however, it isn't necessary to factor the numerator into primes before inverting. This fact makes the Jacobi symbol particularly efficient in evaluating Legendre symbols.

Most of the properties of the Legendre symbol hold for the Jacobi symbol, including the law of reciprocity, but there is one important exception: $(a/m) = 1$ does not imply that $x^2 \equiv a \pmod{m}$ is solvable. (This is the price paid for having the law of reciprocity hold for the Jacobi symbol. If we simply define (a/m) to be 1 or -1 according as $x^2 \equiv a \pmod{m}$ is solvable or not solvable, then the Jacobi symbol would not obey the reciprocity law.)

We next list the definition and the main properties of the Jacobi symbol. (The proofs of these results can be found, for example, in the text by Niven and Zuckerman; see the Bibliography at the end of the book.)

Definition. Let $m = \prod p_i^{k_i}$, where each p_i is an odd prime, and suppose $(a, m) = 1$. Define the *Jacobi symbol* (a/m) by $(a/m) = \prod (a/p_i)^{k_i}$, where the factors (a/p_i) are Legendre symbols.

Theorem. *Let m and n be odd positive integers.*
(i) *If $(a, m) = 1$ and $a \equiv b \pmod{m}$, then $(a/m) = (b/m)$.*
(ii) *If $(a, m) = (b, m) = 1$, then $(ab/m) = (a/m)(b/m)$. In particular, $(a^2/m) = 1$.*
(iii) *If m and n are relatively prime and $(a, m) = (a, n) = 1$, then $(a/mn) = (a/m)(a/n)$.*
(iv) $(-1/m) = 1$ *if and only if* $m \equiv 1 \pmod{4}$.
(v) $(2/m) = 1$ *if and only if* $m \equiv \pm 1 \pmod{8}$.
(vi) *(Reciprocity Law) If $(m, n) = 1$, then $(m/n)(n/m) = (-1)^{\frac{m-1}{2}\frac{n-1}{2}}$.*

5-67. *If (a/m) denotes a Jacobi symbol, give an example to show that $(a/m) = 1$ does not imply that $x^2 \equiv a \pmod{m}$ is solvable.*

Solution. Consider $x^2 \equiv -1 \pmod{21}$. This has no solution, since $x^2 \equiv -1 \pmod{3}$ has no solution (see (5.11)). But $(-1/21) = (-1/3)(-1/7) = (-1)(-1) = 1$.

Note. More generally, the Jacobi symbol (a/m) is equal to 1 as long as an even number of the Legendre symbols (a/p_i) that define (a/m) are equal to -1.

152   CHAPTER 5: QUADRATIC CONGRUENCES

5-68. *Suppose that the Jacobi symbol (a/m) equals -1. Prove that the congruence $x^2 \equiv a \pmod{m}$ is not solvable.*

Solution. If $(a/m) = -1$, then from the definition of the Jacobi symbol, at least one factor (a/p_i) must be -1. Thus $x^2 \equiv a \pmod{p_i}$ has no solution, and therefore $x^2 \equiv a \pmod{m}$ cannot have a solution.

5-69. *Evaluate $(3828/2539)$ with and without the use of Jacobi symbols. (2539 is prime.)*

Solution. Using Jacobi symbols: $(3828/2539) = (-1250/2539) = (-1/2539)(2/2539)(625/2539) = (-1)(-1)(625/2539)$ (since $2539 \equiv 3 \pmod 8$). Now it is obvious that $(625/2539) = 1$, since 625 is a perfect square, but we wish to avoid factoring (except for divisions by 2), since for large numbers, factoring is very slow. Now $(625/2539) = (2539/625) = (39/625) = (625/39) = (1/39) = 1$.

Using Legendre symbols: $(3828/2539) = (4/2539)(3/2539)(11/2539)(19/2539) = [-(2539/3)][-(2539/11)][-(2539/19)] = -(1/3)(9/11)(12/19) = -(3/19) = (1/3) = 1$.

5-70. *Use Jacobi symbols to determine which of the following congruences are solvable: (a) $x^2 \equiv -70 \pmod{709}$; (b) $x^2 \equiv 210 \pmod{263}$; (c) $x^2 \equiv 330 \pmod{997}$. ($263$ and 997 are primes.)*

Solution. (a) $(-70/709) = (-1/709)(2/709)(35/709) = -(35/709)$, since $709 \equiv 1 \pmod 4$ and $709 \equiv 5 \pmod 8$; $-(35/709) = -(709/35) = -(9/35) = -(35/9) = -(-1/9) = -1$, since $9 \equiv 1 \pmod 4$. Thus the congruence is not solvable.

(b) $(210/263) = (2/263)(105/263) = (105/263) = (263/105) = (-52/105) = (13/105) = (105/13) = (1/13) = 1$. Since $(210/263)$ is a *Legendre* symbol (because 263 is prime), it follows that the given congruence is solvable.

(c) $(330/997) = (2/997)(165/997) = -(165/997) = -(997/165) = -(7/165) = -(165/7) = -(4/7) = -1$. Thus the congruence is not solvable.

▷ **5-71.** *(a) Characterize the positive integers m that are relatively prime to 3 and such that 3 is a quadratic nonresidue of m.*

(b) Describe the positive integers m not divisible by 3 such that the Jacobi symbol $(3/m)$ equals 1.

Solution. (a) Let $m = 2^k \prod p_i^{k_i}$, where the p_i are odd primes different from 3. Since the congruence $x^2 \equiv 3 \pmod 4$ does not have a solution, it follows that if $k \geq 2$, then 3 is a quadratic nonresidue of m.

Now suppose that $k = 0$ or $k = 1$. By (5.13.ii), 3 is a quadratic nonresidue of the odd prime p if and only if p is of the form $12k \pm 5$. Thus for m not divisible by 3 or 4, 3 will be a quadratic nonresidue of m if and only if $p_i \equiv \pm 5 \pmod{12}$ for at least one value of i.

(b) The Jacobi symbol (a/b) is not defined when b is even. Suppose $(3, m) = 1$ and $m = \prod p_i^{k_i}$ is odd. According to the definition of the Jacobi symbol, $(3/m)$ is not affected by $(3/p_i)$ if k_i is even. Thus $(3/m) = 1$ if and only if $(3/p_i) = -1$ for an even number (possibly zero) of the p_i for which k_i is odd, i.e., if and only if an even number of the p_i for which k_i is odd are of the form $12k \pm 5$.

EXERCISES FOR CHAPTER 5

1. Evaluate the Legendre symbols (70/97) and (263/331).
2. Compute (14/311), (165/313), and (1891/1999).
3. Calculate $(1/73) + (2/73) + \cdots + (72/73)$.
4. Find the value of $((1 \cdot 2)/73) + ((2 \cdot 3)/73) + \cdots + ((71 \cdot 72)/73)$.
5. Use Gauss's Lemma to calculate (3/31).
6. (a) Use Euler's Criterion to evaluate (37/43).
   (b) Use Gauss's Lemma to compute (13/19).
   (c) Use the Law of Quadratic Reciprocity to find (323/353).
7. Using Euler's Criterion or otherwise, prove Theorem 5.10.
8. Let p be prime. Prove that $(n^2 - 2)(n^2 - 5)(n^2 - 40)$ is divisible by p for infinitely many values of n.
9. (a) Does there exist a positive integer n such that $n^2 - 3$ is a multiple of 313?
   (b) Are there infinitely many n for which $n^2 + 3$ is divisible by 97?
10. Show that the prime divisors of $4n^2 + 28n + 51$ are of the form $8k + 1$ or $8k + 3$.
11. Prove that every odd prime divisor of $n^2 + 100$ is of the form $12k + 1$ or $12k + 5$.
12. Describe the prime divisors of $n^2 + 6$.
13. Characterize the primes p such that -11 is a quadratic residue of p.
14. For which odd primes p is -5 a quadratic residue?
15. Determine the odd primes p that have 11 as a quadratic residue.
16. Characterize the odd primes for which 13 is a quadratic nonresidue.
17. Determine if the congruences $x^2 \equiv \pm 109 \pmod{313}$ have solutions.
18. Decide if the following congruences are solvable:
    (a) $x^2 + 3x + 3 \equiv 0 \pmod{41}$;
    (b) $3x^2 - 4x - 1 \equiv 0 \pmod{1363}$. (Hint. First factor 1363.)
19. Determine if $6x^2 - 15x + 5 \equiv 0 \pmod{749}$ is solvable.
20. Use Euler's Criterion to decide if $5x^2 - 12x + 1 \equiv 0 \pmod{61}$ has solutions.
21. Use the Law of Quadratic Reciprocity to decide if $2x^2 - 6x - 89 \equiv 0 \pmod{1987}$ is solvable. (1987 is prime.)

CHAPTER 5: QUADRATIC CONGRUENCES

22. Determine if there are solutions of $7x^2 - 25x + 1 \equiv 0 \pmod{599}$. (599 is prime.)

23. Is the congruence $x^4 \equiv -1 \pmod{299}$ solvable?

24. Show that $17x^2 + 19x - 2 \equiv 0 \pmod{3493}$ has solutions. How many solutions are there?

25. Find the number of solutions of
    (a) $x^2 \equiv 6 \pmod{175}$;
    (b) $x^2 \equiv 361 \pmod{693}$;
    (c) $x^2 \equiv 41 \pmod{2^6 \cdot 5^3 \cdot 37^2 \cdot 73^3}$.

26. Determine the number of solutions of $x^2 \equiv 57 \pmod{256}$ and $x^2 \equiv 71 \pmod{128}$.

27. How many solutions does $x^2 \equiv -3 \pmod{37^3}$ have?

28. Determine the number of solutions of $x^2 \equiv 69 \pmod{4 \cdot 5^3 \cdot 11^2}$ and $x^2 \equiv 41 \pmod{2^3 \cdot 5^4 \cdot 23^2}$.

29. Find the number of solutions of
    (a) $x^2 \equiv 17 \pmod{2^5 \cdot 13^2 \cdot 19}$;
    (b) $x^2 \equiv 9 \pmod{2^4 \cdot 5^3 \cdot 7^2}$;
    (c) $x^2 \equiv 57 \pmod{2^7 \cdot 7^5 \cdot 59^2}$.

30. Use Problem 5-3 to decide if $x^2 \equiv 3 \pmod{83}$ and $x^2 \equiv 13 \pmod{83}$ are solvable, and find the solutions if they exist.

31. Find all solutions of $x^2 \equiv -1 \pmod{29}$.

32. Use Problem 5-3 to find the solutions of the congruence $9x^2 - 24x + 13 \equiv 0 \pmod{73}$.

33. Find the least positive residue of each solution of
    (a) $9x^2 - 12x - 5 \equiv 0 \pmod{53}$;
    (b) $4x^2 + 47x + 49 \equiv 0 \pmod{59}$.

34. Find all solutions of $5x^2 - 7x - 11 \equiv 0 \pmod{61}$.

35. Use the Law of Quadratic Reciprocity to show that $x^2 \equiv -3 \pmod{79}$ is solvable. Find both solutions.

36. Find all solutions of $2x^2 - 3x - 9 \equiv 0 \pmod{73}$.

37. Prove that $x^2 - 12x + 17 \equiv 0 \pmod{79}$ is solvable, and find the solutions.

38. Find both solutions of (a) $x^2 \equiv 2 \pmod{263}$; (b) $x^2 \equiv -53 \pmod{83}$; (c) $x^2 \equiv 20 \pmod{79}$.

39. Use the Chinese Remainder Theorem and Problem 5-3 to find all solutions of $x^2 \equiv 37 \pmod{77}$.

40. Use the Chinese Remainder Theorem to find the least positive residue of each solution of $25x^2 - 157x + 11 \equiv 0 \pmod{187}$.

41. Find all solutions of $3x^2 - 10x + 7 \equiv 0 \pmod{1547}$. (First factor 1547.)
42. Use the Chinese Remainder Theorem to solve the following congruences:
    (a) $4x^2 - 12x + 5 \equiv 0 \pmod{77}$;
    (b) $2x^2 - x + 7 \equiv 0 \pmod{91}$.
43. Find the solutions of $7x^2 - x + 24 \equiv 0 \pmod{36}$.
44. Solve the congruence $x^2 + 3x - 7 \equiv 0 \pmod{77}$.
45. Determine all solutions of $23x^2 - x - 21 \equiv 0 \pmod{91}$.
46. Find two consecutive quadratic nonresidues of 89.
47. What is the least nonnegative residue of the sum of the quadratic residues of 31?
48. Let N be the number of positive integers less than 16 that are quadratic nonresidues of 31. Show that $15! \equiv (-1)^N \pmod{31}$.
49. What is the least nonnegative residue of the product of the quadratic residues of 59?
50. Prove or disprove: If $x^2 \equiv a \pmod{m}$ is solvable for two different values of a, then each congruence has the same number of solutions.
51. Suppose that $p = 2^{2n} + 1$, where $n \geq 1$. Show that if p is prime, then $3^{(p-1)/2} + 1$ is divisible by p.

NOTES FOR CHAPTER 5

1. In Article 152 of *Disquisitiones Arithmeticae*, Gauss considers briefly the question of finding solutions of $ax^2 + bx + c \equiv 0 \pmod{m}$ and outlines the method of reducing this problem to the study of congruences of the form $y^2 \equiv d \pmod{m}$. In Articles 100–105, Gauss discusses in detail how to reduce this last congruence to congruences of the form $y^2 \equiv d \pmod{p}$.

2. Euler was apparently the first mathematician to define residues and nonresidues and to work systematically with them. But Fermat, a century earlier, knew the primes that have a as a quadratic residue, where $a = -1, 2, 3$, and 5.

3. There is no simple formula, such as the Law of Quadratic Reciprocity, for nth power residues when $n \geq 3$. (There are, however, rather complicated reciprocity laws for such n, the most concrete results occurring for $n = 3$ and $n = 4$.) In Chapter 6, we will give a criterion for determining when an integer is an nth power residue of p^k or $2p^k$, where p is an odd prime.

4. The principal value of the Jacobi symbol (P/Q) occurs when Q is prime, in simplifying and speeding up the calculation of a Legendre symbol.

Jacobi symbol calculations bear a strong formal resemblance to the Euclidean Algorithm, and it is not difficult to see that (P/Q) can be evaluated in roughly the same amount of time as (P, Q). In particular, if we wish to determine the solvability of $x^2 \equiv a \pmod{p}$, where p is a large prime, using the Jacobi symbol and the corresponding reciprocity law is *much* faster than using the Law of Quadratic Reciprocity to evaluate the Legendre symbol (a/p).

When a and p have roughly the same order of magnitude, Jacobi symbol calculations and Euler's Criterion are about equally efficient ways of computing (a/p). If a is very much smaller than p, then using a Jacobi symbol calculation is faster, since after one reciprocity step, we may be dealing with quite small numbers. If we are using a calculator rather than a computer and a prime p with, say, seven digits, then a Jacobi symbol calculation is much easier. The difficulty with using Euler's Criterion is that in computing $a^{(p-1)/2}$ modulo p, we may need to deal with 14-digit numbers.

Biographical Sketches

Ferdinand Gotthold Eisenstein was born in Berlin in 1823. He was frequently sick when young and entered the University of Berlin only in 1843. By this time, he had mastered the techniques of Gauss, Dirichlet, and Jacobi. In 1844, Eisenstein entered explosively on the mathematical scene, publishing 25 short papers in *Crelle's Journal*. Among these were two elegant proofs of the Law of Quadratic Reciprocity, one of which is still reproduced in most texts. The other involved entirely new ideas and enabled him in the same year to prove laws of cubic reciprocity and biquadratic reciprocity. Gauss had sought to prove such a law for many years.

In that same year, Eisenstein visited Gauss in Göttingen for two weeks. Gauss repeatedly expressed his admiration of Eisenstein, calling his talent "one that nature bestows on only a few each century." (There is no evidence for the often-repeated story that Gauss said there had only been three epoch-making mathematicians: Archimedes, Newton, and Eisenstein!) In 1847, Gauss was to write a glowing foreword to a collection of Eisenstein's papers.

Eisenstein continued to do brilliant work on elliptic functions and higher reciprocity laws, despite repeated bouts of illness. There were other difficulties. In 1848, he was involved in revolutionary activity in Berlin. Eisenstein was badly beaten by Prussian soldiers and briefly imprisoned. In the next two years, he wrote papers that were fertile in ideas on quadratic forms, Gaussian sums, and Kummer's ideal theory. In 1852, he was elected to the Berlin Academy, as the successor of Jacobi.

Eisenstein died of tuberculosis in 1852, at the age of 29.

REFERENCES

Harold Davenport, *The Higher Arithmetic* (Sixth Edition), Cambridge University Press, Cambridge, England, 1992.

> This short book is one of the most readable books available and gives a wonderful overview of elementary number theory. Because of the length, Davenport does not prove as many results as in a standard text, but the theorems given cover most of the important areas in number theory and are very nicely motivated. The proofs are detailed and complete, and since they are written in a conversational manner, the notation is not obtrusive. There are not a large number of examples in this book, but the ones included are discussed in detail. All in all, *The Higher Arithmetic* is a very enjoyable book to read, and it is highly recommended for students at any level.

Carl Friedrich Gauss, *Disquisitiones Arithmeticae*, translated by Arthur A. Clarke. (See Chapter 2.)

CHAPTER SIX
Primitive Roots and Indices

In the preceding chapter, we studied the quadratic residues of a positive integer m. In this chapter, we will investigate the *kth power residues* of m for $k \geq 2$ – that is, the numbers a relatively prime to m for which $x^k \equiv a \pmod{m}$ is solvable – and we will give a method for determining the solvability of such congruences.

In order to find the solutions, we will use the existence of a *primitive root* of m and the notion of *indices* to reduce the congruence $x^k \equiv a \pmod{m}$ to one of the form $ky \equiv b \pmod{\phi(m)}$, whose solutions can then be found by any of the methods for linear congruences described in Chapter 2. The properties of indices turn out to be very similar to those of logarithms; the use of indices allows us to reduce a problem involving exponents to one of multiplication, and similarly to reduce a problem of multiplication to one of addition.

While the existence of a primitive root for a given modulus is of theoretical importance and simplifies the study of kth power residues, it is not true that every positive integer has a primitive root. Indeed, the main result in this chapter, proved by Gauss in 1801, is the characterization of which positive integers have primitive roots.

RESULTS FOR CHAPTER 6

The Order of an Integer

We begin with the formal definition of the order of an integer, a concept that was briefly alluded to in Chapter 3 (see (3.9)).

(6.1) Definition. Let m be a positive integer and suppose $(a, m) = 1$. The *order of a modulo m*, denoted by $\operatorname{ord} a$, is the smallest positive integer h such that $a^h \equiv 1 \pmod{m}$.

Notes. 1. The notation $\operatorname{ord} a$ is ambiguous, since the order of an integer also depends on the modulus. It may be clearer to denote the order by $\operatorname{ord}_m a$. However, since the modulus is ordinarily fixed during a calculation, the simpler notation should cause no difficulty.

2. If a is relatively prime to m, then $a^{\phi(m)} \equiv 1 \pmod{m}$ by Euler's Theorem, and hence the order of a is never more than $\phi(m)$. It is easy to see that a smaller exponent may suffice: $\operatorname{ord} 1 = 1$ for every positive integer m, and $\operatorname{ord}(-1) = 2$ if m is greater than 2.

3. In older books, the order of a modulo m is often referred to as *the exponent to which a belongs modulo m*. However, this terminology is rather uncommon now, and we will use *the order of a modulo m* exclusively in this book, a term that is standard in group theory and one that reflects the underlying algebraic structure of a reduced residue system modulo m.

(6.2) Theorem. Let m be a positive integer and suppose that $(a, m) = 1$.

(i) $a^s \equiv 1 \pmod{m}$ *if and only if* $\operatorname{ord} a \mid s$. *In particular,* $\operatorname{ord} a \mid \phi(m)$.

(ii) $a^s \equiv a^t \pmod{m}$ *if and only if* $s \equiv t \pmod{\operatorname{ord} a}$.

Proof. (i) If $s = k \operatorname{ord} a$, then $a^s = (a^{\operatorname{ord} a})^k \equiv 1^k = 1 \pmod{m}$. Conversely, suppose $a^s \equiv 1 \pmod{m}$. By the division algorithm, we have $s = q \operatorname{ord} a + r$, where $0 \le r < \operatorname{ord} a$; thus $1 \equiv a^s = (a^{\operatorname{ord} a})^q a^r \equiv a^r \pmod{m}$. Hence $r = 0$ since, by definition, $a^{\operatorname{ord} a}$ is the smallest positive power of a congruent to 1 modulo m. The second part follows from Euler's Theorem.

(ii) We may suppose that $s \ge t$. If $a^s \equiv a^t \pmod{m}$, then $a^s = a^t a^{s-t} \equiv a^s a^{s-t} \pmod{m}$. Since $(a^s, m) = 1$, it follows from (2.2.vi) that $a^{s-t} \equiv 1 \pmod{m}$. Now apply part (i). Conversely, if $s \equiv t \pmod{\operatorname{ord} a}$, write $s = t + k \operatorname{ord} a$ for some integer k. Then $a^s = a^t (a^{\operatorname{ord} a})^k \equiv a^t \pmod{m}$.

In the case of a prime modulus p, (6.2.i) implies that the order of a is a divisor of $p - 1$. (Euler was the first to publish a proof, in 1736, that if p is prime and d is the smallest positive integer such that $a^d \equiv 1 \pmod{p}$, then d divides $p - 1$, but the result had been stated by Fermat in 1640. See (3.9).)

(6.3) Theorem. Let m be a positive integer and suppose that $(a, m) = 1$.

(i) If $\operatorname{ord} a = d$, then $\operatorname{ord} a^k = d/(k, d)$ for any $k \ge 1$.

(ii) If $\operatorname{ord} a = d$ and e is a positive divisor of d, then $a^{d/e}$ has order e.

Proof. It follows from (6.2.i) that $(a^k)^j \equiv 1 \pmod{m}$ if and only if kj is a multiple of d. Thus a^k has order j if and only if kj is the smallest multiple of k

that is a multiple of d, that is, if and only if kj is the least common multiple of k and d. But this least common multiple is $kd/(k,d)$, and hence $j = d/(k,d)$. Part (ii) follows from part (i) by noting that if $e \mid d$, then $(d/e, d) = d/e$.

The next result shows how to construct an integer whose order is the least common multiple of h and k if we are given elements of order h and k.

(6.4) Theorem. *Suppose $h = \text{ord } a$ and $k = \text{ord } b$. If $(h, k) = 1$, then $\text{ord } ab = hk$. In general, there is an integer c such that the order of c is the least common multiple of h and k.*

Proof. We show first that if $(h, k) = 1$, then ab has order hk. Let $r = \text{ord } ab$. Clearly, $(ab)^{hk} = (a^h)^k (b^k)^h \equiv 1 \pmod{m}$, and hence $r \mid hk$ by (6.2.i). Also, $b^{rh} \equiv (a^h)^r b^{rh} = (ab)^{rh} \equiv 1 \pmod{m}$, and hence $k \mid rh$. Since $(h,k) = 1$, it follows that $k \mid r$. In a similar way, we can show that $h \mid r$, and therefore $hk \mid r$ since $(h,k) = 1$. Thus $r = hk$.

Now suppose $(h, k) > 1$, and let M be the least common multiple of h and k. If $h = p_1^{h_1} \cdots p_t^{h_t}$ and $k = p_1^{k_1} \cdots p_t^{k_t}$, then $M = p_1^{\alpha_1} \cdots p_t^{\alpha_t}$, where $\alpha_i = \max(h_i, k_i)$ for $i = 1, 2, \ldots, t$ (see (1.17)). Let h' be the product of $p_i^{\alpha_i}$ for those i such that $h_i \geq k_i$, and let k' be the product of $p_i^{\alpha_i}$ for values of i where $k_i > h_i$. It is clear that $h' \mid h$, $k' \mid k$, $(h', k') = 1$, and $h'k' = M$.

By (6.3.ii), $a^{h/h'}$ has order h'. Similarly, $b^{k/k'}$ has order k'. Let $c = a^{h/h'} b^{k/k'}$. Since $(h', k') = 1$, it follows from the first part of the proof that c has order $h'k' = M$.

Primitive Roots

It follows from (6.2.i) that $\text{ord } a \leq \phi(m)$ for every a relatively prime to m, and we have seen that the order of a can be strictly less than $\phi(m)$. An obvious question arises: For a given modulus m, does there exist an integer whose order is as large as possible, namely, $\phi(m)$? An integer with this property is called a *primitive root of m*, a term introduced by Euler. We have the following definition.

(6.5) Definition. *Let m be a positive integer, and suppose that $(a, m) = 1$. If the order of a modulo m is $\phi(m)$, then a is called a primitive root of m.*

It is important to note that not every integer has a primitive root. For example, if $m = 8$, then $a^2 \equiv 1 \pmod{m}$ for every odd integer a. Thus $\text{ord } a \leq 2$ for every a relatively prime to 8. But $\phi(8) = 4$, and hence 8 has no primitive roots.

We next show that any prime has a primitive root. This result was first stated in 1769 by J.H. Lambert, in connection with investigations about the decimal expansion of the fraction $1/p$. In 1773, Euler gave an essentially correct, but incomplete, proof that every prime has a primitive root. Legendre showed, in 1785, that if p is an odd prime and d is a divisor of $p-1$, there are precisely $\phi(d)$ incongruent integers of order d modulo p (see (6.14)); thus there exist $\phi(p-1)$ primitive roots of p. Gauss also gave two fully detailed proofs in his *Disquisitiones* (1801). All the proofs, including the one that follows, make use of Lagrange's Theorem on the number of roots of a polynomial congruence. The full characterization of the numbers that have primitive roots will be given in the last section of this chapter.

(6.6) Definition. Let m be a positive integer, and let u be the smallest positive integer such that $a^u \equiv 1 \pmod{m}$ for every a relatively prime to m. Then u is called the *least universal exponent for m*.

Note. By (6.2.i), u is the least common multiple of the numbers $\operatorname{ord} a$, as a ranges over all integers from 1 to m that are relatively prime to m. Thus, applying (6.4) repeatedly, we can find an integer c such that $\operatorname{ord} c = u$. This c has the maximum possible order modulo m. If $u = \phi(m)$, then c is a primitive root of m.

(6.7) Theorem (Legendre). *Every prime has a primitive root.*

Proof. Suppose p is prime. Let u be the least universal exponent for p, and let g be an integer of order u modulo p. Then every integer relatively prime to p is a solution of the congruence $x^u \equiv 1 \pmod{p}$, so the congruence has $p-1$ solutions. But by Lagrange's Theorem, the congruence has no more than u solutions. It follows that $u = p-1$, and hence g is a primitive root of p.

The next result can be quite helpful in showing that g is a primitive root of m.

(6.8) Theorem. *If $(g, m) = 1$, then g is a primitive root of m if and only if $g^{\phi(m)/q} \not\equiv 1 \pmod{m}$ for every prime divisor q of $\phi(m)$.*

Proof. If g is a primitive root of m, then $g^{\phi(m)/q} \not\equiv 1 \pmod{m}$ for any prime q, for if $g^{\phi(m)/q} \equiv 1 \pmod{m}$, then g has order less than $\phi(m)$.

Conversely, suppose that $(g, m) = 1$ and g is not a primitive root of m. Then g has order d for some $d < \phi(m)$, and $d \mid \phi(m)$ by (6.2.i). Let $\phi(m) = dk$, and let q be a prime divisor of k. Then $\phi(m)/q$ is a multiple of d, and since $g^d \equiv 1 \pmod{m}$, it follows that $g^{\phi(m)/q} \equiv 1 \pmod{m}$.

Computational Note. The preceding result gives a fairly efficient way of testing whether g is a primitive root of p when p is a small odd prime. First take $q = 2$. If $g^{(p-1)/2} \equiv 1 \pmod{p}$, then g is *not* a primitive root of p. Thus, in view of Euler's Criterion, *a primitive root g of an odd prime p is always a quadratic nonresidue of p*, and hence $g^{(p-1)/2} \equiv -1 \pmod{p}$.

Having dealt with $q = 2$, calculate $g^{(p-1)/q}$ modulo p for the other prime factors q of $p - 1$. If $g^{(p-1)/q} \not\equiv 1 \pmod{p}$ for all such q, we conclude that g is a primitive root of p. (If p is a large prime, it may be very difficult to find the prime factors of $p - 1$, so the preceding theorem is less useful.)

Example. It is easy to check that 2 is a primitive root of 19. For $\phi(19) = 18$, and the only prime divisors of 18 are 2 and 3. Thus by (6.8), it is enough to show that $2^9 \not\equiv 1 \pmod{19}$ and $2^6 \not\equiv 1 \pmod{19}$. Since 19 is of the form $8k + 3$, 2 is a quadratic nonresidue of 19, and hence $2^9 \equiv -1 \not\equiv 1 \pmod{19}$. Also, $2^3 \equiv 8 \pmod{19}$, so $2^6 \equiv 7 \not\equiv 1 \pmod{19}$.

The following primality test is a partial converse to Fermat's Theorem and uses much the same idea as (6.8). It is used in testing large numbers m for primality in the special case when the prime factorization of $m - 1$ is known.

(6.9) Theorem (Lucas). *Let $m > 1$, and suppose there is an integer a such that $a^{m-1} \equiv 1 \pmod{m}$ and $a^{(m-1)/q} \not\equiv 1 \pmod{m}$ for every prime divisor q of $m - 1$. Then m is prime.*

Proof. By the same reasoning as in the proof of (6.8), we can show that a has order $m - 1$. Since ord $a \le \phi(m) \le m - 1$, it follows that $\phi(m) = m - 1$, and therefore m is prime.

Since a primitive root of m has order $\phi(m)$, (6.2) can be restated as follows.

(6.10) Theorem. *If g is a primitive root of m, then $g^s \equiv g^t \pmod{m}$ if and only if $s \equiv t \pmod{\phi(m)}$. Thus $g^s \equiv 1 \pmod{m}$ if and only if $\phi(m) | s$.*

One of the most important properties of a primitive root of m is that its powers form a reduced residue system modulo m. More precisely, we have the following.

(6.11) Theorem. *The set $g, g^2, g^3, \ldots, g^{\phi(m)}$ is a reduced residue system modulo m if and only if g is a primitive root of m. In particular, $g, g^2, \ldots, g^{p-1}$ are congruent, in some order, to the numbers $1, 2, 3, \ldots, p-1$ if and only if g is a primitive root of the prime p.*

Proof. Let g be a primitive root of m. Since there are $\phi(m)$ numbers in the set $g, g^2, \ldots, g^{\phi(m)}$, it is enough to show that each element is relatively prime to m and that no two of them are congruent modulo m. Since $(g, m) = 1$, it follows that $(g^k, m) = 1$ for each $k \ge 1$. Also, if $g^s \equiv g^t \pmod{m}$, (6.10) implies that $\phi(m) | s - t$. Since s and t are each between 1 and $\phi(m)$, we must have $s = t$. Thus the given set is a reduced residue system modulo m.

PRIMITIVE ROOTS

Conversely, suppose that the set $g, g^2, g^3, \ldots, g^{\phi(m)}$ is a reduced residue system modulo m. If $1 \leq d < \phi(m)$, we must have $g^d \not\equiv g^{\phi(m)} \pmod{m}$, and therefore g has order $\phi(m)$.

There is no easy general way of finding primitive roots of m even when they are known to exist. However, if one primitive root is known, say g, then all of the others can be found with the aid of the previous result, which allows us to write every primitive root of m as a suitable power of g, as follows.

(6.12) Theorem. *Suppose g is a primitive root of m. Then g^k is a primitive root of m if and only if $(k, \phi(m)) = 1$.*

Proof. Since $\text{ord } g = \phi(m)$, it follows from (6.3) that $\text{ord } g^k = \phi(m)/(k, \phi(m))$. Thus g^k will be a primitive root of m, that is, $\text{ord } g^k = \phi(m)$, if and only if $(k, \phi(m)) = 1$.

As the next result shows, if a positive integer has one primitive root, then it generally has quite a few. For example, 19 has six primitive roots, and 125 has 40 primitive roots. (That 125 has primitive roots will follow from (6.25).)

(6.13) Theorem. *Suppose that m has a primitive root. Then m has exactly $\phi(\phi(m))$ incongruent primitive roots.*

Proof. Let g be primitive root of m. Then by (6.11) and (6.12), h is a primitive root of m if and only if $h \equiv g^k \pmod{m}$ for some $k \leq \phi(m)$ with $(k, \phi(m)) = 1$. Since there are clearly $\phi(\phi(m))$ such k, the result follows.

Example. We have already shown that 2 is a primitive root of 19. Since $\phi(19) = 18$, it follows from (6.12) that all of the primitive roots of 19 are given by 2^k, where $1 \leq k \leq 18$ and $(k, 18) = 1$. There are precisely $\phi(18) = 6$ such values of k, namely, $k = 1, 5, 7, 11, 13$, and 17; hence the primitive roots of 19 are 2, 13, 14, 15, 3, and 10.

The idea of Theorems (6.12) and (6.13) can be generalized to characterize and find the number of elements of order d modulo m.

(6.14) Theorem. *Suppose g is a primitive root of m and d is a positive divisor of $\phi(m)$. Then g^k has order d modulo m if and only if k is of the form $j\phi(m)/d$, where $(j, d) = 1$. Thus there are exactly $\phi(d)$ incongruent elements of order d modulo m.*

Proof. By (6.3), the order of g^k is $\phi(m)/(k, \phi(m))$. This order is d if and only if $(k, \phi(m)) = \phi(m)/d$. Let $k = j\phi(m)/d$; then $(k, \phi(m)) = (j, d)\phi(m)/d$. This is equal to $\phi(m)/d$ if and only if $(j, d) = 1$. It is now easy to count the number of elements of order d: Since we can assume $1 \leq k \leq \phi(m)$, it follows that $1 \leq j \leq d$, and there are $\phi(d)$ such j relatively prime to d.

Power Residues and Indices

Suppose that m has a primitive root g. If $(a,m) = 1$, (6.11) implies that there is a unique integer i, with $1 \le i \le \phi(m)$, such that $g^i \equiv a \pmod{m}$. This fact is the basis of a technique that allows us to simplify calculations modulo m that involve only multiplication or exponentiation.

Throughout most of this section, m will denote a fixed positive integer that has primitive roots, and g will denote a fixed primitive root of m.

(6.15) Definition. Let g be a primitive root of m, and suppose $(a,m) = 1$. The smallest positive integer i such that $g^i \equiv a \pmod{m}$ is called the *index of a (to the base g)* and is denoted by $\operatorname{ind} a$.

Strictly speaking, the index of a is depends on both the modulus m and the primitive root g of m. However, since m and g are fixed in any application, the notation $\operatorname{ind} a$ should cause no confusion.

Although the idea of indices goes back to Euler, Gauss was the first mathematician to give a systematic discussion of them. In *Disquisitiones Arithmeticae*, Gauss introduced the term *index* and the notation $\operatorname{ind} a$. The following theorem states the most important properties of indices; the proofs are easy consequences of the definition of index and (6.2).

(6.16) Theorem. *Suppose g is a primitive root of m, and let $\operatorname{ind} a$ denote the index of a to the base g.*

(i) $\operatorname{ind} 1 \equiv 0 \pmod{\phi(m)}$; $\operatorname{ind} g \equiv 1 \pmod{\phi(m)}$.

(ii) $a \equiv b \pmod{m}$ *if and only if* $\operatorname{ind} a \equiv \operatorname{ind} b \pmod{\phi(m)}$.

(iii) $\operatorname{ind} ab \equiv \operatorname{ind} a + \operatorname{ind} b \pmod{\phi(m)}$.

(iv) $\operatorname{ind} a^k \equiv k \operatorname{ind} a \pmod{\phi(m)}$ *for any $k \ge 0$.*

The strong similarity between the properties of indices and the corresponding properties of logarithms is clear, but there is one important difference: The logarithm of a number is unique once the base is specified, whereas the index of a given integer depends also on the modulus m being used. Thus if the modulus is changed, then the indices must be recalculated, and hence a separate table of indices is required for each modulus of interest.

While indices are primarily of theoretical interest, they can be used to solve the polynomial congruences $bx^k \equiv c \pmod{m}$, where $(bc, m) = 1$. By multiplying this congruence by the multiplicative inverse of b modulo m, we can reduce it to an equivalent congruence of the form $x^k \equiv a \pmod{m}$. This leads us to the following definition, which generalizes the notion of quadratic residue.

(6.17) Definition. Let m be a positive integer and suppose $(a, m) = 1$. Then a is called a *kth power residue of m* if the congruence $x^k \equiv a \pmod{m}$ is solvable. If the congruence has no solutions, then a is called a *kth power nonresidue of m*.

The next result provides a way of deciding if a is a kth power residue of m.

(6.18) Theorem. *Let m be a positive integer having a primitive root, and suppose $(a, m) = 1$. Then the congruence $x^k \equiv a \pmod{m}$ has a solution if and only if*

$$a^{\phi(m)/(k,\phi(m))} \equiv 1 \pmod{m}. \tag{1}$$

If the congruence $x^k \equiv a \pmod{m}$ is solvable, then it has exactly $(k, \phi(m))$ incongruent solutions.

Proof. Let g be a primitive root of m, and let $d = (k, \phi(m))$. Taking indices, we see that the congruence $x^k \equiv a \pmod{m}$ holds if and only if $k \operatorname{ind} x \equiv \operatorname{ind} a \pmod{\phi(m)}$. By (2.7), this linear congruence is solvable for $\operatorname{ind} x$ if and only if $d \mid \operatorname{ind} a$, and if solutions exist, then there are exactly d incongruent solutions.

The proof is completed by showing that (1) holds if and only if $d \mid \operatorname{ind} a$. Taking indices, we see that (1) is equivalent to $(\phi(m)/d) \operatorname{ind} a \equiv 0 \pmod{\phi(m)}$, which holds if and only if $d \mid \operatorname{ind} a$.

Since every prime modulus has a primitive root, we have the following result.

(6.19) Corollary. *Suppose p is prime and $(a, p) = 1$. Then a is a kth power residue of p if and only if*

$$a^{(p-1)/(k,p-1)} \equiv 1 \pmod{p}.$$

Computational Note. While the preceding result gives an efficient procedure for determining *whether* a is a kth power residue of p, it is much more difficult to actually *find* a number b such that $b^k \equiv a \pmod{p}$. But if $(k, p - 1) = 1$, the calculation is relatively easy.

Using the Euclidean Algorithm, find positive integers s and t such that $sk = t(p - 1) + 1$. Then $a^{sk} = a^{t(p-1)+1} \equiv a \pmod{p}$. Thus a^s is a solution of the congruence $x^k \equiv a \pmod{p}$. In a similar way, if $d = (k, p - 1)$ and we have found a number b such that $b^d \equiv a \pmod{p}$, it is straightforward to find a solution of $x^k \equiv a \pmod{p}$. Unfortunately, it is not easy in general, given a divisor d of $p - 1$, to solve the congruence $x^d \equiv a \pmod{p}$.

The congruence $x^k \equiv 1 \pmod{m}$ obviously has a solution, and so it follows from (6.18) that if $k \mid \phi(m)$, there are exactly $(k, \phi(m)) = k$ solutions. This gives the following generalization of Corollary 4.8.

(6.20) Theorem. *Suppose that m has a primitive root. If $k \mid \phi(m)$, then the congruence $x^k - 1 \equiv 0 \pmod{m}$ has exactly k solutions.*

(6.21) Corollary. *Suppose that m has a primitive root. Then the number of incongruent kth power residues of m is $\phi(m)/(k, \phi(m))$.*

Proof. By (6.18), a is a kth power residue of m if and only if a is a solution of the congruence $x^{\phi(m)/(k,\phi(m))} \equiv 1 \pmod{m}$. But by (6.20), this congruence has $\phi(m)/(k, \phi(m))$ incongruent solutions.

If the congruence $x^k \equiv a \pmod{m}$ is solvable, indices can be used to find the solutions. To do this, however, we must compute (or have available) a table of indices for the given modulus. (In a supplement to *Disquisitiones Arithmeticae*, Gauss computed tables of indices for all integers less than 100 having primitive roots. In 1839, in *Canon Arithmeticus*, Jacobi published a table of indices for all prime powers less than 1000.)

The following example illustrates this technique.

Example. We will use indices to find all solutions of $7x^{10} \equiv 5 \pmod{13}$. We could first check that the congruence is solvable by using (6.18): Multiply each side by 2 to get the equivalent congruence $x^{10} \equiv 10 \pmod{13}$, and note that $10^{12/(10,12)} = 10^6 \equiv 3^6 = 27^2 \equiv 1 \pmod{13}$. Or we could simply use indices directly; if there are no solutions, this will be evident, since we then obtain a linear congruence that is not solvable.

Check that 2 is a primitive root of 13 (show that 2^4 and 2^6 are not congruent to 1 modulo 13). We set up a table of indices as follows:

a	1	2	3	4	5	6	7	8	9	10	11	12
ind a	12	1	4	2	9	5	11	3	8	10	7	6

Let y denote ind x; hence $x \equiv 2^y \pmod{13}$. Taking indices in the original congruence and using the properties in (6.16), we get the equivalent congruence $\text{ind}(7x^{10}) \equiv \text{ind } 7 + 10 \text{ ind } x \equiv \text{ind } 5 \pmod{12}$, that is, $11 + 10y \equiv 9 \pmod{12}$ or, equivalently, $10y \equiv 10 \pmod{12}$. *Be sure to note that the modulus in the linear congruence is $\phi(13) = 12$.* The congruence $10y \equiv 10 \pmod{12}$ is equivalent to $2y \equiv 2 \pmod{12}$. Now we can divide each side by 2, but the modulus changes to $12/(2, 12) = 6$. We get the equivalent congruence $y \equiv 1 \pmod{6}$, and thus $10y \equiv 10 \pmod{12}$ has the two solutions $y \equiv 1, 7 \pmod{12}$.

Finally, $x \equiv 2^y \equiv 2^1$ or $2^7 \pmod{13}$. Hence the only solutions of the original congruence are 2 and 11.

Notes. 1. It is worth pointing out that we can use the table of indices in this example to find the least positive residue of 2^7 (or indeed 2^j for any j between 1 and 12). The index of 2^7 is clearly 7, and since the table shows that the index of 11 is also 7, it follows that $2^7 \equiv 11 \pmod{13}$. The table of indices can be used, in fact, to find the least positive residue of a^k for any a relatively prime to 13. For example, to find the least positive residue of 5^7, note from the table that 5 has index 9, so the index of 5^7 is congruent to $7 \cdot 9$ modulo 12. Thus 5^7 has index 3. Using the table again, we find that 8 has index 3, so the least positive residue of 5^7 modulo 13 is 8.

2. If we use a different primitive root in the preceding example, the values of the indices will not be the same, but we will still obtain the same solutions.

The Existence of Primitive Roots

We have already seen that every prime has a primitive root, and it is easy to see that 1 and 4 also have primitive roots. To identify the positive integers that have primitive roots, we now consider the problem of finding primitive roots of p^k and $2p^k$, assuming that a primitive root of the odd prime p is known.

(6.22) Theorem. *Suppose that p is an odd prime.*

(i) If g is a primitive root of p and $g^{p-1} \not\equiv 1 \pmod{p^2}$, then g is a primitive root of p^2. If $g^{p-1} \equiv 1 \pmod{p^2}$, then $g + p$ is a primitive root of p^2.

(ii) If $k \geq 2$ and g is a primitive root of p^k, then g is a primitive root of p^{k+1}.

Proof. (i) Let h be the order of g modulo p^2; then $h \mid \phi(p^2) = p(p-1)$. But $g^h \equiv 1 \pmod{p^2}$ implies that $g^h \equiv 1 \pmod{p}$, and since g has order $p-1$ modulo p, $p-1$ must divide h by (6.2.i). Thus $h = p - 1$ or $h = p(p-1)$. If $h = p(p-1)$, then g is a primitive root of p^2. If $h = p - 1$, that is, if $g^{p-1} \equiv 1 \pmod{p^2}$, then g is not a primitive root of p^2. We show that, in this case, $g + p$ is a primitive root of p^2.

Since $g + p$ is congruent to g modulo p, $g + p$ is a primitive root of p. The preceding argument shows that the order of $g + p$ modulo p^2 must be $p - 1$ or $\phi(p^2)$. If the order is $p - 1$, then $(g + p)^{p-1} \equiv 1 \pmod{p^2}$. Using the Binomial Theorem, we get

$$1 \equiv (g+p)^{p-1} \equiv g^{p-1} + (p-1)pg^{p-2} \equiv 1 - pg^{p-2} \pmod{p^2}.$$

Hence $p^2 \mid pg^{p-2}$ and so $p \mid g^{p-2}$, that is, $p \mid g$, a contradiction since $(g, p) = 1$. Thus the order of $g + p$ modulo p^2 is $\phi(p^2)$, and hence $g + p$ is a primitive root of p^2.

(ii) Let h be the order of g modulo p^{k+1}; then $h \mid \phi(p^{k+1}) = p^k(p-1)$. Because $g^h \equiv 1 \pmod{p^{k+1}}$ implies $g^h \equiv 1 \pmod{p^k}$ and g is a primitive root of p^k, $\phi(p^k) = p^{k-1}(p-1)$ must divide h, by (6.2.i). Thus $h = p^{k-1}(p-1)$ or $h = p^k(p-1)$. We will show that $h \neq p^{k-1}(p-1)$.

Let $t = \phi(p^{k-1})$; then $g^t \equiv 1 \pmod{p^{k-1}}$ by Euler's Theorem, and so $g^t = 1 + jp^{k-1}$ for some integer j. If $p \mid j$, we would have $g^t \equiv 1 \pmod{p^k}$, which contradicts the fact that g is primitive root of p^k and therefore has order $\phi(p^k)$ modulo p^k. Thus $p \nmid j$. Since $tp = \phi(p^k)$, the Binomial Theorem implies that

$$g^{tp} = (1 + jp^{k-1})^p \equiv 1 + jp^k \pmod{p^{k+1}}.$$

(Here we use the fact that $p > 2$ and $k \geq 2$. The first neglected term in the binomial expansion is then $(p(p-1)/2)j^2 p^{2k-2}$ and so is divisible by p^{k+1} if $p \cdot p^{2k-2} \geq p^{k+1}$, that is, if $k \geq 2$.)

Thus $g^{\phi(p^k)} \not\equiv 1 \pmod{p^{k+1}}$, since $p \nmid j$. Hence $h \neq p^{k-1}(p-1)$ and therefore $h = p^k(p-1) = \phi(p^{k+1})$, which proves that g is a primitive root of p^{k+1}.

We summarize the previous facts about primitive roots in the following result, which implicitly contains a method for finding primitive roots of any power of an odd prime p if we are given a primitive root of p.

(6.23) Corollary. *Let p be an odd prime.*

(i) *If g is a primitive root of p, then g is a primitive root of p^k for every $k \geq 1$ if $g^{p-1} \not\equiv 1 \pmod{p^2}$. If $g^{p-1} \equiv 1 \pmod{p^2}$, then $g + p$ is a primitive root of p^k for every $k \geq 1$.*

(ii) *If g is primitive root of p^2, then g is a primitive root of p^k for every $k \geq 1$.*

Note. *A primitive root of p is not necessarily a primitive root of p^2.* For example, 14 is a primitive root of 29 but not of 29^2; check that $14^{28} \equiv 1 \pmod{29^2}$. Also, 18 is a primitive root of 37 but not of 37^2, and 19 is a primitive root of 43 but not of 43^2. These are the only examples with $p < 71$. (We are concerned with the primitive roots of p between 1 and $p-1$. Otherwise, there are examples for every odd p; for example, 8 is a primitive root of 3 but not of 9.)

There is a probability of $1 - 1/p$ that a primitive root g of p is a primitive root of p^2 (see Problem 6-73). It is therefore very unlikely, if p is large, that $g^{p-1} \equiv 1 \pmod{p^2}$. Thus *it is usually true that a given primitive root of p is a primitive root of p^2 and hence of p^k for every positive integer k.*

(6.24) Theorem. *Suppose that p is an odd prime, and let g be a primitive root of p^k. If g is odd, then g is also a primitive root of $2p^k$. If g is even, then $g + p^k$ is a primitive root of $2p^k$.*

Proof. If g is odd, then $g^j \equiv 1 \pmod{2}$ for every $j \geq 1$. Thus $g^j \equiv 1 \pmod{2p^k}$ if and only if $g^j \equiv 1 \pmod{p^k}$, and hence the order of g modulo $2p^k$ is equal to the order of g modulo p^k, namely, $\phi(p^k)$. Since $\phi(2p^k) = \phi(p^k)$, g is a primitive root of $2p^k$.

If g is even, then g cannot be a primitive root of $2p^k$, for a primitive root is always relatively prime to the modulus. But $g + p^k$ is odd and is plainly a primitive root of p^k, since it is congruent to g modulo p^k. Thus $g + p^k$ is a primitive root of $2p^k$ by the preceding argument.

The most important results about primitive roots were proved by Gauss, culminating in 1801 with the following characterization of the positive integers that have primitive roots.

(6.25) Theorem. *The only positive integers that have primitive roots are 1, 2, 4, p^k, and $2p^k$, where p is an odd prime.*

Proof. First suppose that $m > 2$ has a primitive root. It follows from (6.18), with $k = 2$, that $x^2 \equiv 1 \pmod{m}$ has only two solutions. Therefore by (5.5), m must be either 4, p^k, or $2p^k$, where p is an odd prime.

To prove the converse, first note that 1, 2, and 4 obviously have primitive roots (1, 1, and 3, respectively). If p is an odd prime, (6.7) implies that p has a primitive root. Thus by (6.23) and (6.24), p^k and $2p^k$ also have primitive roots.

The following is an easy consequence of (5.5) and the preceding theorem.

(6.26) Corollary. *Suppose that $m > 2$. Then the congruence $x^2 \equiv 1 \pmod{m}$ has exactly two solutions (namely, 1 and -1) if and only if m has a primitive root.*

PROBLEMS AND SOLUTIONS

The Order of an Integer

6-1. (a) Use the fact that 6 is a primitive root of 41 to find the least positive residues of all elements of order 8 modulo 41.

(b) *Find all positive integers less than* 61 *having order* 4 *modulo* 61. (2 *is a primitive root of* 61.)

Solution. (a) Since 6 is a primitive root of 41, it has order 40 modulo 41. If $(a, 41) = 1$, then $a \equiv 6^k \pmod{41}$ for some k with $1 \leq k \leq 40$, by (6.11). But (6.3) implies that 6^k has order $40/(k, 40)$. Thus 6^k has order 8 if and only if $(k, 40) = 5$. The only positive $k \leq 40$ for which $(k, 40) = 5$ are 5, 15, 25, and 35. The least positive residues modulo 41 of 6^5, 6^{15}, 6^{25}, and 6^{35} are 27, 3, 14, and 38, respectively.

(b) Argue as in (a) that 2^k has order $60/(k, 60)$ modulo 61. Thus 2^k has order 4 if and only if $(k, 60) = 15$, i.e., $k = 15$ or $k = 45$. The least positive residues of 2^{15} and 2^{45} modulo 61 are 11 and 50.

6-2. *If p is an odd prime, prove that the order of a modulo p is 2 if and only if $a \equiv -1 \pmod{p}$.*

Solution. It is clear that if $a \equiv -1 \pmod{p}$, then a has order 2. Conversely, if a has order 2, then a is a solution of the congruence $x^2 \equiv 1 \pmod{p}$. But this congruence has only two solutions, namely, 1 (which has order 1) and -1.

6-3. *Prove or disprove: Suppose p is an odd prime and $a \not\equiv \pm 1 \pmod{p}$. If $a^r \equiv 1 \pmod{p}$, where $r < p - 1$, then $r | p - 1$.*

Solution. It is easy to find counterexamples to the assertion. For instance, let $p = 13$ and let a be an element of order 4, such as 5. Then $a^4 \equiv 1 \pmod{13}$ and hence $a^8 \equiv 1 \pmod{13}$, but 8 does not divide 12.

Note. Theorem 6.2 asserts only that the *order* of a divides $p - 1$.

6-4. *Let m be a positive integer, and suppose that $ab \equiv 1 \pmod{m}$. Prove that $\operatorname{ord} a = \operatorname{ord} b$.*

Solution. Note that $\operatorname{ord} a$ exists, for since $ab \equiv 1 \pmod{m}$, it follows that $(a, m) = 1$. To show that a and b have the same order, raise each side of the congruence $ab \equiv 1 \pmod{m}$ to the kth power to get $a^k b^k \equiv 1 \pmod{m}$. It follows that $a^k \equiv 1 \pmod{m}$ if and only if $b^k \equiv 1 \pmod{m}$. Thus the *least* positive k such that $a^k \equiv 1 \pmod{m}$ is also the least positive k such that $b^k \equiv 1 \pmod{m}$. Hence $\operatorname{ord} a = \operatorname{ord} b$.

6-5. *Suppose that $n > 1$. Prove that $\phi(2^n - 1)$ is divisible by n. (Hint. Show that 2 has order n modulo $2^n - 1$.)*

Solution. Clearly, $2^n \equiv 1 \pmod{2^n - 1}$, and if $1 \leq k < n$, then $2 \leq 2^k < 2^n$, so $2^k \not\equiv 1 \pmod{2^n - 1}$. It follows from the definition of order that 2 has order n modulo $2^n - 1$, and so (6.2.i) implies that $n | \phi(2^n - 1)$.

6-6. *Let g be a primitive root of the prime p. If $p \equiv 3 \pmod{4}$, prove that the order of $-g$ modulo p is $(p - 1)/2$. (Hint. Use Euler's Criterion.)*

Solution. Note that g cannot be a quadratic residue of p, by Euler's Criterion. Thus $g^{(p-1)/2} \equiv -1 \pmod{p}$, and therefore $-g \equiv g^{(p-1)/2} g \pmod{p}$. If $p = 4k + 3$, we have $-g \equiv g^{2k+2} \pmod{p}$. But $(2k + 2, 4k + 2) = 2$, so it follows from (6.3) that $\operatorname{ord} g^{2k+2} = (p - 1)/2$, and therefore the order of $-g$ is $(p - 1)/2$.

6-7. *Suppose p is a prime of the form $4k+3$. Use (6.4) to show that if $-a$ has order $(p-1)/2$ modulo p, then a is a primitive root of p.*

Solution. Since -1 has order 2, which is relatively prime to $(p-1)/2$, it follows from (6.4) that $a = (-1)(-a)$ has order $p-1$, and so a is a primitive root of p.

6-8. *Let p be an odd prime, and suppose that the order of a modulo p is $2k$. Prove that $a^k \equiv -1 \pmod{p}$.*

Solution. Since $a^{2k} \equiv 1 \pmod{p}$, a^k is a solution of the congruence $x^2 \equiv 1 \pmod{p}$. This congruence has two solutions, 1 and -1. Since a has order $2k$, $a^k \not\equiv 1 \pmod{p}$. It follows that $a^k \equiv -1 \pmod{p}$.

6-9. *Prove or disprove: If $(a,m) = 1$ and a has order $m-1$ modulo m, then m is prime.*

Solution. The result is true. The order of a cannot be larger than $\phi(m)$. Since a has order $m-1$, it follows that $\phi(m) \geq m-1$, and hence $\phi(m) = m-1$. But $\phi(m) = m-1$ implies that m is a prime (see Problem 3-59).

6-10. *Suppose $(a,m) = 1$. If $a^h \equiv 1 \pmod{m}$ and $a^k \equiv 1 \pmod{m}$, prove that $a^{(h,k)} \equiv 1 \pmod{m}$.*

Solution. By (6.2.i), the order of a divides h and k, and hence divides (h,k). Thus $a^{(h,k)} \equiv 1 \pmod{m}$, again by (6.2.i).

6-11. *Prove or disprove: If $d \mid \phi(m)$, then there exists an element of order d modulo m.*

Solution. The assertion is false in general. For example, let $m = 12$. It is easy to see that if $(a, 12) = 1$, then $a^2 \equiv 1 \pmod{12}$, so $\operatorname{ord} a \leq 2$ for all a. In particular, since $\phi(12) = 4$, there is no element of order $\phi(12)$.

6-12. *Show that if a has order $4r$ modulo m, then $-a$ has order $4r$ modulo m.*

Solution. Since $(-a)^{4r} = a^{4r} \equiv 1 \pmod{m}$, (6.2.i) implies that the order of $-a$ divides $4r$. If the order of $-a$ is the even number $2s < 4r$, then $a^{2s} = ((-1)(-a))^{2s} = (-a)^{2s} \equiv 1 \pmod{m}$, contradicting the fact that a has order $4r$. If the order of $-a$ is the odd number t, then $t \mid 4r$ and hence $t \mid r$, and therefore $(-a)^r \equiv 1 \pmod{m}$. It follows that $a^{2r} = ((-1)(-a))^{2r} = (-a)^{2r} \equiv 1 \pmod{m}$, contradicting the fact that a has order $4r$ modulo m.

▷ **6-13.** *Let $a_1, a_2, \ldots, a_r$ be a reduced residue system modulo m, and let $n \geq 1$. Prove that $a_1^n, a_2^n, \ldots, a_r^n$ is a reduced residue system modulo m if and only if $(n, \phi(m)) = 1$.*

Solution. It is clear that each a_i^n is relatively prime to m; thus it is enough to prove that $a_1^n, a_2^n, \ldots, a_r^n$ are incongruent modulo m. We first show that if $(n, \phi(m)) = 1$ and $a^n \equiv b^n \pmod{m}$, then $a \equiv b \pmod{m}$. Let s and t be nonnegative integers such that $ns = t\phi(m) + 1$. Then $a^{ns} \equiv b^{ns}$ (mod m). But $a^{ns} = a^{t\phi(m)+1} \equiv a$ (mod m), and similarly, $b^{ns} \equiv b$ (mod m), and therefore $a \equiv b$ (mod m).

We prove now that if $(n, \phi(m)) \neq 1$, then there exist a and b that are relatively prime to m, incongruent modulo m, but such that $a^n \equiv b^n$ (mod m). Let p be a prime divisor of $(n, \phi(m))$. We show that there is an integer whose order modulo m is p. Obviously, -1 has order 2, so we may assume that p is odd. Let $m = \prod p_i^{a_i}$; then p divides $\phi(p_i^{a_i})$ for some i. If g is a primitive root of $p_i^{a_i}$, then the order of g modulo m is a multiple of the order of g modulo $p_i^{a_i}$, and thus the order of g modulo m is a multiple of p. By (6.3.ii), we can find an integer a whose order modulo m is p. Let $b = 1$; then $a^n \equiv 1 \equiv b^n$ (mod m), but $a \not\equiv b$ (mod m).

Note. If m has a primitive root, then a considerably shorter proof can be given. See Problem 6-65.

Power Residues and Indices

6-14. (a) *Without finding the solutions, show that $x^6 \equiv 8$ (mod 89) is solvable. How many solutions are there?*

(b) *Use the fact that 3 is a primitive root of 89 to find the least positive residues of the solutions of $x^6 \equiv 8$ (mod 89).*

Solution. (a) By (6.18), the congruence is solvable if and only if $8^{44} \equiv 1$ (mod 89). Direct computation of 8^{44} modulo 89 is slightly unpleasant, but it can be bypassed. Since 89 is of the form $8k + 1$, 2 is a quadratic residue of 89, and hence by Euler's Criterion, $2^{44} \equiv 1$ (mod 89). Thus $8^{44} = 2^{44 \cdot 3} \equiv 1$ (mod 89), and it follows that the congruence $x^6 \equiv 8$ (mod 89) has a solution. By (6.18), it has $(6, \phi(89)) = 2$ solutions.

(b) By (6.18), x is a solution of the congruence if and only if $6 \operatorname{ind} x \equiv \operatorname{ind} 8$ (mod 88). We now want to compute $\operatorname{ind} 8 = 3 \operatorname{ind} 2$. It is easy to verify that $\operatorname{ind} 2 = 16$, so we solve the congruence $6k \equiv 3 \cdot 16$ (mod 88); this has the solution $k \equiv 8$ (mod 44). Thus the solutions of the original congruence are $x \equiv 3^8$ (mod 89) and $x \equiv 3^{52}$ (mod 89). The least positive residue of the first solution is 64. But if x is a solution, so is $-x$, so the least positive residue of the second solution is 25.

6-15. *Solve the congruence $x^5 \equiv 2$ (mod 73). (Hint. See the Computational Note following (6.19).)*

Solution. Since 5 is relatively prime to 72, we can find positive integers s and t such that $5s = 72t + 1$, for example, $s = 29$ and $t = 2$. Therefore $2^{5s} = 2^{72t+1} \equiv 2$ (mod 73), and so 2^s is a solution of the congruence $x^5 \equiv 2$ (mod 73). Thus 2^{29} is a solution of the given congruence, and since $(5, 72) = 1$, it is the only solution. Straightforward calculation shows that the least positive residue of 2^{29} modulo 73 is equal to 4.

6-16. *Use a table of indices to find the least positive residues of the solutions of $9x^8 \equiv 8$ (mod 17).*

Solution. It turns out that 3 is a primitive root of 17. This is evident from the following table of indices:

a	1	2	3	4	5	6	7	8	9	10	11	12	13	14	15	16
ind a	16	14	1	12	5	15	11	10	2	3	7	13	4	9	6	8

If we take indices, the congruence $9x^8 \equiv 8 \pmod{17}$ becomes $\text{ind } 9 + 8 \text{ ind } x \equiv \text{ind } 8 \pmod{16}$. From the above table, we get $2 + 8 \text{ ind } x \equiv 10 \pmod{16}$ or, equivalently, $\text{ind } x \equiv 1 \pmod 2$. Thus the solutions are those integers with odd index. From the table, the least positive residues of the solutions are 3, 10, 5, 11, 14, 7, 12, and 6.

6-17. *The congruence $4x^{12} \equiv -23 \pmod{29}$ has four solutions. Use a primitive root of 29 to find the least positive residue of three of them.*

Solution. See Problem 6-34 for a verification that 2 is a primitive root of 29. To avoid the work of preparing a full index table for 29, we do some preliminary manipulation of the given congruence. Since $-23 \equiv 6 \pmod{29}$, we obtain the equivalent congruence $4x^{12} \equiv 6 \pmod{29}$. Divide both sides by 2, replace 3 by -26, divide by 2 again, and replace -13 by 16. We obtain the equivalent congruence $x^{12} \equiv 16 \pmod{29}$. Taking indices, we get the congruence $12 \text{ ind } x \equiv 4 \pmod{28}$ or, equivalently, $3 \text{ ind } x \equiv 1 \pmod 7$. Thus $\text{ind } x \equiv 5 \pmod 7$, and so x has index 5, 12, 19, or 26. An element with index 5 is 2^5, having least positive residue 3. To calculate 2^{12}, note that $2^{12} = 2^5 \cdot 2^7 \equiv 3 \cdot 12 \equiv 7 \pmod{29}$. For the other two solutions, observe that if x is a solution, so is $-x$. So the solutions are 3, 7, 26, and 22. (Note. Each successive solution can be found by multiplying a given solution by $2^7 \equiv 12 \pmod{29}$, since successive exponents differ by 7.)

6-18. *Determine if $x^5 \equiv 6 \pmod{101}$ has solutions. If solutions exist, find them. (Hint. First find the least positive residue of 2^{70} modulo 101.)*

Solution. We calculate the least positive residue of 2^{70} modulo 101, obtaining 6. Thus 2^{14} is a solution of the given congruence, thereby settling the question of existence. If 2 is a primitive root, by taking indices we obtain the equivalent congruence $5 \text{ ind } x \equiv 70 \pmod{100}$, and hence $\text{ind } x$ is 14, 34, 54, 74, or 94. Thus $x = 22, 70, 85, 96,$ or 30. (In fact, we do not need to *know* that 2 is a primitive root. For by using (6.18), we can see that if the congruence has a solution, it has 5 of them, and we have found 5 solutions.)

6-19. *Use the fact that 2 is a primitive root of 59 to find the least positive residues of all solutions of $5x^5 \equiv -16 \pmod{59}$.*

Solution. Since $5 \equiv 64 \pmod{59}$, 5 has index 6. Also, since 2 is a quadratic nonresidue of 59, we have $2^{29} \equiv -1 \pmod{59}$; thus -1 has index 29, and so -16 has index $4 + 29 = 33$. Taking indices, we obtain the equivalent congruence $6 + 5 \text{ ind } x \equiv 33 \pmod{58}$, i.e., $5 \text{ ind } x \equiv 27 \pmod{58}$. Solving, we obtain $\text{ind } x \equiv 17 \pmod{58}$. Thus $x \equiv 2^{17} \pmod{59}$ is the only solution; the least positive residue of the solution x is 33.

6-20. *Show that solutions of $x^5 \equiv 26 \pmod{71}$ exist, and find the least positive residues of the solutions. (Hint. 7 is a primitive root of 71 and $\text{ind } 26 = 45$.)*

Solution. It is easy to verify that $7^{45} \equiv 26 \pmod{71}$. First note that 7 is a quadratic nonresidue of 71, and thus $7^{35} \equiv -1 \pmod{71}$. Now we need only look at 7^{10}.

Since $7^3 \equiv -12 \pmod{71}$, we have $7^6 \equiv 2 \pmod{71}$, $7^9 \equiv -24 \pmod{71}$, and thus $7^{10} \equiv -168 \equiv -26 \pmod{71}$.

Since $7^{45} \equiv 26 \pmod{71}$, it is clear that 7^9 is a solution of the congruence $x^5 \equiv 26 \pmod{71}$. By (6.18), there are 5 solutions. We can use indices to obtain the equivalent congruence $5\operatorname{ind} x \equiv 45 \pmod{70}$, i.e., $\operatorname{ind} x \equiv 9 \pmod{14}$; therefore $\operatorname{ind} x = 9, 23, 37, 51$, or 65. Since $7^{14} \equiv 54 \pmod{71}$, we can calculate fairly quickly that the least positive residues of the solutions are 47, 53, 22, 52, and 39.

6-21. *Use a table of indices for a primitive root of 19 to find the least positive residues of the solutions to the following congruences:*
(a) $8x^4 \equiv 3 \pmod{19}$; (b) $5x^3 \equiv 2 \pmod{19}$; (c) $x^7 \equiv 1 \pmod{19}$.

Solution. We use the primitive root 2 and obtain the following table.

a	1	2	3	4	5	6	7	8	9	10	11	12	13	14	15	16	17	18
$\operatorname{ind} a$	18	1	13	2	16	14	6	3	8	17	12	15	5	7	11	4	10	9

(a) Taking indices, we obtain the equivalent congruence $3+4\operatorname{ind} x \equiv 13 \pmod{18}$, i.e., $2\operatorname{ind} x \equiv 5 \pmod 9$. Thus $\operatorname{ind} x \equiv 7 \pmod 9$, and so $\operatorname{ind} x$ can be 7 or 16. Hence the solutions are 2^7 and 2^{16}, and thus the least positive residues of the solutions are 14 and 5.

(b) We obtain the equivalent congruence $16+3\operatorname{ind} x \equiv 1 \pmod{18}$. Thus $\operatorname{ind} x \equiv 1 \pmod 6$, and so $\operatorname{ind} x = 1, 7$, or 13. This gives solutions 2, 14, and 3.

(c) We obtain the equivalent congruence $7\operatorname{ind} x \equiv 18 \pmod{18}$. Thus x has index 18, giving 1 as the least positive residue of the only solution.

6-22. *Use primitive roots and the properties of indices to find the remainder when 5^{4^3} is divided by 19. (Hint. Use the table in the preceding problem.)*

Solution. To use indices, observe that 2 is a primitive root of 19, and use the table of Problem 6-21. Note that 5 has index 16; if $x = 5^{4^3}$, then $x \equiv 2^{16 \cdot 4^3} \pmod{19}$, and so $\operatorname{ind} x \equiv 16 \cdot 4^3 \pmod{18}$, by (6.16.ii). Since $16 \equiv -2 \pmod{18}$ and $4^2 \equiv -2 \pmod{18}$, we have $\operatorname{ind} x = 16$ and so the required remainder is 5.

6-23. *Suppose p is an odd prime.*
(a) *Prove that the congruence $x^4 \equiv -1 \pmod p$ is solvable if and only if $p \equiv 1 \pmod 8$.*
(b) *Prove that there are infinitely many primes of the form $8k+1$. (Hint. Let $p_1, p_2, \ldots, p_n$ be primes of the form $8k+1$, and consider $(2p_1 p_2 \cdots p_n)^4 + 1$.)*

Solution. (a) By (6.19), the congruence has a solution if and only if $(-1)^{(p-1)/(4, p-1)} \equiv 1 \pmod p$, i.e., if and only if $(p-1)/(4, p-1)$ is even. Let $p = 8k+r$, where $r = 1, 3, 5$, or 7. Then $(p-1)/(4, p-1)$ is equal to $2k, 4k+1, 2k+1$, and $4k+3$ respectively, and so is even if and only if $r = 1$.

(b) Let $p_1, p_2, \ldots, p_n$ be primes of the form $8k+1$, and let $N = (2p_1 p_2 \cdots p_n)^4 + 1$. Let p be a prime divisor of N; since N is odd, p is odd. Then $2p_1 p_2 \cdots p_n$ is a solution of the congruence $x^4 + 1 \equiv 0 \pmod p$, and thus, by part (a), p is of the form $8k+1$.

But clearly p cannot be one of the p_i. Thus for any n, there are more than n primes of the form $8k+1$, and hence there are infinitely many such primes.

Note. Essentially the same proof shows that if p is an odd prime, then the congruence $x^{2^{n-1}} \equiv -1 \pmod p$ is solvable if and only if $p \equiv 1 \pmod{2^n}$. It follows that there are infinitely many primes of the form $2^n k + 1$ for any positive integer n.

6-24. Prove or disprove: *If p is a prime of the form $3k+2$, then a is a cubic residue of p (that is, $x^3 \equiv a \pmod p$ is solvable) for every positive integer a less than p.*

Solution. The statement is true. By (6.19), a is a cubic residue of $p = 3k+2$ if and only if $a^{(p-1)/(3,p-1)} \equiv 1 \pmod p$. But $(3, p-1) = (3, 3k+1) = 1$, and if $(a, p) = 1$, then $a^{p-1} \equiv 1 \pmod p$ by Fermat's Theorem.

6-25. *If $p = 3k+1$ is prime, prove that precisely one-third of the integers $1, 2, \ldots, p-1$ are cubic residues of p.*

Solution. By (6.21), the number of cubic residues of p (in a reduced residue system) is $\phi(p)/(3, \phi(p)) = 3k/(3, 3k) = k$, and k is precisely one-third of $p-1$.

6-26. *If $p = 4k+3$ is prime, show that exactly half of the integers $1, 2, \ldots, p-1$ are fourth power residues of p.*

Solution. By (6.21), the number of fourth power residues of p (in a reduced residue system) is $\phi(p)/(4, \phi(p)) = (4k+2)/2$, which is precisely one-half of $p-1$.

Another proof: We show that every quadratic residue of p is a fourth power residue of p (the converse is obvious). Let a be a quadratic residue of p, and suppose $b^2 \equiv a \pmod p$; it follows that $(-b)^2 \equiv a \pmod p$. But exactly one of b or $-b$ is a quadratic residue of p, since $(-b/p) = (-1/p)(b/p) = -(b/p)$. If $c^2 \equiv \pm b \pmod p$, then $c^4 \equiv a \pmod p$.

6-27. *Let p be a prime of the form $6k-1$. For which values of a does the congruence $x^3 \equiv a \pmod p$ have a unique solution?*

Solution. Clearly, there is a unique solution if $a \equiv 0 \pmod p$. Now suppose $(a, p) = 1$. Let g be a primitive root of p. Then by (6.16), $u^3 \equiv a \pmod p$ if and only if $3 \operatorname{ind} u \equiv \operatorname{ind} a \pmod{p-1}$. But since $(3, p-1) = 1$, the congruence $3y \equiv \operatorname{ind} a \pmod{p-1}$ has a unique solution modulo $p-1$, and thus the congruence $x^3 \equiv a \pmod p$ has a unique solution.

Another proof: There is a solution if $a \equiv 0 \pmod p$. Using (6.21), we can show that there are $p-1$ cubic residues of p. Thus the congruence $x^3 \equiv a \pmod p$ also has a solution whenever $(a, p) = 1$, and it follows that none of the congruences can have more than one solution.

▷ **6-28.** *Suppose g is a primitive root of the odd prime p and $p \nmid a$. Prove that $x^n \equiv a \pmod p$ has solutions if and only if $a \equiv g^{dk} \pmod p$, where $d = (n, p-1)$ and k is some integer with $1 \le k \le (p-1)/d$.*

Solution. By (6.18), a is an nth power residue of p if and only if $a^{(p-1)/d} \equiv 1 \pmod{p}$, where $d = (n, p-1)$. If $a \equiv g^{dk} \pmod{p}$, then $a^{(p-1)/d} \equiv g^{(p-1)k} \equiv 1 \pmod{p}$, so a is an nth power residue of p.

Conversely, suppose that a is an nth power residue of p. Then $a \equiv b^n \pmod{p}$ for some b. Use (6.11) to write $b \equiv g^t \pmod{p}$; then $a \equiv g^{nt} \pmod{p}$. If r is the remainder when nt is divided by $p - 1$, then $a \equiv g^r \pmod{p}$, by Fermat's Theorem. It is easy to see that $d \mid r$. If we set $r = dk$, then $1 \leq k \leq (p-1)/d$. Thus $a \equiv g^{dk} \pmod{p}$ and the result follows.

6-29. *Find all integers x such that (a) $3^x \equiv 7 \pmod{17}$; (b) $3^x \equiv x \pmod{17}$. (Hint. Use the table in Problem 6-16.)*

Solution. (a) We work with the primitive root 3. From the table, we find that 7 has index 11. Thus we obtain the equivalent congruence $x \equiv 11 \pmod{16}$, whose solutions x are all positive integers of the form $11 + 16k$.

(b) The congruence $3^x \equiv x \pmod{17}$ is a bit more complicated. Let $x = r + 16q$, where $0 \leq r \leq 15$. Then since $3^{16} \equiv 1 \pmod{17}$, $3^x \equiv 3^r \pmod{17}$. So our congruence reduces to $3^r - r \equiv 16q \pmod{17}$ or, equivalently, $q \equiv r - 3^r \pmod{17}$. Thus the solutions are all the numbers x of the form $r+16(r-3^r+17n) = 17r-16\cdot 3^r+272n$, where n has to be chosen to make $x \geq 0$. There are 16 families of solutions, corresponding to $r = 0, 1, \ldots, 15$. It is tedious to list them all, so we use one to illustrate matters. Take $r = 5$; then we get $x = 272n - 3803$, where $n > 13$, or, equivalently, $x = 272n - 267$, where n is positive.

6-30. *Does the congruence $x^{13} \equiv 10 \pmod{1323}$ have a solution? (Hint. Use (6.18), but note that 1323 does not have a primitive root.)*

Solution. Since $1323 = 3^3 \cdot 7^2$, we determine whether there are solutions modulo 3^3 and 7^2. By (6.18), since 27 has a primitive root, the congruence $x^{13} \equiv 10 \pmod{27}$ has a solution if only if $10^{18} \equiv 1 \pmod{27}$. (This follows from Euler's Theorem.) Similarly, $x^{13} \equiv 10 \pmod{49}$ has a solution, since $10^{42} \equiv 1 \pmod{49}$. Thus the congruence $x^{13} \equiv 10 \pmod{1323}$ has a solution, by the Chinese Remainder Theorem.

6-31. *Prove the following observation of Gauss: If p and q are prime, with $p = 2q + 1$, then for any positive integer k, the congruence $x^{2k} \equiv 1 \pmod{p}$ has either two solutions or $p - 1$ solutions.*

Solution. By (6.18), the number of solutions of the congruence is $(2k, p-1) = 2(k, q)$. Since q is prime, we have either $(k, q) = 1$ or $(k, q) = q$, and the result follows.

▷ **6-32.** *Let p be an odd prime. Prove that there are infinitely many primes of the form $2kp + 1$. (Hint. Let $a = pp_1p_2\cdots p_n$, where each p_i is of the form $2kp + 1$, and consider prime divisors of $(a^p - 1)/(a - 1)$. First show that $a - 1$ and $(a^p - 1)/(a - 1)$ are relatively prime.)*

Solution. Let $p_1, p_2, \ldots, p_n$ be primes of the form $2kp + 1$; we show that there is a prime of the form $2kp + 1$ different from the p_i. Let $a = pp_1p_2\cdots p_n$, and let $N = (a^p - 1)/(p - 1)$. We show first that $(a - 1, N) = 1$. If q is a prime that divides

$a-1$, then $a \equiv 1 \pmod{q}$ and thus $N \equiv p \pmod{q}$, since $N = a^{p-1} + a^{p-2} + \cdots + a + 1$. Since $p|a$, we have $q \neq p$, and therefore q cannot divide N.

We show next that any prime q that divides N must be of the form $2kp+1$. Since $q | a^p - 1$, we have $a^p \equiv 1 \pmod{q}$. Thus a has order 1 or p modulo q. But ord $a \neq 1$, since $q \nmid a - 1$. Therefore ord $a = p$, and hence, by (6.2.i), p divides $\phi(q) = q - 1$. Thus $q - 1 = mp$ for some integer m, and since N is odd, m must be even. Clearly, q is different from each p_i, for otherwise $q|1$.

Primitive Roots

6-33. *Prove that 3 is a primitive root of 17. Find all incongruent primitive roots of 17 between 1 and 16.*

Solution. We show that 3 has order 16 modulo 17. The order of 3 divides $\phi(17) = 16$, so it is enough to check that $3^8 \not\equiv 1 \pmod{17}$. This can be done by a direct computation or by observing that 3 is a quadratic nonresidue of 17, since $(3/17) = (17/3) = (2/3) = -1$. Thus by Euler's Criterion, $3^8 \equiv -1 \pmod{17}$.

By (6.12), the primitive roots of 17 are congruent modulo 17 to 3^k, where $(k, 16) = 1$. Thus there are 8 primitive roots, namely, $3^1, 3^3, \ldots, 3^{15}$. The least positive residues of these are 3, 10, 5, 11, 14, 7, 12, and 6, respectively.

6-34. *Find all of the primitive roots of 29 between 1 and 28.*

Solution. First note that 2 is a primitive root of 29. To verify this, it is enough, by (6.8), to show that $2^{14} \not\equiv 1 \pmod{29}$ and $2^4 \not\equiv 1 \pmod{29}$. The first congruence can be dealt with easily: Since 29 is of the form $4k+1$, -1 is a quadratic residue of 29, and hence $2^{14} \equiv -1 \pmod{29}$. It is obvious that $2^4 \not\equiv 1 \pmod{29}$. Thus 2 is a primitive root of 29.

By (6.12), the primitive roots of 29 are precisely the numbers congruent modulo 29 to 2^k, where $(k, 28) = 1$. To save some work, we can use the fact that $2^{14} \equiv -1 \pmod{29}$. It follows that the primitive roots of 29 are 2, 8, 3, 19, 18, 14, 27, 21, 26, 10, 11, and 15. (Note that $27 \equiv -2 \pmod{29}$, $21 \equiv -8 \pmod{29}$, and so on.)

6-35. *Which integers $m > 1$ have exactly one primitive root?*

Solution. If m has a primitive root, then by (6.13), m has $\phi(\phi(m))$ primitive roots. Note that the only positive integers k such that $\phi(k) = 1$ are $k = 1$ and $k = 2$. Thus we want to find all integers $m > 1$ such that $\phi(m) = 1$ or $\phi(m) = 2$. It is easy to see that m can only be 2, 3, 4, or 6. For each of these, -1 is the only primitive root.

6-36. *Determine which of the following integers have primitive roots: 198, 199, 200, 201, 202, 203.*

Solution. By (6.25), we need to determine which of the given numbers is of the form p^n or $2p^n$, where p is an odd prime. The only ones are 199 and 202.

6-37. *How many quadratic nonresidues of 47 are not primitive roots of 47?*

Solution. There are 23 quadratic nonresidues of 47, and $\phi(46) = 22$ primitive roots of 47. Since any primitive root must be a quadratic nonresidue, it follows that there is one quadratic nonresidue that is not a primitive root. That quadratic nonresidue is easy to identify: Since 47 is of the form $4k + 3$, -1 is a quadratic nonresidue of 47 but is clearly not a primitive root of 47.

▷ **6-38.** *Let g be a primitive root of m. Prove that the quadratic residues of m are congruent to the even powers of g and the quadratic nonresidues are congruent to the odd powers of g.*

Solution. The result is obvious for $m = 1$ and $m = 2$, so let $m \geq 3$. By (6.11), any element of a reduced residue system is congruent modulo m to a power of g. If $a \equiv g^{2d} \pmod{m}$, then clearly, a is a quadratic residue of m, since $a \equiv (g^d)^2 \pmod{m}$. Conversely, we show that if $a = g^{2d+1}$ is an odd power of g, then a cannot be a quadratic residue of m. For if a is a quadratic residue of m, then $a \equiv b^2 \pmod{m}$ for some b. Suppose that $b \equiv g^e \pmod{m}$. Then $a \equiv g^{2e} \pmod{m}$, and therefore $g^{2d+1} \equiv g^{2e} \pmod{m}$. By (6.2.ii), $\phi(m)$ must then divide $2d + 1 - 2e$. This is impossible, since examination of the usual formula for $\phi(m)$ shows that $\phi(m)$ is even if $m > 2$.

6-39. *If $m > 2$ has a primitive root, show that a quadratic nonresidue a of m satisfies $a^{\phi(m)/2} \equiv -1 \pmod{m}$. (Hint. Use (6.18) and (6.26).)*

Solution. If we take $k = 2$ in (6.18), it follows that $a^{\phi(m)/2} \not\equiv 1 \pmod{m}$. But since $a^{\phi(m)/2}$ is a solution of $x^2 \equiv 1 \pmod{m}$, (6.20) implies that $a^{\phi(m)/2}$ must be congruent to -1 modulo m.

6-40. *Suppose m does not have a primitive root. Use (6.2) and (6.25) to prove that $a^{\phi(m)/2} \equiv 1 \pmod{m}$ for every a relatively prime to m.*

Solution. Let $(a, m) = 1$ and suppose m is odd. By (6.25), m has at least two prime factors. Thus we can write $m = rs$, where $r > 1$, $s > 1$, and $(r, s) = 1$; then $\phi(m) = \phi(r)\phi(s)$. Since $\phi(r)$ and $\phi(s)$ are both even, $\phi(m)/2$ is a multiple of each of them. Therefore by Euler's Theorem, $a^{\phi(m)/2} - 1$ is divisible by r and by s, and since $(r, s) = 1$, it follows that $a^{\phi(m)/2} - 1$ is divisible by m.

Now suppose m is even and let $m = 2^b c$, where c is odd. If $c = 1$, (6.25) implies that $b \geq 3$ and so $\phi(m) = 2^{b-1}$. Since m does not have a primitive root, we have ord $a < \phi(m)$, and it follows from (6.2.i) that ord $a \mid 2^{b-2}$ and hence $a^{\phi(m)/2} \equiv 1 \pmod{m}$. Finally, if $c > 1$, then $\phi(m)/2$ is a multiple of both 2^b and $\phi(c)$, and we again conclude from Euler's Theorem that $a^{\phi(m)/2} \equiv 1 \pmod{m}$.

6-41. *If g^k is a primitive root of m, prove that g is also a primitive root of m.*

Solution. By (6.3), ord $g^k = $ ord $g/(k, \text{ord } g)$; in particular, ord $g^k \leq$ ord g. If g^k is a primitive root of m, then g^k has order $\phi(m)$. It follows that ord $g \geq \phi(m)$, and hence ord $g = \phi(m)$, i.e., g is a primitive root of m.

Alternatively, if g^k is a primitive root of m, then by (6.11), any b relatively prime to m is congruent modulo m to a power of g^k. But then any such b is congruent modulo m to a power of g, and thus g is a primitive root of m.

6-42. *Let q be an odd prime, and suppose that $p = 2q + 1$ is also prime. Prove that there exist q quadratic nonresidues and $q - 1$ primitive roots of p. Thus show that the primitive roots of p are just the quadratic nonresidues of p, with one exception. What is the exception?*

Solution. This is a generalization of Problem 6-37. There are $(p-1)/2 = q$ quadratic nonresidues of p and $\phi(p-1) = \phi(2q) = q - 1$ primitive roots of p. Since every primitive root of p is a quadratic nonresidue of p, exactly one quadratic nonresidue of p fails to be a primitive root of p. Since q is odd, it is of the form $2k + 1$. Thus $p = 2q + 1$ is of the form $4k + 3$, and therefore -1 is a quadratic nonresidue of p. But clearly, -1 has order 2, and so -1 is not a primitive root of p.

Note. It is not known whether there are infinitely many primes of the form $2q + 1$, where q is also a prime. They are usually called *Sophie Germain primes*, after the early nineteenth-century French mathematician who studied them in connection with Fermat's Last Theorem.

6-43. *(a) Let q be a prime of the form $4k + 1$. If $p = 2q + 1$ is prime, prove that 2 is a primitive root of p. (See Problem 6-42.)*

(b) Suppose q is a prime of the form $4k + 3$. If $p = 2q + 1$ is prime, prove that -2 is a primitive root of p.

Solution. (a) Since p is of the form $8k + 3$, 2 is a quadratic nonresidue of p, by (5.12). By Problem 6-42, every quadratic nonresidue of p except -1 is a primitive root of p. Note that $2 \not\equiv -1 \pmod{p}$, since $p > 3$. Thus 2 is a primitive root of p.

(b) Since p is of the form $8k + 7$, 2 is a quadratic residue of p, by (5.12), and -1 is a quadratic nonresidue. Therefore -2 is a quadratic nonresidue. The rest of the proof proceeds as in part (a).

6-44. *Use the previous problem to find the primitive roots of 11 and 23.*

Solution. Let $q = 5$ and $p = 2q + 1 = 11$. Since q is of the form $4k + 1$, 2 is a primitive root of 11. For all the primitive roots, we look at 2^k, where $(k, 10) = 1$, and obtain successively 2, 2^3, 2^7, and 2^9. The least positive residues are 2, 8, 7, and 6.

Let $q = 11$ and $p = 2q + 1 = 23$. Since q is of the form $4k + 3$, -2 is a primitive root of 23. Now look at the numbers $(-2)^k$, where $(k, 22) = 1$. Reducing modulo 23, we obtain the primitive roots 21, 15, 14, 10, 17, 19, 7, 5, 20, and 11.

6-45. *Prove that there are two consecutive positive integers less than 167 that are primitive roots of 167. (Hint. Use Problem 6-42.)*

Solution. Since 83 is prime, Problem 6-42 shows that 167 has 82 primitive roots. But 1 and -1 are not primitive roots; thus there are 82 integers in the interval from 2 to 165 that are primitive roots of 167. The only way to produce 82 integers in this interval with no two integers consecutive is to select 2, 4, ..., 164 or 3, 5, ..., 165. But neither of these is possible, since clearly, 4 and 9 are not primitive roots.

6-46. *Let $p > 2$ be prime. (a) Prove that every quadratic nonresidue of p is a primitive root of p if and only if p is is of the form $2^k + 1$. (b) Prove that if $p > 3$ is a prime of the form $2^k + 1$, then 3 is a primitive root of p.*

Solution. (a) There are $(p-1)/2$ quadratic nonresidues of p and $\phi(p-1)$ primitive roots. Since any primitive root is a quadratic nonresidue, it follows that *every* quadratic nonresidue is a primitive root if and only if $\phi(p-1) = (p-1)/2$. But $\phi(n) = n/2$ if and only if n is of the form 2^k (see Problem 3-65).

(b) By the result of (a), it is enough to show that 3 is a quadratic nonresidue of p. Since $2 \equiv -1 \pmod 3$, $2^k + 1 \equiv 0$ or $2 \pmod 3$. But if $2^k + 1 \equiv 0 \pmod 3$, it cannot be prime, for $p > 3$. Thus $p \equiv 2 \pmod 3$. Because p is of the form $4u + 1$, $(3/p) = (p/3) = (2/3) = -1$; hence 3 is a quadratic nonresidue of p and therefore a primitive root of p.

6-47. *Do there exist solutions of $x^{20} \equiv -1 \pmod{41}$ that are not primitive roots of 41?*

Solution. By Euler's Criterion, the solutions of the given congruence are the 20 quadratic nonresidues of 41. But 41 has only $\phi(40) = 16$ primitive roots, so there are 4 solutions of the congruence that are not primitive roots of 41. Another way to think of it is that we are looking for numbers a of order less than 40 whose order does not divide 20, i.e., numbers a of order 8. By (6.14), there are $\phi(8) = 4$ elements of order 8. It is easy to see that 3 and -3 have order 8 modulo 41, for $3^4 \equiv -1 \pmod{41}$, and so the other elements of order 8 are the multiplicative inverses of 3 and -3 modulo 41, namely, 14 and -14.

6-48. *Let p be a prime of the form $4k + 1$. If g is a primitive root of p, prove that $-g$ is also a primitive root of p. (Hint. Express $-g$ as a power of g, or show that if $-g$ has order less than $p - 1$, so does g.)*

Solution. Since g is a primitive root of p, it is a quadratic nonresidue of p, and therefore, by Euler's Criterion, $g^{2k} \equiv -1 \pmod p$. Thus $-g \equiv g^{2k}g = g^{2k+1} \pmod p$. But clearly, $(2k+1, 4k) = 1$. Thus by (6.12), g^{2k+1} is a primitive root of p, i.e., $-g$ is a primitive root of p.

Alternatively, note that $(-g)^{2k} \equiv -1 \pmod p$, and hence $g \equiv (-g)^{2k+1} \pmod p$. Thus any number that is congruent to a power of g modulo p is also congruent to a power of $-g$. The result now follows from (6.11).

6-49. *Show that -3 is not a primitive root of any prime of the form $4n + 3$.*

Solution. We will show that if $p = 4n + 3$ is prime, then -3 is a quadratic residue of p. Since p is of the form $4k + 3$, $(-1/p) = -1$. By the Law of Quadratic Reciprocity, $(3/p) = -(p/3) = -1$, since $p \equiv 1 \pmod 3$. Thus $(-3/p) = (-1/p)(3/p) = 1$, so -3 is a quadratic residue of p and hence cannot be a primitive root of p.

6-50. *Suppose g is a primitive root of m and $gh \equiv 1 \pmod m$. Prove that h is also a primitive root of m.*

Solution. For every positive integer n, $(gh)^n = g^n h^n \equiv 1 \pmod{m}$. Thus $g^n \equiv 1 \pmod{m}$ if and only if $h^n \equiv 1 \pmod{m}$. In particular, g and h have the same order, and therefore if one is a primitive root of p, then so is the other.

6-51. *Suppose $p > 3$ is prime. Prove that the product of the primitive roots of p between 1 and $p - 1$ is congruent to 1 modulo p. (Hint. Use the preceding problem.)*

Solution. If g is a primitive root of p, then $(g, p) = 1$, and thus g has an inverse h modulo p. By the preceding problem, h is also a primitive root of p. If g is a primitive root of p, then its inverse h is not congruent to g modulo p. Otherwise, we would have $g^2 \equiv 1 \pmod{p}$, giving $g \equiv \pm 1 \pmod{p}$; but neither 1 nor -1 is a primitive root of any prime greater than 3. Thus the primitive roots of p can be arranged in pairs that have a product congruent to 1 modulo p, and hence the product of the primitive roots of p is congruent to 1 modulo p.

Note. In essentially the same way, we can prove that if $m \neq 3$, 4, or 6 and m has a primitive root, then the product of the primitive roots of m is congruent to 1 modulo m.

6-52. *Find the least positive residues of four incongruent primitive roots of 25, of 125, and of 250.*

Solution. The primitive roots of 5 are 2 and 3. By the theory developed in (6.22), if g is a primitive root of 5, then $g + 5k$ is a primitive root of 25 if $(g + 5k)^4 \not\equiv 1 \pmod{25}$. Thus 2, 12, 17, and 22 are primitive roots of 25, as are 3, 8, 13, and 23. By (6.22), these are also primitive roots of 125. According to (6.24), any odd primitive root of 125 is a primitive root of 250. Thus 17, 3, 13, and 23 are primitive roots of 250 (as are $2 + 125 = 127$ and so on).

▷ **6-53.** *Prove that there are infinitely many primes p for which there exists an integer a, with $1 < a < p - 1$, such that neither a nor $-a$ is a primitive root of p.*

Solution. Note that if p is a prime of the form $4k + 1$, then for any a not divisible by p, $(-a/p) = (-1/p)(a/p) = (a/p)$, so a and $-a$ are both quadratic residues or both quadratic nonresidues of p. But by Problem 5-23, there are infinitely many primes of the form $4k+1$, and each such prime $p > 5$ has a quadratic residue a with $1 < a < p-1$ (indeed, we can always take $a = 4$). Then $-a$ is also a quadratic residue of p, and thus neither a nor $-a$ is a primitive root of p.

6-54. *Suppose m has a primitive root, and let $a_1, a_2, \ldots, a_{\phi(m)}$ be a reduced residue system modulo m. Prove that $a_1 a_2 \cdots a_{\phi(m)} \equiv -1 \pmod{m}$. (Note that if m is prime, this is simply Wilson's Theorem.)*

Solution. The result is trivial for $m = 1$ and $m = 2$, so we may take $m \geq 3$. Let g be a primitive root of m, and let $k = \phi(m)$. By (6.11), the elements of a reduced residue system modulo m are, in some order, congruent modulo m to the numbers $g^1, g^2, \ldots, g^k$. Thus their product is congruent modulo m to g^N, where $N = 1+2+\cdots+k$.

By the usual formula for the sum of an arithmetic progression, $2N = k(k+1)$. Note that since $m \geq 3$, k is even and so $k+1$ is odd. Now $g^N = (g^{k/2})^{k+1}$. Since g is a primitive root of m, $g^{k/2} \equiv -1 \pmod{m}$, and thus $g^N \equiv -1 \pmod{m}$.

6-55. *Let q and $p = 4q + 1$ be primes. Prove that 2 is a primitive root of p, and show that 3 is a primitive root of p if $q > 3$. This shows, for example, that 2 is a primitive root of 13 and that 2 and 3 are primitive roots of 29 and 53. (Hint. First show that 2 and 3 are quadratic nonresidues of p.)*

Solution. Clearly $q \neq 2$, so q is of the form $2k+1$, and thus p is of the form $8k+5$. By (5.12), 2 is a quadratic nonresidue of p. We show that 2 has order $p-1$ modulo p. Let $\operatorname{ord} 2 = t$; then $t \mid \phi(p)$ and so $t \mid 4q$. Therefore $t = 1, 2, 4, q, 2q$, or $4q$. Since 2 is a quadratic nonresidue of p, $2^{2q} \equiv -1 \pmod{p}$, and hence t cannot be 1, 2, q, or $2q$. Finally, since $p \geq 13$, p does not divide $2^4 - 1$, and hence $t \neq 4$.

If $q > 3$, then $q \equiv 1 \pmod{3}$, for if $q \equiv 2 \pmod{3}$, then p is divisible by 3. Thus q is of the form $3k+1$, and hence p is of the form $12k+5$. By an easy Legendre symbol calculation, $(3/p) = (p/3) = ((12k+5)/3) = (2/3) = -1$. Let t be the order of 3 modulo p. Exactly as before, t cannot be 1, 2, q, or $2q$. But t cannot be 4, since p does not divide $3^4 - 1$. Thus $t = 4q = \phi(p)$, and hence 3 is a primitive root of p.

6-56. *Let p be a prime other than 2 or 5. Prove that 10 is a primitive root of p if and only if the decimal expansion of $1/p$ has period $p - 1$.*

Solution. Consider the process of "long division" that we use when dividing 1 by p. Immediately after we have obtained the kth digit after the decimal point, we have a "remainder" r_k, which is the ordinary remainder when 10^k is divided by p. The period of the decimal expansion is precisely the same as the period of the sequence of remainders.

Let d be the order of 10 modulo p (this order exists, since $(10, p) = 1$). Then the sequence $r_1, r_2, \ldots$ of remainders has period d, and thus the decimal expansion of $1/p$ has period d. In particular, the decimal expansion has period $p - 1$ if and only if 10 has order $p - 1$ modulo p, i.e., if and only if 10 is a primitive root of p.

6-57. *Show that 3 is a primitive root of 7^k for every $k \geq 1$.*

Solution. By (6.22), if g is a primitive root of the odd prime p and $g^p \not\equiv 1 \pmod{p^2}$, then g is a primitive root of p^k for all $k \geq 1$. Since 3^2 and 3^3 are not congruent to 1 modulo 7, 3 is a primitive root of 7. Since $3^6 \equiv 43 \not\equiv 1 \pmod{49}$, the result follows.

6-58. *Prove or disprove: If p is a prime of the form $4k + 1$, then at least one of the primitive roots of p between 1 and $p - 1$ is odd.*

Solution. If p is of the form $4k+1$ and g is a primitive root of p, then $-g$ is also a primitive root of p (see Problem 6-48). Thus if g is even and between 1 and $p-1$, then $p - g$ is an odd primitive root of p between 1 and $p - 1$. Therefore half of the primitive roots of p between 1 and $p - 1$ are even and half are odd.

6-59. *Prove or disprove: If p is prime and $a^{(p-1)/2} \equiv -1 \pmod{p}$, then a is a primitive root of p.*

Solution. The statement is false. For example, if $p = 7$, then $(-1)^{(p-1)/2} \equiv -1 \pmod{p}$, but -1 is not a primitive root of p.

Note. If $a^{(p-1)/2} \equiv -1 \pmod{p}$, then a is a quadratic nonresidue of p, by Euler's Criterion. Thus the statement will be true only if every quadratic nonresidue of p is a primitive root of p and therefore is false if $\phi(p-1) < (p-1)/2$. If $p = 4k+3$, this inequality always holds. If $p = 4k+1$, $\phi(p-1) < (p-1)/2$ holds if and only if p is not of the form $2^n + 1$.

6-60. *Show that if p is an odd prime, then there is an integer that is a primitive root of p^k and $2p^k$ for every $k \geq 1$. (Hint. If g is a primitive root of p, consider g, $g+p$, $g+p^2$, and $g+p+p^2$.)*

Solution. Let g be a primitive root of p. By (6.22.i), at least one of g and $g+p$ is a primitive root of p^2. By adding p^2 if necessary, we obtain an *odd* primitive root h of p^2; then h is also a primitive root of p, since $h \equiv g \pmod{p}$. It follows from (6.22.i) that h is a primitive root of p^k for every $k \geq 1$. But since h is odd, (6.23) implies that h is also a primitive root of $2p^k$ for all $k \geq 1$.

6-61. *Show that if p is prime and k is positive, then any primitive root of p^{k+1} is a primitive root of p^k.*

Solution. Let g be a primitive root of p^{k+1}. By (6.11), if a is relatively prime to p^{k+1}, then $a \equiv g^j \pmod{p^{k+1}}$ for some positive integer j. Thus $a \equiv g^j \pmod{p^k}$. Since every a relatively prime to p^k is congruent to some power of g modulo p^k, it follows from (6.11) that g is a primitive root of p^k.

▷ **6-62.** *Let g be a primitive root of the odd prime p, and let $k \geq 2$. Prove that g is a primitive root of p^k if and only if $g^{p-1} \not\equiv 1 \pmod{p^2}$. (Hint. See the preceding problem.)*

Solution. If $g^{p-1} \not\equiv 1 \pmod{p^2}$, then by (6.23), g is a primitive root of p^2, and thus g is a primitive root of p^k. Conversely, if g is a primitive root of p^k, then by applying the preceding problem repeatedly, we can see that g is a primitive root of p^j for all positive $j \leq k$, and in particular, g is a primitive root of p^2. Thus g has order $\phi(p^2) = p(p-1)$ modulo p^2, and therefore $g^{p-1} \not\equiv 1 \pmod{p^2}$.

▷ **6-63.** *Suppose p is an odd prime and $k \geq 2$. Prove that there are exactly $(p-1)\phi(p-1)$ primitive roots of p^k that are incongruent modulo p^2.*

Solution. There are $\phi(\phi(p^2)) = (p-1)\phi(p-1)$ primitive roots of p^2, and each of them is a primitive root of p^k by (6.23). But any primitive root of p^k is, by the argument of the previous problem, a primitive root of p^2 and so is congruent modulo p^2 to one of the $(p-1)\phi(p-1)$ primitive roots of p^2.

184   CHAPTER 6: PRIMITIVE ROOTS AND INDICES

6-64. *Show that if $m_1, m_2, \ldots, m_r$ are relatively prime in pairs and each m_i has a primitive root, then there is an integer g that is simultaneously a primitive root of every m_i.*

Solution. Let g_i be a primitive root of m_i, and use the Chinese Remainder Theorem to find an integer g such that $g \equiv g_i \pmod{m_i}$ for every i. Then g is a primitive root of m_i for all i.

6-65. *Suppose m has a primitive root, and let $a_1, a_2, \ldots, a_r$ be a reduced residue system modulo m. Prove that $a_1^n, a_2^n, \ldots, a_r^n$ is a reduced residue system modulo m if and only if $(n, \phi(m)) = 1$.*

Solution. Let g be a primitive root of m. Then the a_i are congruent, in some order, to $g, g^2, \ldots, g^r$, where $r = \phi(m)$. By rearranging the a_i, we can assume that $a_i \equiv g^i \pmod{m}$. Thus $a_i^n \equiv (g^n)^i \pmod{m}$. By (6.12), g^n is a primitive root of m if and only if $(n, \phi(m)) = 1$. It follows that the numbers a_i^n form a reduced residue system modulo m if and only if $(n, \phi(m)) = 1$.

▷ **6-66.** *Let p be an odd prime, and let $S_n = 1^n + 2^n + \cdots + (p-1)^n$. Prove that $S_n \equiv -1 \pmod{p}$ if $p-1$ divides n and $S_n \equiv 0 \pmod{p}$ otherwise. (Compare Problem 3-8.)*

Solution. Let g be a primitive root of p. Then $2, 3, \ldots, p-1$ are congruent modulo p, in some order, to $g, g^2, \ldots, g^{p-2}$. Thus $S_n \equiv 1 + g^n + g^{2n} + \cdots g^{(p-2)n} \pmod{p}$. If $(p-1) \mid n$, let $n = (p-1)t$. Since $g^{kn} = (g^{p-1})^{kt}$, it follows from Fermat's Theorem that $g^{kn} \equiv 1 \pmod{p}$. Thus S_n is a sum of $p-1$ terms each congruent to 1 modulo p, and so $S_n \equiv -1 \pmod{p}$.

Now suppose that $(p-1) \nmid n$. Multiplying each side of the congruence for S_n by $1 - g^n$, we obtain $(1 - g^n)S_n \equiv 1 - g^{(p-1)n} \pmod{p}$. Since $g^{p-1} \equiv 1 \pmod{p}$, it follows that $1 - g^{(p-1)n} \equiv 0 \pmod{p}$. But since $(p-1) \nmid n$ and g has order $p-1$, it follows from (6.2.i) that $g^n \not\equiv 1 \pmod{p}$. Thus $S_n \equiv 0 \pmod{p}$.

Another proof: If $(p-1) \mid n$, let $n = t(p-1)$; then for any a relatively prime to p, $a^n = (a^{p-1})^t \equiv 1 \pmod{p}$ by Fermat's Theorem. Thus S_n is a sum of $p-1$ terms each congruent to 1 modulo p, so $S_n \equiv -1 \pmod{p}$.

Suppose now that $(p-1) \nmid n$, and let a be an integer not divisible by p. Then $a, 2a, \ldots, (p-1)a$ are congruent, in some order, to $1, 2, \ldots, p-1$. Therefore $a^n, (2a)^n, \ldots, ((p-1)a)^n$ are congruent, in some order, to $1^n, 2^n, \ldots, (p-1)^n$. Hence $a^n S_n \equiv S_n \pmod{p}$, and so $(a^n - 1)S_n \equiv 0 \pmod{p}$. In particular, if a is a primitive root of p and $(p-1) \nmid n$, then $a^n \not\equiv 1 \pmod{p}$ and therefore $S_n \equiv 0 \pmod{p}$.

▷ **6-67.** *Show that 7 is a primitive root of any prime of the form $4^{2n} + 1$. (Hint. It is enough to show that $(7/p) = -1$, and then show that a prime of the form $4^{2n} + 1$ is congruent to 3 or 5 modulo 7. See Problem 6-46.)*

Solution. Let $p = 4^{2n} + 1 = 2^{4n} + 1$. By Problem 6-46, the primitive roots of p are precisely the quadratic nonresidues of p, so we need only show that $(7/p) = -1$. It is easy to verify that a power of 2 is always congruent modulo 7 to either 1, 2, or 4. Thus $4^{2n} + 1$ is congruent modulo 7 to one of 2, 3, or 5. We want to rule out the possibility that $4^{2n} \equiv 1 \pmod{7}$. Since the order of 4 modulo 7 is 3, $4^{2n} \equiv 1 \pmod{7}$ only if $3 | 2n$. But if $3 | 2n$, then 4^{2n} is a perfect cube, say x^3, and since $x^3 + 1 = (x+1)(x^2 - x + 1)$, it follows that $4^{2n} + 1$ cannot be prime. Therefore if p is prime, then p is congruent to 3 or 5 modulo 7. By the Law of Quadratic Reciprocity, $(7/p) = (p/7)$, and $(p/7)$ is either $(3/7)$ or $(5/7)$, each of which is -1.

6-68. *Prove directly from the definition of primitive root that if m and n are relatively prime integers greater than 2, then mn does not have a primitive root.*

Solution. We show that if $(a, mn) = 1$, then a has order less than $\phi(mn)$ modulo mn, contradicting the definition of primitive root. Since m and n are greater than 2, $\phi(m)$ and $\phi(n)$ are even; let $\phi(m) = 2s$ and $\phi(n) = 2t$. Then $a^{2st} = (a^{2s})^t \equiv 1 \pmod{m}$ by Euler's Theorem. Similarly, $a^{2st} \equiv 1 \pmod{n}$, so $a^{2st} \equiv 1 \pmod{mn}$. But $2st = \phi(mn)/2$, and hence ord $a \le \phi(mn)/2$. Therefore a cannot be a primitive root of mn.

6-69. *Assume that m has primitive roots. Prove that g is a primitive root of m if and only if g is a qth power nonresidue for every prime divisor q of $\phi(m)$.*

Solution. By (6.18), with $k = q$, g is a qth power nonresidue of m if and only if $g^{(p-1)/q} \not\equiv 1 \pmod{p}$. But (6.8) implies that g is a primitive root of m if and only if $g^{(p-1)/q} \not\equiv 1 \pmod{p}$ for every prime divisor q of $\phi(m)$.

▷ **6-70.** *Suppose that $p \ge 7$ is prime. Prove, using primitive roots, that the sum of the squares of the quadratic residues of p and the sum of the squares of the quadratic nonresidues of p are each divisible by p. (Hint. See Problem 6-38.)*

Solution. Let g be a primitive root of p, and let A denote the sum of the squares of the quadratic residues of p. By Problem 6-38, $A \equiv g^4 + g^8 + \cdots + g^{2(p-1)} \pmod{p}$. Therefore $(g^4 - 1)A \equiv g^4(g^{2(p-1)} - 1) \pmod{p}$. Since the right side is congruent to 0 modulo p, it is enough to show that $g^4 \not\equiv 1 \pmod{p}$. This is clearly true if $p > 5$, since g is a primitive root of p.

Similarly, let B denote the sum of the squares of the quadratic nonresidues of p. Then $B \equiv g^2 + g^6 + \cdots + g^{2(p-2)} \pmod{p}$. Therefore $(g^4 - 1)B \equiv g^2(g^{2(p-1)} - 1) \pmod{p}$. Now we can conclude as in the previous paragraph that $B \equiv 0 \pmod{p}$.

6-71. *Let p be a prime. Use the existence of a primitive root of p to prove directly that if $e | p - 1$, then there exists an element of order e modulo p.*

Solution. A primitive root of p has order $p - 1$ modulo p, and thus by (6.3.ii), there is an element of order e for every positive divisor of $p - 1$.

▷ **6-72.** *Let p be an odd prime, and suppose $k \ge 2$. Prove that the primitive roots of p^{k+1} are precisely (up to congruence modulo p^{k+1}) the numbers of the form $g + jp^k$, where g is a primitive root of p^k, $1 \le g < p^k$, and $0 \le j \le p-1$. (The result is not true for $k = 1$; see the next problem.)*

Solution. Let h be a primitive root of p^{k+1}, with $1 < h < p^{k+1}$. We can find integers g and j, with $1 \le g < p^k$ and $0 \le j \le p-1$, such that $h = g + jp^k$ (simply divide h by

p^k, obtaining quotient j and remainder g). Since h is a primitive root of p^{k+1}, h is a primitive root of p^k, and hence g is a primitive root of p^k, since $g \equiv h \pmod{p^k}$. Thus every primitive root of p^{k+1} is congruent modulo p^{k+1} to a number of the required form. Conversely, any number of the form $g + jp^k$, where g is a primitive root of p^k, is a primitive root of p^{k+1}: Since $g + jp^k \equiv g \pmod{p^k}$, $g + jp^k$ is a primitive root of p^k and hence, by (6.23), also a primitive root of p^{k+1}.

▷ **6-73.** Let p be an odd prime. Prove that there are exactly $\phi(p-1)$ primitive roots of p that are incongruent modulo p^2 and that are not primitive roots of p^2.

Solution. If g is a primitive root of p, then there are p primitive roots of p that are congruent to g modulo p but incongruent modulo p^2, namely, $g, g + p, g + 2p, \ldots, g + (p-1)p$. Since p has $\phi(p-1)$ primitive roots, there are $p\phi(p-1)$ primitive roots of p that are incongruent modulo p^2. But there are $(p-1)\phi(p-1)$ primitive roots of p^2, so the number of primitive roots of p that are incongruent modulo p^2 and that are not primitive roots of p^2 is $p\phi(p-1) - (p-1)\phi(p-1) = \phi(p-1)$.

Another solution: Refer to the proof of (6.22.i). Consider $g+tp$, where $0 \le t \le p-1$. These are the primitive roots of p that are incongruent modulo p^2. Using the Binomial Theorem, we can show that $(g+tp)^{p-1} \equiv g^{p-1} - g^{p-2}tp \pmod{p^2}$. Let $g^{p-1} = 1 + mp$; then $g^{p-1} \equiv 1 \pmod{p^2}$ if and only if $mp - g^{p-2}tp \equiv 0 \pmod{p^2}$, i.e., $g^{p-2}t \equiv m \pmod{p}$. This congruence has a unique solution. Thus $g + tp$ is a primitive root of p for all but one value of t. Since there are $\phi(p-1)$ primitive roots of p, the result follows.

▷ **6-74.** Suppose that $n = 3^k$ for $k \ge 1$. Prove that $3n \mid 2^n + 1$. (Hint. First show that 2 is a primitive root of $3n$.)

Solution. It is easy to see that 2 is a primitive root of 3^2 and hence of 3^{k+1}, by (6.23). Thus 2 has order $\phi(3^{k+1}) = 2 \cdot 3^k = 2n$ modulo 3^{k+1}, and in particular it follows that $2^{2n} \equiv 1 \pmod{3n}$. Therefore 2^n is a solution of the congruence $x^2 \equiv 1 \pmod{3n}$. We cannot have $2^n \equiv 1 \pmod{3n}$, for that would mean that $\text{ord } 2 \le n$. Thus $2^n \equiv -1 \pmod{3n}$, i.e., $3n \mid 2^n + 1$.

▷ **6-75.** (a) Let $m > 1$. Find a positive integer N whose decimal expansion begins with a given digit $b \ge 1$ such that if the b is moved to the end of the decimal expansion of N, the new number is the integer N/m. (For example, if $m = 2$ and $b = 1$, we want a positive integer N that begins with the digit 1 such that if the 1 is moved to the end of the number, we get the integer $N/2$.)

(b) Find the smallest positive solution for (i) $b = 1$, $m = 2$; (ii) $b = 1$, $m = 5$. Justify why the solution obtained is the smallest positive integer with the required property.

Solution. (a) Let $N = a_n 10^n + a_{n-1} 10^{n-1} + \cdots + a_0$, where $a_n = b$. If we take the leading digit of N and move it to the end, we obtain the number $Q = a_{n-1} 10^n + \cdots + 10 a_0 + b$. Thus $10N - Q = b(10^{n+1} - 1)$, and $Q = N/m$ if and only if $Q(10m - 1) = b(10^{n+1} - 1)$. We first show that n can be chosen so that $10m - 1$ divides $b(10^{n+1} - 1)$. To do this, it is enough to ensure that $10m - 1$ divides $10^{n+1} - 1$. Since 10 is relatively prime to $10m - 1$, we can appeal to Euler's Theorem and take $n + 1 = \phi(10m - 1)$. With this choice for n, we get $N = mQ = mb(10^{n+1} - 1)/(10m - 1)$.

(b) (i) By the argument of (a), we need to ensure that $10m - 1$ divides $b(10^{n+1} - 1)$, i.e., that $10^{n+1} \equiv 1 \pmod{19}$. It is easy to verify that 10 is a primitive root of 19, so the smallest possible n is 17. Therefore the smallest N with the required property is $2(10^{18} - 1)/19$. (ii) For $m = 5$, we need to determine the order of 10 modulo 49. Since 10 is a primitive root of 7 and $10^6 \not\equiv 1 \pmod{49}$, 10 is a primitive root of 49 and therefore has order 42 modulo 49. Thus the smallest N with the required property is $5(10^{42} - 1)/49$.

The Least Universal Exponent and the Carmichael Function

The remaining problems in this chapter deal with the least universal exponent of m, that is, the largest possible order of an integer modulo m (see Definition 6.6).

6-76. Show that if a is odd and $n \geq 3$, then $a^{2^{n-2}} \equiv 1 \pmod{2^n}$.

Solution. By (6.25), 2^n does not have a primitive root, and thus $\text{ord } a < \phi(2^n)$. But $\phi(2^n) = 2^{n-1}$, and since $\text{ord } a \mid \phi(2^n)$ by (6.2.i), we conclude that $\text{ord } a \mid 2^{n-2}$. The result now follows from (6.2.i).

6-77. Show that $5^{2^{n-3}} \not\equiv 1 \pmod{2^n}$. (*Hint.* Show by induction that $5^{2^{n-3}} \equiv 1 + 2^{n-1} \pmod{2^n}$.)

Solution. The result is easy to check for $n = 3$. Suppose now that $5^{2^{k-3}} \equiv 1 + 2^{k-1} \pmod{2^k}$; we show that $5^{2^{k-2}} \equiv 1 + 2^k \pmod{2^{k+1}}$. By assumption, $5^{2^{k-3}} = 1 + 2^{k-1} + t2^k$ for some integer t. Squaring both sides and simplifying modulo 2^{k+1}, we find that $5^{2^{k-2}} \equiv (1 + 2^{k-1})^2 = 1 + 2^k + 2^{2k-2} \pmod{2^{k+1}}$. But 2^{2k-2} is divisible by 2^{k+1}, since $2k - 2 \geq k + 1$ when $k \geq 3$, and the result follows.

Note. Together, this problem and the previous one show that the least universal exponent for 2^n is 2^{n-2} if $n \geq 3$.

Define the function $\lambda(m)$ as follows. If $m = 1, 2,$ or 4, let $\lambda(m) = \phi(m)$; if $m > 4$ is a power of 2, then $\lambda(m) = \phi(m)/2$. When m is a power of an odd prime, let $\lambda(m) = \phi(m)$, and if m is a product $P_1 P_2 \cdots P_k$ of powers of distinct primes, then $\lambda(m)$ is the least common multiple of the $\lambda(P_i)$. The function $\lambda(m)$ was introduced by the American mathematician Robert Carmichael in 1908. The idea had been explored some thirty years earlier by François Lucas.

6-78. Show that if a is relatively prime to m, then $a^{\lambda(m)} \equiv 1 \pmod{m}$. (*Hint.* Let m be a product $P_1 P_2 \cdots P_k$ of powers of distinct primes. Calculate modulo the P_i, and use Problem 6-76.)

Solution. It is enough to show that $a^{\lambda(m)} \equiv 1 \pmod{P_i}$ for $1 \leq i \leq k$, since this implies that $a^{\lambda(m)} \equiv 1 \pmod{m}$. If $P_i = 2, 4,$ or a power of an odd prime, $\lambda(P_i) = \phi(P_i)$ by definition, so $a^{\lambda(P_i)} \equiv 1 \pmod{P_i}$. Since $\lambda(P_i)$ divides $\lambda(m)$, $a^{\lambda(m)} \equiv 1 \pmod{P_i}$. If P_i is a power of 2 greater than 4, then $a^{\lambda(P_i)} \equiv 1 \pmod{P_i}$ by Problem 6-76, again giving $a^{\lambda(m)} \equiv 1 \pmod{P_i}$.

6-79. Show that if a is not divisible by 2, 3, 5, 7, or 13, then $a^{12} \equiv 1 \pmod{65520}$.

Solution. $65520 = 2^4 \cdot 3^2 \cdot 5 \cdot 7 \cdot 13$, so $\lambda(65520)$ is the least common multiple of 4, 6, 4, 6, and 12; thus $\lambda(65520) = 12$. Note that $\phi(65520) = 13824$, so $\lambda(m)$ can be very much smaller than $\phi(m)$.

6-80. Let N be the product of the distinct odd primes $p_1, p_2, \ldots, p_k$. Prove that $\lambda(N)/\phi(N) \le 1/2^{k-1}$. *(Thus $\lambda(m)$ can be considerably smaller than $\phi(m)$.)*

Solution. Let $p_i - 1 = 2b_i$ for $1 \le i \le k$. Then $\phi(N) = 2^k P$, where P is the product of the b_i, while $\lambda(N) = 2M$, where M is the least common multiple of the b_i. Since $M \le P$, the result follows.

▷ **6-81.** Show that $\lambda(m)$ is the least universal exponent for m. *(Use Problems 6-77, 6-78, and Theorem 6.25.)*

Solution. By Problem 6-78, $a^{\lambda(m)} \equiv 1 \pmod{m}$ whenever $(a, m) = 1$, and hence ord $a \le \lambda(m)$. It remains to show that there is an element of order exactly $\lambda(m)$. Let $m = P_1 P_2 \cdots P_k$, where the P_i are powers of distinct primes. We show that for every i, there is an integer a_i of order $\lambda(P_i)$ modulo P_i. If $P_i = 2$ or 4, there is obviously such an a_i; if P_i is a power of 2 greater than 4, we can take $a_i = 5$ by Problem 6-77; and if P_i is a power of an odd prime, we let a_i be a primitive root of P_i.

Use the Chinese Remainder Theorem to produce an integer b such that $b \equiv a_i \pmod{P_i}$ for all $i \le k$. Let e be the order of b modulo m. Then $b^e \equiv 1 \pmod{P_i}$ and hence $a_i^e \equiv 1 \pmod{P_i}$. Thus, by (6.2.i), $\lambda(P_i) | e$ for all i, and therefore the least common multiple of the $\lambda(P_i)$, namely, $\lambda(m)$, divides e. Hence ord $b \ge \lambda(m)$, and since ord $b \le \lambda(m)$, we conclude that ord $b = \lambda(m)$.

6-82. Let m be a positive integer and p an odd prime that divides m. If p^k is the largest power of p that divides m, use the preceding problem to show that there is an integer of order $\phi(p^k)$ modulo m.

Solution. It follows from Problem 6-81 and the Note after (6.6) that there is an integer c such that ord $c = \lambda(m)$. By the definition of $\lambda(m)$, we have $\phi(p^k) | \lambda(m)$. Thus by (6.3.ii), $c^{\lambda(m)/\phi(p^k)}$ has order $\phi(p^k)$ modulo m.

EXERCISES FOR CHAPTER 6

1. (a) Find the order of 34 modulo 37.
   (b) What is the order of 2^{12} modulo 37?
2. Find all positive integers less than 37 that have order 6 modulo 37.
3. Prove or disprove: If $(a,m) = 1$ and the order of a modulo m is $m - 1$, then m is prime.
4. Assume that $\text{ord } 9 = 4$ and $\text{ord } 10 = 5$ modulo 41. What is the order of 8 modulo 41?
5. Suppose that the order of a modulo 686 is 42. What is $\text{ord } a^{15}$?
6. Use the fact that 5 is a primitive root of 263 to determine the order of 258 modulo 263.
7. For which primes p will there always exist elements of order 4 modulo p?
8. Find all elements of order 8 modulo 109.
9. Use the fact that 2 is a primitive root of 101 to find the least positive residues of all elements of order 5 modulo 101.
10. Show that 5 is a primitive root of 73, and find all numbers between 1 and 72 that have order 12 modulo 73.
11. Use the fact that 6 is a primitive root of 41 to find the least positive residues of all elements of order 8 modulo 41.
12. Let $n = 2^{45} - 1$. Prove that $\phi(n)$ is a multiple of 45.
13. Determine which of the following integers have primitive roots: 143, 147, 626, 1331.
14. Find all of the primitive roots of 38 between 1 and 37.
15. The integer 5 is a primitive root of 103. Is 25 also a primitive root of 103?
16. Determine the number of primitive roots of (a) 625; (b) 626; (c) 686.
17. Prove that 18 is a primitive root of 37 but not of 37^2.
18. The integer 3 is a primitive root of the prime 199. Is 3 also a primitive root of 199^2?
19. Determine two primitive roots of (a) 17; (b) 289; (c) 578.
20. Find the least positive residues of four incongruent primitive roots of (a) 49; (b) 343; (c) 686.
21. Find a primitive root of 722. How many incongruent primitive roots does 722 have?
22. The integer 6 is a primitive root of 109. Calculate the multiplicative inverse of 6 modulo 109, and use this to find another primitive root of 109.
23. Is every quadratic nonresidue of 257 a primitive root of 257?

24. The integer 7 is a primitive root of the prime 241. Is -7 also a primitive root of 241?

25. Find the least positive residues of three incongruent primitive roots of 7^2, 7^3, and $2 \cdot 7^3$.

26. The integer 2 is a primitive root of 67. Is 34 a primitive root of 67?

▷ 27. Show that $-n^2$ is a primitive root of 83 for every n such that $1 < n < 82$.

28. Does $x^2 \equiv 1 \pmod{5991}$ have solutions other than 1 and -1? Explain.

29. Prove that all but one of the quadratic nonresidues of 227 is a primitive root of 227.

30. How many quadratic nonresidues of 313 are not primitive roots of 313?

31. Find all of the primitive roots of 37 between 1 and 36.

32. Suppose that g is a primitive root of p^3, where p is an odd prime. Is it true that $g^{(p-1)/2} \equiv -1 \pmod{p}$? Explain.

33. Use the fact that 2 is a primitive root of 37 to show that the product of the quadratic residues of 37 between 1 and 36 is congruent to -1 modulo 37. What can be said about the product of the quadratic nonresidues of 37 between 1 and 36?

34. What is the least positive residue of the product of all of the primitive roots of 29 between 1 and 29?

35. Suppose that h is a quadratic nonresidue of 149 whose order is not equal to 4. Is h a primitive root of 149?

36. If 2 is a primitive root of 211, what is the index of -1 to the base 2?

37. Use the table of indices in Problem 6-21 to solve the following congruences:
    (a) $9x \equiv 14 \pmod{19}$;
    (b) $11x^7 \equiv 13 \pmod{19}$;
    (c) $5x^6 \equiv 17 \pmod{19}$;
    (d) $9^x \equiv 7 \pmod{19}$.

38. Prove that 2 is a primitive root of 53. Use this to find the least positive residue of each solution of $45x^{20} \equiv 31 \pmod{53}$.

39. (a) Without finding solutions, show that $x^6 \equiv 8 \pmod{89}$ is solvable. How many solutions are there?
    (b) Use the fact that 3 is a primitive root of 89 to find the solutions of $x^6 \equiv 8 \pmod{89}$.

40. Show that 2 is a primitive root of 67, and use this to determine the least positive residues of all solutions of $x^3 \equiv -3 \pmod{67}$.

41. Find the least positive residue of all solutions of $16^x \equiv 9 \pmod{61}$. (Hint. 2 is a primitive root of 61, and ind $9 = 12$.)

42. (a) Prove that 3 is a primitive root of 43.

(b) Use indices to find the least positive residue of all solutions of $x^9 \equiv 27$ (mod 43).

43. Find the least positive residue of all solutions of $x^3 \equiv 1$ (mod 37). (Hint. First show that 2 is a primitive root of 37 and that ind $11 = 30$.)

44. Use indices to find all solutions of $11x^7 \equiv 7$ (mod 37).

45. Determine the number of solutions of (a) $x^9 \equiv 27$ (mod 43); (b) $x^9 \equiv 27$ (mod 686).

46. Show that solutions of $x^5 \equiv 26$ (mod 71) exist, and find all of the solutions. (Hint. 7 is a primitive root of 71, and ind $26 = 45$.)

47. Determine the number of solutions of the following congruences:
   (a) $x^{12} \equiv 45$ (mod 58);
   (b) $13x^7 \equiv 7$ (mod 58);
   (c) $4x^{20} \equiv 23$ (mod 43).

48. For which values of c is $cx^6 \equiv 17$ (mod 19) solvable?

49. Prove or disprove: If p is a prime of the form $6k - 1$ and $(a, p) = 1$, then $x^3 \equiv a$ (mod p) has a unique solution.

NOTES FOR CHAPTER 6

1. When Gauss began his work on primitive roots (at the age of 16!), he was not aware of earlier work by Lambert, Euler, Legendre, and others. His initial motivation (like that of Lambert) came from investigation of the decimal expansion of fractions a/b, and in particular of $1/p$, where p is a prime other than 2 or 5. Note, for example, that $1/7 = .\overline{142857}$, $1/11 = .\overline{09}$, and $1/13 = .\overline{076923}$. So the decimal expansion of $1/7$ is *periodic* with *period* 6, the expansion of $1/11$ has period 2, while the expansion of $1/13$ has period 6.

If we think about the ordinary "long division" process of finding the decimal expansion of $1/p$, it is fairly easy to see that this expansion is periodic with period k if and only if k is the smallest positive integer such that 10^k leaves a remainder of 1 on division by p, that is, if and only if 10 has order k modulo p. Thus by (6.2.i), the period of the decimal expansion of $1/p$ is a divisor of $p - 1$. Gauss was interested in determining the primes p (for example, $p = 7$) for which the period is $p - 1$. These are precisely the primes p for which 10 is a primitive root (see Problem 6-56).

The literature on periods of decimal expansions is extensive and makes substantial use of the notions developed in this chapter. (See *An Introduction to the Theory of Numbers* by G.H. Hardy and E.M. Wright, and Volume I of *History of the Theory of Numbers* by L.E. Dickson.)

2. Does there exist an integer a such that a is a primitive root of p for infinitely many primes p? Gauss conjectured that 10 is a primitive root of infinitely many primes. In 1927, Emil Artin (1898–1962) made the following broad conjecture: *Every* integer a not equal to -1 or a square is a primitive root of infinitely many primes.

It can be shown (see Problem 6-46) that 3 is a primitive root of every Fermat prime, that is, every prime of the form $F_n = 2^{2^n} + 1$. This would show that 3 is a primitive root of infinitely many primes if it were known that there are infinitely many Fermat primes. Fermat knew 350 years ago that F_4 is prime, but it is not known whether there are any Fermat primes beyond F_4.

By the result of Problem 6-43, if q is a prime of the form $4k + 1$ and $p = 2q + 1$ is also prime, then 2 is a primitive root of p. But it is not known whether there are infinitely many such primes p and q.

There is strong evidence for Artin's Conjecture. Let $P_g(x)$ be the proportion of primes less than or equal to x of which g is a primitive root. It has been conjectured that $\lim_{x \to \infty} P_g(x)$ exists and is positive whenever g is neither -1 nor a perfect square; indeed, there is an explicit conjectured formula for the value of this limit, which agrees very well with the numerical evidence. C. Hooley has shown that this formula is a consequence of the *Extended Riemann Hypothesis*. Without using any unproved hypotheses, R. Gupta, P.M. Ram Murty, D.R. Heath-Brown, and H. Iwaniec have made substantial progress toward proving Artin's Conjecture. It is now known, for example, that all but two prime numbers are primitive roots of infinitely many primes. Oddly enough, however, no *specific* prime is known to be a primitive root of infinitely many primes. For more information, see Ribenboim's *The Book of Prime Number Records*.

3. The existence of a primitive root of m has a particularly nice group-theoretic interpretation. The *multiplicative group* modulo m consists of the integers x relatively prime to m such that $0 \leq x < m$, with multiplication modulo m. The number g is a primitive root of m if and only if g generates this group, in the sense that the elements of the group are all the powers of g. A group which is thus generated by a single element is called a *cyclic group*. In our case, the group has $\phi(m)$ elements. Because it is cyclic, it is abstractly identical to the group whose elements are $0, 1, \ldots, \phi(m) - 1$, under the operation of addition modulo $\phi(m)$. This is what accounts for the logarithm-like character of indices.

In the recent literature on number-theoretic methods in cryptography, if we are given a prime modulus p and a primitive root g of p, the index of a is often referred to as the *discrete logarithm* of a. Considerable effort has been expended in trying to find efficient algorithms for calculating the discrete logarithm. If a is the remainder when g^k is divided by p, then it is fairly easy to calculate a by the method of repeated squaring. But the inverse problem, namely, to find k when we are given a (i.e., the problem of finding the discrete logarithm of a), appears to be computationally very difficult for large primes p.

BIOGRAPHICAL SKETCHES

Adrien-Marie Legendre was born in Paris in 1752, to a wealthy family. After he completed his studies at the Collège Mazarin at the age of 18, his financial independence made it possible to devote several years to pure research. In 1782, he won a prize offered by the Berlin Academy with a paper in ballistics. Legendre was a great mathematician and made lasting contributions to number theory. However, he had the misfortune of living in the era of Lagrange and Gauss and received less recognition than he deserved.

Legendre gave the first complete proof that every prime has a primitive root. He was also the first to determine the number of representations of an integer as a sum of two squares and proved that every odd positive integer which is not of the form $8k + 7$ is a sum of three squares. (Not long after that, Gauss gave a more informative proof.) Legendre conjectured the Prime Number Theorem and the Law of Quadratic Reciprocity but was unable to prove them. He also conjectured that every suitable arithmetic progression has infinitely many primes.

Legendre's contributions to the theory of quadratic forms, like those of Lagrange, were destined to be soon eclipsed by Gauss. Legendre's most beautiful result is probably the complete characterization of when the equation $ax^2 + by^2 + cz^2 = 0$ has a nontrivial solution. In addition to his research in number theory, Legendre worked in more applied areas, such as differential equations and celestial mechanics, and shares credit with Gauss for inventing the method of least squares. Along with Lagrange, Legendre played a leading role in the design of the metric system. In his later years, Legendre's investigations focused on elliptic integrals, and he saw his favorite subject become a major area of research at the hands of Abel and Jacobi.

At the age of 75, Legendre completed an argument of Dirichlet that settled the case $n = 5$ of Fermat's Last Theorem. Legendre died in Paris in 1833.

REFERENCES

Carl Friedrich Gauss, *Disquisitiones Arithmeticae*, translated by Arthur A. Clarke, Yale University Press, New Haven, Connecticut, 1966.

> Gauss gives a very detailed presentation of primitive roots and indices. Most later texts use an approach that is essentially identical to the one in *Disquisitiones*.

CHAPTER SEVEN
Prime Numbers

The existence of infinitely many primes was known to Euclid, as were the essentials of the factorization of a positive integer into a product of primes. In this chapter, we will study some of the deeper properties of primes, including the Prime Number Theorem, which deals with estimating how many primes there are less than a given number, and Dirichlet's Theorem on the infinitude of primes in arithmetic progressions.

The Prime Number Theorem was first formulated around 1800 but not proved until the end of the nineteenth century; the proof of Dirichlet's Theorem was given in 1837. Both results are very difficult to prove and are beyond the scope of this book. It is interesting to note, however, that the standard proof of each involves the use of ideas and techniques from the theory of functions of a complex variable. So even though the the results can be stated quite simply in terms of standard number-theoretic notions, the proofs use concepts that go well beyond those covered in a first course in number theory.

We will also study the existence of primes of the form $2^n - 1$. These numbers are called *Mersenne numbers* and have traditionally generated the largest known primes. (As of this writing, the largest prime known is the Mersenne number $2^{859433} - 1$.) Although Mersenne primes were not studied in detail until the seventeenth century, they are intimately connected with the much older notion of *perfect numbers*, that is, integers which are equal to twice the sum of their positive divisors. The study of perfect numbers goes back to the ancient Greeks, and a method of generating even perfect numbers appears in Euclid's *Elements*.

We conclude the chapter with a number of open questions related to prime numbers. The most celebrated of these is Goldbach's Conjecture, which states that every even integer greater than 2 is the sum of two primes. The conjecture was made in a letter to Euler in 1742; in the ensuing 250 years, much work has been done on this problem, but it remains unsolved.

Because of the nature of the results discussed, much of this chapter is expository. In many cases, either the proofs are too difficult to present in an elementary course in number theory, or the assertions are still open problems.

RESULTS FOR CHAPTER 7

The Sieve of Eratosthenes

In Chapter 1, we presented Euclid's proof that there are infinitely many primes. We discuss next a method for finding all of the primes less than or equal to a given positive integer. The algorithm is known as the *Sieve of Eratosthenes* and is named after the Greek mathematician Eratosthenes (276–194 B.C.), who was librarian of the famous library at Alexandria and is perhaps best remembered for his accurate calculation of the earth's circumference.

Let n be a positive integer. We first list all of the integers from 2 to n. Since 2 is prime, every multiple of 2 greater than 2 is composite, so we cross off our list all multiples of 2 greater than 2. Clearly, the next number in the list that has not been crossed out – namely, 3 – is prime. We now remove all multiples of 3 greater than 3 which have not already been crossed out. The next number in the list after 3 is 5, which must be prime since otherwise, it would be divisible by a prime less than 5 and hence would already be crossed off our list. Remove all multiples of 5 that have not already been crossed out. We proceed in this way: At any given point, after we have crossed out all multiples of a given prime, the next integer in the list that has not been crossed out is a prime, and we then remove all multiples of it (except itself). When the process cannot be carried any further, the numbers that remain represent all of the primes less than or equal to n.

Notes. 1. Eratosthenes observed that *it is necessary only to carry out the process until a prime greater than $\sqrt{n}$ has been found; all of the remaining integers in the list must then be prime.* For if $k \leq n$ is composite, then k has a prime divisor $p \leq \sqrt{n}$ (see Problem 7-1), and hence k would already have been removed from the list.

2. The Sieve of Eratosthenes is not a practical way of determining whether n is prime. A faster way, for example, is to note, as above, that if an integer n is not prime, then it has a prime factor $p \leq \sqrt{n}$. Thus, *to decide if n is prime, it is necessary only to determine if n is divisible by a prime less than or equal to $\sqrt{n}$*. The Sieve of Eratosthenes, however, does have a rather unexpected application: It is commonly used as a benchmark for testing the speed of a computer.

Perfect Numbers

(7.1) Definition. A positive integer n is called a *perfect number* if n is equal to the sum of its positive divisors other than n itself.

Thus n is perfect if and only if $\sigma(n) = 2n$, where $\sigma(n)$ is the sum of all positive divisors of n (including n itself).

The idea of perfect numbers goes back to antiquity, and throughout history, many mystical properties have been attributed to them. Early Greek mathematicians were especially interested in them, although only four perfect numbers were known in Euclid's time (6, 28, 496, and 8128). This sparse information led to the conjecture that even perfect numbers end alternately in 6 or 8. Although this is false (the fifth and sixth perfect numbers are 33,550,336 and 8,589,869,056), it is true that every even perfect number ends in 6 or 8. (See Problem 7-9.)

In the *Elements*, Euclid described the following method for finding even perfect numbers.

(7.2) Theorem (Euclid). *Suppose that $2^n - 1$ is prime. Then $2^{n-1}(2^n - 1)$ is a perfect number.*

Proof. Let $N = 2^{n-1}p$, where $p = 2^n - 1$. Since p is prime, the divisors of $2^{n-1}p$ are clearly of the form 2^i or $2^i p$, where $0 \leq i \leq n-1$. Therefore

$$\sigma(N) = 1 + 2 + \cdots + 2^{n-1} + p + 2p + \cdots + 2^{n-1}p$$
$$= (1+p)(1 + 2 + \cdots + 2^{n-1})$$
$$= (1+p)(2^n - 1) = 2^n(2^n - 1) = 2N.$$

Thus N is perfect.

A natural question to ask is whether the converse of Euclid's result is true: Is every even perfect number of the form given in (7.2)? Almost 2000 years passed before the question was resolved by Euler.

(7.3) Theorem (Euler). *Every even perfect number can be written in the form $2^{n-1}(2^n - 1)$, where $2^n - 1$ is prime.*

Proof. Let N be an even perfect number; then $\sigma(N) = 2N$. Let $N = 2^{n-1}m$, where $n \geq 2$ and m is odd. Because $(2^{n-1}, m) = 1$ and σ is multiplicative (see the Note after (1.20)), we have

$$2^n m = 2N = \sigma(N) = \sigma(2^{n-1})\sigma(m) = (2^n - 1)\sigma(m).$$

Solving the equation $2^n m = (2^n - 1)\sigma(m)$ for $\sigma(m)$, we obtain $\sigma(m) = m + m/(2^n - 1)$. Therefore $m/(2^n - 1)$ is an integer, and hence both m and

$m/(2^n - 1)$ are divisors of m. Since $\sigma(m) = m + m/(2^n - 1)$, it follows that m and $m/(2^n - 1)$ are the *only* positive divisors of m. Thus $m/(2^n - 1) = 1$, that is, $m = 2^n - 1$, and m is prime.

We close this section with two important open questions about perfect numbers. The first is whether there are any odd perfect numbers. It is known that if odd perfect numbers exist, then they must be greater than 10^{300} and have at least eight different prime factors. The numerical evidence would seem to suggest, therefore, that odd perfect numbers do not exist.

The second question is whether there are infinitely many even perfect numbers. Four were known in ancient times, but the fifth was not discovered until the fifteenth century. There are now 33 known even perfect numbers, the last 21 of which were discovered since 1900. To date, the largest is $2^{859432}(2^{859433} - 1)$, an integer with approximately 517,430 digits. However, the existence of infinitely many even perfect numbers remains an open question.

Mersenne Primes

In view of the results of Euclid and Euler, *the existence of an infinite number of even perfect numbers is equivalent to the existence of infinitely many primes of the form* $2^n - 1$. Such primes are known as *Mersenne primes*, after the Franciscan monk who studied them in the seventeenth century.

(7.4) Definition. Let n be a positive integer. The *n*th *Mersenne number* is the integer $M_n = 2^n - 1$. If M_n is prime, then M_n is called a *Mersenne prime*.

More generally, we can ask when a number of the form $a^n - 1$ is prime, where $n \geq 2$. It is not hard to show that a must be 2 and n must be prime (see Problem 7-4). Thus *if M_n is prime, then n must be prime*; hence in looking for primes of this form, it is enough to consider the Mersenne numbers M_p, where p is prime.

In his *Cogitata Physico Mathematica* of 1644, Mersenne conjectured that M_p is composite for all primes $p < 257$, except for the 11 values 2, 3, 5, 7, 13, 17, 19, 31, 67, 127, and 257. In 1772, Euler showed that M_{31} is prime. In 1876, however, Lucas proved that M_{67} is composite, although he could not compute the factors. It required almost three centuries after Mersenne made his conjecture to settle the question completely. Mersenne's guess – for that was probably what it was – proved to be wrong for five prime values less than 257.

In trying to determine factors of Mersenne numbers, Fermat noted that if p is prime, then any prime divisor of $2^p - 1$ leaves a remainder of 1 when divided by p. This observation is contained in the following result, originally proved in Chapter 3. We state it here for completeness. The proof given is essentially the same as that of (3.8), except that it uses properties of the order of an integer developed in the preceding chapter.

(7.5) Theorem. *Let q be a prime divisor of $2^p - 1$, where p is an odd prime. Then q is of the form $2kp + 1$.*

Proof. Since $2^p \equiv 1 \pmod{q}$, (6.2.i) implies that $\text{ord}\,2 \,|\, p$. Thus $\text{ord}\,2 = p$, since p is prime and $\text{ord}\,2$ is plainly not 1. Appealing again to (6.2.i), we find that $p \,|\, q - 1$, and hence $q = mp + 1$ for some m. Since q is odd, m must be even, which proves the result.

Thus in searching for prime factors of M_p, we only have to look at primes of a certain form. Examples of this technique are given in Problems 7-15 to 7-19.

Notes. 1. The largest known primes have traditionally been Mersenne primes. (For an exception, see the Notes at the end of the chapter.) Currently, there are 33 known Mersenne primes (and thus also 33 known even perfect numbers). The largest of these is the number $2^{859433} - 1$, which was discovered in early 1994 after almost 20 hours of computing time on a Cray-2 supercomputer. The next largest known Mersenne prime, $2^{756839} - 1$ was found in 1992 after a similarly lengthy computation. The previous record holder, $2^{216091} - 1$, required "just" 3 hours to find.

2. Perhaps the most important open problem related to Mersenne primes is whether infinitely many such primes exist. There is a probabilistic argument that makes it reasonable to suppose that there are. (See Ribenboim, Chapter 6.)

Fermat Numbers

In looking for primes of the form $a^n - 1$, we have seen that it is enough to consider numbers of the form $2^p - 1$, where p is prime. In a similar way, we can look for odd primes of the form $a^n + 1$, where $n > 1$. For such a number to be an odd prime, a must be even and n must be a power of 2. (See Problem 7-5.) Thus we are led to study numbers of the form $2^{2^n} + 1$.

(7.6) Definition. A *Fermat number* is an integer of the form $F_n = 2^{2^n} + 1$, where $n \geq 0$. If F_n is prime, F_n is called a *Fermat prime*.

For $n = 0, 1, 2, 3$, and 4, F_n assumes the values 3, 5, 17, 257, and 65537, which are all prime. Perhaps on the basis of this somewhat meager evidence, Fermat asserted, in a letter to Frenicle in 1640, that F_n is prime for every $n \geq 0$. As usual, Fermat did not offer a proof, but in 1658, he claimed that he had a proof by the method of infinite descent. The conjecture is in fact false.

Euler showed almost a century later, in 1732, that 641 is a divisor of F_5 (see Problem 2-23).

In response to a question from Frenicle in 1640, Fermat had shown that $2^{37} - 1$ is not prime by showing that it is divisible by 223. Fermat used essentially the same argument as in the proof of (7.5) to conclude that every prime divisor of $2^{37} - 1$ is of the form $74k + 1$.

A similar argument can be used with $F_5 = 2^{32} + 1$. If p is a prime factor of F_5, then p divides $(2^{32} + 1)(2^{32} - 1) = 2^{64} - 1$; by arguing as above, it can be shown that $p = 64k + 1$ for some k. The first few primes of this form are 193, 257, 449, 577, 641, ..., and 641 divides $2^{32} + 1$. In fact, this is how Euler showed that F_5 is not prime, and it is quite surprising that Fermat himself did not apply to $2^{32} + 1$ the method that he had used to factor $2^{37} - 1$.

With a little more work, it can be shown that the prime divisors of $2^{32} + 1$ are actually of the form $128k + 1$, not simply $64k + 1$. We prove next that a similar result holds for the prime divisors of any Fermat number. In obtaining the extra factor of 2, it is necessary to know which primes have 2 as a quadratic residue (see (5.12)).

(7.7) Theorem. *Suppose that $n > 1$. Then any prime divisor of the Fermat number $2^{2^n} + 1$ is of the form $2^{n+2}k + 1$.*

Proof. Let p be a prime divisor of $2^{2^n} + 1$. Then $2^{2^n} \equiv -1 \pmod{p}$, and since 4 divides 2^n, the congruence $x^4 \equiv -1 \pmod{p}$ has a solution. Therefore $(-1)^{(p-1)/(4,p-1)} \equiv 1 \pmod{p}$ by (6.19), and so $(p - 1)/(4, p - 1)$ is even. Thus p is of the form $8k + 1$, and it follows from (5.12) that 2 is a quadratic residue of p. Suppose $s^2 \equiv 2 \pmod{p}$; then s is a solution of the congruence $x^{2^{n+1}} \equiv -1 \pmod{p}$. Appealing once again to (6.19), we conclude as above that $(p - 1)/(2^{n+1}, p - 1)$ is even, and hence 2^{n+2} divides $p - 1$, which proves the theorem.

Notes. 1. Fermat knew that F_0 through F_4 are prime, but no Fermat prime beyond F_4 has yet been found. There are 84 composite Fermat numbers known, the largest of which is F_{23471}.

2. Fermat numbers offer an interesting proof that there are infinitely many primes. See Problem 7-24.

Fermat primes have an unexpected application to the classical problem of deciding when a regular n-sided polygon can be constructed using only a straightedge and compass. In 1801, in the last section of *Disquisitiones Arithmeticae*, Gauss proved that if $n = 2^k$ or $n = 2^k p_1 p_2 \cdots p_m$, where $k \geq 0$ and the p_i are distinct Fermat primes, then the regular n-gon is constructible by straightedge and compass, and he claimed that he had a proof that no other regular polygons were so constructible. (A proof was first published by

Wantzel in 1837.) In particular, a regular 17-sided polygon can be constructed by straightedge and compass. This was the first new construction of a regular polygon since the time of Euclid.

The Prime Number Theorem

We showed in Chapter 1 that there are arbitrarily large gaps in the sequence of primes (see Problem 1-28). It is therefore natural to ask the following questions: Approximately how many primes are less than a given number? Also, how large (approximately) is the nth prime? (Euclid's proof that there exist infinitely many primes shows that $p_n \leq p_1 p_2 \cdots p_{n-1} + 1$, where p_i denotes the ith prime.) Both questions are answered by the Prime Number Theorem.

(7.8) Prime Number Theorem. *Let x be a positive real number, and let $\pi(x)$ denote the number of primes that are less than or equal to x. Then the ratio $\pi(x)/(x/\log x)$ can be made arbitrarily close to 1 by taking x sufficiently large.*

More informally, *the Prime Number Theorem states that the number of primes less than a given positive number x is approximately $x/\log x$ if x is sufficiently large.* (Here, $\log x$ denotes the *natural logarithm* of x, that is, the logarithm to the base e.) This had been conjectured by Gauss in 1793 and by Legendre in 1798, on the basis of values of $\pi(x)$ for $x < 10000$ obtained from tables of primes that were then available.

The Prime Number Theorem is a very deep result; indeed, it was not proved until almost a century later. In 1896, independent proofs of the Prime Number Theorem were given by Jacques Hadamard (1865–1963) and Charles-Jean de la Vallée-Poussin (1866–1962), using techniques from the theory of functions of a complex variable suggested by the work of Riemann (1826–1866). It is now known that, in fact, $\pi(x) \geq x/\log x$ if $x \geq 17$.

If p_n denotes the nth prime, then clearly $\pi(p_n) = n$. Thus, by taking $x = p_n$ in (7.8) and using the fact that $\log(x/\log x)/\log x$ is nearly 1 when x is large, it can be shown that $n \log n / p_n$ approaches 1 as n approaches infinity. We therefore have the following equivalent formulation of the Prime Number Theorem.

(7.9) Prime Number Theorem. *If p_n denotes the nth prime, then $p_n/(n \log n)$ approaches 1 as a limit as n approaches infinity.*

Note. It is known, in fact, that $p_n > n \log n$ for all n. The "error" $p_n - n \log n$ can be quite large, but if n is large, the error is much smaller than $n \log n$.

It follows from the Prime Number Theorem that if $x > 1$, then $\pi(x)/x < C/\log x$ for some constant C; hence the ratio $(x - \pi(x))/x$ approaches 1 as x

approaches infinity. This can be interpreted in the following way. Since $n-\pi(n)$ is the number of positive integers less than or equal to n that are not prime, the ratio $(n-\pi(n))/n$ represents the proportion of composite numbers among the first n integers. Thus, in the sense that this ratio tends to 1, the Prime Number Theorem implies that "almost all" positive integers are composite.

In a certain sense, however, primes are not particularly scarce, since the function $\log x$ grows very slowly. For example, $\log 10^{100}$ is roughly 230. Thus if we choose an integer N "at random" in the neighborhood of 10^{100}, the probability that N is prime is roughly $1/230$. If we make sure that N is not divisible by 2, 3, or 5, the probability that N is prime grows to about $1/61$, and we can raise it significantly by sieving out other small primes. This means that if we use an efficient primality test, it is perfectly feasible to find very large primes.

Notes. 1. One consequence of the Prime Number Theorem is that if a and b are positive real numbers, with $a < b$, then there is at least one prime between ax and bx if x is sufficiently large. Taking $a = 1$ and $b = 1+\epsilon$, where ϵ is any positive real number, we have the following interesting result: *There is at least one prime between n and $n(1+\epsilon)$ for all sufficiently large values of n.*

2. Edmund Landau proved that $\pi(2n) < 2\pi(n)$ if n is sufficiently large. It follows from this that there are more primes between 1 and n than between n and $2n$ for all sufficiently large values of n.

3. Here is another interesting consequence of the Prime Number Theorem: The set of all numbers of the form $\pm p/q$, where p and q are primes, is dense in the set R of real numbers. Equivalently, for any real number x, there is a number of the form $\pm p/q$ that is as close to x as desired.

We close this section with a result, known as *Bertrand's Postulate*, that guarantees the existence of a prime between n and $2n$ for any $n \geq 2$. Joseph Bertrand made his conjecture in 1845, having verified it for all values of n up to 3,000,000, but he was unable to prove his assertion. The first proof was given seven years later by the Russian mathematician Pavnuty Chebyshev (1821–1894). Although the proof is much less difficult than the proof of the Prime Number Theorem, it is nonetheless beyond the scope of this book.

(7.10) Bertrand's Postulate. *For every integer $n > 1$, there is a prime between n and $2n$.*

Notes. 1. A stronger result is true: If $n > 5$, then there are at least two different primes between n and $2n$. One easy consequence of this is that $p_{n+2} < 2p_n$, and thus $p_{n+2} < p_n + p_{n+1}$.

2. In 1892, Bertrand's Postulate was generalized by James Joseph Sylvester, as follows: If m and n are positive integers with $m > n$, then at least one of the numbers $m, m+1, m+2, \ldots, m+n-1$ has a prime divisor greater than n. (Bertrand's Postulate follows by taking $m = n + 1$.)

3. There is a similar long-standing open question concerning primes between consecutive squares: *Is there always a prime between n^2 and $(n+1)^2$?*

Dirichlet's Theorem

Dirichlet's Theorem deals with the following problem: If a and b are relatively prime (with $a \neq 0$), do there exist infinitely many primes of the form $ak + b$? (Clearly, if a and b have a common factor $d > 1$, then d will divide $ak + b$ for any value of k, and so $ak + b$ is prime for at most one value of k.) We have already seen a number of special cases of this problem; for example, it was shown in Chapter 1 (see Problem 1-31) that there are infinitely many primes of the form $4k + 3$, and in Chapter 5, the Law of Quadratic Reciprocity was used to show, for example, the existence of infinitely many primes of the form $8k + 3$, $8k + 5$, and $8k + 7$. (See Problems 5-23, 5-48, 5-50, and 5-54.) The proofs of these results clearly use techniques that are specific to the given values of a and b.

The general problem received a good deal of attention in the early part of the nineteenth century. In fact, in 1785, Legendre claimed to have proved the existence of infinitely many primes of the form $ak + b$ when a and b are relatively prime. The problem was finally settled in 1837 by P. G. Lejeune Dirichlet (1805–1859).

(7.11) Dirichlet's Theorem. *Suppose that a is positive and $(a, b) = 1$. Then $ak + b$ is prime for infinitely many values of k.*

Dirichlet's original proof uses tools from the theory of functions of a complex variable, specifically, properties of the *Riemann zeta-function*, defined for suitable complex numbers s by the formula $\zeta(s) = \sum_1^\infty 1/n^s$. Dirichlet's Theorem represents the first significant application of techniques of analysis to number theory. Dirichlet's Theorem and the Prime Number Theorem are, in fact, the two most important results in elementary number theory whose proofs involve analytic methods. There are "elementary" proofs of both results, that is, proofs that do not use deep properties of functions of a complex variable, but they are even more difficult.

We mention finally a result which combines the ideas of Dirichlet's Theorem and the Prime Number Theorem.

(7.12) Theorem (de la Vallée-Poussin). *Suppose that a is positive and $(a, b) = 1$, and let $\pi_{a,b}(x)$ be the number of primes of the form $ak + b$ that are less than or equal to x. Then the ratio $\pi_{a,b}(x)/\pi(x)$ can be made arbitrarily close to $1/\phi(a)$ by take x sufficiently large.*

Notes. 1. The limit $1/\phi(a)$ is independent of the choice of b; as long as a and b are relatively prime, the ratio approaches a value that depends only on a.

2. The following result is a consequence of (7.12): If $d_1, d_2, \ldots, d_m$ and $e_1, e_2, \ldots, e_n$ are any set of digits, where e_n is odd and $e_n \neq 5$, then there exist infinitely many primes that begin with the digits $d_1, d_2, \ldots, d_m$ and end in the digits $e_1, e_2, \ldots, e_n$ (see Problem 7-41).

Theorem 7.12 has a very nice interpretation, which may not be obvious from the above statement. For example, if we take $a = 4$, (7.12) implies (since $\phi(4) = 2$) that one-half of all primes are of the form $4k+1$ and one-half are of the form $4k+3$. (More precisely, the ratio $\pi_{4,1}(x)/\pi(x)$ can be made arbitrarily close to $1/2$ by taking x sufficiently large.) If $a = 8$, then $\phi(8) = 4$, and we conclude that one-fourth of all primes are of the form $8k+1$, one-fourth are of the form $8k+3$, one-fourth are of the form $8k+5$, and one-fourth are of the form $8k+7$.

In general, for a given choice of $a \neq 0$, if $b' \equiv b \pmod{a}$, then n is of the form $ak + b'$ if and only if n is of the form $ak + b$. Thus there are only $\phi(a)$ essentially different values of b such that a and b are relatively prime. Theorem 7.12 implies that for each allowable value of b, the sequence $a, a+b, a+2b, \ldots$ contains its fair share of primes, that is, a fraction of primes equal to $1/\phi(a)$ of the total number of primes.

Goldbach's Conjecture

In a letter to Euler in 1742, Christian Goldbach (1690–1764) speculated that every even integer greater than 2 is the sum of two primes. This famous assertion is known as *Goldbach's Conjecture* and has been the subject of a great deal of study for the past two and a half centuries. The famous English mathematician G. H. Hardy (1877–1947) described Goldbach's Conjecture as one of the most difficult unsolved problems in mathematics.

(7.13) Goldbach's Conjecture. *Every even integer greater than 2 is the sum of two primes.*

The numerical evidence seems clearly to suggest that Goldbach's Conjecture is true and that there are many ways to represent an even integer as a sum of two primes if the integer is large. The conjecture has been verified for every even integer up to $2 \cdot 10^{10}$. In 1973, J.R. Chen showed that every sufficiently large even integer can be represented as $p + m$, where p is prime and either m is prime or m is the product of two primes.

Goldbach also conjectured that every odd integer greater than 7 is the sum of three odd primes. Although this remains an open problem, I. M. Vinogradov (1891–1983) proved in 1937 that all sufficiently large odd positive integers are the sum of three odd primes. In fact, it has been shown that Vinogradov's result holds for every odd integer greater than $3^{3^{15}}$; thus in theory, the result could be checked for all odd positive integers. However, even using the fastest supercomputer, this is completely impractical: $3^{3^{15}}$ is an integer with almost seven million digits!

Other Open Problems

Another famous conjecture states that there are infinitely many pairs of *twin primes*, that is, pairs of consecutive odd integers that are both prime. Leopold Kronecker (1823–1891) remarked, but did not prove, that every even integer can in fact be represented in infinitely many ways as the difference of two primes, which would imply, in particular, that there are an infinite number of twin primes.

Although no proof has been given, the numerical data indicate that the twin-prime conjecture is true. Indeed, it is also likely that there are an infinite number of prime triplets, that is, integers p, $p+2$, $p+6$ which are all prime. (The only three *consecutive* odd integers that are prime are 3, 5, and 7; see Problem 7-3.)

In 1737, Euler proved that the sum $\sum 1/p$ of the reciprocals of the primes is infinite (see Problem 7-8). An interesting way to approach the problem of twin primes is to consider the sum $\sum 1/p + 1/(p+2)$, where the sum is taken over all primes p such that $p+2$ is also prime. If this sum were infinite, that is, if the series diverged, then this would prove the conjecture (for a finite number of prime pairs would clearly imply that the sum is finite). It is known, however, that this series *converges*.

Finally, we mention the following question: *Is there a simple formula that yields every prime, or at least only primes?* No practical formula is known, although Fermat thought (incorrectly) that $2^{2^n} + 1$ is always prime. In this connection, there is the curious result proved by W. H. Mills in 1947: A positive real number a exists such that $[a^{3^n}]$ is prime for all positive integers n ($[x]$ denotes the greatest integer function). However, this is not a practical way to generate primes, since to use this result, the number a must be known to an arbitrarily high degree of accuracy. (It is known that a can be taken to be approximately 1.3064.)

There are, however, well-known examples of polynomials that take on many prime values, at least initially. In a letter to Johann Bernoulli in 1772, Euler pointed out that the quadratic polynomial $x^2 - x + 41$ is prime for $x = 0, 1, \ldots, 40$ (it is clearly composite for $x = 41$), but it is not even known if this polynomial assumes infinitely many prime values.

Goldbach, in correspondence with Euler in 1742, observed that a nonconstant polynomial with integer coefficients cannot assume *only* prime values.

(This is not hard to prove; see Problem 7-47.) More generally, we can ask the question: *Is there a nonlinear polynomial in one variable with integer coefficients that assumes an infinite number of prime values?* No such polynomial is known. (We restrict our attention to nonlinear polynomials, since, for example, the polynomials $4x+1$, $4x-1$, $8x+1$, ... all generate infinitely many primes.) In the case of quadratic polynomials, there is the following long-standing open problem: *Do there exist infinitely many primes of the form $n^2 + 1$?*

For related results concerning these problems, as well as other open questions related to primes, see the book by Ribenboim mentioned in the references at the end of the chapter.

PROBLEMS AND SOLUTIONS

7-1. *If n is composite, prove that n has a prime divisor $p \leq \sqrt{n}$.*

Solution. If n is composite, write $n = ab$, where $a > 1$ and $b > 1$; then a and b cannot both be greater than $\sqrt{n}$, for otherwise, $ab > n$. Therefore one of a or b, and hence n, has a prime factor less than or equal to $\sqrt{n}$.

7-2. *Let p be an odd prime. Prove that every prime divisor of $2^p - 1$ is of the form $8k + 1$ or $8k - 1$.*

Solution. We show in fact that if n is odd, then every prime divisor of $2^n - 1$ is of the required form. Suppose that the prime q (necessarily odd) divides $2^n - 1$. Then $2^n \equiv 1 \pmod{q}$, and hence $2^{n+1} \equiv 2 \pmod{q}$. If $a = 2^{(n+1)/2}$, then $a^2 \equiv 2 \pmod{q}$, and therefore 2 is a quadratic residue of q. By (5.12), it follows that q is of the form $8k \pm 1$.

7-3. *Show that the only three consecutive odd numbers that are all prime are 3, 5, and 7.*

Solution. Let the consecutive odd numbers be n, $n + 2$, and $n + 4$. These are all incongruent modulo 3, so exactly one of them is divisible by 3. Since the numbers are all prime, one of them – clearly n – must be 3.

7-4. *If $n > 1$ and a is a positive integer such that $a^n - 1$ is prime, prove that $a = 2$ and n is prime. (Hint. Use the identity $x^j - 1 = (x-1)(x^{j-1} + x^{j-2} + \cdots + x + 1)$.)*

Solution. Using the identity in the Hint with $x = a$ and $j = n$, we find that $a - 1$ divides $a^n - 1$. Clearly, $a - 1 < a^n - 1$, and therefore $a - 1$ is a nontrivial divisor of $a^n - 1$ unless $a - 1 = 1$. Thus if $a^n - 1$ is prime, then $a = 2$.

Next we show that if $2^n - 1$ is prime, then n is prime. If n is not prime, let $n = st$ with $1 < s < n$. In the identity of the Hint, take $j = t$ and $x = 2^s$. It follows that $2^s - 1$ divides $2^n - 1$, and hence if n has a nontrivial divisor, then $2^n - 1$ cannot be prime.

7-5. If $a \geq 2$ and $a^n + 1$ is prime, show that a is even and n is a power of 2. (Hint. If j is odd, then $x^j + 1 = (x+1)(x^{j-1} - x^{j-2} + \cdots - x + 1)$.)

Solution. The identity of the Hint is easily verified by direct multiplication. If $a^n + 1$ is prime, with $a \geq 2$, then certainly a must be even, for $a \geq 2$ odd implies that $a^n + 1$ is even and greater than 3 and so cannot be prime.

We next show that if $a^n + 1$ is prime, then n is a power of 2. If not, let $n = jk$ with j odd and $1 < j \leq n$. If $x = a^k$, then $a^n + 1 = x^j + 1$. By the identity in the Hint, $a^n + 1$ is then divisible by $a^k + 1$, which is obviously larger than 1 but less than $a^n + 1$. Thus if n is not a power of 2, then $a^n + 1$ cannot be prime.

7-6. Show that there are infinitely many composite positive integers which cannot be expressed as the sum of two primes.

Solution. We will look for *odd* composite integers with the required property, since the existence of infinitely many even examples would strongly violate the Goldbach Conjecture. Let $n = 6k + 5$. If n is the sum of two primes, then n must be of the form $p + 2$, where p is prime. But $n - 2$ is obviously divisible by 3, and thus n is not the sum of two primes if $k > 0$.

▷ **7-7.** If k is a positive integer, let p_k denote the kth prime, and let $\alpha_k(x)$ be the number of positive integers not exceeding x all of whose prime divisors are less than or equal to p_k. Prove that $\alpha_k(x) \leq 2^k \sqrt{x}$. (Hint. Show first that there are no more than 2^k such integers which are square-free.)

Solution. Let $a \leq x$ be a square-free positive integer that is not divisible by any prime greater than p_k, and let $a = 2^{a_1} 3^{a_2} \cdots p_k^{a_k}$. Since each a_i is 0 or 1, there are at most 2^k possibilities for the exponents and therefore at most 2^k possibilities for a. For any a, there are at most $\sqrt{x/a}$ positive integers n such that $n^2 a \leq x$, so for any a, there are no more than $\sqrt{x}$ possibilities for n. It follows that $\alpha_k(x) \leq 2^k \sqrt{x}$.

▷ **7-8.** Use the preceding problem to show that the sum $\sum 1/p$ of the reciprocals of the primes diverges.

Solution. Let p_i denote the ith prime. If the series converges, then there is a positive integer k such that $1/p_{k+1} + 1/p_{k+2} + \cdots < 1/2$. The number of positive integers less than a given x that are divisible by the prime p is no more than x/p. Using the notation of the preceding problem, $x - \alpha_k(x)$ represents the number of positive integers not exceeding x that are divisible by at least one prime greater than p_k, and so $x - \alpha_k(x) < x/p_{k+1} + x/p_{k+2} + \cdots < x/2$. By Problem 7-7, we then have $x/2 < \alpha_k(x) \leq 2^k \sqrt{x}$, and thus $x < 2^{2k+2}$. This inequality is obviously false if $x \geq 2^{2k+2}$, and therefore we conclude that the series diverges.

Perfect Numbers

7-9. Prove that every even perfect number ends in the digit 6 or 8.

Solution. By (7.3), every even perfect number N is of the form $N = 2^{p-1}(2^p - 1)$, where p is prime. If $p = 2$, then $N = 6$; thus we may suppose that p is odd. If

$p = 4k + 1$, then $2^{p-1} = 2^{4k} = 16^k \equiv 6 \pmod{10}$ and $2^p - 1 = 2 \cdot 2^{p-1} - 1 \equiv 1 \pmod{10}$; thus $N \equiv 6 \cdot 1 = 6 \pmod{10}$. If $p = 4k + 3$, then $2^{p-1} = 2^{4k+2} = 4 \cdot 16^k \equiv 4 \pmod{10}$, and $2^p - 1 = 2 \cdot 2^{p-1} - 1 \equiv 7 \pmod{10}$; hence $N \equiv 4 \cdot 7 \equiv 8 \pmod{10}$.

7-10. Suppose that $N > 6$ is an even perfect number. Prove that N is of the form $9k + 1$.

Solution. Every even perfect number N is of the form $N = 2^{p-1}(2^p - 1)$, where p is prime. In particular, if $N > 6$, then p is odd. We show that $2^{n-1}(2^n - 1) \equiv 1 \pmod{9}$ for every odd integer n.

If $n = 1, 3, 5, 7, \ldots$, then 2^{n-1} is congruent in turn to $1, 4, 7, 1, 4, 7, \ldots$ modulo 9, while $2^n - 1$ is congruent in turn to $1, 7, 4, 1, 7, 4, \ldots$ modulo 9. Multiplying, we find that $2^{n-1}(2^n - 1)$ is congruent in turn to $1, 1, 1, 1, \ldots$ modulo 9, and hence $N \equiv 1 \pmod{9}$.

7-11. Prove that if p is prime, then p^k cannot be a perfect number.

Solution. The sum of the divisors of p^k is $1 + p + \cdots + p^k$. This cannot be equal to $2p^k$, or indeed to *any* multiple of p, for it is clear that p cannot divide $1 + p + \cdots + p^k$.

Another proof: Since $1 + p + \cdots + p^{k-1} = (p^k - 1)/(p - 1) < p^k$, it follows that $1 + p + \cdots + p^k < 2p^k$, and so p^k is not perfect.

7-12. Prove that n is a perfect number if and only if $\sum_{d|n} 1/d = 2$.

Solution. As d ranges over the positive divisors of n, n/d also ranges over the positive divisors of n. Thus $\sum_{d|n} n/d = \sigma(n)$, so n is perfect if and only if $\sum_{d|n} n/d = 2n$. Dividing each side by n now yields the required result.

7-13. Use the preceding problem to show that no proper divisor of a perfect number is itself perfect.

Solution. Let n be perfect, and let k be a proper divisor of n. By the preceding problem, $\sum_{d|n} 1/d = 2$. Now every divisor of k is a divisor of n, but n has at least one additional divisor, namely, n itself. Thus $\sum_{d|k} 1/d < 2$, and hence k is not perfect.

7-14. If p and q are distinct odd primes, prove that pq is not a perfect number.

Solution. By the Note after (1.20), $\sigma(pq) = \sigma(p)\sigma(q) = (p+1)(q+1)$. This cannot be equal to $2pq$, for $(p+1)(q+1)$ is divisible by 4 but $2pq$ is not.

Note. The same argument can be used to show that a product of any number of distinct odd primes cannot be perfect. Thus an odd perfect number (if one exists) cannot be square-free.

Mersenne Primes and Fermat Numbers

7-15. *Use Theorem 7.5 to show that $2^7 - 1$ is prime.*

Solution. By (7.5), every prime divisor of $2^7 - 1 = 127$ is of the form $14k + 1$. We only have to check primes of this form up to $\sqrt{127}$. Since there aren't any, we conclude that 127 is prime.

7-16. *Use Theorem 7.5 to determine if $2^{11} - 1$ is prime.*

Solution. In view of (7.5) and Problem 7-1, we check for prime divisors q of the form $22k + 1$, where $q < \sqrt{2^{11} - 1} \approx 45$. The only such q is 23, and it is easily checked that $23 | 2^{11} - 1$, so $2^{11} - 1$ is composite.

7-17. *Determine if $2^{29} - 1$ is prime.*

Solution. The only possible divisors are of the form $58k + 1$, namely, 59, 117, 175, 233, and so on. It is easily checked that 59 is not a divisor; 117 and 175 are not primes and need not be checked, since the smallest divisor of $2^{29} - 1$ greater than 1 must be prime. It is easy to verify that 233 is prime and divides $2^{29} - 1$, so $2^{29} - 1$ is composite.

7-18. *Prove that $2^{16}(2^{17} - 1)$ is a perfect number.*

Solution. In view of (7.2), it suffices to show that $2^{17} - 1$ is prime. It is necessary only to check for prime divisors up to $\sqrt{2^{17} - 1}$. By (7.5), prime factors of M_{17} are of the form $34k + 1$; thus we have to check only 103, 137, 239, and 307. Since none of these divides M_{17}, we conclude that $2^{17} - 1$ is prime.

7-19. *(a) Use (7.5) to decide if $M_{23} = 2^{23} - 1$ is prime.*
  (b) Use the fact that 2 is a quadratic residue of 47 to determine if M_{23} is prime.

Solution. (a) By (7.5), any prime divisor of M_{23} must be of the form $46k + 1$: 47, 139, 277, ... It is easy to verify that $47 | M_{23}$.
  (b) Since 2 is a quadratic residue of 47, $2^{23} \equiv 1 \pmod{47}$ by Euler's Criterion, and hence $47 | 2^{23} - 1$.

7-20. *Let q be a prime of the form $4k + 3$, and suppose that $2q + 1$ is prime. Show that $2q + 1$ is a divisor of M_q. Use this to conclude that if q is such a prime and $q > 3$, then M_q is composite. (Hint. Show first that 2 is a quadratic residue of $2q + 1$.)*

Solution. Since $2q + 1$ is a prime of the form $8k + 7$, it follows from (5.12) that 2 is a quadratic residue of $2q + 1$. Thus $2^q \equiv 1 \pmod{2q + 1}$ by Euler's Criterion, and hence $2q + 1$ divides $2^q - 1$. Therefore $2q + 1$ is a proper divisor of M_q unless $2q + 1 = M_q$. If $q = 3$, then $2q + 1 = M_q = 7$, and we do not obtain a proper divisor of M_q, but it is easy to verify that if $n > 3$, then $2n + 1 < 2^n - 1$.

PROBLEMS AND SOLUTIONS 209

7-21. *Let m be an odd positive integer. Prove that m divides infinitely many Mersenne numbers M_n. (Hint. Use Euler's Theorem.)*

Solution. By Euler's Theorem, $2^{\phi(m)} \equiv 1 \pmod{m}$, and thus $2^{k\phi(m)} \equiv 1 \pmod{m}$ for any positive integer k. Therefore m divides M_n whenever n is a positive multiple of $\phi(m)$.

7-22. *Prove that $(2^m - 1, 2^n - 1) = 2^{(m,n)} - 1$. (Hint. Let $d = (2^m - 1, 2^n - 1)$, and consider the order of 2 modulo d. Also recall that $2^r - 1$ divides $2^s - 1$ if r divides s.)*

Solution. If $d = (2^m - 1, 2^n - 1)$, then $2^m \equiv 1 \pmod{d}$, and so $\text{ord}\, 2 \mid m$ by (6.2.i); similarly, $\text{ord}\, 2 \mid n$. Thus $\text{ord}\, 2 \mid (m, n)$ and hence $2^{(m,n)} \equiv 1 \pmod{d}$, that is, d divides $2^{(m,n)} - 1$. But $2^{(m,n)} - 1$ is a common divisor of $2^m - 1$ and $2^n - 1$ and so must divide d. We therefore conclude that $d = 2^{(m,n)} - 1$.

Another proof: We can also give an induction argument that uses only simple properties of the greatest common divisor. Let d be fixed. We show that if the result holds for all s and t such that $0 < s \le t < n$ with $(s, t) = d$, then it holds for all s, t with $0 < s \le t \le n$ and $(s, t) = d$. This is obvious if $s = t$, so we may assume that $s < t$. Using the fact that $(a, b) = (a, b - a)$, we find that

$$(2^s - 1, 2^t - 1) = (2^s - 1, 2^t - 2^s) = (2^s - 1, 2^s(2^{t-s} - 1)) = (2^s - 1, 2^{t-s} - 1).$$

Since $(s, t - s) = d$ and s and $t - s$ are both less than n, we have $(2^s - 1, 2^{t-s} - 1) = 2^d - 1$ by the induction hypothesis, and therefore $(2^s - 1, 2^t - 1) = 2^d - 1$.

Note. A similar argument shows that $(a^m - 1, a^n - 1) = a^{(m,n)} - 1$ for any $a \ge 2$.

7-23. *Show that if p and q are odd primes and $p \mid 2^q + 1$, then either $p = 3$ or p is of the form $2kq + 1$. Use this to find two prime divisors of $2^{13} + 1$.*

Solution. First we prove the general assertion by arguing as in the proof of (7.5). If $p \mid 2^q + 1$, then $2^q \equiv -1 \pmod{p}$ and thus $2^{2q} \equiv 1 \pmod{p}$. Therefore the order of 2 modulo p divides $2q$ and hence is 2 or $2q$. (It cannot be q, since $2^q \equiv -1 \pmod{p}$.) The integer 2 has order 2 only when $p = 3$. In all other cases, $2q \mid p - 1$, i.e., p is of the form $2kq + 1$.

Since $2^2 \equiv 1 \pmod{3}$, it follows that $2^{13} \equiv 2 \pmod{3}$, so $3 \mid 2^{13} + 1$. Since $(2^{13} + 1)/3 = 2731$, the other prime factors of $2^{13} + 1$ will be of the form $26k + 1$ and must divide 2731. But there are no primes of this form less than $\sqrt{2731}$. Thus 2731 is prime, and so 3 and 2731 are the only prime divisors of $2^{13} + 1$.

▷ **7-24.** *(a) Let F_k denote the kth Fermat number. Prove that $(F_m, F_n) = 1$ if $m \ne n$. (Hint. Suppose $p \mid F_m$; then $2^{2^m} \equiv -1 \pmod{p}$.)*
(b) Use part (a) to give another proof that there exist infinitely many primes.

Solution. (a) We may suppose that $m < n$. Let p be any prime divisor of F_m; then $F_m - 1 = 2^{2^m} \equiv -1 \pmod{p}$. Since 2^{2^n} is 2^{2^m} raised to an even power, $F_n \equiv 2 \pmod{p}$. In particular, since p is odd, p cannot divide F_n.

(b) Each F_k has some prime divisor p_k (possibly F_k itself). Since $(F_m, F_n) = 1$ if $m \neq n$, it follows that $p_m \neq p_n$ if $m \neq n$. Thus distinct values of k gives rise to distinct primes p_k, and hence there are infinitely many primes.

Note. We do not need to know whether infinitely many of the F_k are primes; indeed, this is still an open question.

7-25. Let $c_0 = 2$ and $c_{n+1} = c_n^2 - c_n + 1$ for $n \geq 0$. Show that if $m \neq n$, then c_m and c_n are relatively prime. (Hint. We may assume that $m < n$. Show that if the prime p divides c_m, then p does not divide $c_{m+1}, c_{m+2}, \ldots$.)

Solution. We show by induction on i that if p divides c_m, then $c_{m+i} \equiv 1 \pmod{p}$ for every $i \geq 1$. Let $P(x) = x^2 - x + 1$. If $i = 1$, then $c_{m+i} = P(c_m) \equiv 0^2 - 0 + 1 = 1 \pmod{p}$. Now suppose that $c_{m+i} \equiv 1 \pmod{p}$; then $c_{m+i+1} = P(c_{m+i}) \equiv 1^2 - 1 + 1 = 1 \pmod{p}$.

Note. This gives still another proof that there are infinitely many primes. There is a close relationship between this problem and the preceding problem. If F_n is the nth Fermat number, it is easy to verify that $F_{n+1} = F_n^2 - 2F_n + 2 = Q(F_n)$, where $Q(x) = x^2 - 2x + 2$. The argument we have used can be easily adapted to show that if the prime p divides F_m, then $F_{m+i} \equiv 2 \pmod{p}$ for every positive integer i. Since every Fermat number is odd, this shows that any two distinct Fermat numbers are relatively prime. Similar results about Fermat-like numbers can be proved by using related polynomials.

7-26. Prove that $F_4 = 2^{16} + 1$ is prime. (Hint. Use (7.7).)

Solution. By (7.7), any prime divisor of F_4 must be of the form $2^6 k + 1$. We examine all the primes less than or equal to $\sqrt{F_4}$ which are of this form. Only 193 qualifies, and it does not divide F_4; thus F_4 is prime.

7-27. If F_n denotes the nth Fermat number, prove that $F_n + 2$ is composite for infinitely many n. (Hint. Work modulo 7.)

Solution. Note that $2^{2^{n+1}}$ is the square of 2^{2^n}; thus, beginning with $n = 0$, 2^{2^n} is in turn congruent to $2, 4, 2, 4, \ldots$ modulo 7. It follows that if n is odd, then $F_n + 2 \equiv 4 + 1 + 2 \equiv 0 \pmod{7}$. Thus $F_n + 2$ is composite for all odd $n > 1$.

7-28. In a letter to Frenicle, Fermat conjectured that if a is even, then $N = a^{2^n} + 1$ is prime unless N is divisible by some Fermat number F_k. Disprove this conjecture. (Hint. Take $a = 12$.)

Solution. Let $a = 12$ and $n = 2$. Then $N = 89 \cdot 233$, 89 and 233 are prime, and neither is a Fermat number.

7-29. If $n > 1$, prove that the Fermat number F_n ends in the digit 7.

Solution. We show that if $n > 1$, then $F_n \equiv 2 \pmod{5}$. Since F_n is odd, it will follow that F_n ends in the digit 7. If $n > 1$, then $4 | 2^n$; let $2^n = 4k$. Then $F_n = (2^4)^k + 1 \equiv 1^k + 1 = 2 \pmod{5}$.

7-30. (Pepin's Test.) *Let $n \geq 1$. Prove that the Fermat number F_n is prime if and only if $3^{(F_n-1)/2} \equiv -1 \pmod{F_n}$. (Hint. Use (6.9).)*

Solution. Suppose first that F_n is prime, where $n \geq 1$. It is easy to verify that 3 is a quadratic nonresidue of F_n: $F_n \equiv 1 \pmod 4$ and $F_n \equiv 2 \pmod 3$, so $(3/F_n) = (F_n/3) = (2/3) = -1$. Hence $3^{(F_n-1)/2} \equiv -1 \pmod{F_n}$ by Euler's Criterion.

The converse is an immediate consequence of (6.9), the Lucas primality test. For if $3^{(F_n-1)/2} \equiv -1 \pmod{F_n}$, then $3^{F_n-1} \equiv 1 \pmod{F_n}$. Since the only prime divisor of $F_n - 1$ is $q = 2$ and $3^{(F_n-1)/2} \not\equiv 1 \pmod{F_n}$, it follows from (6.9) that F_n is prime.

The Prime Number Theorem and Bertrand's Postulate

7-31. *Let p_n denote the nth prime. Use Bertrand's Postulate to show that $p_n < 2^n$ if $n \geq 2$.*

Solution. We prove the result by induction on n. The result is clearly true when $n = 2$, for $p_2 = 3 < 2^2$. Now suppose the result holds for $n = k$, i.e., $p_k < 2^k$; we show that $p_{k+1} < 2^{k+1}$. By Bertrand's Postulate, there is a prime between p_k and $2p_k$, and thus $p_{k+1} < 2p_k$. But by the induction hypothesis, $p_k < 2^k$ and therefore $p_{k+1} < 2p_k < 2 \cdot 2^k = 2^{k+1}$. Thus the result holds for $n = k+1$.

7-32. *Let p_k denote the kth prime. Using the fact that for $n \geq 6$ there are always at least two primes between n and $2n$, prove that $p_{k+2} \leq p_k + p_{k+1}$.*

Solution. For $p_k = 2, 3,$ or 5, it is easy to verify the assertion directly. If $p_k > 5$, then there are at least two primes between p_k and $2p_k$, so $p_{k+2} < 2p_k$. But clearly, $2p_k < p_k + p_{k+1}$ and hence $p_{k+2} < p_k + p_{k+1}$.

7-33. *Use Bertrand's Postulate to prove that if $m \geq 2$ and $m! = p_1^{a_1} \cdots p_r^{a_r}$, then $a_i = 1$ for at least one value of i.*

Solution. The result is clearly true if $m = 2$. For $m > 2$, let $m = 2k$ or $m = 2k + 1$, according as m is even or odd. If p is a prime between k and $2k$, then $2p > m$ and therefore the exponent of p in the prime factorization of $m!$ is equal to 1.

7-34. *Suppose that $m > 1$. Use the previous problem to show that $m!$ is never a kth power for any $k \geq 2$.*

Solution. By the previous problem, there exists a prime p such that the exponent of p in the prime factorization of $m!$ is equal to 1. But the exponent of any prime in the prime factorization of a kth power is a multiple of k, so $m!$ cannot be a kth power for any $k > 1$.

▷ **7-35.** *Let $s(x)$ denote the number of positive perfect squares less than or equal to x. Use the Prime Number Theorem to show that $s(x)/\pi(x)$ can be made as small as desired by taking x sufficiently large. (This shows that there are* many more primes than squares in the sense that for large x, there are many more primes than squares in the interval $1 \leq u \leq x$.)

Solution. Suppose $x > 1$. Let $g(x) = x/\log x$; then $s(x)/\pi(x) = (s(x)/g(x))(g(x)/\pi(x))$. But $s(x) \leq \sqrt{x}$, so $s(x)/g(x) \leq (\log x/\sqrt{x})(g(x)/\pi(x))$. Now by the Prime Number Theorem, $g(x)/\pi(x)$ approaches 1 as x gets large. Also, by standard techniques from calculus, $\log x/\sqrt{x}$ approaches 0 as x gets large. It follows that $s(x)/\pi(x)$ approaches 0 as x gets large.

▷ **7-36.** A theorem of Chebyshev states that there are positive constants c and C such that $c(x/\log x) < \pi(x) < C(x/\log x)$ for all $x \geq 2$. Use this to give another proof of the fact that there are arbitrarily long gaps in the sequence of primes.

Solution. Suppose to the contrary that there are not arbitrarily large gaps between primes. Then there is an integer d such that the difference between consecutive primes never exceeds d, and thus for any number $x \geq 2$, there is a prime p with $x < p \leq x+d$. It follows that there are at least $(x-2)/d$ primes less than or equal to x, and so $\pi(x) \geq (x-2)/d$. But by Chebyshev's Theorem, $\pi(x) < Cx/\log x$ for some positive number C. Hence for all $x > 2$, $(x-2)/d \leq \pi(x) < Cx/\log x$. This implies that $\log x < Cdx/(x-2)$ for all $x > 2$, which is obviously false since $\log x$ can be made arbitrarily large by taking x large enough.

Dirichlet's Theorem

7-37. Suppose that whenever c and d are positive and relatively prime, there is at least one prime of the form $ck + d$. If a and b are positive and $(a, b) = 1$, prove that there are infinitely many primes of the form $ak + b$. (*Hint.* Show that there is a prime of the form $a^n k + b$ for any $n \geq 1$.)

Solution. We can assume $a > 1$, since for $a = 1$ the sequence $ak + b$ is just $b + 1, b + 2, b + 3, \ldots$, which clearly contains infinitely many primes. Since $(a, b) = 1$, we also have $(a^n, b) = 1$ for any $n \geq 1$; hence by assumption, there is a prime p of the form $a^n k + b$, and therefore of the form $am + b$. Since $p > a^n$, it follows that for any $n \geq 1$, we can find a prime of the form $am + b$ that exceeds a^n, and hence there exist infinitely many such primes.

7-38. Let N be a positive integer, and suppose that $(a, b) = 1$. Without using Dirichlet's Theorem, prove that there are infinitely many numbers of the form $ak + b$ which are relatively prime to N. (*Hint.* Let P be the product of the primes that divide N but do not divide a, and let $k = s + tP$ for a suitable s.)

Solution. Let P be defined as in the hint. Since $(a, P) = 1$, the congruence $ax + b \equiv 1 \pmod{P}$ has a solution s; then $s + tP$ is also a solution for any integer t. Let $k = s + tP$. We show that $ak + b$ is relatively prime to N by showing that if the prime p divides N, then p does not divide $ak + b$.

If p divides a, then p cannot divide $ak+b$, for a and b are relatively prime. If p does not divide a, then p divides P, and hence $as + b \equiv 1 \pmod{p}$. Thus $ak + b \equiv as + b \equiv 1$

(mod p), so again p does not divide $ak+b$. It follows that $ak+b$ and N are relatively prime. Since there are no restrictions on t, there exist infinitely many numbers of the form $ak+b$ that are relatively prime to N.

▷ **7-39.** *Find a prime p of the form $8k+1$ such that 3, 5, 7, and 11 are quadratic residues of p. (Hint. Use the Law of Quadratic Reciprocity and Dirichlet's Theorem.)*

Solution. Since p is to be of the form $8k+1$, the Law of Quadratic Reciprocity implies that we want $(p/3) = (p/5) = (p/7) = (p/11) = 1$. This imposes certain conditions on the remainders when p is divided by 3, 5, 7, and 11. We need $p \equiv 1 \pmod{3}$, $p \equiv 1$ or $4 \pmod{5}$, $p \equiv 1, 2,$ or $4 \pmod{7}$, and $p \equiv 1, 3, 4, 5,$ or $9 \pmod{11}$ (in addition to $p \equiv 1 \pmod{8}$). By combining the possibilities for the remainders in any of the $2 \cdot 3 \cdot 5 = 30$ possible ways, we obtain 30 different systems of five congruences. For any of these systems, we can use the Chinese Remainder Theorem to obtain an equivalent congruence of the form $p \equiv b \pmod{N}$, where $N = 8 \cdot 3 \cdot 5 \cdot 7 \cdot 11$ and b takes on any one of 30 values, which are all easily seen to be relatively prime to N. Thus there are 30 arithmetic progressions in which we can look for primes, and by Dirichlet's Theorem, each will contain infinitely many primes. The simplest case is obtained by taking $p \equiv 1$ modulo 8, 3, 5, 7, and 11; in this case, we look for primes of the form $1 + Nk$. We get lucky immediately: $N = 9240$, and it is not difficult to show that 9241 is prime.

Alternatively, we can combine just the congruences modulo 8, 3, and 5 and therefore look for primes of the form $120k+1$ or $120k+49$ which satisfy additional congruences modulo 7 and 11. If we proceed in this way, we find fairly quickly that $p = 2689$ has the required properties, being congruent to 1 modulo 3, 7, and 8, to 4 modulo 5, and to 5 modulo 11. This is the smallest example.

Note. In general, if we want a prime p of the form $8k+1$ such that the first $n-1$ odd primes $p_2, p_3, \ldots, p_n$ are quadratic residues of p, we can take p to be any prime in the sequence $N+1, 2N+1, 3N+1, \ldots,$ where $N = 8p_2p_3 \cdots p_n$.

▷ **7-40.** *Use Dirichlet's Theorem to prove that for any number x, there are infinitely many primes p such that the smallest positive primitive root of p is greater than x. (Hint. Use the preceding Note.)*

Solution. Let $p_1 < p_2 < \cdots < p_n$ be the primes less than or equal to x, and let p be a prime of the form $8k+1$ such that $p_2, p_3, \ldots, p_n$ are quadratic residues of p; by Dirichlet's Theorem, there are infinitely many such p (see the preceding Note). Since p is of the form $8k+1$, 2 is also a quadratic residue of p. If m is any integer whose prime factorization uses at most the primes $p_1, p_2, p_3, \ldots, p_n$, then m is a quadratic residue of p, since a product of quadratic residues is a quadratic residue. Thus any positive integer less than or equal to x is a quadratic residue of p. Now let g be a positive primitive root of p. Since g is a quadratic nonresidue of p, it follows that g must exceed x.

▷ **7-41.** *Use Theorem 7.12 to show that there are infinitely many primes whose decimal representation begins with 1 and ends with 7.*

Solution. Denote by S_n the set of all numbers of the form $10k+7$ that lie between 10^n and $2 \cdot 10^n$. The decimal representation of any number in S_n begins with 1 and ends with 7.

Let $f(x)$ be the number of primes of the form $10k+7$ that are no greater than x. If $y = 10^n$, then the number of primes in S_n is $f(2y) - f(y)$. Since $\phi(10) = 4$, (7.12) implies that the ratio $f(x)/(x/\log x)$ can be made arbitrarily close to $1/4$ by taking x large enough. Take n so large that for all $x > 10^n$, the ratio lies between $1/5$ and $1/3$. Thus there are at least $g(y) = (1/5)2y/\log(2y) - (1/3)y/\log y$ primes in S_n. Calculation shows that $g(y) = y(\log y - 5\log 2)/15 \log y(\log 2 + \log y)$. It is easy to show that $g(y)$ can be made arbitrarily large by taking y large enough; in particular, $g(y)$ can be made greater than 1. Thus for any sufficiently large integer n, there is at least one prime in S_n.

Goldbach's Conjecture and Other Open Problems

7-42. *Prove that Goldbach's Conjecture implies that every odd integer greater than 7 is the sum of three odd primes.*

Solution. If n is an odd integer greater than 7, then $n-3$ is even and greater than 4. Thus, by Goldbach's Conjecture, $n-3$ is the sum of two primes p and q. Moreover, p and q are odd, since only 4 is the sum of two even primes. It follows that $n = 3+p+q$, the sum of three odd primes.

7-43. *It is known that every sufficiently large odd positive integer is a sum of three odd primes (Vinogradov). Use this to prove that every sufficiently large positive integer is a sum of at most four primes.*

Solution. We show, using Vinogradov's result, that every sufficiently large *even* integer is a sum of four primes. Let n be even and so large that the odd number $n-3$ is a sum of the three odd primes p, q, and r. Then $n = 3 + p + q + r$.

7-44. *Prove that Goldbach's Conjecture is equivalent to the following statement: Every $n > 5$ is a sum of three primes.*

Solution. First assume that Goldbach's Conjecture is true. We show that every $n > 5$ is a sum of three primes. If $n > 5$ is even, then $n-2$ is even and at least 4, so $n-2$ is the sum $p+q$ of two primes. Therefore $n = 2+p+q$ is the sum of three primes. If $n > 5$ is odd, then $n-3$ is even and at least 4, so $n-3$ is the sum of two primes, and hence n is the sum of three primes.

Now assume that every $n > 5$ is the sum of three primes. We show that Goldbach's Conjecture holds. Let $n \geq 4$ be even; then $n+2$ is a sum of three primes. But since $n+2$ is even, at least one of these primes must be equal to 2 (if they were all odd, then $n+2$ would be odd). Thus $n+2 = 2+p+q$ for some primes p and q, and therefore $n = p+q$ is the sum of two primes.

7-45. (a) Show that if $x-1$ and $x+1$ are prime, then x^2-1 has precisely four positive divisors. (b) Show that if x is positive and x^2-1 has precisely four positive divisors, then either $x=3$ or $x-1$ and $x+1$ are both prime.

Solution. (a) Let $p = x-1$ and $q = x+1$. Then $x^2-1 = pq$, and it is clear that 1, p, q, and pq are the only positive divisors of pq.

(b) The numbers 1, $x-1$, $x+1$, and x^2-1 are divisors of x^2-1 and are all distinct, since clearly $x > 2$. Thus they are the only divisors of x^2-1. In particular, $x-1$ cannot have positive divisors other than 1 and $x-1$, and therefore $x-1$ is prime. Also, $x+1$ cannot have positive divisors other than 1, $x+1$, and (possibly) $x-1$. Thus, if $x-1$ does not divide $x+1$, then $x+1$ is prime, and if $x > 2$ and $x-1$ divides $x+1$, then $x = 3$.

7-46. (Schinzel, 1958.) (a) Let m and n be positive integers. Prove that there is an integer c such that $(c, m) = (2n - c, m) = 1$. (Hint. First consider the case when m is prime, and then use the prime factorization of m.)
(b) Use the result of part (a) and Dirichlet's Theorem to show that given any modulus m and any even number $2n$, there are infinitely many primes p and q such that $2n \equiv p + q \pmod{m}$.

Solution. (a) Suppose m has the prime factorization $m = \prod_1^t p_i^{a_i}$. We show that for any prime p_i, there is an integer c_i such that $(c_i, p_i) = (2n - c_i, p_i) = 1$. If $p_i = 2$, set $c_i = 1$. If p_i is an odd prime, then p_i cannot divide both $2n-1$ and $2n+1$. Therefore set $c_i = 1$ if $(2n-1, p_i) = 1$ and $c_i = -1$ otherwise. Now use the Chinese Remainder Theorem to find an integer c such that $c \equiv c_i \pmod{p_i}$ for $1 \le i \le t$. It is clear that $(c, m) = (2n - c, m) = 1$.

(b) Use part (a) to find an integer c such that $(c, m) = (2n - c, m) = 1$. Let p and q, respectively, be primes of the form $mk + c$ and $mk + 2n - c$. (There are infinitely many primes of each form, by Dirichlet's Theorem.) It is obvious that $p + q \equiv 2n \pmod{m}$.

Note. Part (b) is a *very* weak version of Goldbach's Conjecture, since the primes p and q depend on m.

7-47. Prove that there is no nonconstant polynomial $f(x)$ with integer coefficients such that $f(n)$ is prime for every positive integer n.

Solution. Suppose that $f(1) = p$, where p is prime. Since $1 + kp \equiv 1 \pmod{p}$, it follows from (2.3.iii) that $f(1 + kp) \equiv f(1) \equiv 0 \pmod{p}$ for any integer k. Thus if $f(n)$ is to be prime for every positive integer n, we must have $f(1 + kp) = p$ for *every* integer k. Let $g(x) = f(x) - p$. If $f(1 + kp) = p$ for all k, then the equation $g(x) = 0$ has infinitely many roots. This is impossible unless $g(x)$ is the zero polynomial, i.e., $f(x)$ is the constant polynomial p.

▷ **7-48.** Let $f(x)$ be a nonconstant polynomial with integer coefficients. Show that there are infinitely many primes p such that the congruence $f(x) \equiv 0 \pmod{p}$ has a solution. (Hint. We may assume that $a_0 \ne 0$, where a_0 is the constant term of $f(x)$. Consider $f(a_0 x) = a_0(1 + xg(x))$.)

Solution. If $a_0 = 0$, then $f(0) = 0$, and thus $f(0)$ is divisible by every prime. If $a_0 \neq 0$, then $f(a_0 x) = a_0(1 + xg(x))$ for some polynomial g with integer coefficients. We show that given any integer n, we can find a prime $p > n$ such that the congruence $f(x) \equiv 0$ (mod p) has a solution.

The equations $1 + xg(x) = 1$ and $1 + xg(x) = -1$ have only finitely many solutions each. Let $b = N!$, where $N \geq n$ is chosen so that b is larger than any of these solutions. Then $1 + bg(b)$ cannot be equal to 1 or -1 and therefore is divisible by some prime p. If $p \leq N$, then $p \mid b$ and so p cannot divide $1 + bg(b)$. Thus $p > N \geq n$. Since $p \mid 1 + bg(b)$, it follows that $f(a_0 b) \equiv 0$ (mod p).

7-49. Use the previous problem to show that if $f(x)$ is a nonconstant polynomial, then for any k, there is a number a such that $f(a)$ is divisible by at least k primes.

Solution. By the previous problem, for any k, we can find k primes $p_1, p_2, \ldots, p_k$ such that the congruence $f(x) \equiv 0$ (mod p_i) has a solution, say a_i. Use the Chinese Remainder Theorem to find an integer a such that $a \equiv a_i$ (mod p_i) for all $i \leq k$. Then $f(a) \equiv 0$ (mod p_i) for all $i \leq k$, and thus $f(a)$ is divisible by at least k primes.

EXERCISES FOR CHAPTER 7

1. Prove or disprove: There are values of $n > 1$ such that $n!$ is a perfect square.
2. What is the first even perfect number greater than 10^6?
3. What are the last two digits of the perfect number $2^{11212}(2^{11213} - 1)$?
4. Prove or disprove: If $2^p - 1$ is prime, then $n = 2^{p-1} + 2^p + 2^{p+1} + \cdots + 2^{2(p-1)}$ is a perfect number.
5. Show that every perfect number that ends in 8 ends in 28.
6. Prove or disprove: If r, s, and t are distinct even perfect numbers, then $\phi(rst) = 4\phi(r)\phi(s)\phi(t)$.
7. Show that any odd perfect number is of the form pm^2, where p is prime and $p \equiv 1$ (mod 4). (See Problem 7-14.)
8. The positive integer n is called *abundant* if $\sigma(n) > 2n$. (a) Find the smallest odd abundant number. (It was once believed that there were none.) (b) If n is abundant, show that any positive multiple of n is abundant.
9. Find the greatest common divisor of $2^{30} - 1$ and $2^{54} - 1$.
10. Let N be an integer whose decimal expansion consists entirely of 1's. Show that if N is prime, then the number of 1's is prime.
11. Factor $2^{26} + 1$ and $2^{34} + 1$. (Hint. $4n^4 + 1 = (2n^2 + 2n + 1)(2n^2 - 2n + 1)$.)
12. If F_n denotes the nth Fermat number, prove that $F_n + 4$ is never prime if $n \geq 1$.

13. Show that $2^{2^n}+7$ is composite for infinitely many values of n. (Hint. Work modulo 11.)

14. Prove that if $n \geq 1$, then the Fermat number $2^{2^n}+1$ is of the form $9k-1$ or $9k-4$.

15. Show that $\phi(n)$ is a power of 2 if and only if n is of the form $2^s p_1 p_2 \cdots p_t$, where $s, t \geq 0$ and the p_i are distinct Fermat primes. (Note that an empty product is defined to be 1.)

16. Prove that n and $2^{2^n}+1$ are relatively prime for every $n \geq 1$. (Hint. Use the fact that any prime divisor of $2^{2^n}+1$ is of the form $2^{n+2}k+1$.)

17. Use Bertrand's Postulate to show that for every integer $k > 2$, there is a prime p such that $p < k < 2p$.

18. Use Dirichlet's Theorem to show that there exist infinitely many primes whose decimal representation ends in 1111.

19. Use Dirichlet's Theorem to show that for any positive integer m, there exist infinitely many primes p and q such that $q \equiv p+2 \pmod{m}$. (Hint. See Problem 7-46.)

20. Assume that Goldbach's Conjecture is true. Show that any odd integer n can be expressed in infinitely many different ways as $n = p+q-r$, where p, q, and r are prime.

NOTES FOR CHAPTER 7

1. The function $x/\log x$ is by no means the best reasonably simple approximation to $\pi(x)$. By the age of 15, Gauss had somehow guessed that an excellent approximation is given by $\operatorname{li}(x)$, the *logarithmic integral*, defined by $\operatorname{li}(x) = \int_2^x (1/\log t)\, dt$.

A vast amount of effort has gone into estimating the size of the error term, that is, in studying the behavior of the function $\operatorname{li}(x) - \pi(x)$. It is known that this function changes sign infinitely often, and there are grounds for conjecturing that for large enough x, we have $|\operatorname{li}(x) - \pi(x)| < x^{1/2} \log x$, but the problem is very difficult. The question of the size of the error term is intimately connected with one of the most famous open problems in all of mathematics, the *Riemann Hypothesis*. This asserts that all the zeros of the function $\sum_{n=1}^{\infty} 1/n^s$, where s is a complex number, are of the form $s = 1/2 + t\sqrt{-1}$, where t is real. There is a very large amount of evidence that the Riemann Hypothesis is true, but a proof has eluded determined attacks for more than a hundred years.

It is known that $|\operatorname{li}(x) - \pi(x)| > \sqrt{x}$ for infinitely many x. That is something we should remember if we interpret the Prime Number Theorem as saying that $\pi(x)$ is "approximately" $\operatorname{li}(x)$ (or "approximately" $x/\log x$). This is true only in the sense that for large x, the error term is very much smaller than $\pi(x)$, but that does not mean that the error is small in absolute terms. For example, it is known that if $x = 4 \cdot 10^{16}$, then $\pi(x)$ is exactly equal to 1,075,292,778,753,150. The number $x/\log x$ is too small by about 28,929,900,579,950, a 2.7% error, whereas li(x) is too large by approximately 5,538,861, an error of about five parts per billion.

2. Let $\omega(n)$ be the number of *distinct* prime factors of n. The function $\omega(n)$ behaves quite irregularly, but its *average* size can be determined. If $f(x) = \sum_{n \leq x} \omega(n)$, then the ratio $f(x)/x \log(\log x)$ can be made arbitrarily close to 1 by taking x sufficiently large. Thus, for example, the positive integers up to approximately one billion have, on average, only three distinct prime factors. A related result is even more surprising. Let $\Omega(n)$ be the *total* number of prime factors of n; thus if $n = p_1^{a_1} p_2^{a_2} \cdots p_k^{a_k}$, then $\Omega(n) = a_1 + a_2 + \cdots + a_k$. If $g(x) = \sum_{n \leq x} \Omega(n)$, then again the ratio $g(x)/x \log(\log x)$ can be made arbitrarily close to 1 by taking x sufficiently large. For more details, see Hardy and Wright's *An Introduction to the Theory of Numbers*.

3. The largest known primes have almost always been Mersenne primes. The most recent exception was a short period from August 1989 to March 1992 when $391581 \cdot 2^{216193} - 1$ was the largest prime known; it was supplanted in March 1992 by the Mersenne prime $2^{756839} - 1$.

BIOGRAPHICAL SKETCHES

Pavnuty Chebyshev was born in 1821 and became by far the most important Russian mathematician of his generation. Under his guidance, St. Petersburg became, for the first time since the age of Euler, a center of mathematical activity. He made significant contributions to many branches of mathematics, including probability and numerical analysis. In his work in numerical analysis, Chebyshev used the continued fraction expansion of a function extensively. Unlike many nineteenth-century mathematicians, Chebyshev paid careful attention to error bounds.

Beside obtaining good upper and lower bounds on the ratio of $\pi(x)$ to $x/\log x$, Chebyshev showed in 1852 that if the ratio had a limit, that limit must be 1. He also pointed out that Bertrand's postulate follows easily from his estimates for $\pi(x)$. Chebyshev's other main contribution to number theory is in the area of Diophantine approximation (which is, roughly speaking, the study of minimum values of functions as the variables range over the integers).

Chebyshev died in St. Petersburg in 1894.

Peter Gustav Lejeune Dirichlet was born in 1805, near the city of Cologne, Germany. His high school mathematics teacher was Georg Ohm, now famous for Ohm's law. In 1822, Dirichlet went to Paris, at that time the world center for the mathematical sciences. Dirichlet was strongly influenced by Fourier but even more by reading Gauss's *Disquisitiones Arithmeticae*. After a brief period in Breslau, he accepted a position at the University of Berlin, where he stayed almost to the end of his life, among colleagues and students who included such mathematicians as Jacobi, Kummer, Eisenstein, Kronecker, Dedekind, and Riemann. Jacobi in particular was a close personal friend.

Dirichlet made major advances in complex analysis and Fourier series. In number theory, his first significant result came in 1828, when he settled, together with Legendre, the case $n = 5$ of Fermat's Last Theorem. Dirichlet's main number-theoretic achievement is probably the proof that all suitable arithmetical progressions have infinitely many primes; in the proof, he introduced analytic methods that, when suitably refined by others, were to lead to many other fundamental results, including the Prime Number Theorem. Dirichlet also made an important contribution to the theory of algebraic integers by showing that the units could be generated in a relatively simple way. His *Vorlesungen über Zahlentheorie* was the standard advanced text in number theory for more than fifty years.

Dirichlet died in 1859, four years after succeeding Gauss as Professor of Mathematics at Göttingen.

Georg Bernhard Riemann was born in 1826 and studied at Göttingen with Gauss, receiving his doctorate in 1851. During his short life, Riemann revolutionized mathematics, making ground-breaking contributions in differential geometry and complex analysis. He wrote only one paper in number theory, in which he sketched a method for proving the Prime Number Theorem using properties of the function now known as the Riemann zeta-function. The sketch, tremendously fertile in ideas, turned out to be the key to the proof found later by Hadamard and de la Vallée-Poussin, though many technical difficulties had to be overcome.

Riemann died of tuberculosis in 1866, at the age of 39.

REFERENCES

Tom M. Apostol, *Introduction to Analytic Number Theory*, Springer-Verlag, New York, 1976.

> This text is an excellent introduction to number theory for students who have a strong background in analysis. In particular, the book provides detailed proofs of the Prime Number Theorem and Dirichlet's Theorem on primes in arithmetic progressions. There are many other estimates, usually with error bounds, of the average size of the important number-theoretic functions.

David M. Bressoud, *Factorization and Primality Testing*. (See Chapter 3.)

G.H. Hardy and E.M. Wright, *An Introduction to the Theory of Numbers* (Fourth Edition), The Clarendon Press, Oxford, England, 1971.

This is the great classic among number theory books in English. Many sections are elementary, but the book contains a complete proof of the Prime Number Theorem, as well as a discussion of topics such as partitions and the geometry of numbers that are rarely found in introductory books.

Paulo Ribenboim, *The Book of Prime Number Records* (Second Edition), Springer-Verlag, New York, 1989.

This is an encyclopedic overview of what was known about prime numbers in 1989, with a large amount of historical information and a first-rate bibliography. The open problems mentioned in this chapter, and many more, are discussed thoroughly, with progress reports and detailed numerical evidence. This book is highly recommended for the insight it gives about the vast sweep and range of research about prime numbers.

CHAPTER EIGHT

Some Diophantine Equations and Fermat's Last Theorem

The term *Diophantine equation* originally referred to an algebraic equation for which only positive rational solutions were sought, although it now commonly applies to equations where solutions are restricted to integers. Some problems of this type can be found in the mathematics of the Babylonians (roughly 2000 B.C. to 1600 B.C.), but the first systematic treatment appears in the *Arithmetica* of Diophantus (c. 250 A.D.). This work, only a part of which survives, is a collection of over 250 problems, involving mainly equations of the second degree. Diophantus is ordinarily content to produce a single rational solution to each problem.

One of the oldest Diophantine equations deals with right triangles whose sides are of integral length, sometimes called *Pythagorean triangles*. The corresponding equation is $x^2 + y^2 = z^2$. Fifteen solutions, some quite large, such as $(4961, 6480, 8161)$, appear on a Babylonian clay tablet, so there is reason to believe that a systematic method was known for generating solutions. This equation received much attention from the early Greek mathematicians, especially Pythagoras (c. 570 B.C.), who is generally given credit for a formula that yields infinitely many solutions.

Much later, the eleventh-century Arab mathematician al-Karaji expanded on the methods of Diophantus. In the Middle Ages, Leonardo of Pisa (c. 1175–1250), better known as Fibonacci, brought the algebra of the Arab world, and thus indirectly Diophantus's work, to the attention of European mathematicians.

In the seventeenth century, Fermat studied Diophantine equations extensively. In 1637, in his copy of Bachet's Latin translation of Diophantus's *Arithmetica*, Fermat made one of the most famous conjectures in mathematics: If $n \geq 3$, the equation $x^n + y^n = z^n$ has no solution in nonzero integers x, y, and z. This conjecture, known as *Fermat's Last Theorem*, captured the imagination of generations of mathematicians and defied proof for over three and a half centuries. A complete proof, over 200 pages long, was finally given in October, 1994.

Another famous problem, which goes back to Diophantus, is to describe the positive integers that can be represented as a sum of two squares. Although he provided no proofs, Fermat gave the correct answer to this problem, and discussed the number of such representations, in correspondence with Mersenne and Roberval in 1640.

A related problem deals with the question of which positive integers can be expressed as a sum of four squares. Bachet had conjectured that every positive integer has such a representation. In his copy of Diophantus, Fermat wrote that he had a proof using his *method of infinite descent*, a technique that he had also used in the two-squares problem. Euler made important contributions to the solution of this problem, but the first complete proof was given by Lagrange in 1770.

In that same year, in trying to generalize the four-squares result, the Englishman Edward Waring (1734–1798) conjectured that every number can be written as a sum of 4 squares, 9 cubes, 19 fourth powers, and, in general, a sum of a fixed number of nonnegative kth powers for any positive integer k. Since then, Waring's Problem has been the subject of much attention. It was finally settled in 1909 by the German mathematician David Hilbert (1862–1943), but related questions remain the subject of vigorous research.

RESULTS FOR CHAPTER 8

We begin our discussion of Diophantine equations with a classification of all right triangles having sides of integral length.

The Equation $x^2 + y^2 = z^2$

The well-known Pythagorean Theorem states that in a right triangle, the square of the hypotenuse is equal to the sum of the squares of the other two sides. We are therefore led to the equation $x^2 + y^2 = z^2$. The problem of finding integer solutions of this equation goes back almost 4000 years.

(8.1) Definition. Suppose x, y, and z are positive integers. If $x^2 + y^2 = z^2$, then (x, y, z) is called a *Pythagorean triple*. (The triples (x, y, z) and (y, x, z) are not considered to be different.) If, in addition, x, y, and z have no positive factor in common except 1, then (x, y, z) is said to be a *primitive triple*.

We will also refer to a right triangle with integral sides as a *Pythagorean triangle*.

THE EQUATION $x^2 + y^2 = z^2$

Note. If an integer k divides two of x, y, and z, then it divides the third. For example, if k divides x and y, then $k^2 \mid x^2 + y^2$; hence $k \mid z$. Similarly, if $k \mid y$ and $k \mid z$, then $k^2 \mid z^2 - y^2$, so $k \mid x$. Thus, if one of (x, y), (x, z), or (y, z) is equal to 1, then they are all equal to 1.

If (x, y, z) is a Pythagorean triple and d is the greatest common divisor of x, y, and z, then $(x/d, y/d, z/d)$ is clearly a primitive Pythagorean triple. Thus any Pythagorean triple is a multiple of a *primitive* triple, and so it suffices to look for primitive solutions.

Pythagoras is credited with the geometric equivalent of the following formula for generating Pythagorean triples: $x = k$, $y = (k^2 - 1)/2$, and $z = (k^2 + 1)/2$, where $k > 1$ is odd. Plato (c. 380 B.C.) reputedly possessed a similar rule, namely, $x = 2k$, $y = k^2 - 1$, $z = k^2 + 1$. Neither formula, however, generates all primitive Pythagorean triples. (For example, the triple $(20, 21, 29)$ is not of either form.) The form of the general solution, which we give next, appears in Diophantus's *Arithmetica* and in Euclid's *Elements*. The proof that it is indeed the general solution came much later.

Note. If (x, y, z) is a Pythagorean triple, then x and y cannot both be odd, for otherwise, $x^2 \equiv y^2 \equiv 1 \pmod{4}$ and hence $z^2 \equiv 2 \pmod{4}$, which is impossible.

The following gives a way of generating all Pythagorean triples by showing how to generate all of the primitive triples.

(8.2) Theorem. *A Pythagorean triple (x, y, z), with y even, is primitive if and only if it is of the form*

$$x = a^2 - b^2, \qquad y = 2ab, \qquad z = a^2 + b^2,$$

where a and b are positive integers of opposite parity with $a > b$ and $(a, b) = 1$. Every Pythagorean triple is a multiple of a primitive Pythagorean triple.

Proof. If x, y, and z are defined as above, it is easy to check that $x^2 + y^2 = z^2$. Note that $z + x = 2a^2$ and $z - x = 2b^2$. Thus if p is a common prime divisor of x and z, then $p \mid 2a^2$ and $p \mid 2b^2$; since p is odd, it follows that $p \mid a$ and $p \mid b$. Since $(a, b) = 1$, it follows that $(x, z) = 1$, and therefore (x, y, z) is a primitive triple.

Conversely, if (x, y, z) is a primitive triple, then clearly, x and y cannot both be even. Since we are looking for solutions with y even, x must be odd and hence z is odd. Thus $z + x$ and $z - x$ are even. Let $r = (z + x)/2$ and $s = (z - x)/2$. Since $z = r + s$ and $x = r - s$, any common divisor of r and s also divides (z, x), and thus $(r, s) = 1$. Since $y^2 = z^2 - x^2 = (z + x)(z - x)$, we have $(y/2)^2 = rs$; but r and s are relatively prime, so r and s must each be perfect squares. (This follows easily by considering the prime factorization of r and s; see Problem 1-33.) If $r = a^2$ and $s = b^2$, then $a > b$ and $x = r - s = a^2 - b^2$, $z = r + s = a^2 + b^2$, and $y^2 = 4rs = 4a^2b^2$, that is, $y = 2ab$. It is clear that a and b are of opposite parity, for otherwise x would be even. Finally, note that $(a, b) = 1$, since any common divisor of a and b clearly divides x, y, and z.

Notes. 1. It is important to note that even if no restrictions are placed on a and b (except that $a > b$), the formulas in (8.2) will *not* generate *all* Pythagorean triples. (See Problem 8-5.) The triples that can be obtained in this way are characterized in Problem 8-6.

2. In Chapter 11, we will give another proof of (8.2), using Gaussian integers.

Fermat's Last Theorem

Some time after 1630, Fermat wrote the following in the margin of his copy of Bachet's 1621 Latin translation of Diophantus's *Arithmetica*:

> No cube can be split into two cubes, nor any biquadrate into two biquadrates, nor generally any power beyond the second into two of the same kind. I have discovered a truly remarkable proof for this, but the margin is too narrow to contain it.

This conjecture, one of the most famous problems in mathematics, has come to be known as *Fermat's Last Theorem*. In modern terms, it can be stated as follows.

(8.3) Fermat's Last Theorem. *The equation $x^n + y^n = z^n$ has no solution in nonzero integers if $n \geq 3$.*

Of the results that Fermat wrote in the margins of *Arithmetica*, all were later proved to be true, with the exception of his Last Theorem. A great many mathematicians tried, unsuccessfully, to prove this conjecture. In June 1993, it was announced that after three and a half centuries, a proof of Fermat's Last Theorem had been found. The proof, by Andrew Wiles, was very long (over 200 pages) and used many deep results of algebraic geometry. However, there were some gaps in the original argument; these were corrected in October 1994 by Wiles and Richard Taylor.

The proof that Fermat had in mind was likely one using the *method of infinite descent*, the main technique that he used in his proofs and one to which he often referred in correspondence (almost always without details). This method can be characterized as follows. Assuming that a certain statement holds for a given positive integer, we show that the statement is then true for a smaller positive integer. Proceeding this way eventually leads to a contradiction, since we cannot indefinitely produce successively smaller positive integers. Hence we conclude that the statement holds for no positive integer.

We will use the method of infinite descent to prove Fermat's Last Theorem for the case $n = 4$. In fact, we prove the somewhat stronger result given below. Fermat himself gave, in 1659, a proof of a closely related result using infinite descent. It is the *only* proof Fermat left for any of his number-theoretic results. The first recorded proof of the next theorem is due to Frenicle, but there is strong reason to believe that the argument originates with Fermat.

(8.4) Theorem. *The equation $x^4 + y^4 = z^2$ has no solution in nonzero integers.*

Proof. We may assume that any solution involves positive integers, since a change in sign of x, y, or z still yields a solution. The proof will be by Fermat's *method of infinite descent*. More precisely, if we assume that the equation has some solution in positive integers, then there is a positive solution having a smallest value of z; we then produce another positive solution with a smaller value of z, and this contradiction will establish the result.

Thus suppose x, y, z is a positive solution, where z is minimal. If x and y have a common factor greater than 1, then there is a prime p dividing x and y. Since $p^4 | x^4 + y^4$, we have $p^4 | z^2$ and so $p^2 | z$. Thus $(x/p)^4 + (y/p)^4 = (z/p^2)^2$ and we have produced a positive solution with a smaller value of z. Therefore x and y can have no factor greater than 1 in common; in particular, (x^2, y^2, z) is a primitive Pythagorean triple. Hence x^2 and y^2 are of opposite parity; we will assume that x^2 is odd and y^2 is even. Thus by (8.2), there exist relatively prime u and v such that

$$x^2 = u^2 - v^2, \qquad y^2 = 2uv, \qquad z = u^2 + v^2.$$

In particular, (x, v, u) is a primitive Pythagorean triple with x odd. Therefore there exist relatively prime integers s, t such that

$$x = s^2 - t^2, \qquad v = 2st, \qquad u = s^2 + t^2.$$

Since $y^2 = 2uv = 4ust$ and s, t, and u are pairwise relatively prime, they are all perfect squares (see Problem 1-33). Hence there exist integers a, b, c such that $s = a^2$, $t = b^2$, and $u = c^2$. Since $u = s^2 + t^2$, it follows that $a^4 + b^4 = c^2$. Note that $c^2 = u$ and $z = u^2 + v^2$; therefore $c < z$. Thus a, b, c is a solution of the original equation with $c < z$, which contradicts the minimality of z. This completes the proof.

The next result can be proved using a similar argument. For the details, see Problem 8-18.

(8.5) Theorem. *The equation $x^4 - y^4 = z^2$ has no solution in nonzero integers.*

Notes. 1. In view of (8.4), Fermat's Last Theorem follows for every $n \geq 3$ if it can be shown to be true for *prime* exponents. (See Problem 8-14.)

2. We have called the proof a proof by infinite descent, since that is the term Fermat used. It is in fact a proof by induction, for what is shown is that if the equation has no solution with z positive and less than an integer w, then it has no solution with $z = w$. (The proof shows that if we had a solution with $z = w$, there would be a solution with $z < w$.)

In 1770, Euler gave an incomplete proof of Fermat's Last Theorem for the case $n = 3$, also using infinite descent. (Euler's proof relied on unique factorization of integers in the algebraic number field $Q(\sqrt{-3})$. This topic will be discussed in Chapter 11.) The case $n = 5$ was dealt with by Dirichlet and by Legendre in 1825. (Both proofs used infinite descent.) Lamé settled the case $n = 7$ in 1839.

The most important nineteenth-century advance in proving Fermat's Last Theorem was made by the German mathematician Ernst Eduard Kummer (1810–1893). A natural approach to Fermat's Last Theorem is to express the polynomial $z^n - x^n$ as a product of linear factors with coefficients in a suitable extension of the integers. In 1847, Lamé claimed to have a complete proof based on this idea. His proof, however, assumed (incorrectly) that unique factorization into a product of primes holds in more general algebraic number fields. (See the Notes at the end of the chapter.) At almost the same time, to restore unique factorization, Kummer was developing his theory of ideals and was able to prove Fermat's Last Theorem for all primes less than 100 except for 37, 59, and 67 and sketched out an essentially correct argument for the remaining cases under 100. A complete proof, however, eluded him.

For a historical treatment of Fermat's Last Theorem and a detailed discussion of related ideas, see the books by H. Edwards and P. Ribenboim given in the references at the end of this chapter.

Sums of Two Squares

Several problems in Diophantus's *Arithmetica* raise the question of representing a positive integer as a sum of two squares. The problem was taken up by Fermat in the seventeenth century. In a letter to Mersenne dated Christmas Day, 1640, Fermat stated, without proof, that every prime of the form $4k + 1$ has a unique representation as a sum of two squares. He later indicated that he had used his method of infinite descent to prove this; in fact, this is how Euler proved the result over a hundred years later, in 1745.

The complete characterization of precisely which integers are sums of two squares is given in Theorem 8.9 and is generally credited to Fermat; the first proof was published by Euler in 1749. It is easy to see that no prime – indeed, no integer – of the form $4k+3$ has such a representation. For suppose $n = a^2 + b^2$. Since the square of any integer is congruent to 0 or 1 modulo 4, it is clear that n must be congruent to 0, 1, or 2 modulo 4. But a stronger result is true.

(8.6) Theorem. *Suppose that n is divisible by a prime q of the form $4k+3$.* (i) *If $n = a^2 + b^2$, then $q|a$ and $q|b$.* (ii) *If n is a sum of two squares, then q must appear to an even power in the prime factorization of n. In particular, no prime of the form $4k+3$ is a sum of two squares.*

Proof. (i) If $q \nmid a$, then there is an s such that $sa \equiv 1 \pmod{q}$. But if $n = a^2 + b^2$ and $q|n$, then $a^2 + b^2 \equiv 0 \pmod{q}$. Multiplying by s^2, we find that $(sb)^2 \equiv -1 \pmod{q}$, which is impossible since -1 is not a quadratic residue of any prime of the form $4k+3$. Thus q divides a, and, by symmetry, q also divides b.

(ii) It follows from part (i) that if $a^2 + b^2 = n$, then $q^2 | n$. We can then divide the equation $a^2 + b^2 = n$ by q^2 to get $n/q^2 = (a/q)^2 + (b/q)^2$. Let $n = q^2 n_1$; if $q|n_1$, then the above argument shows that $q^2|n_1$. Proceeding this way, we see that n must be divisible precisely by an *even* number of factors of q.

The converse of this result is also true: *If every $4k+3$ prime that divides n appears with an even exponent in the prime factorization of n, then n can be represented as a sum of two squares.* The standard proof of this uses an identity which was implicitly used by Diophantus and which appears in Fibonacci's *Liber Quadratorum* (1225). A more general identity was used by Brahmagupta in seventh-century India.

(8.7) Lemma. *If m and n are each a sum of two squares, then their product mn is also a sum of two squares. In particular, if $m = a^2 + b^2$ and $n = c^2 + d^2$, then*

$$mn = (ac + bd)^2 + (ad - bc)^2 = (ac - bd)^2 + (ad + bc)^2.$$

Note. The conclusion of (8.7) easily extends (by induction) to the product of any finite number of terms, each expressible as a sum of two squares.

Suppose now that every $4k+3$ prime divisor of n appears to an even power in the prime decomposition of n. Write $n = c^2 m$, where m is *square-free*; then m is clearly not divisible by any prime of the form $4k+3$. Therefore m is of the form $2^t \prod p_i$, where t is 0 or 1 and the p_i are primes of the form $4k+1$. Note that $2 = 1^2 + 1^2$; thus if we can prove that every $4k+1$ prime is a sum of two squares, then repeated application of (8.7) implies that the product $m = 2^t \prod p_i$ is also such a sum. Finally, if $m = a^2 + b^2$, then $n = (ca)^2 + (cb)^2$, and therefore n is expressible as a sum of two squares.

Thus the proof of Fermat's characterization can be completed by showing that every prime of the form $4k+1$ is a sum of two squares. In Problem 8-48, we give a proof of this result that is essentially the one given by Euler using the method of infinite descent. The following argument, however, is much shorter.

(8.8) Theorem. *Every prime of the form $4k+1$ can be expressed as a sum of two relatively prime squares.*

Proof. If p is a prime of the form $4k+1$, (5.11) implies that -1 is a quadratic residue of p, so there exists an integer s such that $s^2 \equiv -1 \pmod{p}$. Consider the integers $sx - y$, where $0 \leq x, y < \sqrt{p}$. There are $[\sqrt{p}] + 1$ choices for each of x and y, and since $([\sqrt{p}] + 1)^2 > (\sqrt{p})^2 = p$, at least two of the values of $sx - y$ must be congruent modulo p, say, $sx_1 - y_1 \equiv sx_2 - y_2 \pmod{p}$. Let $x = x_1 - x_2$ and $y = y_1 - y_2$; then x and y are not both zero since the pairs x_1, y_1 and x_2, y_2 are distinct. Clearly, $sx \equiv y \pmod{p}$ and therefore $s^2 x^2 \equiv y^2 \pmod{p}$, that is, $-x^2 \equiv y^2 \pmod{p}$. Thus $x^2 + y^2$ is a multiple of p, and since $0 < x^2 + y^2 < 2(\sqrt{p})^2 = 2p$, it follows that $x^2 + y^2 = p$.

To show that x and y are relatively prime, note that if $d \mid x$ and $d \mid y$, then $d^2 \mid x^2 + y^2$; since $x^2 + y^2$ is prime, it follows that $d^2 = 1$.

Note. A proof of this result, using continued fractions, is given in Problem 9-10, and another, based on the existence of solutions to Pell's Equation, can be found in Theorem 10.20. Both give a method for finding the squares needed to represent a $4k+1$ prime. Still another proof, using properties of Gaussian primes, is given in the discussion that precedes (11.12).

We summarize the preceding results in the following theorem.

(8.9) Theorem (Fermat; Euler, 1749). *Let n be a positive integer. Then n is a sum of two squares if and only if every prime divisor of n of the form $4k+3$ appears to an even power in the prime factorization of n.*

Finally, we consider the problem of determining the *number* of representations of an integer as a sum of two squares. In Fermat's letter to Mersenne on Christmas Day, 1640, Fermat calculated the number of representations of p^m, where p is a prime of the form $4k+1$, and more generally of any integer n that can be written as a sum of two squares.

(8.10) Definition. *Let $N(n)$ denote the total number of representations of n as a sum of two squares. Representations are counted as distinct even if they differ only in the sign of the terms or in the order of the terms.*

For example, $N(13) = 8$, since $13 = (\pm 2)^2 + (\pm 3)^2 = (\pm 3)^2 + (\pm 2)^2$. Thus there are eight representations obtained by changing the sign and order of the numbers used.

A proof of the next result can be found in Problems 8-69 to 8-72. The result is also an easy consequence of properties of Gaussian integers and will be proved again in Chapter 11.

(8.11) Theorem. *Suppose n can be represented as a sum of two squares, and write $n = 2^a \prod p_i^{a_i} \prod q_i^{b_i}$, where the p_i are $4k+1$ primes and the q_i are $4k+3$ primes (the b_i are necessarily even). Then*

$$N(n) = 4\prod(a_i + 1).$$

(A product of no terms is interpreted to be 1.) In particular, if p is a prime of the form $4k+1$, then $N(p^m) = 4(m+1)$.

There is another useful characterization of $N(n)$, due to Carl Gustav Jacobi (1804–1851), which we give next. For a proof, see Problems 8-73 to 8-75.

(8.12) Theorem. *Let n be a positive integer, and let D_1 and D_3 denote, respectively, the number of positive divisors of n of the form $4k+1$ and $4k+3$. Then $N(n) = 4(D_1 - D_3)$.*

Notes. 1. A representation of n as a sum of two squares that are nonzero and have different absolute values generates eight different representations (counting itself), obtained by varying the order and sign of the terms. If n is itself a square, say $n = m^2$, we can write $n = m^2 + 0^2$; this generates the four representations $m^2 + 0^2$, $(-m)^2 + 0^2$, $0^2 + m^2$, and $0^2 + (-m)^2$. These are all counted as distinct in the formula for $N(n)$. If $n = 2m^2$, four representations are generated, namely, $n = (\pm m)^2 + (\pm m)^2$.

2. If p is a prime of the form $4k+1$, then $N(p) = 8$. Hence, except for the order and sign of the terms, *any prime of the form $4k+1$ can be represented in one and only one way as a sum of two squares.*

The number of *essentially distinct* representations (that is, representations where we do not count variations obtained by changing the order and sign of the terms) is discussed in Problem 8-45.

Sums of Two Relatively Prime Squares

In this section, we consider the problem of representing an integer as a sum of relatively prime squares. We begin with a definition.

(8.13) Definition. *The representation $n = a^2 + b^2$ is called a* positive representation *if a and b are positive. The representation is* primitive *if a and b are relatively prime. (Note that a primitive representation need not be a positive representation.)*

Most of the theorems on primitive representations follow from the next result. The argument used in its proof is very similar to that of (8.8).

(8.14) Theorem. *Let $n > 1$. For every positive primitive representation $n = a^2 + b^2$, there is a unique solution s of $x^2 \equiv -1 \pmod{n}$ such that $sa \equiv b \pmod{n}$, and different positive primitive representations correspond to different s.*

Conversely, any solution s of $x^2 \equiv -1 \pmod{n}$ determines a unique positive primitive representation $n = a^2 + b^2$ such that $sa \equiv b \pmod{n}$.

Proof. Suppose $n = a^2 + b^2$, with a and b positive and $(a, b) = 1$. Then $(a, n) = 1$, and so by (2.8), there is a unique s modulo n such that $sa \equiv b \pmod{n}$. Since $a^2 + b^2 \equiv 0 \pmod{n}$, we have $a^2 \equiv -b^2 \equiv -(sa)^2 \equiv -s^2 a^2 \pmod{n}$, and dividing by a^2 gives $s^2 \equiv -1 \pmod{n}$. Assume now that the representation $c^2 + d^2$ also corresponds to s, that is, $sc \equiv d \pmod{n}$. Since $a^2 + b^2$ and $c^2 + d^2$ are positive representations, we have $1 \leq a < \sqrt{n}$ and $1 \leq d < \sqrt{n}$; thus $1 \leq ad < n$. Similarly, $1 \leq bc < n$. Since $bc \equiv (sa)c \equiv ad \pmod{n}$, it follows that $bc = ad$. Thus b divides ad, and since $(a, b) = 1$, we have $b \mid d$. By symmetry, we also have $d \mid b$. Therefore $b = d$ and hence $a = c$.

Now suppose $s^2 \equiv -1 \pmod{n}$, and consider the set of all ordered pairs (j, k), where $0 \leq j, k < \sqrt{n}$ if n is not a square, and $0 \leq j \leq \sqrt{n}$ and $0 \leq k < \sqrt{n}$ if n is a square. There are then, respectively, $(1 + [\sqrt{n}])^2$ or $(1 + \sqrt{n})\sqrt{n}$ such ordered pairs, and therefore more than n in either case. Now consider $sj - k$ as (j, k) ranges over these ordered pairs. Since any integer is congruent modulo n to one of $0, 1, \ldots, n - 1$, the numbers $sj - k$ cannot all be distinct modulo n. Thus there exist two pairs (j_1, k_1) and (j_2, k_2) such that $sj_1 - k_1 \equiv sj_2 - k_2 \pmod{n}$. Let $A = j_1 - j_2$ and $B = k_1 - k_2$; then $|A| \leq \sqrt{n}$, $|B| < \sqrt{n}$, and $sA \equiv B \pmod{n}$. Since the ordered pairs (j_1, k_1) and (j_2, k_2) are distinct, it follows that at least one of A or B is nonzero; thus $A^2 + B^2 \geq 1$. Also, $sA \equiv B \pmod{n}$ implies $s^2 A^2 \equiv B^2 \pmod{n}$, that is, $-A^2 \equiv B^2 \pmod{n}$. Therefore $A^2 + B^2$ is a multiple of n. Since $A^2 + B^2$ is positive and less than $2n$, we must have $A^2 + B^2 = n$.

We show next that A and B are relatively prime. Let $d = (A, B)$; since $B \equiv sA \pmod{n}$, it follows that $B/d \equiv sA/d \pmod{n/d}$, and therefore $(B/d)^2 \equiv s^2(A/d)^2 \equiv -(A/d)^2 \pmod{n/d}$. Hence $n/d^2 = (A/d)^2 + (B/d)^2 \equiv (A/d)^2 - (A/d)^2 = 0 \pmod{n/d}$. Thus n/d divides n/d^2, and clearly, this can happen only if $d = 1$.

Finally, if A and B are both positive or both negative, set $a = |A|$ and $b = |B|$; then, for example, if $A < 0$ and $B < 0$, $sa = s(-A) = -sA \equiv -B = b \pmod{n}$. Otherwise, if one of A and B is positive and the other is negative, let $a = |B|$ and $b = |A|$; thus, for example, if $A > 0$ and $B < 0$, $sa = -sB \equiv -s^2 A \equiv A = b \pmod{n}$. In each case, $a^2 + b^2$ is a positive primitive representation of n such that $sa \equiv b \pmod{n}$.

Fermat stated in 1640, and Euler proved in 1742, that if $n = a^2 + b^2$ with

$(a, b) = 1$, then all odd prime divisors of n are of the form $4k + 1$. This is contained in the next result, which gives a complete description of the positive integers that have primitive representations.

(8.15) Theorem. *A positive integer n has primitive representations as a sum of two squares if and only if n is not divisible by 4 or by any prime of the form $4k + 3$. Thus n can be represented as a sum of two relatively prime squares if and only if $n = p_1^{a_1} \cdots p_r^{a_r}$ or $n = 2p_1^{a_1} \cdots p_r^{a_r}$, where each p_i is a prime of the form $4k + 1$.*

Proof. If n has primitive representations, then the congruence $x^2 \equiv -1$ (mod n) is solvable, by (8.14). But if $4 \mid n$ and $x^2 \equiv -1$ (mod n), then $x^2 \equiv -1$ (mod 4) must be solvable, which it clearly is not; and if $q \mid n$, where q is prime and $q \equiv 3$ (mod 4), then $x^2 \equiv -1$ (mod q) must be solvable, which contradicts (5.11).

Conversely, if n is not divisible by 4 or by any $4k + 3$ prime, then either $n = p_1^{a_1} \cdots p_r^{a_r}$ or $n = 2p_1^{a_1} \cdots p_r^{a_r}$, where each prime p_i is of the form $4k+1$. By (5.11), $x^2 \equiv -1$ (mod p_i) is solvable for each i, and therefore (5.4.i) implies that $x^2 \equiv -1$ (mod $p_i^{a_i}$) has solutions as well. Thus $x^2 \equiv -1$ (mod n) is also solvable, by (5.5), and it follows from (8.14) that n can be represented as a sum of relatively prime squares.

We turn now to an investigation of the *number* of primitive representations a positive integer can have. It follows from Theorem 8.14 that there is a one-to-one correspondence between solutions of $x^2 \equiv -1$ (mod n) and positive primitive representations of n. In counting positive primitive representations in (8.14), the representation $a^2 + b^2$ of n is viewed as distinct from $b^2 + a^2$; that is, they arise from *different* solutions of $x^2 \equiv -1$ (mod n). For suppose the solution s determines the representation $a^2 + b^2$; then $sa \equiv b$ (mod n). Since $-s$ is also a solution, $-s$ determines a unique positive primitive representation $c^2 + d^2$ such that $(-s)c \equiv d$ (mod n). Since $(-s)b \equiv (-s)sa \equiv -s^2 a \equiv -(-1)a = a$ (mod n), it follows from (8.14) that $-s$ must correspond to the representation $b^2 + a^2$. Thus, *the pair of solutions s and $-s$ generate positive primitive representations which differ only in the order of the summands.*

It is clear that a given positive primitive representation generates a total of four primitive representations by changing the sign (but not the order) of the terms: $n = a^2 + b^2 = a^2 + (-b)^2 = (-a)^2 + b^2 = (-a)^2 + (-b)^2$. We therefore have the following result.

(8.16) Theorem. *Suppose $n > 1$, and let S denote the number of solutions of $x^2 \equiv -1$ (mod n). Then n has precisely S positive primitive representations, and the total number of primitive representations is $4S$.*

Example. (Refer to Theorem 8.14.) Suppose $p = 13$; then $x^2 \equiv -1$ (mod 13) has the two solutions ± 5, that is, 5 and 8. The root 5 corresponds to the representation $3^2 + 2^2$, since $5 \cdot 3 \equiv 2$ (mod 13), and by varying the sign of the terms, we obtain the representations $(3, -2)$, $(-3, 2)$, and $(-3, -2)$. Since $8 \cdot 2 \equiv 3$ (mod 13), the root 8 determines the representation $2^2 + 3^2$, and variations of this give the representations $(2, -3)$, $(-2, 3)$ and $(-2, -3)$. Thus all eight variations obtained by changing the order and sign of the terms are accounted for.

The next result answers the question of which integers can be expressed in an essentially unique way as a sum of two relatively prime squares.

(8.17) Theorem. *An integer $n > 2$ has a unique primitive representation (apart from the order and sign of the terms) if and only if $n = p^m$ or $n = 2p^m$, where p is a prime of the form $4k + 1$ and $m \geq 1$.*

Proof. Since changing the order and sign of the terms produces eight variations for a given representation, n will have an essentially unique representation as a sum of two relatively prime squares if and only if n has a total of eight primitive representations. By (8.16), this happens if and only if $x^2 \equiv -1$ (mod n) has exactly two solutions. The conclusion now follows from (5.5).

In general, we can use (8.16) to determine the number of primitive representations when such representations exist.

(8.18) Theorem. *Suppose $n > 2$ and n has primitive representations; thus by (8.15), $n = p_1^{a_1} \cdots p_r^{a_r}$ or $n = 2p_1^{a_1} \cdots p_r^{a_r}$, where $p_1, \ldots, p_r$ are primes of the form $4k + 1$. Then n has precisely 2^{r+2} primitive representations overall and 2^{r-1} essentially distinct primitive representations (that is, representations where variations in the order and sign of terms are disregarded.)*

Proof. By (8.16), the number of primitive representations of n is equal to four times the number of solutions of $x^2 \equiv -1$ (mod n). If n is of either form described in the theorem, it follows from (5.5) that this congruence has precisely 2^r solutions; hence n has exactly $4 \cdot 2^r = 2^{r+2}$ primitive representations. Since each primitive representation involves two distinct nonzero squares, there are eight variations obtained by changing the order and sign of the terms. Thus there are $2^{r+2}/8 = 2^{r-1}$ essentially distinct primitive representations of n.

Finally, we note that even if n has primitive representations, not all representations of n need be primitive. In fact, *if $n > 2$ is representable as a sum of two squares, then every representation of n is primitive if and only if $n = p_1 p_2 \cdots p_r$ or $n = 2p_1 p_2 \cdots p_r$, where each p_i is a prime of the form $4k+1$.* (See Problem 8-60.)

Sums of Four Squares

In view of (8.6), it is clear that not all integers are the sum of two squares; similarly, not all numbers can be written as a sum of three squares (see the following section).

In 1621, Bachet stated, without proof, that every positive integer is a sum of at most four squares. In 1636, in a note in his copy of Diophantus, Fermat claimed to have proved this result using his method of infinite descent.

Euler, the most prominent mathematician of the eighteenth century, first became interested in this problem in 1730 and made important contributions in a series of papers from 1747 to 1751, but he was unable to give a complete proof. However, he played a major role in the eventual solution of the problem, which was given by Lagrange in 1770 and was based on Euler's work. Two years later, Euler gave an elegant proof of the Four-Squares Theorem that is very similar to his proof of the two-squares result for primes of the form $4k + 1$. (See Problem 8-95.) Critical to the arguments of Lagrange and Euler is a remarkable identity of Euler (see the proof of the following result). It was communicated in a letter to Goldbach in 1748 and allows for a proof by infinite descent.

(8.19) Lemma (Euler). *If m and n are each a sum of four squares, then their product mn is also a sum of four squares.*

Proof. Suppose that $m = a^2 + b^2 + c^2 + d^2$ and $n = A^2 + B^2 + C^2 + D^2$. Then $mn = r^2 + s^2 + t^2 + u^2$, where $r = aA + bB + cC + dD$, $s = aB - bA + cD - dC$, $t = aC - bD - cA + dB$, and $u = aD + bC - cB - dA$. This can be verified by multiplying out each side.

Every integer $n > 1$ is a product of primes. Thus by repeated application of (8.19), it will follow that every positive integer is a sum of four squares (where some of the squares may be 0) if we can show that every prime is a sum of four squares. It is clear that 2 is such a sum, as is any prime of the form $4k + 1$, by (8.8). Hence it is enough to prove that any prime of the form $4k + 3$ is a sum of four squares.

This is the standard approach used in proving the Four-Squares Theorem (see Problem 8-95). The proof we give, however, does not make use of Euler's formula and is very similar to the argument used to prove the Two-Squares Theorem (8.8). We require the following lemmas.

(8.20) Lemma (Sylvester, 1847). *Suppose that $3m$ is a sum of four squares. Then m is also a sum of four squares.*

Proof. Let $3m = s^2 + t^2 + u^2 + v^2$. Since every square is congruent to 0 or 1 modulo 3, at least one of s, t, u, and v must be divisible by 3. Thus we may suppose that s is a multiple of 3. Since $t^2 + u^2 + v^2 \equiv 0 \pmod{3}$, we can also assume, for a suitable choice of the signs of t, u, and v, that $t \equiv u \equiv v \pmod{3}$. By expanding and simplifying, we find that m is the sum of the squares of the integers $(t+u+v)/3$, $(s+u-v)/3$, $(s-t+v)/3$, and $(s+t-u)/3$.

(8.21) Lemma. *Let n be a square-free integer. Then there exist integers a and b such that $a^2 + b^2 \equiv -1 \pmod{n}$.*

Proof. We first show that the result holds when n is a prime p. This is obvious when $p = 2$. For primes p of the form $4k + 1$, it follows from (5.11) that there is a number a such that $a^2 \equiv -1 \pmod{p}$; thus we can take $b = 0$. We give, however, a separate proof that works for all odd primes.

Let $q = (p-1)/2$, and define A to be the set $\{0^2, 1^2, \ldots, q^2\}$ and B the set $\{-1 - 0^2, -1 - 1^2, \ldots, -1 - q^2\}$. Any two elements of A are incongruent modulo p. For let $0 \le i < j \le q$. If $i^2 \equiv j^2 \pmod{p}$, then $p \mid j^2 - i^2$, and therefore $p \mid j+i$ or $p \mid j-i$. But this is impossible, since j and i are nonnegative and less than p. Similarly, any two elements of B are incongruent modulo p. Thus A and B each contain $(p+1)/2$ incongruent integers. It follows that there is an element a^2 of A and an element $-1 - b^2$ of B such that $a^2 \equiv -1 - b^2 \pmod{p}$, for otherwise, we would have $2(p+1)/2 = p + 1$ incongruent integers modulo p, which is impossible. With this choice of a and b, we have $a^2 + b^2 \equiv -1 \pmod{p}$.

It is easy to extend the result to numbers n of the form $n = \prod p_i$, where the p_i are distinct primes. For each p_i, choose a_i, b_i such that $a_i^2 + b_i^2 \equiv -1 \pmod{p_i}$. Use the Chinese Remainder Theorem to find numbers a and b such that $a \equiv a_i \pmod{p_i}$ and $b \equiv b_i \pmod{p_i}$ for all i. Then $a^2 + b^2 \equiv -1 \pmod{p_i}$ for all i, and hence $a^2 + b^2 \equiv -1 \pmod{n}$.

We are now ready to prove the Four-Squares Theorem. Let n be a positive integer. Then n can be expressed in the form $n = k^2 m$, where m is square-free. If we can prove that m is the sum of four squares, say $m = s^2 + t^2 + u^2 + v^2$, then it will follow that $n = (ks)^2 + (kt)^2 + (ku)^2 + (kv)^2$ is a sum of four squares.

(8.22) Theorem (Lagrange, 1770; Euler, 1772). *Every positive integer is a sum of four squares.*

Proof. In view of the preceding comments, it is enough to show that if the positive integer n is square-free, then n is a sum of four squares; clearly, we can take $n > 1$. Let a and b be integers such that $a^2 + b^2 \equiv -1 \pmod{n}$. Consider all ordered pairs $(as + bt - u, bs - at - v)$, where s, t, u, and v range over all integers from 0 to $[\sqrt{n}]$. There are $(1 + [\sqrt{n}])^4 > n^2$ choices for s, t, u, v. But since there are only n^2 distinct ordered pairs modulo n, there exist distinct ordered quadruples (s_1, t_1, u_1, v_1) and (s_2, t_2, u_2, v_2), with all entries lying in the interval from 0 to $\sqrt{n}$, such that $as_1 + bt_1 - u_1 \equiv as_2 + bt_2 - u_2$ (mod n) and $bs_1 - at_1 - v_1 \equiv bs_2 - at_2 - v_2$ (mod n).

Let $s = s_1 - s_2$, $t = t_1 - t_2$, $u = u_1 - u_2$, and $v = v_1 - v_2$. Then $as + bt \equiv u$ (mod n) and $bs - at \equiv v$ (mod n). Therefore $(as + bt)^2 + (bt - as)^2 \equiv u^2 + v^2$ (mod n). But $(as + bt)^2 + (bt - as)^2 = (s^2 + t^2)(a^2 + b^2)$, and since $a^2 + b^2 \equiv -1$ (mod n), it follows that $-(s^2 + t^2) \equiv u^2 + v^2$ (mod n), so $s^2 + t^2 + u^2 + v^2 = kn$ for some k. Clearly, s, t, u, and v are all less than or equal to $[\sqrt{n}]$. They are not all 0, since the ordered quadruples (s_1, t_1, u_1, v_1) and (s_2, t_2, u_2, v_2) are distinct. Also, $s^2 + t^2 + u^2 + v^2 \leq 4[\sqrt{n}]^2 < 4n$ and thus $k = 1, 2$, or 3.

If $k = 1$, we are done, while if $k = 3$, then n itself is a sum of four squares by (8.20). Now suppose $k = 2$; since $2n$ is even, either zero, two, or four of the numbers s, t, u, and v must be even. If exactly two are even, we may assume that they are s and t. Then in each case, $s \pm t$ and $u \pm v$ are even, and hence n is the sum of the squares of $(s + t)/2$, $(s - t)/2$, $(u + v)/2$, and $(u - v)/2$. This completes the proof.

It is not true that every positive integer – or even every sufficiently large positive integer – is a sum of four *nonzero* squares. In fact, the positive integer n is a sum of four nonzero squares if and only if $n \neq 1, 3, 5, 9, 11, 17, 29, 41, 2 \cdot 4^k, 6 \cdot 4^k, 14 \cdot 4^k$. In particular, there are infinitely many even integers, but only a finite number of odd integers, that cannot be represented as a sum of four nonzero squares. However, the following result holds. (See Problem 8-80 for a proof.)

(8.23) Theorem. *Every sufficiently large positive integer is a sum of five nonzero squares.*

Note. The only positive integers that are *not* the sum of five nonzero squares are 1, 2, 3, 4, 6, 7, 9, 10, 12, 15, 18, and 33.

Finally, we have the following theorem of Jacobi.

(8.24) Theorem. *Let $M(n)$ denote the total number of representations of n as a sum of four squares (including changes in the order or signs of the terms). Then $M(n)$ is 8 times the sum of the positive divisors of n that are not divisible by 4.*

Sums of Three Squares

The product of two numbers, each the sum of three squares, need not be itself a sum of three squares (see Problem 8-82). Thus in trying to determine which numbers are representable as a sum of three squares, we cannot use an identity similar to (8.7) or (8.19). Hence the question of which integers are sums of three squares is much more difficult than the analogous question for two squares and four squares.

An easy congruence argument shows that no number of the form $8k+7$ is such a sum, nor in general is any number of the form $4^m(8k+7)$ (see Problem 8-83). They are in fact the *only* numbers that cannot be written as a sum of three squares. This was first noted by Fermat, probably around 1630. In 1798, Legendre gave an obscure and somewhat incomplete proof. Gauss was the first to give a complete proof, in *Disquisitiones Arithmeticae*, together with a formula for the number of representations, but the argument is difficult and uses his theory of quadratic forms. The characterization is as follows.

(8.25) Theorem (Legendre, Gauss). *A positive integer can be expressed as a sum of three squares if and only if it is not of the form $4^m(8k+7)$.*

Note. It is not hard to show, using a proof similar to that of (8.8), that any prime of the form $8k+1$ or $8k+3$ is of the form $2a^2+b^2$ (see Problem 8-90). In particular, any such prime is a sum of three squares.

Waring's Problem

Lagrange's result of 1770 shows that every positive integer is a sum of four squares. In the same year, Waring made a broader conjecture, stating that every number is a sum of 4 squares, 9 cubes, 19 fourth powers, and in general, a sum of a fixed number of kth powers for any positive integer k. The following is a precise statement of Waring's conjecture.

(8.26) Waring's Problem. *For any $k \geq 2$, there is a positive integer s (which depends only on k) such that every positive integer can be expressed as a sum of s nonnegative kth powers.*

There are really two questions in Waring's Problem. The first is whether the integer s exists for every $k \geq 2$. By the end of the nineteenth century, this was known to be true for $k \leq 8$. The question was settled completely in 1909 by David Hilbert; his proof, however, is an existence argument and gives no method of determining s for a particular value of k.

The second question is to find, for a given k, the *smallest* value of s. For $k \geq 2$, define $g(k)$ to be the smallest number s such that every positive integer can be represented as the sum of s nonnegative kth powers. For $k=2$, it is not hard to see that $g(2) = 4$: Lagrange's theorem implies that no more than four squares are needed for any integer, and it follows from (8.25) that every number of the form $8k+7$ requires exactly four squares.

For $k > 2$, the problem is considerably harder. For almost one hundred years, very little progress was made. In 1859, Liouville proved that $g(4) \leq 53$, and Wieferich showed in 1909 that $g(3) = 9$. This problem has captured the attention of many famous mathematicians, among them Hardy, Littlewood, and Vinogradov. It is now known that $g(4) = 19$ (this was finally proved in 1986) and that $g(5) = 37$ (proved in 1964). Except for the cases $k = 2$ and $k = 3$, the results use analytic techniques and rely on the theory of functions of a complex variable.

In 1772, Johannes Euler (son of the famous Leonhard Euler) noted that $g(k) \geq 2^k + [(3/2)^k] - 2$ for every $k \geq 2$ (see Problem 8-96). To prove this inequality, it is enough to find some positive integer requiring this number of kth powers to be represented. In a sense, this is the worst possible case, and it has been conjectured – and is very likely true – that in fact $g(k) = 2^k + [(3/2)^k] - 2$ for every value of k. This equality is now known to hold for every $k \leq 471,600,000$ and also for all sufficiently large k.

PROBLEMS AND SOLUTIONS

Note. The following problem is very old. Tradition says that it is about Diophantus, but the straightforward linear equation is not at all characteristic of Diophantus's work.

8-1. *His boyhood lasted one-sixth of his life. His beard grew after one-twelfth more. After one-seventh more, he married, and his son was born five years later; the son lived to half his father's age, and the father died four years after his son. How old was he when he died?*

Solution. If he lived to age x, we have $(1/6)x + (1/12)x + (1/7)x + 5 + (1/2)x + 4 = x$. Thus $x = 84$.

The Equation $x^2 + y^2 = z^2$

8-2. *Find all Pythagorean triples (x, y, z) with $40 \leq z \leq 50$.*

Solution. First we list all primitive triples by expressing z in the form $a^2 + b^2$, where a and b are relatively prime and of opposite parity. The equation $41 = 16 + 25$ produces the triple $(9, 40, 41)$; no other odd number in our range is the sum of two relatively prime squares.

In general, we try to express z in the form $z = tz_1$, where z_1 is the hypotenuse of a primitive triangle. For $z = 40$, we can take $z_1 = 5$, which yields the triple $(24, 32, 40)$. For $z = 45$, again take $z_1 = 5$; this gives the triple $(27, 36, 45)$. For $z = 50$, we can take $z_1 = 5$ or $z_1 = 25$, producing the triples $(30, 40, 50)$ and $(14, 48, 50)$.

8-3. *Find all Pythagorean triangles having a side of length 18.*

Solution. Using (8.2), we first find the *primitive* Pythagorean triangles that have a side which divides 18. Clearly, neither 1 nor 2 is the side of a Pythagorean triangle, and 3 is a side only of the $(3, 4, 5)$ triangle; thus 18 is a side of the $(18, 24, 30)$ triangle. Since 6 is even, to make 6 a side of a primitive Pythagorean triangle, 6 must be expressed as $2ab$, where a and b have opposite parity; clearly, this cannot be done. Similarly, 18 is not a side of a primitive triangle. Finally, express 9 in the form $a^2 - b^2$ (9 cannot be $a^2 + b^2$ or $2ab$); thus $a = 5$ and $b = 4$. This yields the $(9, 40, 41)$ triangle; multiplying by 2, we obtain the triangle $(18, 80, 82)$.

8-4. *Find all Pythagorean triples (x, y, z) with $x < y$ and $z = 481$. Which of these triples are primitive? (Hint. Factor 481, and use (8.7) to write 481 as a sum of two squares.)*

Solution. Since $481 = 13 \cdot 37 = (3^2 + 2^2)(6^2 + 1^2) = 20^2 + 9^2 = 16^2 + 15^2$, we can take $a = 20$, $b = 9$ in (8.2) to get the primitive triple $(319, 360, 481)$ and $a = 16$, $b = 15$ to get the primitive triple $(31, 480, 481)$. Now look for triples of the form (kx, ky, kz), where $kz = 481$, $k > 1$, and (x, y, z) is primitive. Thus k must be 13 or 37. If $k = 13$, then $a^2 + b^2 = 37$ implies that $a = 6$ and $b = 1$, which gives the primitive triple $(35, 12, 37)$ and hence, multiplying by 13, the nonprimitive triple $(156, 455, 481)$ (swapping x and y to ensure $x < y$). Similarly, for $k = 37$, we have $a^2 + b^2 = 13$, and so $a = 3$, $b = 2$. This gives $(5, 12, 13)$, and, multiplying by 37, we obtain the nonprimitive triple $(185, 444, 481)$.

8-5. *Theorem 8.2 guarantees that every primitive Pythagorean triple is of the form $x = a^2 - b^2$, $y = 2ab$, $z = a^2 + b^2$, where $(a, b) = 1$. Is it true that every Pythagorean triple is of this form, where a and b are not necessarily relatively prime?*

Solution. No. For example, the triple $(9, 12, 15)$ is not of this form, since 15 cannot be expressed as the sum of two squares.

The following problem describes the Pythagorean triples that can be obtained from the formulas in (8.2) when a and b are not necessarily relatively prime or of opposite parity.

▷ **8-6.** *Suppose that $r = a^2 - b^2$, $s = 2ab$, and $t = a^2 + b^2$, where a and b are positive integers with $a > b$. Let $d = (a, b)$ and set $A = a/d$, $B = b/d$.*

(a) If exactly one of A and B is odd, prove that the Pythagorean triple (r, s, t) is of the form $(d^2 x, d^2 y, d^2 z)$, where (x, y, z) is a primitive triple with y even.

(b) If A and B are both odd, prove that (r, s, t) is of the form $(2d^2 x, 2d^2 y, 2d^2 z)$, where (x, y, z) is a primitive triple with x even.

Solution. (a) It is easily checked that (r, s, t) is a Pythagorean triple. Since A and B are of opposite parity and $(A, B) = 1$, (8.2) implies that $(A^2 - B^2, 2AB, A^2 + B^2)$ is a primitive triple, and clearly, $r = d^2(A^2 - B^2)$, $s = d^2(2AB)$, $t = d^2(A^2 + B^2)$.

(b) Since A and B are odd, $x = (A + B)/2$ and $y = (A - B)/2$ are integers. Then $(x, y) = 1$, for if a prime p divides x and y, then $p \mid x + y$ and $p \mid x - y$, i.e., $p \mid A$ and $p \mid B$, which contradicts $(A, B) = 1$. Also, x and y are of opposite parity, since $x + y = A$ and A is odd.

Since $(x, y) = 1$ and x and y are of opposite parity, (8.2) implies that $(2xy, x^2 - y^2, x^2 + y^2)$ is a primitive triple (where we list the even member first). Note that $4xy = A^2 - B^2$; since $A = x + y$ and $B = x - y$, we also have $2AB = 2(x^2 - y^2)$ and $A^2 + B^2 = 2(x^2 + y^2)$. Thus $r = d^2(A^2 - B^2) = 2d^2(2xy)$, $s = d^2(2AB) = 2d^2(x^2 - y^2)$, and $t = d^2(A^2 + B^2) = 2d^2(x^2 + y^2)$.

8-7. Let (x, y, z) be a Pythagorean triple. (a) Prove that at least one of x and y is divisible by 3. (b) Prove that at least one of x, y, and z is divisible by 5.

Solution. (a) It is easy to see that any square is congruent to 0 or 1 modulo 3. Thus if neither x nor y is divisible by 3, then $x^2 + y^2 \equiv 2 \pmod{3}$; in particular, $x^2 + y^2$ cannot be a perfect square. (b) Any square is congruent to 0, 1, or -1 modulo 5. Thus if neither x nor y is divisible by 5, then $x^2 + y^2$ is congruent to 0, 2, or -2 modulo 5. But if $x^2 + y^2 \equiv \pm 2 \pmod 5$, then $x^2 + y^2$ cannot be a perfect square. Thus $x^2 + y^2 \equiv 0 \pmod 5$, and hence either x or y (or both) is divisible by 5, or z is divisible by 5.

8-8. If (x, y, z) is a Pythagorean triple, prove that xyz is divisible by 60.

Solution. By Problem 8-7, xyz is divisible by 3 and by 5. We need only show that xyz is divisible by 4. It is clearly enough to deal with primitive triples. By (8.2), one of x and y, say y, is even, and then y has the form $y = 2ab$, where a and b have opposite parity. Thus y is divisible by 4.

8-9. Suppose $n \geq 3$. Prove that there is a Pythagorean triple with n as one of its members.

Solution. Suppose first that $n \geq 3$ is odd, say, $n = 2k + 1$. Then $n = (k+1)^2 - k^2$. Letting $a = k+1$ and $b = k$ in (8.2), we find that n is a member of the primitive triple $(a^2 - b^2, 2ab, a^2 + b^2)$.

Now let n be even and suppose that $n = 2^t m$, where m is odd. If $t \geq 2$, choose $a = 2^{t-1}m$ and $b = 1$ in (8.2). Then n is a member of the primitive triple $(a^2 - b^2, 2ab, a^2 + b^2)$. If $t = 1$, then $n = 2m$. Since m is odd, there is a primitive triple (x, y, z) with $x = m$. Therefore $(2x, 2y, 2z)$ is a triple with n as its first member.

8-10. (a) Prove that $(3, 4, 5)$ is the only primitive Pythagorean triple whose terms are in arithmetic progression.

(b) Prove that $(3k, 4k, 5k)$, with $k \geq 1$, are the only Pythagorean triples whose terms are in arithmetic progression.

Solution. (a) (b) Let (x, y, z) be a triple whose members are in arithmetic progression. Then for some integers a and k, we have $x = a - k$, $y = a$, and $z = a + k$. From the equation $x^2 + y^2 = z^2$, we obtain $a^2 = 4ak$, and therefore $a = 4k$. Thus the triple must be $(3k, 4k, 5k)$. If the triple is to be primitive, we must have $k = 1$.

8-11. *Prove or disprove: There are infinitely many primitive Pythagorean triples (x, y, z) such that y is even and y is a perfect square.*

Solution. This is true. The argument of Problem 8-9 shows that for every such y, there is a primitive Pythagorean triple (x, y, z), since y is divisible by 4.

8-12. *Determine all right triangles with sides of integral length whose areas equal their perimeters.*

Solution. Any Pythagorean triple (x, y, z) is of the form $(k(a^2 - b^2), k(2ab), k(a^2 + b^2))$ for some positive integers k, a, and b. The condition that the area is numerically equal to the perimeter is $xy/2 = x + y + z$, i.e., $k^2 ab(a^2 - b^2) = k(2a^2 + 2ab)$. Dividing both sides by $ka(a + b)$, we obtain the equivalent condition $kb(a - b) = 2$. If $k = 1$, either $b = 1$, $a - b = 2$, so that $a = 3$, giving the triple $(8, 6, 10)$; or $b = 2$, $a - b = 1$, so that $a = 3$, which yields $(5, 12, 13)$. If $k = 2$, then $b(a - b) = 1$, and so $b = 1$, $a = 2$, giving $(6, 8, 10)$. Thus apart from interchanging x and y, the only solutions are $(6, 8, 10)$ and $(5, 12, 13)$.

▷ **8-13.** *Let (x, y, z) be a primitive Pythagorean triple. Can $x - y$ be a square greater than 1?*

Solution. Suppose y is even; then $x = a^2 - b^2$ and $y = 2ab$, with a and b of opposite parity and relatively prime. Then $x - y = (a - b)^2 - 2b^2$. Thus we want a, b, c such that $(a - b)^2 - 2b^2 = c^2$, i.e., $(a - b)^2 - c^2 = 2b^2$. Factoring gives $(a - b - c)(a - b + c) = 2b^2$. To find solutions, set $a - b - c = 2$, $a - b + c = b^2$; subtracting, we get $2c = b^2 - 2$. Thus take b even and solve for a. For example, $b = 4$ implies $c = 7$, so $a = c + b + 2 = 13$; hence $x = 153$, $y = 104$, and $x - y = 7^2$. Similarly, $b = 6$ implies $c = 17$, so $a = 25$; then $x = 589$, $y = 300$, and $x - y = 17^2$.

Notes. 1. The problem of characterizing *all* triples for which $x - y = k^2$ ($k \geq 1$) is more difficult, and solutions exist only for certain values of k. The general solution is given by
$$a = e^2 + (e + d)^2, \qquad b = 2de, \qquad k = 2e^2 - d^2,$$
where d is odd and $(d, e) = 1$. For example, if $d = 3$ and $e = 4$, then $k = 23$, $b = 24$, $a = 65$; hence $x = 3649$, $y = 3120$, and $x - y = 23^2$.

2. If we want triples (x, y, z) with $y > x$, y even, and $y - x$ a perfect square, the solution is more complicated and best handled by using the technique for solving Pell's Equation (see Problem 10-40). For $y - x = 1$, this approach yields the following three solutions (from among infinitely many): $(3, 4, 5)$, $(119, 120, 169)$, and $(4059, 4060, 5741)$.

Fermat's Last Theorem

8-14. *Show that to prove Fermat's Last Theorem, it is enough to prove it for the case where the exponent is an odd prime. (Hint. Use (8.4).)*

Solution. If the exponent n is a multiple of 4, Fermat's Last Theorem follows from (8.4). For if $n = 4k$ and $a^n + b^n = c^n$, then $(a^k)^4 + (b^k)^4 = (c^{2k})^2$, contradicting the fact that the equation $x^4 + y^4 = z^2$ has no solution in nonnegative integers. Suppose now that n has an odd prime divisor p, and let $n = kp$. If Fermat's Last Theorem failed for the exponent n, then it would also fail for the exponent p, since if $x^n + y^n = z^n$, then $(x^k)^p + (y^k)^p = (z^k)^p$.

8-15. Show that a Pythagorean triple cannot be a solution of $x^n + y^n = z^n$ if $n \geq 3$.

Solution. Suppose $x^2 + y^2 = z^2$ with $y > x$. Then $x^n + y^n = x^2 x^{n-2} + y^2 y^{n-2} < (x^2 + y^2) y^{n-2} = z^2 y^{n-2} < z^2 z^{n-2} = z^n$.

8-16. Show that if $x^5 + y^5 = z^5$, then at least one of x, y, and z is divisible by 5. (*Hint.* Work modulo 25.)

Solution. We list all possible values of a^5 modulo 25. Since $(a+5k)^5 \equiv a^5 \pmod{25}$, it is enough to consider $a = 0, \pm 1$, and ± 2. Thus a^5 is congruent to either $0, \pm 1$, or ± 7 modulo 25. Simple examination of cases now shows that if $x^5 + y^5 = z^5$ and neither x nor y is divisible by 5, then z must be divisible by 5.

8-17. Use (8.4) to prove that $x^4 - 4y^4 = z^2$ has no solution in positive integers. (*Hint.* If there is a solution, square each side and rearrange terms to get a solution of $x^4 + y^4 = z^2$.)

Solution. Suppose there are positive integers r, s, t such that $r^4 - 4s^4 = t^2$. Squaring each side, we can rewrite the resulting equation as $(2rs)^4 + t^4 = (r^4 + 4s^4)^2$. Thus there is also a solution of $x^4 + y^4 = z^2$ in positive integers, contradicting (8.4).

Note. We have shown that from a supposed solution of $x^4 - 4y^4 = z^2$ in positive integers, we can produce a solution of $x^4 + y^4 = z^2$ in positive integers. The argument is easily reversed to produce a solution of $x^4 - 4y^4 = z^2$ from a solution of $x^4 + y^4 = z^2$. The equations $x^4 + 4y^4 = z^2$ and $x^4 - y^4 = z^2$ are also equivalent in this sense.

▷ **8-18.** Show that the equation $x^4 - y^4 = z^2$ has no solution in positive integers.

Solution. If there is a positive solution, let x be the smallest positive integer for which there are positive y and z such that $x^4 - y^4 = z^2$. As in the proof of (8.4), if x, y, and z are not pairwise relatively prime, we can obtain a smaller positive solution, contradicting the minimality of x. Thus (y^2, z, x^2) is a primitive Pythagorean triple.

If y is odd, then by (8.2), there exist positive integers a and b such that $x^2 = a^2 + b^2$ and $y^2 = a^2 - b^2$; then $a^4 - b^4 = (xy)^2$. Since $a^2 < x^2$, this contradicts the minimality of x.

If y is even, then there exist relatively prime integers a and b of opposite parity such that $x^2 = a^2 + b^2$ and $2ab = y^2$. We can assume that a is odd and b is even. Thus (a, b, x) is a primitive Pythagorean triple with b even, and hence there exist relatively prime s and t such that $a = s^2 - t^2$ and $b = 2st$. But since $2ab = y^2$, it follows that a is a perfect square and b is twice a perfect square. Let $a = c^2$ and $b = 2d^2$. Since $b = 2st = 2d^2$ and $(s,t) = 1$, s and t are each perfect squares. If $s = u^2$ and $t = v^2$, then $c^2 = a = s^2 - t^2 = u^4 - v^4$. Since clearly $u < x$, this again contradicts the minimality of x.

8-19. Prove that at most one member of a Pythagorean triple can be a perfect square. (*Hint.* Use (8.4) and (8.5).)

CHAPTER 8: DIOPHANTINE EQUATIONS

Solution. Let (x, y, z) be a Pythagorean triple. If x and y are perfect squares, then z^2 can be expressed as a sum of two fourth powers, contradicting (8.4). If y and z are perfect squares, then x^2 can be expressed as a difference of two fourth powers, which contradicts (8.5).

8-20. Prove that there are no positive integers a and b such that $a^2 + b^2$ and $a^2 - b^2$ are both perfect squares.

Solution. If $a^2 + b^2$ and $a^2 - b^2$ are perfect squares, then so is their product $a^4 - b^4$, contradicting (8.5).

8-21. Can the area of a Pythagorean triangle be a perfect square? Twice a perfect square? (Hint. Use Problems 8-20 and 8-18.)

Solution. Let (x, y, z) be a Pythagorean triple, and suppose that the area $xy/2$ of the associated triangle is a perfect square, say a^2. Then $(x + y)^2 = z^2 + (2a)^2$ and $(x - y)^2 = z^2 - (2a)^2$, a contradiction in view of Problem 8-20.

Similarly, if $xy/2 = 2a^2$, we get $(x + y)^2 = z^2 + 8a^2$ and $(x - y)^2 = z^2 - 8a^2$. Thus $z^4 - 4(2a)^4 = (x^2 - y^2)^2$, contradicting Problem 8-18.

8-22. (a) Prove that $x^4 - 2y^2 = 1$ has no solution in positive integers. In particular, $x^4 - 2y^4 = 1$ is not solvable.

(b) Prove that $x^4 - 2y^2 = -1$ has no solution in positive integers except $x = y = 1$. (Thus $x^4 - 2y^4 = -1$ has only one positive solution.)

Solution. (a) If $x^4 - 2y^2 = 1$, then $2y^2 = x^4 - 1$ and hence $4y^4 = x^8 - 2x^4 + 1$. Add $4x^4$ to each side to get $4(y^4 + x^4) = (x^4 + 1)^2$; then $x^4 + 1$ is even. If $z = (x^4 + 1)/2$, it follows that $y^4 + x^4 = z^2$, which contradicts (8.4).

(b) The same argument used in (a) now yields $y^4 - x^4 = z^2$, which has only the trivial solutions $x = \pm 1$, $y = \pm 1$, and $z = 0$. Thus the only solution in positive integers to the original equation is $x = y = 1$.

▷ **8-23.** Prove that $x^2 - 2y^4 = 1$ has no solution in positive integers. (Hint. Factor $x^2 - 1$ and use part (a) of the preceding problem.)

Solution. If $x^2 - 2y^4 = 1$, then $(x - 1)(x + 1) = 2y^4$. By changing the sign of x if necessary, we can assume $x \equiv 1 \pmod 4$, i.e., $x = 4k + 1$. Hence the last equation can be written as $4k(2k + 1) = y^4$, and so y is even, say $y = 2z$. Then $k(2k + 1) = 4z^4$, and since $2k + 1$ is odd, we must have $4 \mid k$. Thus, letting $k = 4s$, we obtain $s(8s + 1) = z^4$. It follows that $s = u^4$ and $8s + 1 = v^4$, since s and $8s + 1$ are clearly relatively prime. Hence $v^4 - 8u^4 = 1$, i.e., $v^4 - 2(2u^2)^2 = 1$. By part (a) of the preceding problem, this equation has only the trivial solution $v = \pm 1$, $u = 0$. Thus $s = 0$, which implies that $k = 0$ and thus $x = 1$. Therefore the original equation has no solution in positive integers.

Sums of Two Squares

8-24. *Determine which of the following integers is a sum of two squares: 98, 343, 735, 1428, and 4680.*

Solution. By (8.8), n is a sum of two squares if and only if any prime of the form $4k+3$ appears to an even power in the prime factorization of n. This is clearly true for $98 = 2 \cdot 49$ and false for $343 = 7^3$. By summing the digits of 735, we see that 735 is divisible by 3 but not by 9, so 735 is not a sum of two squares. For the same reason, 1428 is not the sum of two squares. Since $4680 = 2^3 \cdot 3^2 \cdot 5 \cdot 13$, it is a sum of two squares.

8-25. *Find all of the essentially different representations of 5525 as a sum of two squares.*

Solution. Since $5525 = 5^2 \cdot 13 \cdot 17$, (8.11) implies that $N(5525) = 48$. Since 5525 is neither a perfect square nor twice a perfect square, it follows that 5525 has six essentially different representations as a sum of two squares. First we find the representations of $13 \cdot 17 = 221$ as a sum of two squares. There are two essentially different such representations, which can be found by inspection or by using (8.7). To apply (8.7), note that $13 = 2^2 + 3^2$ and $17 = 1^2 + 4^2$. The formula $(a^2+b^2)(c^2+d^2) = (ac-bd)^2 + (ad+bc)^2$, with $a=2$, $b=3$, $c=1$, and $d=4$, yields (after we change the sign) $221 = 10^2 + 11^2$. Interchanging the roles of a and b, we obtain $221 = 5^2 + 14^2$. Multiplying by 25 then gives $5525 = 50^2 + 55^2$ and $5525 = 25^2 + 70^2$.

Using the same technique and the fact that $25 = 3^2 + 4^2$, we get that $25 \cdot 13 = 6^2 + 17^2$ and $25 \cdot 13 = 1^2 + 18^2$. Combining these two representations with $17 = 1^2 + 4^2$ gives $5525 = 62^2 + 41^2 = 7^2 + 74^2 = 71^2 + 22^2 = 14^2 + 73^2$.

8-26. *Find all of the essentially different representations of 14365 as a sum of two squares.*

Solution. Using the fact that $14365 = 5 \cdot 13^2 \cdot 17$, we conclude, exactly as in the previous problem, that 14365 has six essentially different representations as the sum of two squares. They are $26^2 + 117^2$, $78^2 + 91^2$, $98^2 + 69^2$, $21^2 + 118^2$, $54^2 + 107^2$, and $37^2 + 114^2$.

8-27. *Find the two essentially different representations of 229320 as a sum of two squares.*

Solution. Since $229320 = 2^3 \cdot 3^2 \cdot 7^2 \cdot 5 \cdot 13$, we will first represent $2 \cdot 5 \cdot 13 = 130$. We could use the machinery of (8.7) for this, but the solutions can be easily found by inspection: $130 = 3^2 + 11^2 = 7^2 + 9^2$. Multiplying by $2^2 \cdot 3^2 \cdot 7^2$ yields the representations $126^2 + 462^2$ and $294^2 + 378^2$.

8-28. *Let q be a prime of the form $4k+3$. Prove that q^2 is not the sum of two nonzero squares.*

Solution. Suppose $q^2 = a^2 + b^2$. By (8.6.i), q divides a. Thus if a is nonzero, then $a^2 \geq q^2$ and therefore $b = 0$.

244   CHAPTER 8: DIOPHANTINE EQUATIONS

Another proof: We have four obvious representations, namely, $0^2 + (\pm q)^2$ and $(\pm q)^2 + 0^2$. But $N(q^2) = 4$, by (8.11), and therefore these are the only representations.

8-29. Prove that the prime p is of the form $4k + 1$ if and only if p divides $n^2 + (n+1)^2$ for some $n \geq 1$. (Hint. Note that $(n, n+1) = 1$.)

Solution. First suppose p divides $n^2 + (n+1)^2$. If p is of the form $4k+3$, then by (8.6.i), we have $p \mid n$ and $p \mid n+1$, which is clearly impossible. Since p is odd, it follows that p is of the form $4k+1$.

Now suppose $p \equiv 1 \pmod 4$. We want a value of n such that $n^2 + (n+1)^2 \equiv 0 \pmod p$, i.e., $2n^2 + 2n + 1 \equiv 0 \pmod p$. Multiplying by 2 and completing the square, we can rewrite this congruence as $(2n+1)^2 \equiv -1 \pmod p$. Since -1 is a quadratic residue of p, there is an x such that $x^2 \equiv -1 \pmod p$. If x is odd, let $n = (x-1)/2$, and if x is even, let $n = (p-x-1)/2$.

8-30. (a) Let $n = 1769625$. Find the number of primitive representations of n as a sum of two squares. (b) Find all essentially distinct representations of n as a sum of two squares. (c) Find one representation of $8n$ as a sum of two squares.

Solution. (a) Factor n as $n = 3^2 \cdot 5^3 \cdot 11^2 \cdot 13$. Since n is divisible by a $4k+3$ prime, it follows from (8.15) that n has no primitive representations.

(b) By (8.11), $n = 33^2 \cdot 5^3 \cdot 13$ has $4 \cdot 4 \cdot 2 = 32$ representations, and thus n has $32/8 = 4$ essentially distinct representations. To find them, first write $5^3 \cdot 13$ as a sum of two squares. Clearly, the only representation of 13 is $3^2 + 2^2$. Since $5 = 1^2 + 2^2$, (8.7) implies that $5^2 = (1^2+2^2)(1^2+2^2) = 5^2 + 0^2 = 3^2 + 4^2$, and so $5^3 = (1^2+2^2)(5^2+0^2) = 5^2 + 10^2$ and $5^3 = (1^2+2^2)(3^2+4^2) = 11^2 + 2^2$. Applying (8.7) to the various combinations for $5^3 \cdot 13$ then gives

$$(5^2 + 10^2)(3^2 + 2^2) = (15 \pm 20)^2 + (10 \mp 30)^2 = 35^2 + 20^2 = 5^2 + 40^2;$$
$$(11^2 + 2^2)(3^2 + 2^2) = (33 \pm 4)^2 + (22 \mp 6)^2 = 37^2 + 16^2 = 29^2 + 28^2.$$

Thus the representations of n itself are obtained by multiplying each representation of $5^3 \cdot 13$ by 33^2, yielding $1155^2 + 660^2$, $165^2 + 1320^2$, $1221^2 + 528^2$, and $957^2 + 924^2$.

(c) Since $8 = 2^2 + 2^2$, we can use (8.7) with each of the representations obtained in (b) to find all of the (essentially distinct) representations of $8n$: $3630^2 + 990^2$, $2970^2 + 2310^2$, $3498^2 + 1386^2$, and $3762^2 + 66^2$.

8-31. Find an integer n such that n has exactly three essentially distinct representations as a sum of two squares.

Solution. We use (8.11). If $n = p^2 q$, where p and q are distinct primes of the form $4k+1$, then $N(n) = 4 \cdot 3 \cdot 2 = 24$. Any representation of n as the sum of two distinct nonzero squares yields 8 variations obtained by changing the sign or order of the terms. Thus $p^2 q$ has $24/8 = 3$ essentially distinct representations as a sum of two squares. For example, $n = 5^2 \cdot 13$ has the required property.

Alternatively, if p is a prime of the form $4k+1$, then $N(p^m) = 4(m+1)$. By the same argument as above, we find that if $m+1 = 6$, then p^m has exactly three essentially distinct representations as a sum of two squares.

Another example is $n = p^4$, where p is of the form $4k+1$. By (8.11), we have $N(n) = 20$. Clearly, n has the representation $n = (p^2)^2 + 0^2$, which yields a total of four representations if we count order and sign. The remaining 16 representations split into two families, each with 8 variations. Thus p^4 has exactly three essentially distinct representations.

Note. It is not hard to show that *all* solutions can be obtained by multiplying the above solutions by a number of the form m^2 or $2m^2$, where m has no prime divisors of the form $4k+1$.

8-32. *Prove or disprove: At least one of every four consecutive positive integers is not a sum of two squares.*

Solution. This is true. One of the numbers must be congruent to 3 modulo 4 and hence cannot be a sum of two squares.

8-33. *Let $n = 2 \cdot 3^2 \cdot 5^3 \cdot 13^5$. Is there a Pythagorean triangle with hypotenuse n? Does there exist a primitive Pythagorean triangle with hypotenuse n?*

Solution. We ask whether n^2 can be represented as the sum of two nonzero squares. Certainly, n^2 is a sum of two squares, so by (8.11), we obtain $N(n^2) = 4 \cdot (6+1)(10+1)$. Remove the four trivial representations of n^2 as a sum of two squares, one of which is 0. We thus find that n^2 has $4 \cdot (77-1)$ representations as the sum of nonzero squares. It follows that there are in fact $(4 \cdot 76)/8 = 38$ different Pythagorean triangles with hypotenuse n.

There are no primitive triangles with hypotenuse n. This follows from (8.15), since n^2 is divisible by a prime of the form $4k+3$.

8-34. *Prove that n is a sum of two squares if and only if $n = m^2 N$, where N is square-free and every odd prime divisor of N is of the form $4k+1$.*

Solution. This is a simply a restatement of (8.9). If n is of the form specified, then by (8.9), N is a sum of two squares. If $N = a^2 + b^2$, then $n = (ma)^2 + (mb)^2$. Conversely, if n is a sum of two squares, then by (8.9), any prime of the form $4k+3$ occurs to an even power in the prime factorization of n. Thus, if we express n as $m^2 N$, where N is square-free, then no prime of the form $4k+3$ can divide N, and therefore every odd prime divisor of N is of the form $4k+1$.

8-35. (Fermat, 1636.) *Show that 21 cannot be expressed as the sum of the squares of two rational numbers.*

Solution. Suppose to the contrary that $21 = r^2 + s^2$, where r and s are rational. Write $r = a/c$, $s = b/c$, where a, b, and $c > 0$ are integers. Then $21c^2 = a^2 + b^2$. But since the prime 3 appears to an odd power in the prime factorization of $21c^2$, (8.6) implies that $21c^2$ cannot be the sum of the squares of two integers.

8-36. *Show that the positive integer n is a sum of two nonzero squares if and only if every prime divisor of n of the form $4k+3$ appears to an even power in the prime factorization of n and either the exponent of 2 is odd or n has at least one prime factor of the form $4k+1$.*

Solution. If n is of the specified form, then, by (8.9), n is a sum of two squares, and these must be nonzero unless n is a perfect square. Suppose n is a perfect square and let $n = m^2 p^2$, where p is of the form $4k+1$. We show that p^2 is a sum of nonzero squares. Write $p = a^2 + b^2$; then by (8.7), $p^2 = (a^2+b^2)(a^2+b^2) = (2ab)^2 + (a^2-b^2)^2$. Clearly, neither a nor b is 0 and $a \neq b$, so we have expressed p^2 as the sum of two nonzero squares.

Conversely, if n is the sum of two nonzero squares, then all primes of the form $4k+3$ must appear to an even power in the prime factorization of n. Thus let $n = s^2 m$, where m is not divisible by any prime of the form $4k+3$. If m is divisible by a prime of the form $4k+1$, we are done. Otherwise, m is a power of 2, and it is easy to see that m cannot be an even power of 2. Hence n is of the form $2^t s^2$, where t is odd and all prime divisors of s are of the form $4k+3$.

Note. A similar argument shows the following: n is a sum of two *different* nonzero squares if and only if every prime divisor of n of the form $4k+3$ appears to an even power and n has at least one prime divisor of the form $4k+1$.

8-37. *Prove that a positive integer n is the hypotenuse of a Pythagorean triangle if and only if n is divisible by at least one prime of the form $4k+1$.*

Solution. The positive integer n is the hypotenuse of some Pythagorean triangle if and only if n^2 has a representation as the sum of two nonzero squares. Suppose first that $p|n$, where p is a prime of the form $4k+1$, and let $n^2 = k^2 p^2$. Then by (8.15), p^2 has a representation $p^2 = a^2 + b^2$ as a sum of two relatively prime squares; clearly, neither a nor b can be 0. Thus $n^2 = (ka)^2 + (kb)^2$, and so n is the hypotenuse of a Pythagorean triangle.

Conversely, suppose that n is the hypotenuse of a Pythagorean triangle; then $n^2 = a^2 + b^2$ for some positive integers a and b. Let $d = (a,b)$. Since $(n/d)^2 = (a/d)^2 + (b/d)^2$, n/d is the hypotenuse of a primitive Pythagorean triangle. Therefore by (8.15), n/d is not divisible by any prime of the form $4k+3$. Since n/d is odd and not equal to 1, it follows that n/d, and hence n, is divisible by a prime of the form $4k+1$.

Alternate proof: Suppose first that n, and hence n^2, has no prime divisor of the form $4k+1$. Then by (8.11), $N(n^2) = 4$. Since n^2 has four obvious representations as a sum of two squares, one of which is 0, n^2 has no representations as the sum of two nonzero squares.

Conversely, suppose that n has a prime divisor of the form $4k+1$. Then (8.11) implies that $N(n^2) \geq 12$. It follows that n^2 must have at least one representation as a sum of nonzero squares.

8-38. *For a given positive integer z, prove that the number of Pythagorean triples (x,y,z), with $0 < x < y$, is $(N(z^2) - 4)/8$, where $N(z^2)$ is the (total) number of representations of z^2 as a sum of two squares.*

Solution. Each Pythagorean triple (x, y, z) corresponds to a representation of z^2 as a sum of two positive squares, and this representation is uniquely determined if we assume that $0 < x < y$. (We cannot have $x = y$, for then $z^2 = 2x^2$, which is impossible since z^2 has an even number of factors of 2 and $2x^2$ has an odd number.) Each representation of z^2 as a sum of two nonzero squares has eight variations if we allow for sign changes and the order of the summands; also, z^2 has the representation $0^2 + z^2$, which allows for four variations and clearly does not correspond to a Pythagorean triple. If we disregard these four representations, the remaining $N(z^2) - 4$ representations correspond to $(N(z^2) - 4)/8$ representations of the form $x^2 + y^2$, where $0 < x < y$.

8-39. *In a letter to Mersenne dated December 25, 1640, Fermat asked for a number n which is the hypotenuse of precisely 367 right triangles with integer sides. Use the preceding problem to answer Fermat's question.*

Solution. Let $n = m \prod p_i^{a_i}$, where the p_i are primes of the form $4k + 1$, m is not divisible by any prime of the form $4k + 1$, and any prime divisor of m of the form $4k + 3$ occurs to an even power. Then (8.11) implies that $N(n^2) = 4 \prod(2a_i + 1)$, and hence, by the preceding problem, there are $(4 \prod(2a_i + 1) - 4)/8$ different Pythagorean triangles with hypotenuse n. It is now easy to find n such that there are as many Pythagorean triangles with hypotenuse n as we wish. For example, let $n = 5^k$. Then there are $(4(2k + 1) - 4)/8 = k$ Pythagorean triangles with hypotenuse n. Thus if $n = 5^{367}$, then n is the hypotenuse of precisely 367 right triangles with integer sides. (There are many other solutions; the smallest is $n = 5^{14} \cdot 13^2 \cdot 17$.)

8-40. *Prove or disprove: Every positive integer n has at least as many divisors of the form $4k + 1$ as of the form $4k + 3$.*

Solution. This is true and follows at once from (8.12), since $N(n) \geq 0$ for any n.

8-41. *Let $n = 2^3 \cdot 5^4 \cdot 17 \cdot 3^6 \cdot 11^4$. Prove that n has exactly 10 more divisors of the form $4k + 1$ than of the form $4k + 3$. (Hint. First show that n has representations as a sum of two squares.)*

Solution. By (8.9), n has a representation as a sum of two squares, since every prime of the form $4k + 3$ occurs to an even power in the prime factorization of n. Therefore by (8.11), $N(n) = 4 \cdot (4 + 1) \cdot (1 + 1) = 40$. But by (8.12), $N(n) = 4(D_1 - D_3)$, so $D_1 - D_3 = 10$.

8-42. *Suppose $n = q_1^{b_1} \cdots q_r^{b_r}$, where each q_i is a prime of the form $4k + 3$. If $b_1 + b_2 + \cdots + b_r$ is odd, prove that n has as many divisors of the form $4k + 3$ as of the form $4k + 1$.*

Solution. Since the sum of the b_i is odd, at least one of the b_i is odd. It follows from (8.9) that n has no representations as a sum of two squares, and thus by (8.12), $D_1 - D_3 = 0$, i.e., $D_1 = D_3$.

248   CHAPTER 8: DIOPHANTINE EQUATIONS

8-43. *Let $n = 670761000$. How many more divisors of the form $4k + 1$ does n have compared to the number of divisors of the form $4k + 3$?*

Solution. Factoring, we obtain $n = 2^3 \cdot 3^4 \cdot 7^2 \cdot 5^3 \cdot 13^2$, and therefore $N(n) = 4(3 + 1)(2 + 1) = 48 = 4(D_1 - D_3)$ by (8.11) and (8.12); hence $D_1 - D_3 = 12$.

8-44. *Show that 98049603 has the same number of divisors of the form $4k + 3$ as of the form $4k + 1$. (It is not necessary to factor this number.)*

Solution. Note that 3 divides 98049603 but 9 does not; hence $N(n) = 0$, by (8.9). It follows from (8.12) that $D_1 - D_3 = 0$.

▷ **8-45.** *(a) If $8 | N(n)$, prove that there are exactly $N(n)/8$ esssentially distinct representations of n as a sum of two squares.*

(b) If $8 \nmid N(n)$, show that $N(n) = 8k + 4$ for some integer k and that there are exactly $k + 1$ essentially distinct representations of n.

Solution. (a) Each representation of n as a sum of two different nonzero squares has eight trivial variations (see the Example preceding (8.17)); if one of the squares is 0 or if the squares are equal, there are only four variations. But n can have at most one representation of the form $m^2 + 0^2$ and at most one of the form $m^2 + m^2$; it cannot have representations of both these forms, since n cannot be simultaneously a perfect square and twice a perfect square. Thus $N(n)$ is of the form $8k$ if n is neither a perfect square nor twice a perfect square, and n is of the form $8k + 4$ otherwise. If $8 | N(n)$, then every representation involves two unequal nonzero squares, and hence there will be precisely $N(n)/8$ essentially distinct representations.

(b) If $8 \nmid N(n)$, then n must be a square or twice a square, say, $n = m^2$ or $n = 2m^2$. If $n = m^2$, let k be the number of essentially distinct representations of n involving two nonzero squares. These generate a total of $8k$ variations. In addition, n has the representation $m^2 + 0^2$, with 4 variations. Thus $N(n) = 8k + 4$, and there is a total of $k + 1$ essentially distinct representations. The argument for $n = 2m^2$ is essentially the same.

Note. The preceding argument shows that $N(n)$ is a multiple of 8 if and only if n is neither a square nor twice a square.

8-46. *Let $n = 5^4 \cdot 13^2$; then $N(n) = 4 \cdot 5 \cdot 3 = 60$. Using the preceding problem, what can you conclude about the number of essentially distinct representations of n as a sum of two squares?*

Solution. In view of the preceding problem, since $60 = 7 \cdot 8 + 4$, n has 7 essentially distinct representations as a sum of two distinct nonzero squares, as well as one representation where one square is 0^2, namely, $n = (5^2 \cdot 13)^2 + 0^2$.

8-47. *Find a positive integer which is divisible by 14 and has precisely five essentially distinct representations as a sum of two nonzero squares.*

Solution. We will look for a number $n = 2 \cdot 7^2 \cdot m$, where every prime that divides m is of the form $4k+1$. Let $m = \prod p_i^{a_i}$; then $N(n) = 4\prod(a_i+1)$. If we make sure that m is not a perfect square, then n has representations only as a sum of unequal nonzero squares, and thus the number of essentially distinct representations is $(4\prod(a_i+1))/8$. We can make this equal to 5 by letting $a_1 = 4$ and $a_2 = 1$; for example, take $n = 2 \cdot 7^2 \cdot 5^4 \cdot 13$. We can also look for solutions which are perfect squares or twice a perfect square. A similar analysis shows that if, for example, $n = 2 \cdot 7^2 \cdot 5^2 \cdot 13^2$, then n has five essentially distinct representations as a sum of two nonzero squares. This is the smallest solution.

▷ **8-48.** *The following is essentially Euler's proof that every prime of the form $4k+1$ is a sum of two squares.*

(a) *Let p be a prime of the form $4k+1$. Prove that there exists $1 \le k < p$ such that kp is a sum of two squares.*

(b) *Let m be the smallest positive integer such that mp is a sum of two squares. Prove that $m = 1$. Thus p is itself a sum of two squares.*

Solution. (a) By (5.11), -1 is a quadratic residue of p, so there exists an integer s between $-(p-1)/2$ and $(p-1)/2$ such that $s^2 \equiv -1 \pmod{p}$. Thus $s^2 + 1 = kp$ for some $k \ge 1$, and $kp < (p/2)^2 + 1 < p^2$, whence $1 \le k < p$.

(b) Assume that $m > 1$ and write $mp = a^2 + b^2$. Let A, B be numbers of minimal absolute value that are congruent modulo m to a and b; thus $|A| \le m/2$ and $|B| \le m/2$. Then
$$A^2 + B^2 \equiv a^2 + b^2 \equiv 0 \pmod{m},$$
and so
$$A^2 + B^2 = km, \quad \text{with} \quad 0 \le km \le 2(m/2)^2 < m^2.$$

Thus $0 \le k < m$. If $k = 0$, then $A = B = 0$ and m divides a and b; hence $m^2 \mid mp$, that is, m divides p, a contradiction since $m < p$. Therefore $k > 0$. By (8.7),
$$m^2kp = (mp)(km) = (a^2 + b^2)(A^2 + B^2) = r^2 + s^2, \tag{1}$$
where $r = aB - Ab$ and $s = aA + bB$. Since $A \equiv a \pmod{m}$ and $B \equiv b \pmod{m}$, we have $r \equiv ab - ab \equiv 0 \pmod{m}$ and $s \equiv a^2 + b^2 \equiv 0 \pmod{m}$; thus r and s are each divisible by m. Dividing (1) by m^2, we get $kp = (r/m)^2 + (s/m)^2$, with $0 < k < m$. This contradicts the fact that mp is the smallest positive multiple of p that is a sum of two squares. Thus $m = 1$, and therefore p is itself a sum of two squares.

8-49. *If p is prime and $2p - 1$ is a perfect square, prove that $p = k^2 + (k+1)^2$ for some $k \ge 1$.*

Solution. Let $2p - 1 = n^2$; clearly, n is odd. If $k = (n-1)/2$, then $k + 1 = (n+1)/2$ and $p = k^2 + (k+1)^2$.

8-50. *The Fermat number $F_5 = 2^{32} + 1$ can be written as $20449^2 + 62264^2$. Use this fact to explain why F_5 cannot be prime.*

Solution. We also have $F_5 = (2^{16})^2 + 1^2$, so F_5 has at least two essentially distinct representations as the sum of two squares. Thus, by Note 2 following (8.12), F_5 is composite.

▷ **8-51.** *Suppose that n has (at least) two essentially distinct representations as a sum of two squares. Specifically, let $n = s^2 + t^2 = u^2 + v^2$, where $s \geq t \geq 0$, $u \geq v \geq 0$, and $s > u$. If $d = (su - tv, n)$, show that d is a proper divisor of n. (Hint. First show that $s^2u^2 \equiv t^2v^2 \pmod{n}$.)*

Solution. We have $s^2 \equiv -t^2 \pmod{n}$ and $u^2 \equiv -v^2 \pmod{n}$. Multiplying, we obtain $s^2u^2 \equiv t^2v^2 \pmod{n}$, and so $(su + tv)(su - tv)$ is divisible by n. Since $n^2 = (s^2 + t^2)(u^2 + v^2) = (su + tv)^2 + (sv - tu)^2$ and $sv > tu$, it follows that $su + tv < n$. Because $0 < su - tv < n$ and $(su + tv)(su - tv)$ is divisible by n, we conclude that $su - tv$ and n have a nontrivial common divisor, and therefore $(su - tv, n)$ is a proper divisor of n.

Note. If we apply this idea to the representations of F_5 as a sum of two squares given in the preceding problem, with $s = 2^{16}$, $t = 1$, $u = 62264$, and $v = 20449$, we find that $(su - tv, F_5) = 641$, and hence F_5 is divisible by 641.

Sums of Two Relatively Prime Squares

8-52. *Can 194922 be expressed as a sum of two squares? Does 194922 have a representation as a sum of two relatively prime squares?*

Solution. Let $n = 194922$. Then $n = 2 \cdot 3^2 \cdot 7^2 \cdot 13 \cdot 17$. In particular, all primes of the form $4k + 3$ occur to an even power in the prime factorization of n, and therefore by (8.9), n can be expressed as a sum of two squares. Since n is divisible by a prime of the form $4k + 3$, (8.15) implies that n is not the sum of two relatively prime squares.

8-53. *Express 332514 as a sum of two squares. Can 332514 be represented as a sum of two relatively prime squares?*

Solution. Since $332514 = 2 \cdot 3^2 \cdot 7^2 \cdot 13 \cdot 29$, it has prime divisors of the form $4k + 3$ and so, by (8.15), cannot be represented as the sum of relatively prime squares. To represent it as a sum of squares, observe that $26 = 1^2 + 5^2$ and $29 = 2^2 + 5^2$. Thus using (8.7), we have $2 \cdot 13 \cdot 29 = 27^2 + 5^2 = 23^2 + 15^2$. Multiplying by $3^2 \cdot 7^2$, we obtain $332514 = 567^2 + 105^2 = 483^2 + 315^2$. (These are the only solutions.)

8-54. *Find all representations of 11050 as a sum of two relatively prime squares.*

Solution. We have $11050 = 2 \cdot 5^2 \cdot 13 \cdot 17$. Thus by (8.18), 11050 has $2^{3-1} = 4$ essentially distinct representations as the sum of two relatively prime squares. It is not difficult to find them all. Using the fact that $50 = 1^2 + 7^2$, $13 = 2^2 + 3^2$, and $17 = 1^2 + 4^2$, and applying (8.7), we get $50 \cdot 13 = 23^2 + 11^2 = 19^2 + 17^2$, and hence $11050 = 67^2 + 81^2 = 21^2 + 103^2 = 87^2 + 59^2 = 49^2 + 93^2$.

8-55. *Find the six smallest integers that have exactly four essentially distinct representations as a sum of two squares, with at least one representation primitive.*

Solution. In view of (8.15), any solution n has at most one factor of 2 and has no prime divisors of the form $4k + 3$. We also require $N(n) = 3 \cdot 8 + 4 = 28$ or $N(n) = 4 \cdot 8 = 32$

(see Problem 8-45); thus if $n = 2^a p_1^{a_1} \cdots p_r^{a_r}$ with each p_i a $4k+1$ prime, (8.11) implies that $(a_1 + 1) \cdots (a_r + 1) = 7$ or 8. Thus n must be one of the following forms: p_1^6, $p_1 p_2 p_3$, $p_1 p_2^3$, or p_1^7, or twice any of these numbers. The six smallest numbers of these forms are $5 \cdot 13 \cdot 17 = 1105$, $5^3 \cdot 13 = 1625$, $5 \cdot 13 \cdot 29 = 1885$, $5^3 \cdot 17 = 2125$, $2 \cdot 5 \cdot 13 \cdot 17 = 2210$, and $5 \cdot 17 \cdot 29 = 2465$.

8-56. *Find a positive integer that has five essentially distinct representations as a sum of two squares, with none of these representations primitive.*

Solution. We look at solutions of the form N^2. To ensure that no representation is primitive, let $N = nm$, where $n > 1$ and n has no prime divisors of the form $4k + 1$, while m is divisible only by primes of the form $4k + 1$. Thus we want m^2 to have five essentially distinct representations. The obvious representation $m^2 = 0^2 + m^2$ has four variations, so we want $N(m^2) = 36$, since then, apart from the obvious representation, there will be $(36 - 4)/8 = 4$ other, essentially distinct representations. It follows from (8.11) that $m = p^4$, where p is a prime of the form $4k + 1$, or $m = pq$, where p and q are distinct primes of the form $4k + 1$. Thus, for example, we can take $N = 2 \cdot 5^4$. Similar reasoning shows that to obtain a solution of the form $2N^2$, we must choose n and m as above.

For solutions that are not of the form N^2 or $2N^2$, let $x = n^2 m$ or $2n^2 m$, where $n > 1$ and n has no prime divisors of the form $4k + 1$, while all prime divisors of m are of the form $4k + 1$. We then want $N(m) = 40$. If follows from (8.11) that $m = p^9$ or $m = p^4 q$, where p and q are distinct primes of the form $4k + 1$.

8-57. *Prove or disprove: If m and n each have a primitive representation as a sum of two squares, then mn also has a primitive representation.*

Solution. A counterexample is easy to find. Let $m = n = 2$; then m and n have a primitive representation, but mn does not. We can in fact characterize all the counterexamples. By (8.15), an integer has a primitive representation if and only if it is not divisible by 4 or by any prime of the form $4k + 3$. Thus if m and n have a primitive representation, but mn does not, it must be because mn is divisible by 4. Therefore m and n are each twice a product of primes of the form $4k + 1$.

8-58. *Suppose m and n each have primitive representations as a sum of two squares, where $m > 2$ and $n > 2$. If $(m, n) = 1$, prove that mn has at least two essentially distinct primitive representations.*

Solution. Suppose that m has r distinct prime divisors of the form $4k + 1$ and that n has s such divisors. Because m and n have primitive representations, it follows that $r \geq 1$ and $s \geq 1$. Since $(m, n) = 1$, mn has $r + s$ different prime divisors of the form $4k + 1$, and it follows from (8.18) that mn has 2^{r+s-1} essentially distinct primitive representations. It is clear that this number is at least 2.

8-59. *Let n be an integer of the form $4k + 1$, with $k > 1$. Prove that n is prime if and only if n has exactly one representation (apart from the order and sign of the terms) as a sum of two squares and that representation is primitive.*

Solution. Suppose that n is a prime of the form $4k + 1$. Then by (8.9), n has a representation as a sum of two squares, and clearly, any representation of n is primitive. By (8.11), $N(n) = 8$, so n has essentially only one representation.

Conversely, suppose n has only primitive representations as a sum of two squares. Then n is square-free, for if $n = c^2m$ with $c > 1$, we can produce a nonprimitive representation of n from a representation of m. (The integer m has a representation by (8.9).) Thus if n is of the form $4k + 1$, it must be a product of distinct primes of the form $4k + 1$. But if such an n is not prime, it has at least two distinct prime factors, and so (8.11) implies that $N(n) \geq 16$. It follows that if n has only primitive representations, and essentially only one, then n is prime.

8-60. *Let $n > 2$ be representable as a sum of two squares. Show that every representation of n is primitive if and only if $n = p_1p_2 \cdots p_r$ or $n = 2p_1p_2 \cdots p_r$, where each p_i is a prime of the form $4k + 1$.*

Solution. We show first that if n has one of the two specified forms, then every representation of n is primitive. For suppose $n = a^2 + b^2$, and let $d = (a, b)$; then $d^2 \mid n$. Since clearly no square greater than 1 divides $p_1p_2 \cdots p_r$ or $2p_1p_2 \cdots p_r$, it follows that $d = 1$.

Suppose now that every representation of n is primitive. By (8.15), n is not divisible by 4 or by any prime of the form $4k + 3$. Also, n is square-free. For if $n = c^2m$, where $c > 1$, then m has a representation $m = a^2 + b^2$ as a sum of two squares, since m has no prime divisors of the form $4k + 3$. But then $n = (ca)^2 + (cb)^2$ is a nonprimitive representation of n. Hence n is a product of distinct primes, and the result follows.

8-61. *Suppose n has primitive representations as a sum of two squares. Prove directly that n cannot be divisible by 4 or by any number of the form $4k + 3$.*

Solution. Let $n = a^2 + b^2$. If n is divisible by 4, then a and b must be even, since the sum of two odd squares is congruent to 2 modulo 4. Thus a and b are not relatively prime, and hence the representation $n = a^2 + b^2$ is not primitive.

If n is divisible by a number of the form $4k + 3$, then n is divisible by a prime p of the form $4k + 3$. But then $p \mid a$ and $p \mid b$ by (8.6.i), and again the representation is not primitive.

8-62. *Prove or disprove: If n is odd or twice an odd integer and has only odd prime divisors of the form $4k + 1$, then all representations of n are primitive.*

Solution. This is false; for example, $25 = 0^2 + 5^2$. In general, if such an n is divisible by a square greater than 1, then n has a representation which is not primitive. For let $n = c^2m$ where $c > 1$, and let $m = a^2 + b^2$ (there is a representation, since m is not divisible by any prime of the form $4k + 3$). Then $n = (ca)^2 + (cb)^2$ is a representation of n which is not primitive.

8-63. *If n has a primitive representation as a sum of two squares, will every multiple of n also have primitive representations?*

Solution. No. In fact, if some prime of the form $4k + 3$ occurs to an odd power in the prime factorization of m, then mn will not have any representation as a sum of two squares. It is also easy to produce an example in which mn has a representation but no primitive representation. For example, by (8.15), we can take $m = 4$ (or $m = 2$ if n is even).

8-64. *Prove or disprove: If n has a primitive representation and $d \mid n$, then d also has a primitive representation.*

Solution. It follows from (8.15) that if n has a primitive representation, then n is not divisible by 4 or by any prime of the form $4k+3$. If $d \mid n$, then d cannot be divisible by 4 or by any prime of the form $4k+3$, and thus by (8.15), d has a primitive representation.

8-65. *Can $N(n) = 44$ for an integer n that has no primitive representations? If so, find such an n. If not, explain why not.*

Solution. First find an integer m such that $N(m) = 44$; in view of (8.11), we can take $m = 5^{10}$. Let $n = 9 \cdot 5^{10}$; then (8.11) implies that $N(n) = 44$, but n has no primitive representations, by (8.15).

8-66. *Let p be a prime of the form $4k + 1$, and suppose $p = a^2 + b^2$. Use (8.7) to prove directly that p^2 has a primitive representation as a sum of two squares.*

Solution. By (8.7), $p^2 = (a^2 + b^2)^2 = (2ab)^2 + (a^2 - b^2)^2$, and it is not hard to show that $(2ab, a^2 - b^2) = 1$. For if $d = (2ab, a^2 - b^2)$, then $d^2 \mid p^2$ and so d is 1 or p. If $d = p$, then $p \mid 2ab$ and hence $p \mid a$ or $p \mid b$. We may suppose that $p \mid a$; since $p \mid a^2 - b^2$, we have $p \mid b^2$ and thus $p \mid b$. This contradicts the fact that $(a, b) = 1$. Therefore $d = 1$.

▷ **8-67.** *Let p be a prime of the form $4k+1$. Use (8.11) to prove that, apart from the order and sign of the terms, p^n has exactly one primitive representation.*

Solution. For $n = 1$, (8.11) implies that p^n has essentially one representation as a sum of two squares; this representation is clearly primitive. Now suppose $n \geq 2$. By (8.11), p^n has a total of $4(n+1)$ representations, and p^{n-2} has $4(n-1)$ representations. Every representation of p^{n-2} generates a representation of p^n; if $p^{n-2} = a^2 + b^2$, then $p^n = (pa)^2 + (pb)^2$. Clearly, none of these representations is primitive. Also, if $p^n = u^2 + v^2$ is a nonprimitive representation of p^n, then the greatest common divisor of u and v divides p^n, and so u and v are multiples of p. Therefore every nonprimitive representation of p^n is of the form $(pa)^2 + (pb)^2$, where $a^2 + b^2 = p^{n-2}$. It follows that p^n has exactly $4(n-1)$ nonprimitive representations. Thus p^n has precisely $4(n+1) - 4(n-1) = 8$ primitive representations, that is, exactly one primitive representation apart from the sign and order of the terms.

8-68. *Prove or disprove: No integer of the form $n^2 - 1$ has primitive representations.*

Solution. This is true. Note that $n^2 - 1 \equiv 0$ or $3 \pmod 4$ according as n is odd or even. Thus $n^2 - 1$ is divisible by 4 if n is odd, and if n is even, then $n^2 - 1$ has at least one prime divisor of the form $4k + 3$. In either case, $n^2 - 1$ has no primitive representations, by (8.15).

254   CHAPTER 8: DIOPHANTINE EQUATIONS

The next four problems give a proof of Theorem 8.11. Recall that $N(n)$ is the total number of representations of n as a sum of two squares.

8-69. Let $n = q^2 m$, where $q = 2$ or q is a prime of the form $4k + 3$. Show directly (without using (8.11)) that $N(n) = N(m)$.

Solution. Suppose first that $q = 2$. If $X^2 + Y^2 = m$ and $x = 2X$, $y = 2Y$, then $x^2 + y^2 = 4m = n$. Moreover, every representation of n as a sum of two squares arises in this way, for if $x^2 + y^2 = n$, then since $n \equiv 0 \pmod{4}$, x and y must be even. Therefore if we let $x = 2X$ and $Y = 2y$, then $X^2 + Y^2 = n/4 = m$. Thus there is a one-to-one correspondence between representations of n and representations of m, and therefore $N(n) = N(m)$. Essentially the same argument works if q is a prime of the form $4k+3$, for if $x^2 + y^2 = n$, then, by (8.6.i), q must divide x and y.

8-70. Let $n = 2m$, where m is odd. Show directly that $N(n) = N(m)$. (*Hint.* Note that $X^2 + Y^2 = m$ if and only if $(X+Y)^2 + (X-Y)^2 = n$.)

Solution. Since $(X+Y)^2 + (X-Y)^2 = 2X^2 + 2Y^2$, every pair X, Y such that $X^2 + Y^2 = m$ determines a pair $X+Y$, $X-Y$ such that $(X+Y)^2 + (X-Y)^2 = n$. Moreover, every pair x, y for which $x^2 + y^2 = n$ arises in this way. For since n is twice an odd integer, we have $x^2 + y^2 \equiv 2 \pmod{4}$, and hence x and y are odd. Thus there exist uniquely determined integers X and Y such that $X+Y = x$ and $X-Y = y$, namely, $X = (x+y)/2$, $Y = (x-y)/2$, and it is easy to verify that $X^2 + Y^2 = m$. Thus there is a one-to-one correspondence between representations of n and representations of m, and hence $N(n) = N(m)$.

▷ **8-71.** If x is a positive integer, let $\omega(x)$ denote the number of distinct primes that divide x. Suppose that n and d are positive integers such that $d^2 \mid n$. Show that the number of ordered pairs x, y of positive integers such that $(x,y) = d$ and $xy = n$ is $2^{\omega(n/d^2)}$. Use this to conclude that the number $\tau(n)$ of divisors of n is $\sum \omega(n/d^2)$, where the sum is taken over all integers d such that $d^2 \mid n$. (*Hint.* See Problem 1-26.)

Solution. If $(x,y) = d$ and $xy = n$, let $X = x/d$ and $Y = y/d$. Then X and Y are relatively prime and $XY = n/d^2$. Conversely, if we are given relatively prime positive integers X and Y such that $XY = n/d^2$, then by letting $x = dX$ and $y = dY$, we find that $(x,y) = d$ and $xy = n$. Thus we want to count the number of relatively prime pairs X, Y such that $XY = n/d^2$. By Problem 1-26, the number of such pairs is $2^{\omega(n/d^2)}$.

Finally, since $2^{\omega(n/d^2)}$ counts the pairs x, y such that $(x,y) = d$ and $xy = n$, summing over all d such that $d^2 \mid n$ counts the total number of ordered pairs x, y such that $xy = n$, i.e, the number of divisors of n.

▷ **8-72.** Let $\omega(x)$ be as in the preceding problem. Suppose that m is a product of $4k+1$ primes. Show that the number of ordered pairs x, y of integers such that $x^2 + y^2 = m$ is $4 \sum \omega(m/d^2)$, where the sum is taken over all integers d such that $d^2 \mid m$. Use this result and the preceding problem to prove Theorem 8.11.

PROBLEMS AND SOLUTIONS 255

Solution. Suppose $d^2 \mid m$. We obtain the solutions of $x^2 + y^2 = m$ such that $(x, y) = d$ by taking relatively prime integers X, Y for which $X^2 + Y^2 = m/d^2$ and letting $x = dX$, $y = dY$. But by (8.18), there are $4 \cdot 2^{\omega(m/d^2)}$ such ordered pairs. Summing over all d gives a count of the total number of representations of m as a sum of two squares. Thus $N(n) = 4 \sum \omega(m/d^2)$, where the sum is taken over all positive integers d such that $d^2 \mid m$.

By the preceding problem, $4 \sum \omega(m/d^2) = 4\tau(m)$. Suppose n can be represented as a sum of two squares, and write $n = 2^a \prod p_i^{a_i} \prod q_i^{b_i}$, where the p_i are $4k + 1$ primes and the q_i are $4k + 3$ primes (so each b_i is even). Applying Problem 8-69 repeatedly and Problem 8-70 if necessary, we find that $N(n) = N(m)$, where $m = \prod p_i^{a_i}$. Thus $N(n) = 4\tau(m) = 4 \prod (a_i + 1)$.

The three problems that follow provide a proof of Theorem 8.12.

▷ **8-73.** Let $f(n) = D_1(n) - D_3(n)$, where $D_1(n)$ and $D_3(n)$ are, respectively, the number of divisors of n of the form $4k + 1$ and $4k + 3$. Show directly from the definition of $D_1(n)$ and $D_3(n)$ that $f(n)$ is a multiplicative function.

Solution. We need to show that if $(m, n) = 1$, then $f(mn) = f(m)f(n)$. Every positive divisor r of mn can be expressed uniquely in the form $r = de$, where $d \mid m$ and $e \mid n$. Also, r is of the form $4k + 1$ if and only if d and e are both of the form $4k + 1$ or both of the form $4k + 3$. Therefore $D_1(mn) = D_1(m)D_1(n) + D_3(m)D_3(n)$. Similarly, $D_3(mn) = D_1(m)D_3(n) + D_3(m)D_1(n)$. Hence

$$f(mn) = D_1(mn) - D_3(mn)$$
$$= D_1(m)D_1(n) + D_3(m)D_3(n) - D_1(m)D_3(n) - D_3(m)D_1(n)$$
$$= (D_1(m) - D_3(m))(D_1(n) - D_3(n)) = f(m)f(n).$$

▷ **8-74.** For $n \geq 1$, let $g(n) = N(n)/4$. Using (8.11), show that $g(n)$ is a multiplicative function.

Solution. We need to show that if $(m, n) = 1$, then $g(mn) = g(m)g(n)$. Suppose first that some prime of the form $4k + 3$ appears to an odd power in the prime factorization of either m or n, say m; then $g(m) = 0$. Since $(m, n) = 1$, the same prime appears to an odd power in the prime factorization of mn. Therefore $g(mn) = 0$ and hence $g(mn) = g(m)g(n)$.

If no prime of the form $4k + 3$ appears to an odd power in the prime factorization of m or n, then m, n, and mn are representable as a sum of two squares. Let m_1 and n_1 be, respectively, the product of the primes of the form $4k + 1$ that divide m and n. By (8.11), $g(m) = g(m_1)$, $g(n) = g(n_1)$, and $g(mn) = g(m_1 n_1)$. Let $m_1 = \prod p_i^{a_i}$ and $n_1 = \prod q_j^{b_j}$ be the prime factorizations of m_1 and n_1. Since m_1 and n_1 are relatively prime, the primes p_i and q_j are distinct. It then follows from (8.11) that $g(m_1) = \prod (a_i + 1)$, $g(n_1) = \prod (b_j + 1)$, and $g(m_1 n_1) = \prod (a_i + 1) \prod (b_j + 1)$. Thus $g(m_1 n_1) = g(m_1)g(n_1)$, and hence $g(mn) = g(m)g(n)$.

8-75. *Prove Theorem 8.12. (Hint. Use the two preceding problems.)*

Solution. Let $f(n) = D_1(n) - D_3(n)$ and $g(n) = N(n)/4$. We will show that $f(n) = g(n)$ for any $n \geq 1$. Let $n = \prod p_i^{a_i}$ be the prime factorization of n. By the two preceding problems, f and g are multiplicative functions, and hence

$$f(n) = \prod f(p_i^{a_i}), \qquad g(n) = \prod g(p_i^{a_i}).$$

We complete the proof by showing that $f(p^a) = g(p^a)$ for any prime power p^a.

If $p = 2$, then $f(p^a) = 1$ and $g(p^a) = 1$. If p is of the form $4k+1$, then $f(p^a) = a+1$, and, by (8.11), $g(p^a) = a+1$. Finally, let p be of the form $4k+3$. If a is odd, then p^a has $(a+1)/2$ divisors of the form $4k+1$ and $(a+1)/2$ divisors of the form $4k+3$, and hence $f(p^a) = 0$. Similarly, if a is even, then $f(p^a) = 1$. But if a is odd, then $N(p^a) = 0$, while if a is even, then $N(p^a) = 4$, so again $f(p^a) = g(p^a)$.

Sums of Three and Four Squares

8-76. *Determine the total number of representations of 360 as a sum of four squares.*

Solution. We could attack the problem directly by listing the representations, but it is easier to use the theorem of Jacobi ((8.24)). Since $360 = 2^3 \cdot 3^2 \cdot 5$, $\sigma(n) = (1 + 2 + 4 + 8)(1 + 3 + 9)(1 + 5) = 1170$. The sum of the divisors of 360 that are divisible by 4 is $4\sigma(90) = 936$. So the sum of the divisors of 360 that are not divisible by 4 is 234, and hence 360 has $8 \cdot 234 = 1872$ representations as a sum of four squares.

8-77. *Find a positive integer which has exactly five essentially distinct representations as a sum of four nonzero squares.*

Solution. Jacobi's formula (8.24) is helpful in finding the right range, but the count is complicated by the fact that the number of inessential variations of a representation depends on how many equal squares are used. A bit of experimentation shows that, for example, 100 has five essentially different representations as a sum of nonzero squares: $81 + 9 + 9 + 1$, $64 + 16 + 16 + 4$, $49 + 49 + 1 + 1$, $49 + 25 + 25 + 1$, and $25 + 25 + 25 + 25$.

8-78. (a) *Suppose $n = a^2 + b^2 + c^2 + d^2$, where a, b, c, and d are distinct nonzero integers. How many variations can be obtained by changing the sign and order of the terms?* (b) *What if a, b, c, and d are nonzero but not all distinct?* (c) *What if a, b, c, and d are distinct and exactly one is 0?*

Solution. (a) Without changing order, we can produce 2^4 variations with just sign changes. For each sign change variation, there are 4! variations produced by rearrangement, giving a total of $16 \cdot 24 = 384$ variations.

(b) Suppose now that a, b, c, and d are not all distinct. If exactly two are equal, say, $a = b$, there are $2^4 \cdot 2\binom{4}{2} = 192$ variations, while if $a = b$ and $c = d$ but $a \neq c$, there are $2^4 \binom{4}{2} = 96$ variations. If exactly three are equal, there are $2^4 \binom{4}{1} = 64$ variations. And if $a = b = c = d$, then there are just $2^4 = 16$ variations.

(c) If a, b, c, and d are distinct and exactly one is 0, then there are $2^3 \cdot 4! = 192$ variations.

▷ **8-79.** *Find a positive integer which can be represented as a sum of four nonzero squares in only two ways (ignoring sign changes and the order of the summands). (Hint. Use (8.24) and the preceding problem.)*

Solution. Since we will ignore sign changes and order when we write $n = a^2+b^2+c^2+d^2$, we can assume that $a \geq b \geq c \geq d \geq 0$. To simplify matters, we look for an n not divisible by 4; thus, by (8.24) the total number of representations is 8 times the sum of the positive divisors of n, i.e., $M(n) = 8\sigma(n)$. In looking for representations of n, we must allow for the fact that some of the squares may be 0 or some of the squares may repeat. We will not investigate all the possibilities, but we will look at the following cases.

If $M(n) = 768$, this allows for the possibility that (i) all of the representations of n use distinct positive squares (each representation having 384 variations, by the preceding problem); (ii) exactly two of four nonzero squares repeat; (iii) exactly one square is 0 and the other three are distinct (each representation in (ii) and (iii) generating 192 variations). Since $M(n) = 8\sigma(n)$, we must have $\sigma(n) = 96$. Suppose $n = rs$, with $(r,s) = 1$. Since σ is multiplicative (see the Note after (1.20)), we have $\sigma(r)\sigma(s) = 96$. There are several possibilities.

First, if we write $96 = 8 \cdot 12$, then we need $\sigma(r) = 8$, $\sigma(s) = 12$, i.e., $r = 7$, $s = 11$, and hence $n = 77$. It is easy to write down the representations of 77 as a sum of four squares: $77 = 8^2 + 3^2 + 2^2 + 0^2 = 6^2 + 6^2 + 2^2 + 1^2 = 6^2 + 5^2 + 4^2 + 0^2 = 6^2 + 4^2 + 4^2 + 3^2$, each representation having 192 variations by the preceding problem. Thus 77 has exactly two essentially different representations as a sum of four nonzero squares. Writing $96 = 3 \cdot 32$, we find that $r = 2$, $s = 31$, and so $n = 62$. Check that $62 = 6^2 + 5^2 + 1^2 + 0^2 = 7^2 + 3^2 + 2^2 + 0^2 = 6^2 + 4^2 + 3^2 + 1^2$. The first two representations have 192 variations, and the last has 384 (for a total of 768), but 62 has only one representation as a sum of four nonzero squares. And $96 = 4 \cdot 24$ yields $r = 3$, $s = 23$, and hence $n = 69 = 8^2 + 2^2 + 1^2 + 0^2 = 7^2 + 4^2 + 2^2 + 0^2 = 6^2 + 5^2 + 2^2 + 2^2 = 6^2 + 4^2 + 4^2 + 1^2$, each representation having 192 variations. Thus 69 has exactly two essentially different representations as a sum of four nonzero squares.

Now consider the case where $M(n) = 8\sigma(n) = 384 + 192 = 576$; hence $\sigma(n) = 72$. Writing $72 = 3 \cdot 24$, we have $r = 2$, $s = 23$, i.e., $n = 46$. Since $46 = 5^2+4^2+2^2+1^2 = 6^2+3^2+1^2+0^2$ (the first with 384 variations and the second with 192), it follows that 46 has only one representation as a sum of nonzero squares. Now write $72 = 4 \cdot 18$; then $r = 3$, $s = 17$, and so $n = 51$. Check that $51 = 5^2+4^2+3^2+1^2 = 7^2+1^2+1^2+0^2 = 5^2+5^2+1^2+0^2$, the first having 384 variations and the other two 96 variations (2^3 sign changes and $2\binom{4}{2} = 12$ rearrangements of terms). Thus 51 is not a solution. Finally, write $72 = 6 \cdot 12$; then $r = 5$, $s = 11$ and so $n = 55$. Note that $55 = 7^2 + 2^2 + 1^2 + 1^2 = 5^2 + 5^2 + 2^2 + 1^2$ (each with 192 variations) and $55 = 6^2 + 5^2 + 2^2 + 0^2$ (with 192 variations: 2^3 sign changes and 4! rearrangements of terms). Hence 55 has exactly two essentially distinct representations as a sum of four nonzero squares.

▷ **8-80.** (a) *Prove Theorem 8.23: Every $n > 169$ is a sum of five nonzero squares. (Hint. Use the fact that 169 is a square and also the sum of two, three, or four nonzero squares.)*

(b) *Show that there exist infinitely many positive integers which are not the* sum of four nonzero squares. (Hint. Show that if $8 \mid n$ and n is a sum of four nonzero squares, so is $n/4$.)

Solution. (a) Let $k = n - 169$. Then by Lagrange's Theorem, $k = a^2 + b^2 + c^2 + d^2$ for some integers a, b, c, d. If these are all nonzero, then $n = a^2 + b^2 + c^2 + d^2 + 169$ has been written as a sum of five nonzero squares. If exactly one of a, b, c, d is zero (say d), then since $169 = 25 + 144$, we have $n = a^2 + b^2 + c^2 + 25 + 144$, a sum of five nonzero squares. Similarly, we can use $169 = 9 + 16 + 144$ and $169 = 16 + 16 + 16 + 121$ to deal with the cases where two or three of a, b, c, d are zero.

(b) Suppose that $n \equiv 0 \pmod 8$ and $n = a_1^2 + a_2^2 + a_3^2 + a_4^2$ is a sum of four nonzero squares. Since the square of an odd integer is congruent to 1 modulo 8, we have $\sum a_i^2 \equiv 4 \pmod 8$ if the a_i are all odd. If exactly two of the a_i are odd, then $\sum a_i^2 \equiv 2 \pmod 4$. Thus, if $n \equiv 0 \pmod 8$ and $n = a_1^2 + a_2^2 + a_3^2 + a_4^2$, the a_i must all be even. Let $a_i = 2A_i$. Then $n/4 = A_1^2 + A_2^2 + A_3^2 + A_4^2$, and so $n/4$ is a sum of four nonzero squares. Since clearly 2 is not a sum of four nonzero squares, it follows that 8, 32, 128, ... cannot be expressed as the sum of four nonzero squares. Similarly, since 6 is not the sum of four nonzero squares, neither are 24, 96, 384, ...

8-81. Use Euler's identity (8.19) to express 8050 as a sum of four squares.

Solution. We have $8050 = 2 \cdot 5^2 \cdot 7 \cdot 23$. It is clearly enough to express $2 \cdot 7 \cdot 23$ as a sum of four squares. Now $14 = 0^2 + 1^2 + 2^2 + 3^2$ and $23 = 1^2 + 2^2 + 3^2 + 3^2$. Use Euler's identity with (in the notation of (8.19)) $a = 0$, $b = 1$, $c = 2$, $d = 3$, and $A = 1$, $B = 2$, $C = 3$, $D = 3$. Then $r = 17$, $s = -4$, $t = 1$, and $u = -4$. This gives $8050 = 85^2 + 20^2 + 5^2 + 20^2$. There are many other solutions.

8-82. Prove that the product of two integers, each the sum of three squares, need not itself be a sum of three squares. Show that this happens infinitely often. (Hint. Use (8.25).)

Solution. Note, for example, that 3 and 5 are each the sum of three squares but 15 is not. In general, if $m \equiv 3 \pmod 8$ and $n \equiv 5 \pmod 8$, then (8.25) implies that m and n are both sums of three squares, but their product, which is of the form $8k + 7$, is not. (Other solutions: $m = 2^a(8s + 3)$, $n = 2^b(8t + 5)$, where $a + b$ is even; and $m = 2^a(8s + 1)$, $n = 2^b(8t + 7)$, with a and b odd.)

▷ **8-83.** Show, without using (8.25), that an integer of the form $8k + 7$ cannot be written as a sum of three squares. Use this to show that if n is of the form $4^m(8k + 7)$, then n is not a sum of three squares.

Solution. Any square is congruent to one of 0, 1, or 4 modulo 8. No combination of three of these yields a sum congruent to 7 modulo 8. It follows that an integer of the form $8k + 7$ cannot be a sum of three squares.

We show by induction on m that $4^m(8k + 7)$ cannot be a sum of three squares. The calculation above dealt with the case $m = 0$. We show now that if $4^s(8k + 7)$ is not the sum of three squares, then $4^{s+1}(8k + 7)$ is not the sum of three squares. Suppose to the contrary that $4^{s+1}(8k + 7) = a^2 + b^2 + c^2$. Then, in particular, $a^2 + b^2 + c^2 \equiv 0 \pmod 4$. It is easy to see that a, b, and c must all be even. Let $a = 2A$, $b = 2B$, and

$c = 2C$. Then $4^{s+1}(8k+7) = 4A^2 + 4B^2 + 4C^2$, and therefore $4^s(8k+7) = A^2 + B^2 + C^2$, contradicting the assumption that $4^s(8k+7)$ is not the sum of three squares.

8-84. (Fermat, 1636.) *Show that no integer of the form $8k+7$ can be expressed as the sum of three squares of rational numbers.*

Solution. Suppose that n is of the form $8k+7$ and $n = r^2 + s^2 + t^2$, where r, s, and t are rational. Let $r = a/d$, $s = b/d$, and $t = c/d$, where a, b, c, and $d > 0$ are integers. Then $nd^2 = a^2 + b^2 + c^2$. Let $d = 2^e q$, where q is odd; then $nd^2 = 2^{2e} q^2$. Since $q^2 \equiv 1 \pmod{8}$, we have $nq^2 \equiv 7 \pmod{8}$; also, 2^{2e} is a power of 4. Thus nd^2 cannot be expressed as a sum of the squares of three integers (see (8.25) and Problem 8-83), contradicting the fact that $nd^2 = a^2 + b^2 + c^2$.

8-85. *Prove or disprove: If $n \geq 1$, then either n or $2n$ is a sum of three squares.*

Solution. Clearly, at least one of n or $2n$ is not of the form $4^m(8k+7)$. But by (8.25), any number not of this form is a sum of three squares, so the statement is true.

8-86. *Use Gauss's Theorem on sums of three squares (Theorem 8.25) to prove that every odd positive integer can be written as a sum of four squares, one of which is 0 or 1.*

Solution. Let n be an odd positive integer. If n is of the form $4k+1$, then (8.25) implies that n is a sum of three squares and hence a sum of four squares, one of which is 0. If n is of the form $4k+3$, (8.25) implies that $n-1$ is a sum of three squares, and hence n is a sum of four squares, one of which is 1.

8-87. *Use Theorem 8.25 to prove that every odd positive integer is of the form $r^2 + s^2 + 2t^2$. (Hint. First show that for any n, $4n+2$ is a sum of three squares, exactly two of which are odd.)*

Solution. Let n be a positive integer; then $4n+2$ is clearly not of the form $4^m(8k+7)$. Hence by (8.25), we can write $4n+2 = a^2 + b^2 + c^2$. The integers a, b, and c cannot all be even, since $4n+2$ is not divisible by 4; hence exactly two of a, b, and c are odd. Therefore suppose that a and b are odd and c is even, say, $c = 2t$; thus $a+b$ and $a-b$ are even. If $a+b = 2r$ and $a-b = 2s$, then $a = r+s$ and $b = r-s$. It follows that $4n+2 = (r+s)^2 + (r-s)^2 + 4t^2 = 2r^2 + 2s^2 + 4t^2$, and hence $2n+1 = r^2 + s^2 + 2t^2$.

8-88. *Use Gauss's Theorem on sums of three squares to prove that every odd positive integer can be written as $r^2 + s^2 + t^2 + (t+1)^2$. (Hint. For $n \geq 1$, show that $4n+1 = a^2 + b^2 + c^2$ with only c odd, say, $c = 2t+1$. Now express $2n+1$ in terms of a, b, and t.)*

Solution. For every $n \geq 0$, $4n+1$ is a sum of three squares, by (8.25). Let $4n+1 = a^2 + b^2 + c^2$; then exactly one of a, b, and c is odd, since every square is congruent to 0 or 1 modulo 4. Suppose $a = 2u$, $b = 2v$, $c = 2t+1$; then it is easily checked that the odd number $2n+1$ is the sum of the squares of $u+v$, $u-v$, t, and $t+1$.

8-89. *Prove that there exist infinitely many primes p such that p can be represented as $a^2 + b^2 + c^2 + 1$.*

Solution. There are infinitely many primes p of the form $4k + 3$. If p is such a prime, then $p - 1$ is of the form $4k + 2$ and hence is not of the form $4^m(8k + 7)$. Therefore by (8.25), $p - 1$ is a sum of three squares, which proves the result.

▷ **8-90.** (Euler, 1763.) *Suppose p is a prime of the form $8k + 1$ or $8k + 3$. Use an argument similar to the proof of (8.8) to show that there exist integers a, b such that $p = 2a^2 + b^2$. In particular, p is the sum of three squares. (Hint. Note that $(-2/p) = 1$ if and only if $p \equiv 1, 3 \pmod 8$.)*

Solution. By (5.13.i), -2 is a quadratic residue of p, and so there is an integer s such that $s^2 \equiv -2 \pmod p$. Consider the integers $sx - y$, where $0 \leq x, y < \sqrt{p}$. There are $[\sqrt{p}] + 1$ choices for each of x and y, and since $([\sqrt{p}] + 1)^2 > (\sqrt{p})^2 = p$, at least two of the values of $sx - y$ must be congruent modulo p, say, $sx_1 - y_1 \equiv sx_2 - y_2 \pmod p$. Let $a = x_1 - x_2$ and $b = y_1 - y_2$; then a and b cannot both be 0 since the ordered pairs (x_1, y_1) and (x_2, y_2) are distinct. Since $sa \equiv b \pmod p$, it follows that $s^2 a^2 \equiv b^2 \pmod p$, i.e., $-2a^2 \equiv b^2 \pmod p$. Hence $2a^2 + b^2$ is a positive multiple of p, and since $2a^2 + b^2 < 3(\sqrt{p})^2 = 3p$, we must have $2a^2 + b^2 = p$ (in which case we are finished) or $2a^2 + b^2 = 2p$. In the latter case, b must be even; if $b = 2c$, then $2a^2 + (2c)^2 = 2p$, that is, $a^2 + 2c^2 = p$.

8-91. *Prove or disprove: If the odd prime p is of the form $2a^2 + b^2$, then $p \equiv 1 \pmod 8$ or $p \equiv 3 \pmod 8$.*

Solution. This is true. If $2a^2 + b^2$ is odd, then b is odd, so $b^2 \equiv 1 \pmod 8$. Also, since any square is congruent to 0, 1, or 4 modulo 8, it follows that $2a^2 \equiv 0$ or $2 \pmod 8$. Thus $2a^2 + b^2 \equiv 1$ or $3 \pmod 8$.

8-92. *Use Gauss's result on sums of three squares to prove that every positive integer is the sum of four squares.*

Solution. In view of (8.25), it is enough to show that if $n = 4^m(8k + 7)$, then n is a sum of four squares. Now $n - 4^m = 4^m(8k + 6)$, so $n - 4^m$ is a sum of three squares, by (8.25). Let $n - 4^m = a^2 + b^2 + c^2$; then $n = a^2 + b^2 + c^2 + (2^m)^2$.

8-93. *Prove the following generalization of (8.21): The congruence $x^2 + y^2 + 1 \equiv 0 \pmod m$ is solvable if and only if $4 \nmid m$.*

Solution. If $4 \mid m$ and $a^2 + b^2 + 1 \equiv 0 \pmod m$, then $a^2 + b^2 + 1 \equiv 0 \pmod 4$, which is impossible since any square is congruent to 0 or 1 modulo 4.

Now suppose $4 \nmid m$. Write $m = 2^{k_0} p_1^{k_1} \cdots p_r^{k_r}$, where $k_0 = 0$ or 1 and the p_i are distinct odd primes; if $k_0 = 1$, let $p_0 = 2$. We show that for each i, there exist a_i and b_i such that $a_i^2 + b_i^2 + 1 \equiv 0 \pmod{p_i^{k_i}}$. The result is obvious for p_0, so it is enough to show that for any odd prime p_i, the congruence $x^2 + y^2 + 1 \equiv 0 \pmod{p_i^{k_i}}$ has a solution. By (8.21), there exist integers a_i and y_i such that $a_i^2 + y_i^2 + 1 \equiv 0 \pmod{p_i}$.

Since the congruence $y^2 \equiv -(a_i^2+1)$ (mod p_i) has a solution, it follows from (5.4) that the congruence $y^2 \equiv -(a_i^2 + 1)$ (mod $p_i^{k_i}$) has a solution for every i, and in particular there exists an integer b_i such that $a_i^2 + b_i^2 + 1 \equiv 0$ (mod $p_i^{k_i}$).

Now, as in the proof of (8.21), use the Chinese Remainder Theorem to find integers a and b such that $a \equiv a_i$ (mod $p_i^{k_i}$) and $b \equiv b_i$ (mod $p_i^{k_i}$) for each i. Then $a^2+b^2+1 \equiv 0$ (mod m).

The next two problems provide another proof that every positive integer is a sum of four squares. As indicated in the paragraph following (8.19), it is enough to show that *any prime q of the form $4k + 3$ is a sum of four squares.* The idea of the proof is to show that *some* positive multiple of q is a sum of four squares and then to argue that the *least* positive multiple of q which is a sum of four squares is q itself. Note the similarity with the proof of the Two-Squares Theorem given in Problem 8-48.

▷ **8-94.** *Suppose q is a prime of the form $4k + 3$. Prove that there exist integers a and b such that $a^2 + b^2 + 1 \equiv 0$ (mod q) by letting $a^2 \equiv h - 1$ (mod q), where h is the least positive quadratic nonresidue of q. In particular, show that there exists an integer r, with $0 < r < q$, such that rq is a sum of four squares.*

Solution. Let h be the least positive quadratic nonresidue of q. Then $h \geq 2$, and since $h - 1$ is a quadratic residue of q, there exists an integer a, with $0 < a < q/2$, such that $a^2 \equiv h - 1$ (mod q). By (5.11), -1 is a quadratic nonresidue of q; hence $(-h/q) = (-1/q)(h/q) = 1$. Thus there is an integer b, with $0 < b < q/2$, such that $b^2 \equiv -h$ (mod q). It follows that $a^2+b^2+1 \equiv (h-1)+(-h)+1 \equiv 0$ (mod q). Therefore $a^2 + b^2 + 1 = rq$ for some integer r, and so rq is a sum of four (in fact, three) squares. Since $0 < a^2 + b^2 + 1 < 2(q/2)^2 + 1 < q^2$, we conclude that $0 < r < q$.

▷ **8-95.** *Show that every prime of the form $4k + 3$ is a sum of four squares.*

Solution. Suppose q is a prime of the form $4k + 3$. Let m be the *least* positive integer such that mq is a sum of four squares, say, $mq = a^2 + b^2 + c^2 + d^2$. Then $0 < m < q$ by Problem 8-94; we will show that $m = 1$.

If m is even, then either 0, 2, or 4 of the numbers a, b, c, d are even; if exactly two are even, we may suppose that they are a and b. Then in all three cases, $a \pm b$ and $c \pm d$ are even; hence

$$(m/2)q = ((a+b)/2)^2 + ((a-b)/2)^2 + ((c+d)/2)^2 + ((c-d)/2)^2,$$

contradicting the minimality of m. Therefore m is not even.

Suppose now that m is odd; if m is not 1, then $3 \leq m < q$. The integers a, b, c, d are congruent modulo m to integers A, B, C, D chosen from the complete residue system consisting of all integers x such that $-(m-1)/2 \leq x \leq (m-1)/2$; thus A, B, C, and D are less than $m/2$ in absolute value. Then

$$A^2 + B^2 + C^2 + D^2 \equiv a^2 + b^2 + c^2 + d^2 \equiv 0 \pmod{m},$$

and so

$$A^2 + B^2 + C^2 + D^2 = km, \quad \text{with} \quad 0 \leq km < 4(m/2)^2 = m^2.$$

Thus $0 \leq k < m$. If $k = 0$, then $A = B = C = D = 0$, and m divides a, b, c, and d; hence $m^2 | mq$, that is, m divides q, a contradiction since $m < q$. Therefore $k > 0$. Hence

$$m^2 kq = (mq)(km) = (a^2 + b^2 + c^2 + d^2)(A^2 + B^2 + C^2 + D^2) = r^2 + s^2 + t^2 + u^2, \quad (1)$$

where r, s, t, and u are defined as in the proof of (8.19). Since $A \equiv a$, $B \equiv b$, $C \equiv c$, and $D \equiv d \pmod{m}$, it follows that r, s, t, and u are each divisible by m. (For example, $t = aC - bD - cA + dB \equiv ac - bd - ca + db = 0 \pmod{m}$.) Dividing (1) by m^2, we get

$$kq = (r/m)^2 + (s/m)^2 + (t/m)^2 + (u/m)^2,$$

with $0 < k < m$. This contradicts the fact that mq is the smallest multiple of q that is a sum of four squares. Thus $m = 1$, and the proof is complete.

Waring's Problem

▷ **8-96.** (Johannes Euler, 1772.) *Let $g(k)$ be the least number of kth powers needed to represent every positive integer. Prove that $g(k) \geq 2^k + [(3/2)^k] - 2$ for every $k \geq 2$. (Hint. Let $n = 2^k [(3/2)^k] - 1$, and find the number of kth powers required to represent n.)*

Solution. Let $q = [(3/2)^k]$. Since $q < (3/2)^k$, it follows that $n = 2^k q - 1 < 3^k$. Thus in representing n as a sum of kth powers, we can use only 1^k and 2^k. Clearly, the minimum number of kth powers needed requires the use of as many copies of 2^k as possible, namely, $q - 1$. This leaves $n - 2^k(q - 1) = 2^k - 1$ to represent, which obviously requires $2^k - 1$ copies of 1^k. Therefore we need at least $(q-1) + (2^k - 1) = 2^k + [(3/2)^k] - 2$ kth powers to represent n.

8-97. *Prove that $g(3) \geq 4$ by showing that if $n \equiv 4 \pmod{9}$ or $n \equiv 5 \pmod{9}$, then n has no representation as a sum of fewer than four cubes.*

Solution. By looking at $0^3, 1^3, \ldots, 8^3$, we can verify that any cube is congruent to 0, 1, or -1 modulo 9. A sum of cubes that uses a cubes congruent to 1 modulo 9, b cubes congruent to -1, and possibly some cubes congruent to 0 will be congruent to $a - b$ modulo 9. It is easy to see that 4 is the smallest value of $a + b$ for which $a - b \equiv 4 \pmod{9}$ and also the smallest value for which $a - b \equiv 5 \pmod{9}$.

8-98. *Prove that there are infinitely many positive integers which cannot be represented as the sum of fewer than 15 fourth powers. (Hint. Consider the residues of fourth powers modulo 16.)*

Solution. It is clear that if x is even, then $x^4 \equiv 0 \pmod{16}$; if x is odd, then $x^4 \equiv 1 \pmod{16}$. This can be verified in several ways, for example, by expanding $(2y+1)^{16}$ using the Binomial Theorem, or more simply by noting that 1^4, 3^4, 5^4, and 7^4 are each conguent to 1 modulo 16.

Suppose that we have expressed n as a sum of fourth powers, using exactly k fourth powers of odd integers (and possibly fourth powers of some even integers). Then $n \equiv k$ (mod 16). In particular, if $n \equiv 15$ (mod 16), then n cannot be represented using fewer than 15 fourth powers.

Notes. 1. The result above can be improved slightly. We can show by induction on m that if $n = 31 \cdot 16^m$, then n requires 16 fourth powers. This is clearly true for $m = 0$. We show that if it is true for $m = k$, then it is true for $m = k+1$. Let $n = 31 \cdot 16^{k+1}$. If we use any odd numbers in representing n, we will have to use at least 16. If we use only even numbers, their fourth powers are all divisible by 16. Thus we obtain a representation of $31 \cdot 16^k$ as a sum of fourth powers, and by the induction assumption, such a representation requires at least 16 fourth powers.

2. The value of $g(k)$ is unduly affected by "accidental" properties of small integers. Define $G(k)$ as the smallest integer s such that every *large enough* integer can be expressed as the sum of s or fewer positive kth powers. Clearly, $G(k) \leq g(k)$. It turns out that, in general, $G(k)$ is substantially smaller than $g(k)$. The argument above shows that $G(4) \geq 16$. Davenport showed in 1939 that, in fact, $G(4) = 16$.

▷ **8-99.** (a) *Show that if $p > 3$ is prime, then p^n cannot be the sum of two positive cubes for any $n \geq 1$. (b) What happens when $p = 2$ or $p = 3$?*

Solution. (a) The proof is by the method of descent (induction). If there is a power of p which is the sum of two positive cubes, then there is a smallest power of p with this property. Let it be p^n, and suppose $p^n = a^3 + b^3$. Since $a^3 + b^3 = (a+b)(a^2 - ab + b^2)$, it follows that $a + b = p^k$ and $a^2 - ab + b^2 = p^{n-k}$ for some integer k. Then $k > 0$ since a and b are positive, and $k < n$, since $a^3 + b^3 > a + b$ unless $a = b = 1$.

Since $3ab = (a+b)^2 - (a^2 - ab + b^2) = p^{2k} - p^{n-k}$ and $0 < k < n$, it follows that p divides $3ab$. Since $p > 3$, p divides either a or b, say a; then $a + b = p^k$ implies that p divides b. Let $a = pA$ and $b = pB$; then it is easy to see that $A^3 + B^3 = p^{n-3}$. This contradicts the choice of n as the smallest integer such that p^n is a sum of two positive cubes.

(b) Suppose $p = 2$. We can show in the same way that $a^3 + b^3 = 2^n$ implies that a and b are divisible by 2 unless $a = b = 1$, and if A and B are defined as above, then $A^3 + B^3 = 2^{n-3}$. Thus if a and b are positive and $a^3 + b^3 = 2^n$, then n must be of the form $3k + 1$ and $a = b = 2^k$.

For $p = 3$, there is a complication, since from $p \mid 3ab$ we cannot conclude that $p \mid a$ or $p \mid b$. But it is easy to see that $a^3 + b^3 > 3(a+b)$ unless the larger of a and b, say a, is 2 or less, and thus 3^{n-k} is divisible by 9. We conclude that a and b are divisible by 3 except in the case $a = 2$, $b = 1$. Thus if a and b are positive and $a^3 + b^3 = 3^n$ with $a \geq b$, then n is of the form $3k + 2$ and $a = 2 \cdot 3^k$, $b = 3^k$.

EXERCISES FOR CHAPTER 8

1. Find all Pythagorean triples which contain 15 as one of their members.
2. Determine all Pythagorean triples (x, y, z) with $x < y$ and $z = 377$.

264   CHAPTER 8: DIOPHANTINE EQUATIONS

3. Let $N = 2 \cdot 3^2 \cdot 5^3 \cdot 13^5$. Is there a Pythagorean triangle with hypotenuse N? Does there exist a primitive Pythagorean triangle with hypotenuse N?

4. How many Pythagorean triples (x, y, z) have $z = 1885$?

5. Prove or disprove: There exists a Pythagorean triple with hypotenuse 77.

6. Let n be a positive integer. Prove that if there is at least one primitive Pythagorean triangle where one side is n units less than the hypotenuse, then there are infinitely many.

7. Find all of the essentially different representations of 150280 and 707850 as a sum of two squares.

8. Use a divisibility test to determine if 76549317 has any representations as a sum of two squares.

9. Is 646^{41} a sum of two squares?

10. Give an example of a positive integer that is a multiple of 21 and has exactly six more divisors of the form $4k + 1$ than of the form $4k + 3$.

11. Find all of the representations of 33323400 as a sum $a^2 + b^2$, where a and b are positive integers and $a > b$.

12. Prove that $2^3 \cdot 3^5 \cdot 5^3 \cdot 7^2 \cdot 11^4 \cdot 13$ has exactly the same number of divisors of the form $4k + 3$ as of the form $4k + 1$.

13. Without factoring, decide if 768312741 or 351694843 can be represented as a sum of two squares.

14. Find a positive integer that is a multiple of 33 and has precisely six essentially different representations as a sum of two squares.

15. Prove that if $n \equiv 3 \pmod{9}$, then n has the same number of divisors of the form $4k + 3$ as of the form $4k + 1$.

16. Determine all of the essentially different representations of 434826 and 2533986 as a sum of two squares.

17. The integer 99221 can be represented as $50^2 + 311^2$ and $25^2 + 314^2$. Use this to show that 99221 cannot be prime.

18. Find all primitive representations of 6409 as a sum of two squares.

19. Prove or disprove: If q is a prime of the form $4k + 3$ and q divides $c^2 + d^2$, then q^2 divides $c^2 + d^2$.

20. Determine all of the essentially different representations of 169000 as a sum of two squares. Are there any primitive representations?

21. Find a positive integer that has precisely six essentially distinct representations as a sum of two nonzero squares, with none of the representations primitive.

22. Calculate all of the primitive representations of 11050 as a sum of two squares.

23. What are the five smallest positive integers that have exactly three essentially different representations as a sum of two squares, with at least one representation primitive?

24. Prove or disprove: If $p_1, p_2, \ldots, p_n$ are primes of the form $4k + 1$, then $p_1 p_2 \cdots p_n$ has only primitive representations as a sum of two squares.

25. If n is square-free (that is, divisible by no square greater than 1), is every representation of n primitive?

26. Suppose $n = a^2 + b^2$, where $(a, b) = 1$. If p is an odd prime divisor of n, prove that $p \equiv 1 \pmod 4$ by showing that $(-1/p) = 1$. (Hint. Use $a^2 \equiv -b^2 \pmod p$ to show that $x^2 \equiv -1 \pmod p$ is solvable.)

27. Show that an integer can be expressed as the sum of the squares of two rational numbers if and only if it can be expressed as the sum of the squares of two integers.

28. Give an example of an integer greater than 20,000,000 which is not the sum of four nonzero squares.

29. Is 9952 a sum of three squares? Is 12! a sum of three squares?

30. Show that a power of 2 cannot be the sum of three positive squares.

31. Determine which of the following integers is a sum of four nonzero squares: 12345; 73728; 98304.

32. Find the first four integers greater than 1,000,000 that are not the sum of four nonzero squares.

33. What are the first two integers greater than 98303 that are the sum of four nonzero squares?

34. Prove that 2 is the only prime which is a sum of two positive cubes. (Hint. $x^3 + y^3 = (x + y)((x - y)^2 + xy)$.)

35. Let p be an odd prime. Prove that p has a unique representation of the form $a^2 - b^2$, where a and b are positive integers.

36. Let n be an odd positive integer. Show that if n has exactly one representation of the form $a^2 - b^2$, where a and b are positive integers, then n is either prime or the square of a prime.

37. Prove that every positive integer is of the form $r^2 + s^2 + t^2$ or $r^2 + s^2 + 2t^2$.

NOTES FOR CHAPTER 8

1. Euler conjectured in 1778 that if $n \geq 3$, then no positive nth power is a sum of fewer than n positive nth powers. Fermat's Last Theorem would of course follow at once from this. Euler's conjecture is true for $n = 3$, for as Euler himself proved, $x^3 + y^3 = z^3$ has no solutions in positive integers.

However, in 1966, the conjecture was shown to be false by L.J. Lander and T.R. Parkin:
$$144^5 = 27^5 + 84^5 + 110^5 + 133^5.$$
In 1988, N. Elkies showed that the conjecture is also false for $n = 4$; the smallest counterexample is
$$95800^4 + 217519^4 + 414560^4 = 422481^4.$$
Euler's conjecture is still open for $n \geq 6$.

2. The gap in Lamé's sketch of a proof of Fermat's Last Theorem was basically the same as the problem in Euler's proof of Fermat's Last Theorem for $n = 3$: Each assumed without justification that "integers" in more general number fields can always be factored uniquely into a product of primes. (This problem is discussed in Chapter 11.) There is, in fact, unique factorization for the integers in $Q(\omega)$, where $\omega = (-1 + \sqrt{-3})/2$, the case that is relevant for $n = 3$, but Cauchy showed that the Fundamental Theorem of Arithmetic does not hold, for example, for the "integers" that arise when $n = 23$.

3. There are two cases of Fermat's Last Theorem: Case I (the easier case): p does not divide any of x, y, or z; and Case II (the harder case): p divides one of x, y, or z. Before Wiles's work, Fermat's Last Theorem was known to be true in Case I for all exponents that have an odd prime factor less than 10^{17} and was known to be true in Case II for all primes less than 10^6.

To obtain the estimate in Case I: In 1909, Wieferich proved that if Case I is false for a prime exponent p, then p satisfies $2^{p-1} \equiv 1 \pmod{p^2}$. Such primes are known as *Wieferich primes*, and the only Wieferich primes less than $6 \cdot 10^9$ are 1093 and 3511. In 1913, Mirimanoff showed similarly that if Case I is false for p, then $3^{p-1} \equiv 1 \pmod{p^2}$. The primes 1093 and 3511 do not satisfy this congruence, and hence Case I is true for all $p < 6 \cdot 10^9$.

This approach was extended in the following way: If Case I is false for a prime exponent p, then $q^{p-1} \equiv 1 \pmod{p^2}$ for every prime $q \leq 89$. This showed, until Wiles proved Fermat's Last Theorem with no restrictions, that Case I held for all primes less than 714,591,416,091,389; later analysis took this bound up to 10^{17}. In addition, Case I had been shown to hold for any Mersenne prime.

4. There are a number of easy-to-state problems related to Pythagorean triangles that remain unsolved despite considerable effort. Here are some examples.

(a) Is there a Pythagorean box? That is, is there a box whose sides, diagonals of the faces, and main diagonal all have integral length?

(b) Are there a square $ABCD$ and a point P such that the distances AB, PA, PB, PC, and PD are all integers?

(c) Are there infinitely many Pythagorean triples such that the hypotenuse and one of the other sides are both prime numbers?

5. We quote in full Fermat's proof (as translated by Heath) of the fact that the area of a Pythagorean triangle cannot be a square or, equivalently, that the equation $x^4 - y^4 = z^2$ has no positive solutions:

> If the area of a right triangle were a square, there would be two biquadrates whose difference is a square, and hence two squares whose sum and difference are squares. Thus there would be a square equal to the sum of a square and the double of a square, such that the sum of the two component squares is a square. But if a square is the sum of a square and the double of a square, its root is likewise the sum of a square and the double of a square, which I can easily prove. It follows that this root is the sum of the two legs of a right triangle, one of the squares forming the base and the double of the other square forming the height. This right triangle will therefore be formed from two squares whose sum and difference are squares. But both of these squares can be shown to be smaller than the squares of which it was assumed that the sum and difference are squares. Similarly, we would have smaller and smaller integers satisfying the same conditions. But this is impossible, since there is not an infinitude of positive integers smaller than a given one. The margin is too narrow for the complete demonstration and all its developments.

BIOGRAPHICAL SKETCHES

Diophantus (c. 250 A.D.) introduced algebraic methods into Greek mathematics. In particular, he introduced notation for the unknown, its square, and its cube. However, there was no direct way for Diophantus to deal simultaneously with several unknowns. His *Arithmetica* is a collection of over 250 problems, many with clever solutions, which deal with algebraic equations where the solutions are required to be rational numbers. Usually, Diophantus is content to produce a single numerical answer, but often it is clear that he is in possession of a general method.

Essentially nothing is known about Diophantus's life. If we accept the tradition that Problem 8-1 is about him, then Diophantus died at the age of 84.

Carl Gustav Jacobi was born in Potsdam in 1804. A child prodigy, he was largely self-taught, learning his mathematics from the works of Euler and Lagrange. In 1821, he entered the University of Berlin; he received his first degree in the same year and his doctorate in 1825, with a thesis on continued fractions. He taught for eighteen years at Königsberg, but his quickly growing fame gained him election to the Prussian Academy of Sciences, and he spent much time in Berlin, moving there permanently in 1844. Jacobi is mainly known for his work in the theory of elliptic functions and was not primarily a number theorist. But he showed how identities involving elliptic functions could be used to solve problems in number theory, in particular problems about the number of representations of an integer as a sum of a specified number of squares. Related ideas were to become one of the main themes of number theory from the middle of the nineteenth century to today. Jacobi, unlike most of his colleagues, was lively, argumentative, and extroverted. He had an encyclopedic knowledge of mathematics and was known as a great teacher and prodigiously hard worker.

Jacobi died of smallpox in Berlin in 1851.

REFERENCES

Leonard Eugene Dickson, *History of the Theory of Numbers*, Volume II. (See Chapter 3.)

Harold M. Edwards, *Fermat's Last Theorem: A Genetic Introduction to Algebraic Number Theory*, Springer-Verlag, New York, 1977.

> Edwards describes how algebraic number theory arose naturally from attempts to settle Fermat's Last Theorem and other Diophantine problems, mainly involving quadratic forms. The book is indispensable for understanding the mathematical context within which the ideas arose. The first three chapters are quite accessible and have valuable information on sums of squares and Pell's Equation. Edwards gives a large amount of historical detail; in particular, it was he who showed that there seems to be no basis in fact for the often-repeated assertion that Kummer claimed to have proved Fermat's Last Theorem but had mistakenly assumed a unique factorization property.

Sir Thomas L. Heath, *Diophantus of Alexandria*, Dover, New York, 1964.

> In this book, first published in 1885, Heath gives a modern translation of Tannery's definitive edition of the part of Diophantus's *Arithmetica* that was available at the time. He follows this with a description of results of Fermat and Euler that were inspired by Diophantus, including the famous marginal notes of Fermat. Heath's book is in the great tradition of nineteenth-century scholarship. It is thorough but accessible to anyone with knowledge of high school algebra. Among Heath's other books are translations of Euclid, Apollonius, and Archimedes.

Paulo Ribenboim, *The Book of Prime Number Records* (Second Edition). (See Chapter 7.)

Paulo Ribenboim, *13 Lectures on Fermat's Last Theorem*, Springer-Verlag, New York, 1979.

Jacques Sesiano, *Books IV to VII of Diophantus' Arithmetica*, Springer-Verlag, New York, 1982.

> In the early 1970s, it became known that a hitherto unknown part of the *Arithmetica*, in an Arabic translation, was among a collection of mathematical manuscripts in the Mashhad Shrine Library. This is a critical edition of the text, together with an English translation and scholarly and mathematical commentary.

André Weil, *Number Theory: An approach through history from Hammurapi to Legendre*. (See Chapter 3.)

CHAPTER NINE
Continued Fractions

In this chapter, we describe a technique for writing any real number as an iterated sequence of quotients. (For example, 158/49 can be written in the form 3+1/(4+1/(2+1/5)).) Such an expression, called a *continued fraction*, can be obtained from the Euclidean Algorithm when the given number is rational, but continued fraction expansions can be derived for irrational numbers as well.

Because of their close connection with the Euclidean Algorithm, it is difficult to date the origin of continued fractions precisely. In more or less modern form, continued fractions first appear in the work of the Italian mathematicians Rafael Bombelli (1526–1573) and Pietro Cataldi (1548–1626), where they were used to approximate square roots. The Dutch physicist Christiaan Huygens (1629–1695), in his *Descriptio Automati Planetarii* (1703), studied continued fractions in connection with the design of a mechanical model of the planets. For example, the ratio of the rotation period of Saturn and Earth was known to be approximately $\alpha = 77708431/2640858$. Huygens needed to construct two gears with the number of teeth p and q reasonably small and such that p/q is approximately α. He found that $p = 206$, $q = 7$ gave an error of less than 40 minutes of arc per century. More than 1100 years earlier, Aryabhata had used a related technique to approximate the period of Jupiter.

Continued fractions and their connection to quadratic irrationals appear in Euler's work around 1730, in his letters to Goldbach. Euler noted the connection between continued fractions and approximations by rational numbers, and he observed that a rational number has a finite continued fraction expansion that can be derived from the Euclidean Algorithm. In 1759, Euler used continued fractions to solve equations of the form $x^2 - dy^2 = 1$. This equation, known as *Pell's Equation*, has a long history and will be studied in detail in the next chapter. The fundamental results about periodic continued fractions were first proved by Lagrange, beginning in 1766.

Generalizations of continued fractions have many applications in analysis,

FINITE CONTINUED FRACTIONS

first explored by Euler and Lagrange. For example, the algorithms used by scientific calculators to approximate functions such as e^x and $\sin x$ have their roots in continued fractions.

RESULTS FOR CHAPTER 9

Finite Continued Fractions

We begin by examining the connection between the Euclidean Algorithm and the continued fraction expansion of a rational number. If we apply the Euclidean Algorithm to 158 and 49, we obtain

$$158 = 3 \cdot 49 + 11, \quad \text{and thus} \quad 158/49 = 3 + 11/49;$$
$$49 = 4 \cdot 11 + 5, \quad \text{and thus} \quad 49/11 = 4 + 5/11;$$
$$11 = 2 \cdot 5 + 1, \quad \text{and thus} \quad 11/5 = 2 + 1/5.$$

We can therefore write the rational number 158/49 as follows:

$$158/49 = 3 + \frac{11}{49} = 3 + \cfrac{1}{4 + \cfrac{5}{11}} = 3 + \cfrac{1}{4 + \cfrac{1}{2 + \cfrac{1}{5}}}$$

An expression of this type is called a *continued fraction*, which we will denote by $\langle 3, 4, 2, 5 \rangle$. The numbers 3, 4, 2, and 5 are called the *terms*, or the *partial quotients*, of the continued fraction, since they are the quotients that occur in the Euclidean Algorithm.

(9.1) Definition. Let $a_0, a_1, a_2, \ldots, a_n$ be real numbers, all positive except possibly a_0. Define the continued fraction $\langle a_0, a_1, a_2, \ldots, a_n \rangle$ by

$$a_0 + \cfrac{1}{a_1 + \cfrac{1}{a_2 + \cfrac{1}{\ddots + \cfrac{1}{a_{n-1} + \cfrac{1}{a_n}}}}}$$

If each a_i is an integer, the continued fraction is called *simple*.

In the above example, since $1/5 = 1/(4+1/1)$, we also have $158/49 = \langle 3,4,2,4,1\rangle$. In fact, $\langle 3,4,2,5\rangle$ and $\langle 3,4,2,4,1\rangle$ are the only representations of $158/49$ as a finite simple continued fraction. As the following result shows, any rational number has precisely two simple continued fraction expansions.

(9.2) Theorem. *Every rational number r has precisely two finite continued fraction expansions. When r is an integer, the representations are $\langle r\rangle$ and $\langle r-1,1\rangle$. If r is not an integer and $\langle a_0,a_1,\ldots,a_n\rangle$ is the continued fraction obtained from the Euclidean Algorithm, then $a_n > 1$ and hence the other representation of r is $\langle a_0,a_1,\ldots,a_n-1,1\rangle$.*

Proof. It is clear that $1 = \langle 1\rangle = \langle 0,1\rangle$. In Problem 9-5, we prove that a rational number can have only one finite continued fraction expansion where the last partial quotient is greater than 1. It follows that if $r \neq 1$, then r has a unique continued fraction expansion where the last partial quotient is 1. (Otherwise, we can produce two different expansions ending in an integer greater than 1, since $\langle a_0,a_1,\ldots,a_n\rangle = \langle a_0,a_1,\ldots,a_n-1,1\rangle$ if $a_n > 1$.) Therefore if r is an integer, then $\langle r-1,1\rangle$ and $\langle r\rangle$ are the only expansions of r.

Now suppose that $r = a/b$ is not an integer. Apply the Euclidean Algorithm to get $a = q_1 b + r_1$, $b = q_2 r_1 + r_2$, $r_1 = q_3 r_2 + r_3$, $\ldots$, $r_{n-2} = q_n r_{n-1} + r_n$, $r_{n-1} = q_{n+1} r_n$; thus $r_1 > r_2 > \cdots > r_{n-1} > r_n > 0$. (Since r is not an integer, $r_1 > 0$ and hence $n \geq 1$.) Then, as in the preceding example, $a/b = \langle q_1, q_2, \ldots, q_{n+1}\rangle$. Clearly, q_{n+1} is a positive integer. If $q_{n+1} = 1$, then $r_{n-1} = q_{n+1} r_n = r_n$, contradicting the fact that $r_{n-1} > r_n$. Hence we must have $q_{n+1} \geq 2$. Since $q_{n+1} = (q_{n+1}-1) + 1/1$ and $q_{n+1} - 1$ is a positive integer, a/b also has the expansion $\langle q_1, q_2, \ldots, q_{n+1}-1, 1\rangle$. By Problem 9-5, it follows that these are the only continued fraction expansions of a/b.

We turn now to the problem of finding an efficient way to evaluate a given simple continued fraction $\langle a_0, a_1, \ldots, a_n\rangle$.

(9.3) Definition. Let $a_0, a_1, a_2, \ldots$ be a sequence of integers, all positive except possibly a_0. Define sequences (p_k) and (q_k) as follows:

$$p_{-1} = 1, \quad p_0 = a_0, \quad \text{and} \quad p_k = a_k p_{k-1} + p_{k-2}$$
$$q_{-1} = 0, \quad q_0 = 1, \quad \text{and} \quad q_k = a_k q_{k-1} + q_{k-2}$$

for $k \geq 1$.

Note. From the definition, $q_0 = 1$; since a_k is a positive integer if $k \geq 1$, we have $1 = q_0 \leq q_1 < q_2 < \cdots$. In particular, $q_k \geq k$ for every $k \geq 0$. In fact, the q_k grow at an exponential rate; it is not hard to show that $q_k \geq 2^{k/2}$ if $k \geq 2$ (see Problem 9-7).

The next result will be used throughout this chapter. For a proof, see Problem 9-6.

FINITE CONTINUED FRACTIONS   273

(9.4) Theorem. *Let $a_0, a_1, a_2, \ldots$ be a sequence of integers, all positive except possibly a_0. Then for any positive real number x and any $k \geq 1$,*

$$\langle a_0, a_1, \ldots, a_{k-1}, x \rangle = (xp_{k-1} + p_{k-2})/(xq_{k-1} + q_{k-2}).$$

In particular,

$$\langle a_0, a_1, a_2, \ldots, a_k \rangle = p_k/q_k.$$

(9.5) Definition. *The number $c_k = \langle a_0, a_1, a_2, \ldots, a_k \rangle$, where $k \leq n$, is called the kth* convergent *of the continued fraction $\langle a_0, a_1, a_2, \ldots, a_n \rangle$. Thus by (9.4), $c_k = p_k/q_k$.*

(9.6) Example. We can use (9.3) and (9.4) to evaluate $\langle -4, 2, 3, 1, 2 \rangle$. The calculations are more easily carried out by using the following table:

k	-1	0	1	2	3	4
a_k		-4	2	3	1	2
p_k	1	-4	-7	-25	-32	-89
q_k	0	1	2	7	9	25

(The three bold numbers are the same for *any* such calculation.) The entries are computed according to the formulas given in (9.3). For example, to find $p_3 = a_3 p_2 + p_1$, multiply $a_3 = 1$ by the last computed p-value p_2 (namely, -25) and add the preceding term p_1: $p_3 = 1(-25) + (-7) = -32$. Finally, note that $\langle -4, 2, 3, 1, 2 \rangle = p_4/q_4 = -89/25$.

The following theorem plays a role in many of the subsequent results; it can be used, for example, to find solutions of the linear equation $ax + by = c$. (See the next section.)

(9.7) Theorem. *The equation*

$$p_k q_{k-1} - p_{k-1} q_k = (-1)^{k-1}$$

holds for every $k \geq 0$. In particular, $(p_k, q_k) = 1$.

Proof. The proof is by induction. The result is clear for $k = 0$. Now suppose the equation holds for some integer k. Then

$$p_{k+1} q_k - p_k q_{k+1} = (a_{k+1} p_k + p_{k-1}) q_k - p_k (a_{k+1} q_k + q_{k-1})$$
$$= -(p_k q_{k-1} - p_{k-1} q_k) = -(-1)^{k-1} = (-1)^k,$$

which is the result with k replaced by $k + 1$.

(9.8) Corollary. *The equation $p_k q_{k-2} - p_{k-2} q_k = (-1)^k a_k$ holds for every $k \geq 1$.*

Proof. Note that
$$p_k q_{k-2} - p_{k-2} q_k = (a_k p_{k-1} + p_{k-2}) q_{k-2} - p_{k-2}(a_k q_{k-1} + q_{k-2})$$
$$= a_k(p_{k-1} q_{k-2} - p_{k-2} q_{k-1}) = (-1)^{k-2} a_k = (-1)^k a_k.$$

The next result shows that the odd-numbered convergents of a continued fraction $r = \langle a_0, a_1, a_2, \ldots, a_n \rangle$ are successively closer to r and hence give better and better approximations to r. A similar result is true for even-numbered convergents.

(9.9) Theorem. *Let c_k be the kth convergent of $r = \langle a_0, a_1, \ldots, a_n \rangle$. Then*
$$c_0 < c_2 < c_4 < \cdots \quad \cdots < c_5 < c_3 < c_1.$$
In particular, any even-numbered convergent is less than any odd-numbered convergent.

Proof. Since $c_k = p_k/q_k$, the equation in (9.8) can be divided by $q_k q_{k-2}$ to give $c_k - c_{k-2} = (-1)^k a_k / q_k q_{k-2}$. Since $q_i > 0$ for every $i \geq 0$ and $a_i \geq 1$ if $i > 0$, we have $c_k - c_{k-2} > 0$ for every even $k \geq 2$. Similarly, if $k \geq 3$ is odd, then $c_k - c_{k-2} < 0$. To prove that $c_{2m} < c_{2n+1}$ for any choice of nonnegative integers m and n, first note that if each side of the equation in (9.7) is divided by $q_k q_{k-1}$, we obtain $c_k - c_{k-1} = (-1)^{k-1}/q_k q_{k-1}$. Hence $c_k > c_{k-1}$ if k is odd. Thus $c_{2m} < c_{2m+2n} < c_{2m+2n+1} < c_{2n+1}$.

An Application: Solutions of $ax + by = c$

Let a/b be a rational number in lowest terms, where we may suppose that $b > 0$. Use the Euclidean Algorithm to write $a/b = \langle a_0, a_1, \ldots, a_n \rangle$. By (9.5), $a/b = p_n/q_n$; since $(a, b) = (p_n, q_n) = 1$ and $q_n > 0$, we have $a = p_n$ and $b = q_n$. Thus by (9.7), $aq_{n-1} - bp_{n-1} = (-1)^{n-1}$. We therefore have the following result.

(9.10) Theorem. *Suppose $a/b = \langle a_0, a_1, \ldots, a_n \rangle$, with $(a, b) = 1$ and $b > 0$. If n is odd, the equation $ax + by = c$ has the solution $x = cq_{n-1}, y = -cp_{n-1}$. If n is even, then $x = -cq_{n-1}, y = cp_{n-1}$ is a solution.*

(9.11) Example. We can use (9.10) to find a solution of $89x - 25y = 3$. Rewrite as $-89x + 25y = -3$; here, $a = -89$, $b = 25$, and $c = -3$. By (9.6), $-89/25 = \langle -4, 2, 3, 1, 2 \rangle$ and so n, the subscript on the last partial quotient, is 4. Thus (9.10) implies that one solution is

$$x = -cq_{n-1} = 3q_3 = 3(9) = 27, \quad y = cp_{n-1} = -3p_3 = -3(-32) = 96.$$

Infinite Continued Fractions

In this section, we define an *infinite* simple continued fraction and show that the value is always an irrational number.

(9.12) Theorem. *Let $a_0, a_1, a_2, \ldots$ be an infinite sequence of integers, all positive except possibly a_0, and suppose $c_k = \langle a_0, a_1, \ldots, a_k \rangle$. Then $\lim_{k \to \infty} c_k$ exists. Also, $c_{2i} < \lim_{k \to \infty} c_k < c_{2j+1}$ for any $i \geq 0$ and $j \geq 0$.*

Proof. By (9.9), (c_{2k}) is an increasing sequence that is bounded above (for example, by c_1), and therefore $\lim_{k \to \infty} c_{2k}$ exists. Similarly, (c_{2k+1}) is a decreasing sequence bounded below (by c_0), and so $\lim_{k \to \infty} c_{2k+1}$ also exists. We will show that these two limits are equal.

As in the proof of (9.9), we have $c_{2k+1} - c_{2k} = 1/q_{2k+1}q_{2k}$, which has limit 0 as $k \to \infty$ since $q_{2k} \geq 2k$ (see the Note following (9.3)). Therefore $\lim_{k \to \infty} c_{2k} = \lim_{k \to \infty} c_{2k+1}$. Let $\alpha = \lim_{k \to \infty} c_{2k}$; then it is easily checked that $\lim_{k \to \infty} c_k$ exists and equals α. Finally, it follows that $c_{2i} < \alpha < c_{2j+1}$ for every choice of i and j, since (c_{2k}) is strictly increasing and (c_{2k+1}) is strictly decreasing.

Note. It is worth emphasizing that *if $\alpha = \lim_{k \to \infty} c_k$, then every odd-numbered convergent is greater than α and every even-numbered convergent is less than α.* This fact will be used in many of the problems. Also, note that *the odd-numbered convergents are successively closer to α, and similarly, the even-numbered convergents are successively closer to α.*

(9.13) Definition. Let $a_0, a_1, a_2, \ldots$ be an infinite sequence of integers, all positive except possibly a_0. If $c_k = \langle a_0, a_1, \ldots, a_k \rangle$, define the value of the infinite simple continued fraction $\langle a_0, a_1, a_2, \ldots \rangle$ to be $\lim_{k \to \infty} c_k$.

We will shortly prove that two distinct infinite continued fractions converge to different values. We require the following lemma.

(9.14) Lemma. *Let $\alpha = \langle a_0, a_1, a_2, \ldots \rangle$. Then $a_0 = [\alpha]$ and $\alpha = a_0 + 1/\langle a_1, a_2, \ldots \rangle$.*

Proof. It is clear from (9.9) and (9.13) that $a_0 = c_0 < \alpha < c_1 = a_0 + 1/a_1$. Thus $a_0 < \alpha < a_0 + 1$ since $a_1 \geq 1$, and therefore $[\alpha] = a_0$. If $k \geq 1$, then $c_k = \langle a_0, a_1, \ldots, a_k \rangle = a_0 + 1/\langle a_1, a_2, \ldots, a_k \rangle$. Taking the limit as $k \to \infty$, we obtain $\alpha = a_0 + 1/\langle a_1, a_2, \ldots \rangle$.

(9.15) Theorem. *Suppose that the two infinite continued fractions $\langle a_0, a_1, a_2, \ldots \rangle$ and $\langle b_0, b_1, b_2, \ldots \rangle$ converge to the same value. Then $a_i = b_i$ for every $i \geq 0$.*

Proof. Suppose each continued fraction converges to α. It follows from (9.14) that $a_0 + 1/\langle a_1, a_2, \ldots \rangle = b_0 + 1/\langle b_1, b_2, \ldots \rangle$. Since $a_0 = b_0 = [\alpha]$, we have $\langle a_1, a_2, \ldots \rangle = \langle b_1, b_2, \ldots \rangle$. The same argument then shows that $a_1 = b_1$. Proceeding by induction, we conclude that $a_i = b_i$ for every $i \geq 0$.

We show next that any infinite continued fraction converges to an irrational number.

(9.16) Theorem. *The value of any infinite continued fraction is irrational.*

Proof. Let $\alpha = \langle a_0, a_1, a_2, \ldots \rangle$. If α is rational, then $\alpha = m/n$, where m and n are integers and $n > 0$. If k is even, then $c_k < \alpha < c_{k+1}$ by (9.12). Thus

$$0 < \frac{m}{n} - \frac{p_k}{q_k} < \frac{p_{k+1}}{q_{k+1}} - \frac{p_k}{q_k} = \frac{1}{q_k q_{k+1}}.$$

Multiplying by nq_k, we obtain $0 < mq_k - np_k < n/q_{k+1}$, and therefore $0 < mq_k - np_k < 1$ if k is even and sufficiently large. This is impossible, since $mq_k - np_k$ is an integer, so we conclude that α is irrational.

The Infinite Continued Fraction of an Irrational Number

We have shown, in (9.16), that an infinite continued fraction represents an irrational number. Is it true that *every* irrational number has an infinite continued fraction expansion? The answer is yes, and in this section, we give a method for determining the infinite continued fraction of a given irrational number.

(9.17) Discussion. Suppose α is irrational, and let $\alpha_0 = \alpha$. Define $a_0 = [\alpha_0]$; then $0 < \alpha_0 - a_0 < 1$, since α_0 is irrational. Let $\alpha_1 = 1/(\alpha_0 - a_0)$; then α_1 is irrational, $\alpha_1 > 1$, and $\alpha_0 = a_0 + 1/\alpha_1$, that is, $\alpha_0 = \langle a_0, \alpha_1 \rangle$. Similarly, set $a_1 = [\alpha_1]$ and let $\alpha_2 = 1/(\alpha_1 - a_1)$; then $\alpha_0 = \langle a_0, a_1, \alpha_2 \rangle$. Proceeding this way, if $\alpha_1, \alpha_2, \ldots, \alpha_k$ and $a_0, a_1, \ldots, a_k$ have been defined, let

$$\alpha_{k+1} = \frac{1}{\alpha_k - a_k} \quad \text{and} \quad a_{k+1} = [\alpha_{k+1}].$$

Then $\alpha_0 = \langle a_0, a_1, \ldots, a_k, \alpha_{k+1} \rangle$ for any $k \geq 0$.

The next result gives an explicit formula for the difference between α and its kth convergent p_k/q_k.

INFINITE CONTINUED FRACTION OF AN IRRATIONAL NUMBER 277

(9.18) Theorem. Let α be irrational, and let α_i and a_i be defined as in (9.17). Then

$$\alpha - \frac{p_k}{q_k} = \frac{(-1)^k}{q_k(\alpha_{k+1}q_k + q_{k-1})}.$$

Proof. Since $\alpha = \langle a_0, a_1, \ldots, a_k, \alpha_{k+1}\rangle$ for $k \geq 0$, (9.4) implies that $\alpha = (\alpha_{k+1}p_k + p_{k-1})/(\alpha_{k+1}q_k + q_{k-1})$. Subtracting p_k/q_k from both sides and simplifying, we obtain

$$\alpha - \frac{p_k}{q_k} = \frac{p_{k-1}q_k - p_k q_{k-1}}{q_k(\alpha_{k+1}q_k + q_{k-1})},$$

and the result follows from (9.7).

We now prove that every irrational number has a continued fraction expansion.

(9.19) Theorem. Let α be irrational, and let α_k and a_k be defined as in (9.17). Then $\alpha = \langle a_0, a_1, a_2, \ldots\rangle$. In general, $\alpha_k = \langle a_k, a_{k+1}, a_{k+2}, \ldots\rangle$ for every $k \geq 0$.

Proof. It is clear that α_k is irrational for all k. Thus $0 < \alpha_k - a_k < 1$, and therefore $\alpha_{k+1} > 1$ and $a_{k+1} \geq 1$ for every $k \geq 0$. It follows from (9.18) that $\lim_{k\to\infty} p_k/q_k = \alpha$, and hence by definition, $\alpha = \langle a_0, a_1, a_2, \ldots\rangle$.

To prove that $\alpha_k = \langle a_k, a_{k+1}, a_{k+2}, \ldots\rangle$ for any $k \geq 1$, simply begin the process with α_k instead of $\alpha = \alpha_0$ and apply the first result.

(9.20) Example. We will use the method described above to calculate the continued fraction expansion of $\alpha = \sqrt{15}$. Proceeding as in (9.17), we obtain

$a_0 = [\sqrt{15}] = 3,\qquad \alpha_1 = 1/(\alpha_0 - a_0) = 1/(\sqrt{15} - 3) = (\sqrt{15} + 3)/6;$
$a_1 = [\alpha_1] = 1,\qquad \alpha_2 = 1/(\alpha_1 - a_1) = 1/((\sqrt{15} - 3)/6) = \sqrt{15} + 3;$
$a_2 = [\alpha_2] = 6,\qquad \alpha_3 = 1/(\alpha_2 - a_2) = 1/(\sqrt{15} - 3).$

Since $\alpha_3 = \alpha_1$, it is clear that the continued fraction expansion of $\sqrt{15}$ is periodic, that is, the partial quotients repeat, and we have

$$\sqrt{15} = \langle 3, 1, 6, 1, 6, 1, 6, \ldots\rangle.$$

Note. If we express $\alpha_1, \alpha_2, \alpha_3, \ldots$ *algebraically*, using, for example, the exact expression $\alpha_1 = (\sqrt{15}+3)/6$ rather than a numerical approximation, then it is clear when the α's begin to repeat. However, if we evaluate expressions involving $\sqrt{15}$ on a calculator, getting an *approximation* to the actual value, then we have not *proved*, for example, that $\alpha_3 = \alpha_1$; we have shown only that the two numbers are nearly equal. In the next section, we will give a much better method for finding the continued fraction expansion of certain irrational numbers, using only exact integer arithmetic.

Periodic Continued Fractions

We study now the continued fraction expansion of numbers of the form $(a+\sqrt{d})/b$, where a and b are integers and d is a positive integer which is not a perfect square. (Note that $\sqrt{d}$ is irrational, by Problem 1-35.) We will show that the continued fraction expansions of such numbers are periodic and that these are the *only* numbers that have a periodic expansion.

(9.21) Definition. The infinite simple continued fraction

$$\langle c_0, c_1, \ldots, c_n, \overline{a_0, a_1, \ldots, a_{m-1}} \rangle$$

is called *periodic*. The bar indicates that the sequence $a_0, a_1, \ldots, a_{m-1}$ repeats indefinitely. The smallest m with this property is called the *length of the period* or, more simply, the *period*. If the continued fraction is of the form $\langle \overline{a_0, a_1, \ldots, a_{m-1}} \rangle$, it is called *purely periodic*.

For example, $\sqrt{15} = \langle 3, 1, 6, 1, 6, 1, 6, \ldots \rangle = \langle 3, \overline{1,6} \rangle$. (See (9.20).)

(9.22) Definition. The real number α is a *quadratic irrational* if α is irrational and is the root of a quadratic polynomial with integer coefficients, that is, if α is a solution of the equation $ax^2 + bx + c = 0$, where a, b, c are integers with $a \neq 0$.

(9.23) Theorem. *If α is a quadratic irrational, then α can be written in the form $\alpha = (r + \sqrt{d})/s$, where d is a positive integer that is not a perfect square and r and s are integers such that $s \mid d - r^2$. Moreover, if we are given integers d, r and s such that $\alpha = (r + \sqrt{d})/s$ but $s \nmid d - r^2$, then we can find integers D, R, and S such that $\alpha = (R + \sqrt{D})/S$ and $S \mid D - R^2$.*

Proof. If α is a quadratic irrational, then by definition, α satisfies an equation of the form $ax^2 + bx + c = 0$, with $a \neq 0$; thus $\alpha = (-b \pm \sqrt{b^2 - 4ac})/2a$. Since α is irrational, $b^2 - 4ac$ is not a perfect square. Set $d = b^2 - 4ac$. If $\alpha = (-b + \sqrt{d})/2a$, take $r = -b$ and $s = 2a$; if $\alpha = (-b - \sqrt{d})/2a$, let $r = b$ and $s = -2a$. In both cases, $s \mid d - r^2$.

If $\alpha = (r + \sqrt{d})/s$ and $s \nmid d - r^2$, write $\alpha = (R + \sqrt{D})/S$, where $D = ds^2$, $R = r|s|$, and $S = s|s|$. Then $S \mid D - R^2$.

We next give an algorithm that can be used to find the continued fraction expansion of a quadratic irrational. The method is due to Euler, but it is essentially the same as a technique given by Lord William Brouncker (1620–1684) a century earlier, although Brouncker did not use continued fractions explicitly. The method is also closely related to the "cyclic method" of Bhaskara, more than five centuries before Brouncker.

(9.24) Theorem. *Let α be a quadratic irrational, and set $\alpha_0 = \alpha$, $a_0 = [\alpha]$. Use (9.23) to write $\alpha_0 = (r_0 + \sqrt{d})/s_0$, where d is not a perfect square and $s_0 \mid d - r_0^2$. Recall that $\alpha_{k+1} = 1/(\alpha_k - a_k)$ and $a_{k+1} = [\alpha_{k+1}]$ (see (9.17)). For $k \geq 0$, define*

$$r_{k+1} = a_k s_k - r_k, \qquad s_{k+1} = (d - r_{k+1}^2)/s_k.$$

Then the following are true for every $k \geq 0$:

(i) *r_k and s_k are integers, with $s_k \neq 0$;*

(ii) *$s_k \mid d - r_k^2$;*

(iii) *$\alpha_k = (r_k + \sqrt{d})/s_k$, and hence $a_k = [(r_k + \sqrt{d})/s_k]$.*

Proof. The proof is by induction. The results are clearly true for $k = 0$. Now suppose the statements hold for some positive integer k. It is obvious from the definition that r_{k+1} is an integer and $r_{k+1} \equiv -r_k \pmod{s_k}$. Thus

$$d - r_{k+1}^2 \equiv d - r_k^2 \equiv 0 \pmod{s_k}.$$

From the definition of s_{k+1}, it follows that s_{k+1} is an integer, and nonzero since d is not a perfect square. Also, $d - r_{k+1}^2 = s_k s_{k+1}$, and therefore $s_{k+1} \mid d - r_{k+1}^2$. Finally, since $\alpha_k = (r_k + \sqrt{d})/s_k$, we have

$$\alpha_{k+1} = \frac{1}{\alpha_k - a_k} = \frac{s_k}{r_k + \sqrt{d} - a_k s_k} = \frac{s_k}{-r_{k+1} + \sqrt{d}}$$

$$= \frac{s_k(r_{k+1} + \sqrt{d})}{d - r_{k+1}^2} = \frac{r_{k+1} + \sqrt{d}}{s_{k+1}}.$$

Computational Note. For a given quadratic irrational $\alpha = (r_0 + \sqrt{d})/s_0$, the initial partial quotient a_0 is simply $[\alpha]$. The integers r_1 and s_1 are then computed from the above formulas, and the next partial quotient a_1 is given by $a_1 = [\alpha_1] = [(r_1 + \sqrt{d})/s_1]$. In general, if r_k and s_k have been found, then $a_k = [\alpha_k] = [(r_k + \sqrt{d}/s_k]$. If $s_k \geq 1$, the calculation can be simplified.

$$a_k = [(r_k + a_0)/s_k],$$

where $a_0 = [\sqrt{d}]$ (see Problem 1-59). Thus all we need to know about $\sqrt{d}$ is its integer part. The task of computing the continued fraction expansion of $\sqrt{d}$ can therefore be accomplished using exact integer arithmetic.

The algorithm described in (9.24) can be speeded up by noting that $s_{k+1} = s_{k-1} + a_k(r_k - r_{k+1})$ (see Problem 9-28).

(9.25) Example. We will use the above method to find the continued fraction expansion of $\alpha = (4 - \sqrt{3})/3$. If we write $\alpha = \alpha_0$ as $(r + \sqrt{d})/s = (\sqrt{3} - 4)/(-3)$, then $s \nmid d - r^2$. So we must multiply numerator and denominator by $|s| = |-3| = 3$, getting $\alpha_0 = (r_0 + \sqrt{d})/s_0 = (-12 + \sqrt{27})/(-9)$, where $s_0 | d - r_0^2$. Now construct the following table, using the formulas given in (9.24):

k	0	1	2	3	4	5
r_k	−12	12	1	5	5	5
s_k	−9	13	2	1	2	1
a_k	0	1	3	10	5	

Note that $a_k = [\alpha_k] = [(r_k + 5)/s_k]$ for each k, since $[\sqrt{27}] = 5$. The table clearly repeats beginning with $k = 5$, since $r_5 = r_3$ and $s_5 = s_3$. Therefore $(4 - \sqrt{3})/3 = \langle 0, 1, 3, \overline{10, 5} \rangle$.

It is also true that $\alpha_k = (r_k + \sqrt{d})/s_k = \langle a_k, a_{k+1}, a_{k+2}, \ldots \rangle$. Thus, for example, $\langle 3, \overline{10, 5} \rangle = \alpha_2 = (1 + \sqrt{27})/2$ and $\langle \overline{10, 5} \rangle = \alpha_3 = 5 + \sqrt{27}$.

(9.26) Definition. Let $\alpha = (a + b\sqrt{d})/c$, where a, b, and c are integers, with $c \neq 0$, and d is an integer which is not a perfect square. Then α', the *conjugate* of α, is defined by $\alpha' = (a - b\sqrt{d})/c$.

The operation of taking the conjugate has a number of simple properties that will be used frequently. For proofs, see Problem 9-29.

(9.27) Lemma. *Suppose that α and β are each of the form $(a + b\sqrt{d})/c$ for a fixed d. Then $(\alpha + \beta)' = \alpha' + \beta'$, $(\alpha\beta)' = \alpha'\beta'$, and $(\alpha/\beta)' = \alpha'/\beta'$.*

(9.28) Theorem (Lagrange). *The continued fraction expansion of any quadratic irrational is periodic.*

Proof. Let α be a quadratic irrational, and carry out the algorithm described in (9.24). First we show that there is an integer m such that $\alpha'_m < 0$, where α'_m denotes the conjugate of α_m. Recall from (9.18) that $\alpha - p_k/q_k = (-1)^k/q_k(\alpha_{k+1}q_k + q_{k-1})$. Taking conjugates of both sides and multiplying, we find

$$\left(\alpha - \frac{p_k}{q_k}\right)\left(\alpha' - \frac{p_k}{q_k}\right) = \frac{1}{q_k^2(\alpha_{k+1}q_k + q_{k-1})(\alpha'_{k+1}q_k + q_{k-1})}.$$

Since $\alpha' \neq \alpha$ and p_k/q_k converges to α, there is a value of k such that p_k/q_k lies between α and α' (see the Note after (9.12)). For this value of k, the left side of the above equation is negative. Thus $\alpha'_{k+1}q_k + q_{k-1} < 0$; hence if we take $m = k + 1$, then $\alpha'_m < 0$.

In fact, if $\alpha'_m < 0$ for $m > 0$, then $-1 < \alpha'_n < 0$ for every $n > m$. To prove this, it is enough to show that $-1 < \alpha'_{m+1} < 0$. Since $\alpha_{m+1} = 1/(\alpha_m - a_m)$, by taking conjugates, we obtain $\alpha'_{m+1} = 1/(\alpha'_m - a_m) < 0$, and so $-1 < \alpha'_{m+1}$ since $a_m \geq 1$. Now let $k \geq 1$ be large enough that $-1 < \alpha'_k < 0$. Then $1 < \alpha_k - \alpha'_k = 2\sqrt{d}/s_k$ and hence $0 < s_k < 2\sqrt{d}$. Since $\alpha_k = (r_k + \sqrt{d})/s_k$, we have $\alpha_k \alpha'_k = (r_k^2 - d)/s_k^2$. It follows that $r_k^2 < 0$, that is, $-\sqrt{d} < r_k < \sqrt{d}$. In fact, $r_k > 0$ since $0 < \alpha_k + \alpha'_k = 2r_k/s_k$.

Thus for large k, the pair r_k, s_k can take on at most $2d$ different values, and hence there are integers m and n, with $m < n$, such that $r_n = r_m$ and $s_n = s_m$. The recurrence relations for r_k and s_k given in (9.24) then imply that $a_{m+t} = a_{n+t}$ for every positive integer t, and therefore the continued fraction expansion of α is periodic.

Note. The converse of Lagrange's Theorem also holds: *The value of any periodic continued fraction is a quadratic irrational* (see Problem 9-30).

We give a concrete example to show how to carry out the calculations.

(9.29) Example. To find the value of the periodic continued fraction $\alpha = \langle 4, \overline{3, 2} \rangle$, let $\beta = \langle \overline{3, 2} \rangle$. Then $\beta = 3 + 1/(2 + 1/\beta) = 3 + \beta/(2\beta + 1) = (7\beta + 3)/(2\beta + 1)$, and hence $2\beta^2 - 6\beta - 3 = 0$. From the quadratic formula, we then have $\beta = (3 + \sqrt{15})/2$, since β is clearly positive. Thus $\alpha = 4 + 1/\beta = (9 + \sqrt{15})/3$.

Purely Periodic Continued Fractions

In the previous section, we showed that every quadratic irrational has a periodic continued fraction expansion. We now consider the problem of determining which quadratic irrationals have a *purely periodic* expansion, that is, an expansion of the form $\langle \overline{a_0, a_1, a_2 \ldots, a_m} \rangle$.

(9.30) Definition. A quadratic irrational $\alpha = (r + \sqrt{d})/s$ is *reduced* if $\alpha > 1$ and its conjugate $\alpha' = (r - \sqrt{d})/s$ satisfies $-1 < \alpha' < 0$.

The next result characterizes the irrational numbers that have a purely periodic expansion. The characterization is already implicit in the work of Lagrange, but the first proof was given in 1828 by Évariste Galois (1811–1832).

(9.31) Theorem. *The quadratic irrational α has a purely periodic continued fraction if and only if α is reduced. Also, if $\alpha = \langle \overline{a_0, a_1, \ldots, a_{m-1}} \rangle$, then $-1/\alpha' = \langle \overline{a_{m-1}, a_{m-2}, \ldots, a_1, a_0} \rangle$.*

Proof. Suppose first that $\alpha = \langle a_0, a_1, \ldots \rangle$ is reduced, and let α_k be defined as in (9.17). From $\alpha_{k+1} = 1/(\alpha_k - a_k)$, we obtain $\alpha'_{k+1} = 1/(\alpha'_k - a_k)$ by taking conjugates. Since $a_k \geq 1$, it follows that if $\alpha'_k < 0$, then $-1 < \alpha'_{k+1} < 0$. Thus if α_k is reduced, then so is α_{k+1}. But $\alpha = \alpha_0$ is reduced, and therefore α_n is reduced for all n.

The equation $\alpha'_{k+1} = 1/(\alpha'_k - a_k)$ can be rewritten as $\alpha'_k = a_k + 1/\alpha'_{k+1}$. Letting $\beta_i = -1/\alpha'_i$, we obtain $\beta_{k+1} = a_k + 1/\beta_k$. Note that $0 < \beta_k < 1$, and hence $[\beta_{k+1}] = a_k$.

The continued fraction expansion of α is periodic, say, with period m. Thus $\alpha_{m+j} = \alpha_j$ for some j, and taking conjugates, we obtain $\beta_{m+j} = \beta_j$. If $j > 0$, then

$$\beta_j = a_{j-1} + \frac{1}{\beta_{j-1}} \quad \text{and} \quad \beta_{m+j} = a_{m+j-1} + \frac{1}{\beta_{m+j-1}}.$$

Since $[\beta_j] = a_{j-1}$ and $[\beta_{m+j}] = a_{m+j-1}$, it follows that $a_{j-1} = a_{m+j-1}$, and therefore $1/\beta_{j-1} = 1/\beta_{m+j-1}$. Inverting and taking conjugates now yields $\alpha_{j-1} = \alpha_{m+j-1}$; if $j - 1 > 0$, we repeat the process. Ultimately, we find that $\alpha_0 = \alpha_m$, that is, α has a purely periodic continued fraction expansion.

To prove the converse, let $\alpha = \langle \overline{a_0, a_1, \ldots, a_{m-1}} \rangle$. The case $m = 1$ is trivial, so assume that $m > 1$. Since $\alpha_0 = \alpha_m > 1$, we have $a_0 \geq 1$, and thus all of the a_i are positive. Letting $k = m - 1$ in the equation $\beta_{k+1} = a_k + 1/\beta_k$, we obtain $\beta_0 = \beta_m = a_{m-1} + 1/\beta_{m-1} = \langle a_{m-1}, \beta_{m-1} \rangle$. But $\beta_{m-1} = a_{m-2} + 1/\beta_{m-2}$, and hence $\beta_0 = \langle a_{m-1}, a_{m-2}, \beta_{m-2} \rangle$. Continuing in this way, we obtain

$$\beta_0 = \langle a_{m-1}, a_{m-2}, \ldots, a_0, \beta_0 \rangle,$$

and therefore

$$-1/\alpha' = \beta_0 = \langle \overline{a_{m-1}, a_{m-2}, \ldots, a_0} \rangle.$$

In particular, $-1/\alpha' > 1$, and hence α is reduced.

(9.32) Example. If $\alpha = 5 + \sqrt{27}$, then $\alpha' = 5 - \sqrt{27}$ and $-1 < \alpha' < 0$, so α is reduced. By (9.25), α has the purely periodic expansion $\langle \overline{10, 5} \rangle$.

Rational Approximations to Irrational Numbers

If α is irrational and p_k/q_k is the kth convergent of the continued fraction expansion of α, then the sequence (p_k/q_k) converges to α, that is, p_k/q_k can be made arbitrarily close to α by making k sufficiently large. Thus the convergents of α provide arbitrarily close approximations to α. The next result gives an estimate of just how good these approximations are.

(9.33) Theorem. Let p_k/q_k be the kth convergent of the irrational number α. Then

$$\left| \alpha - \frac{p_k}{q_k} \right| < \frac{1}{q_k q_{k+1}}$$

for every $k \geq 0$. In particular, $|\alpha - p_k/q_k| < 1/q_k^2$ for every $k \geq 0$.

Proof. By (9.18), $|\alpha - p_k/q_k| = 1/q_k(\alpha_{k+1}q_k + q_{k-1})$. Since $\alpha_{k+1} > a_{k+1}$ and $a_{k+1}q_k + q_{k-1} = q_{k+1}$, the result follows.

(9.34) Theorem. *Let α be irrational and suppose that $\alpha = \langle a_0, a_1, a_2, \ldots \rangle$, with kth convergent $p_k/q_k = \langle a_0, a_1, \ldots, a_k \rangle$. Then the convergents of α are successively closer to α; that is,*

$$|\alpha - p_{k+1}/q_{k+1}| < |\alpha - p_k/q_k| \tag{1}$$

for every $k \geq 0$. In fact, we have the stronger inequality

$$|q_{k+1}\alpha - p_{k+1}| < |q_k\alpha - p_k|. \tag{2}$$

Proof. Using (9.33), with $k+1$ instead of k, we obtain $|q_{k+1}\alpha - p_{k+1}| < 1/q_{k+2}$. By (9.18), $|q_k\alpha - p_k| = 1/(\alpha_{k+1}q_k + q_{k-1})$. Since $[\alpha_{k+1}] = a_{k+1}$, we have $q_k\alpha_{k+1} + q_{k-1} < q_k a_{k+1} + q_k + q_{k-1} = q_{k+1} + q_k$. But $q_{k+2} = a_{k+2}q_{k+1} + q_k \geq q_{k+1} + q_k$, and inequality (2) follows. Inequality (1) is an easy consequence of (2).

The next result shows that the convergents of α play a very special role in finding integers s and t such that $|t\alpha - s|$ is small.

(9.35) Theorem. *Let α be irrational, and suppose that s/t (with t positive) has the property that $|t\alpha - s| < |v\alpha - u|$ whenever $1 \leq v < t$. Then s/t is a convergent of α.*

Proof. Let $c_k = p_k/q_k$, where p_k/q_k is the kth convergent of α. Suppose first that $s/t < c_0$. Then $|t\alpha - s| \geq |\alpha - s/t| > |\alpha - c_0| = |q_0\alpha - p_0|$ (since $q_0 = 1$), contradicting the hypothesis on s/t.

Suppose now that $s/t > c_1$; then $|\alpha - s/t| > |c_1 - s/t|$. Multiply through by t to obtain $|t\alpha - s| > |tp_1 - sq_1|/q_1 \geq 1/q_1$. But by (9.33), we have $|q_0\alpha - p_0| < 1/q_1$, again contradicting the hypothesis on s and t.

Thus s/t lies between c_{n-1} and c_{n+1} for some integer n. If s/t is either c_{n-1} or c_{n+1}, we are finished. Otherwise, note that these two convergents are on the same side of α, and c_n lies on the other side. It follows that $|s/t - c_{n-1}| < |c_n - c_{n-1}|$. Multiplying both sides by tq_nq_{n-1}, we obtain $q_n|sq_{n-1} - tp_{n-1}| < t|p_nq_{n-1} - p_{n-1}q_n| = t$. Since $sq_{n-1} - tp_{n-1}$ is a nonzero integer, it follows that $q_n < t$.

We also have $|\alpha - s/t| > |c_{n+1} - s/t|$. If we multiply through by t, we find that $|t\alpha - s| > |tp_{n+1} - sq_{n+1}|/q_{n+1} \geq 1/q_{n+1}$. But $|q_n\alpha - p_n| < 1/q_{n+1}$ by (9.33). Since $q_n < t$, this contradicts the assumption on s/t, completing the proof.

(9.36) Corollary. *Let α be irrational, and suppose that c and d are integers, with d positive. If $|d\alpha - c| < |q_k\alpha - p_k|$, then $d \geq q_{k+1}$.*

Proof. Let t be the *smallest* positive integer for which there is an s such that $|t\alpha - s| < |q_k\alpha - p_k|$. Then by (9.35), s/t is a convergent of α; let $s/t = p_m/q_m$. Since convergents of α are successively closer to α, it follows that $m \geq k+1$, and hence $t \geq q_{k+1}$. Since clearly $d \geq t$, the result follows.

The next result shows that the kth convergent p_k/q_k of the irrational number α is in fact the best rational approximation to α among all rational numbers whose denominators are less than or equal to q_k.

(9.37) Theorem. *If $|\alpha - c/d| < |\alpha - p_k/q_k|$ for some $k \geq 1$, then $d > q_k$.*

Proof. Suppose to the contrary that $d \leq q_k$. Multiplying on the left by d and on the right by q_k, we obtain $|d\alpha - c| < |q_k\alpha - p_k|$. Thus by (9.36), we have $d \geq q_{k+1}$. This contradicts the assumption that $d \leq q_k$ unless $q_k = q_{k+1}$, which cannot happen if $k \geq 1$.

Note. In fact, a stronger result holds if $q_{k+1} > 2q_k$: If $|\alpha - c/d| < |\alpha - p_k/q_k|$, then $d > q_{k+1}/2$. For a proof, see Problem 9-46.

The following more general question is harder to answer: *If s is a positive integer, what is the best approximation to the irrational α among all rational numbers with denominators less than or equal to s?* (The previous result answers this question only for the case where $s = q_k$ for some k.) The solution involves the notion of *secondary convergents* of α, which are discussed in the last section of the Problems. The complete answer is given in Problem 9-66: For any positive integer s, the best rational approximation to α with denominator not exceeding s is either a convergent or a secondary convergent of α.

(9.38) Example. The preceding results can be used to find the "best" rational approximations to π. Using the technique of (9.17), we can evaluate the first nine partial quotients of π: $\pi = \langle 3, 7, 15, 1, 292, 1, 1, 1, 2, \ldots \rangle$. (See Problem 9-57. The accuracy of these values depends on the accuracy of the calculator being used.) Set up the following table:

k	−1	0	1	2	3	4
a_k		3	7	15	1	292
p_k	1	3	22	333	355	103993
q_k	0	1	7	106	113	33102

The convergent p_1/q_1 is just the familiar approximation $22/7$, first given by Archimedes around 250 B.C.; it is the best approximation among all rationals with denominator not exceeding 7. The approximation $333/106 = p_2/q_2$ first

appeared in Europe in the sixteenth century, and the remarkable approximation 355/113 was known to Chinese mathematicians in the fifth century. By (9.33),
$$|\pi - 355/113| < 1/113 \cdot 33102 < .0000003,$$
and so 355/113 is accurate to six decimal places. In fact, 355/113 is the best approximation to π among all rationals with denominators not exceeding 16603 (Problem 9-70). For further results concerning π, see Problems 9-54 to 9-59 and Problem 9-69.

The following result shows that if a rational approximates an irrational number α "sufficiently" closely, then the rational number must be one of the convergents of the infinite continued fraction expansion of α. (For a proof, see Problem 9-53.)

(9.39) Theorem. *Suppose α is irrational. If c/d is a rational number, with $d \geq 1$, such that $|\alpha - c/d| < 1/2d^2$, then c/d is one of the convergents of the infinite continued fraction expansion of α.*

The preceding theorem will be used in Chapter 10 to exhibit the connection between continued fractions and Pell's Equation.

Note. It is not true that every convergent of α satisfies the inequality in (9.39) (which is clearly stronger than the second inequality in (9.33)). However, *of any two successive convergents, at least one, say p/q, will satisfy the inequality* $|\alpha - p/q| < 1/2q^2$. (For a proof, see Problem 9-52.)

We mention finally that with minor exceptions, the preceding approximation results also hold when α is *rational*.

An Application: Calendars

The earth orbits the sun approximately once every 365 days, 5 hours, 48 minutes, and 46 seconds. Thus the actual length of a year exceeds 365 days by 20926/86400 of a day. How can a calendar based on a year of 365 days be modified to give a more accurate approximation to the true value? The answer lies in finding a good rational approximation to 20926/86400, one that is easy to implement.

We begin by using the Euclidean Algorithm to find the continued fraction expansion of 20926/86400, namely, $\langle 0, 4, 7, 1, 3, 5, 64 \rangle$. Now set up the following table:

k	-1	0	1	2	3	4	5
a_k		0	4	7	1	3	5
p_k	1	0	1	7	8	31	163
q_k	0	1	4	29	33	128	673

In 45 B.C., Julius Caesar used the correction $1/4$ (which is the convergent p_1/q_1) by adding one day – February 29 – every four years, that is, $1/4$ of a day for every year. The use of a leap year made the Julian calendar accurate to within 11 minutes a year, which amounts to an error of approximately 10 days in 1500 years. This necessitated a further modification by Pope Gregory XIII in 1582, as described in the next paragraph.

In the eleventh century, the Persian astronomer, mathematician, and poet al-Khayyami (1050–1123), known in the West as Omar Khayyam, proposed the correction $8/33$ (the convergent p_3/q_3), which would give an error of only 19 seconds a year. The calendar that we use today, which is known as the *Gregorian calendar*, has an extra day in every year divisible by 4, except years that are divisible by 100 but not by 400. The Gregorian calendar is accurate to within 26 seconds a year. This correction corresponds to approximating $20926/86400$ by $97/400$, since there are 97 leap years every 400 years. Although $97/400$ is not one of the convergents of $20926/86400$, it is easier to implement the correction just described than to use the more accurate correction $8/33$.

Finally, we point out that the fourth convergent $31/128$ (which could be implemented, for example, by skipping one leap year every 128 years) gives an extremely accurate correction: 365 31/128 agrees with the actual length of a year to four decimal places. In fact, the error amounts to about one second a year, or only about one and a half minutes a century.

PROBLEMS AND SOLUTIONS

Finite Continued Fractions

9-1. *Find the two continued fraction expansions of each of the following rational numbers: (a) $355/113$; (b) $5/32$; (c) $4756/1121$.*

Solution. (a) Apply the Euclidean Algorithm to get $355 = 3 \cdot 113 + 16$, $113 = 7 \cdot 16 + 1$, $16 = 16 \cdot 1$. Thus $355/113 = \langle 3, 7, 16 \rangle = \langle 3, 7, 15, 1 \rangle$. (b) Similarly, the Euclidean Algorithm gives $5/32 = \langle 0, 6, 2, 2 \rangle = \langle 0, 6, 2, 1, 1 \rangle$. (c) $4756 = 4 \cdot 1121 + 272$, $1121 = 4 \cdot 272 + 33$, $272 = 8 \cdot 33 + 8$, $33 = 4 \cdot 8 + 1$, $8 = 8 \cdot 1$. Hence $4756/1121 = \langle 4, 4, 8, 4, 8 \rangle = \langle 4, 4, 8, 4, 7, 1 \rangle$.

Note. We could also find the continued fraction expansion of $4756/1121$ using the technique described in (9.17), although α here is *rational*. (It should be clear that the method still produces the continued fraction of α even when α is rational.) In this case, with $\alpha_0 = 4756/1121$, we have $a_0 = [\alpha_0] = 4$, $\alpha_1 = 1/(\alpha_0 - a_0) = 4.1213\ldots$, and so on. This is very easy to implement on a calculator: If α_{k-1} has been found, subtract its integral part (which is a_{k-1}) and invert, and repeat this process until some α_k is an integer. Then $a_k = [\alpha_k]$ is the last partial quotient.

But beware: the lack of enough accuracy in the calculator will usually give a number that is not an exact integer in what should be the final step. In this example, the last partial quotient on the calculator used was actually 7.999994954, an approximation to the correct value of 8. If we had not assumed that this should be 8 and repeated the process by subtracting 7 and inverting, we would have obtained a partial quotient of 1, and one more step would yield the completely erroneous value 198159!

9-2. *Determine the rational numbers with the following continued fraction expansions: (a)* $\langle 1,2,3 \rangle$*; (b)* $\langle -3,4,2,5 \rangle$*; (c)* $\langle 1,1,1,1,1,1 \rangle$.

Solution. In each case, the best way is to set up a table as in (9.6):

k	-1	0	1	2
a_k		1	2	3
p_k	1	1	3	10
q_k	0	1	2	7

k	-1	0	1	2	3
a_k		-3	4	2	5
p_k	1	-3	-11	-25	-136
q_k	0	1	4	9	49

Thus by (9.4), $\langle 1,2,3 \rangle = p_2/q_2 = 10/7$ and $\langle -3,4,2,5 \rangle = p_3/q_3 = -136/49$. In a similar way, we can show that $\langle 1,1,1,1,1,1 \rangle = 13/8$.

9-3. *Set up a table, and compute the convergents of* $776/247$.

Solution. Apply the Euclidean Algorithm to obtain $776/247 = \langle 3,7,17,2 \rangle$. (If the method described in the Note after Problem 9-1 is used, be sure to recognize 1.999999898 as 2.) Now set up the following table:

k	-1	0	1	2	3
a_k		3	7	17	2
p_k	1	3	22	377	776
q_k	0	1	7	120	247

Thus the convergents of $776/247$ are 3, $22/7$, $377/120$, and $776/247$.

9-4. *Use induction to prove that a finite simple continued fraction represents a rational number.*

Solution. We show by induction on n that any finite simple continued fraction with n partial quotients represents a rational number. For $n = 0$, the result is obvious. Now suppose that any continued fraction with k partial quotients represents a rational number. Then if $\langle a_0, a_1, \ldots, a_k \rangle$ is a continued fraction with $k+1$ partial quotients, we have $\langle a_0, a_1, \ldots, a_k \rangle = a_0 + 1/\langle a_1, a_2, \ldots, a_k \rangle$, which is also rational, since the induction hypothesis implies that $\langle a_1, a_2, \ldots, a_k \rangle$ is rational.

288   CHAPTER 9: CONTINUED FRACTIONS

9-5. Suppose that $r = \langle a_0, a_1, \ldots, a_n \rangle = \langle b_0, b_1, \ldots, b_m \rangle$, where a_i and b_i are integers, all positive except possibly a_0 and b_0. If $a_n > 1$ and $b_m > 1$, prove that $m = n$ and $a_i = b_i$ for $i = 0, 1, 2, \ldots, n$.

Solution. If r is an integer, then $n = 0$; otherwise, $a_0 < r = \langle a_0, a_1, \ldots, a_n \rangle = a_0 + 1/\langle a_1, a_2, \ldots, a_n \rangle < a_0 + 1$, which is impossible. Similarly, we have $m = 0$, and hence $a_0 = b_0 = r$. Now suppose that r is not an integer. Then $n \geq 1$ and $m \geq 1$, and we have $r = a_0 + 1/\langle a_1, a_2, \ldots, a_n \rangle = b_0 + 1/\langle b_1, b_2, \ldots, b_m \rangle$. Clearly, $1/\langle a_1, a_2, \ldots, a_n \rangle > 0$ and $1/\langle b_1, b_2, \ldots, b_m \rangle > 0$. Since $a_n > 1$, we have $a_1 > 1$ or $n \geq 2$. In either case, $1/\langle a_1, a_2, \ldots, a_n \rangle < 1$; similarly, $1/\langle b_1, b_2, \ldots, b_m \rangle < 1$. Thus $a_0 < r < a_0 + 1$ and $b_0 < r < b_0 + 1$, and therefore $a_0 = b_0 = [r]$. Hence $\langle a_1, a_2, \ldots, a_n \rangle = \langle b_1, b_2, \ldots, b_m \rangle$. Without loss of generality, suppose that $n \geq m$. Applying the above argument repeatedly, we eventually get $a_1 = b_1, a_2 = b_2, \ldots, a_{m-1} = b_{m-1}$, and $\langle a_m, a_{m+1}, \ldots, a_n \rangle = \langle b_m \rangle = b_m$. Since b_m is an integer, it follows from the first part of the proof that $n = m$ and $a_m = b_m$.

9-6. Prove (9.4): Let $a_0, a_1, a_2, \ldots$ be a sequence of integers, all positive except possibly a_0. Prove that for any positive real number x and any $k \geq 1$,

$$\langle a_0, a_1, \ldots, a_{k-1}, x \rangle = (xp_{k-1} + p_{k-2})/(xq_{k-1} + q_{k-2}).$$

In particular, $\langle a_0, a_1, a_2, \ldots, a_k \rangle = p_k/q_k$. (Hint. Use induction.)

Solution. For $k = 1$, we have $(xp_0 + p_{-1})/(xq_0 + q_{-1}) = (x \cdot a_0 + 1)/(x \cdot 1 + 0) = a_0 + 1/x = \langle a_0, x \rangle$. Now suppose the result holds for $\langle a_0, a_1, \ldots, a_{k-1}, x \rangle$. Then

$$\langle a_0, a_1, \ldots, a_k, x \rangle = \langle a_0, a_1, \ldots, a_{k-1}, a_k + 1/x \rangle$$
$$= ((a_k + 1/x)p_{k-1} + p_{k-2})/((a_k + 1/x)q_{k-1} + q_{k-2})$$
$$= (x(a_k p_{k-1} + p_{k-2}) + p_{k-1})/(x(a_k q_{k-1} + q_{k-2}) + q_{k-1})$$
$$= (xp_k + p_{k-1})/(xq_k + q_{k-1}),$$

using Definition 9.3.

9-7. Let q_k be defined as in (9.3). Show by induction that $q_k \geq 2^{k/2}$ if $k \geq 2$.

Solution. It is easy to see that the result holds for $k = 2$ and $k = 3$. Suppose now that we know that $q_n \geq 2^{n/2}$; we show that $q_{n+2} \geq 2^{(n+2)/2}$. Obviously, $q_{n+1} > q_n$ and thus $q_{n+2} = a_{n+2}q_{n+1} + q_n > 2q_n$. Since $q_n \geq 2^{n/2}$ by assumption, it follows that $q_{n+2} > 2 \cdot 2^{n/2} = 2^{(n+2)/2}$.

9-8. Let p_k/q_k be the kth convergent of $\langle a_0, a_1, a_2, \ldots, a_n \rangle$. If $a_0 \geq 1$, prove that $p_k/p_{k-1} = \langle a_k, a_{k-1}, \ldots, a_1, a_0 \rangle$ and $q_k/q_{k-1} = \langle a_k, a_{k-1}, \ldots, a_2, a_1 \rangle$. (Hint. Use (9.3) to show that $p_k/p_{k-1} = a_k + 1/(p_{k-1}/p_{k-2})$.)

Solution. It is easily verified that $p_k/p_{k-1} = a_k + 1/(p_{k-1}/p_{k-2})$. Since $p_{k-2} < p_{k-1}$ (see (9.3)), it follows that the first partial quotient of p_k/p_{k-1} is equal to $[p_k/p_{k-1}] = a_k$. Thus $p_k/p_{k-1} = \langle a_k, p_{k-1}/p_{k-2} \rangle$. Repeating the argument shows that the second partial quotient is a_{k-1}, the third is a_{k-2}, and so on. In the kth step, we have $p_k/p_{k-1} = \langle a_k, a_{k-1}, a_{k-2}, \ldots, a_1, p_0/p_{-1} \rangle = \langle a_k, a_{k-1}, \ldots, a_1, a_0 \rangle$, since $p_0 = a_0$ and $p_{-1} = 1$. For q_k/q_{k-1}, use the fact that $q_k = a_k q_{k-1} + q_{k-2}$ and argue as above. In the $(k-1)$th step, we then have $q_k/q_{k-1} = \langle a_k, a_{k-1}, \ldots, a_2, q_1/q_0 \rangle = \langle a_k, a_{k-1}, \ldots, a_2, a_1 \rangle$, since $q_1 = a_1$ and $q_0 = 1$ (see (9.3)).

▷ **9-9.** Let a/b be a rational number in lowest terms, and suppose $a/b = \langle a_0, a_1, \ldots, a_n \rangle$, where each a_i is a positive integer. Prove that if n is odd, the continued fraction is symmetric (that is, $a_0 = a_n$, $a_1 = a_{n-1}$, ...) if and only if $a \mid b^2 + 1$, while if n is even, the continued fraction is symmetric if and only if $a \mid b^2 - 1$. (Hint. Use the preceding problem.)

Solution. If $n = 0$, the result is obvious, since then a/b is an integer. Thus we may suppose that $n \geq 1$. If the continued fraction is symmetric, it follows from Problem 9-8 and (9.4) that $p_n/p_{n-1} = \langle a_0, a_1, \ldots, a_n \rangle = p_n/q_n = a/b$. Since $(p_n, q_n) = 1$, we have $p_{n-1} = q_n = b$ and $p_n = a$; thus by (9.7), $p_n q_{n-1} - p_{n-1} q_n = a q_{n-1} - b^2 = (-1)^{n-1}$ and hence $a \mid b^2 + (-1)^{n-1}$. Now suppose that $a \mid b^2 + (-1)^{n-1}$. Then, by (9.7), $(-1)^{n-1} = p_n q_{n-1} - p_{n-1} q_n = a q_{n-1} - b p_{n-1}$, and thus a divides $b^2 + (-1)^{n-1} - (b p_{n-1} + (-1)^{n-1})) = b(b - p_{n-1})$. Because $(a, b) = 1$, we must have $a \mid b - p_{n-1}$. Note that $n \geq 1$ implies that $a/b > 1$; thus $a > b > b - p_{n-1}$ (since $p_i > 0$ if $i \geq 0$). Also, since $p_{n-1} < p_n$ and $p_n = a$, we have $p_{n-1} < a < a + b$, and so $b - p_{n-1} > -a$. Thus $-a < b - p_{n-1} < a$, and since $a \mid b - p_{n-1}$, we must have $b - p_{n-1} = 0$, i.e., $b = p_{n-1}$. Hence $\langle a_0, a_1, \ldots, a_n \rangle = a/b = p_n/p_{n-1} = \langle a_n, a_{n-1}, \ldots, a_1, a_0 \rangle$, by Problem 9-8.

▷ **9-10.** Let $n > 1$, and suppose that $s^2 \equiv -1 \pmod{n}$. Let p_k/q_k be the convergent of the continued fraction expansion of s/n such that $q_k < \sqrt{n} \leq q_{k+1}$. Show that $(sq_k - np_k)^2 + q_k^2 = n$.

Solution. Since $s^2 \equiv -1 \pmod{n}$, it is easy to verify that $(sq_k - np_k)^2 + q_k^2 \equiv 0 \pmod{n}$. But $|s/n - p_k/q_k| \leq 1/q_k q_{k+1}$ by (9.33), and hence $|sq_k - np_k| \leq n/q_{k+1}$. Since $q_{k+1} \geq \sqrt{n}$, it follows that $|sq_k - np_k| \leq \sqrt{n}$. By the choice of q_k, we have $q_k < \sqrt{n}$. Hence $(sq_k - np_k)^2 + q_k^2 < 2n$. Since $q_k \geq 1$, it follows that $(sq_k - np_k)^2 + q_k^2$ is a positive multiple of n which is less than $2n$ and therefore is equal to n.

Note. It is not hard to verify that the above representation of n as a sum of two squares is *primitive*, and so we have another proof for the main part of Theorem 8.14. Thus once we have a solution s of the congruence $x^2 \equiv -1 \pmod{n}$, the continued fraction procedure gives a very efficient way of constructing a primitive representation of n as a sum of two squares.

The Equation $ax + by = c$

9-11. *Use the method of continued fractions to find a solution of* $377x - 120y = -3$. *(See Problem 9-3.)*

Solution. To apply (9.10), we need $b > 0$; hence we could write the equation as $-377x + 120y = 3$ and proceed as in (9.11). Or we could simply find solutions of the equation $377x + 120y = -3$ and replace y by $-y$. By the Euclidean Algorithm, $377/120 = \langle 3, 7, 17 \rangle$. Set up the following table:

k	-1	0	1	2
a_k		3	7	17
p_k	1	3	22	377
q_k	0	1	7	120

Now apply (9.10). Since $n = 2$ is even, $377x + 120y = -3$ has the solution $x = -cq_{n-1} = 3 \cdot 7 = 21$, $y = cp_{n-1} = -3 \cdot 22 = -66$. Thus $x = 21$, $y = 66$ is a solution of $377x - 120y = -3$.

9-12. *Use continued fractions to express the greatest common divisor of 377 and 120 as a linear combination of these numbers. (Hint. Use (9.10) and the preceding problem.)*

Solution. Check that $(377, 120) = 1$. By Problem 9-11, $377/120 = \langle 3, 7, 17 \rangle$. Therefore (9.10) implies that $377x + 120y = 1$ has the solution $x = -q_1, y = p_1$, and thus from the table in the preceding problem, $x = -7$ and $y = 22$.

9-13. *Use continued fractions to find a solution of each of the following equations: (a) $98x + 263y = 1$; (b) $1255x + 177y = -1$.*

Solution. (a) $263/98 = \langle 2, 1, 2, 6, 5 \rangle$. Set up the following table:

k	-1	0	1	2	3	4
a_k		2	1	2	6	5
p_k	1	2	3	8	51	263
q_k	0	1	1	3	19	98

Applying (9.10) gives the solution $x = 51$, $y = -19$.
(b) $1255/177 = \langle 7, 11, 16 \rangle$. Construct the following table:

k	-1	0	1	2
a_k		7	11	16
p_k	1	7	78	1255
q_k	0	1	11	177

Then (9.10) yields the solution $x = 11$, $y = -78$.

Infinite Continued Fractions

9-14. *A certain number α has a continued fraction expansion whose first three partial quotients are 1, 2, and 3, and α has at least one more partial quotient* (the continued fraction could be finite or infinite). What can we say about the size of α?

Solution. We have $\alpha = \langle 1, 2, 3, \beta \rangle$, where $\beta \geq 1$. Evaluating the finite continued fraction, we obtain $\alpha = (10\beta + 3)/(7\beta + 2) = 10/7 + 1/(49\beta + 14)$. Thus $\alpha > 10/7$ and $\alpha \leq 10/7 + 1/63 = 13/9$.

9-15. Suppose $\alpha = \langle a_0, a_1, \ldots, a_{n-1}, a_n, \ldots \rangle$ and $\beta = \langle b_0, b_1, \ldots, b_{n-1}, b_n, \ldots \rangle$ are finite or infinite continued fractions that agree in their first n partial quotients but not in the next; thus $a_i = b_i$ if $i \leq n - 1$ and $a_n \neq b_n$. Prove that
(a) if n is even, then $\alpha > \beta$ if and only if $a_n > b_n$;
(b) if n is odd, then $\alpha > \beta$ if and only if $b_n > a_n$.

Solution. As usual, let $\alpha_k = \langle a_k, a_{k+1}, \ldots \rangle$ and $\beta_k = \langle b_k, b_{k+1}, \ldots \rangle$. Then $\alpha > \beta$ if and only if $a_0 + 1/\alpha_1 > b_0 + 1/\beta_1$ if and only if $\beta_1 > \alpha_1$. Similarly, $\beta_1 > \alpha_1$ if and only if $a_1 + 1/\alpha_2 > b_1 + 1/\beta_2$ if and only if $\alpha_2 > \beta_2$. Proceeding this way, we show that, if n is even, $\alpha > \beta$ if and only if $\alpha_n > \beta_n$ if and only if $a_n + 1/\alpha_{n+1} > b_n + 1/\beta_{n+1}$. Since $1/\alpha_{n+1} < 1$ and $1/\beta_{n+1} < 1$, the last inequality holds if and only if $a_n > b_n$. If n is odd, we have $\alpha > \beta$ if and only if $\beta_n > \alpha_n$ if and only if $b_n > a_n$.

9-16. Suppose $\alpha = \langle a_0, a_1, \ldots, a_{n-1} \rangle$ and $\beta = \langle a_0, a_1, \ldots, a_{n-1}, b_n, \ldots \rangle$. Use the preceding problem to prove that $\alpha > \beta$ if and only if n is even.

Solution. If $a_{n-1} = 1$, write $\alpha = \langle a_0, a_1, \ldots, a_{n-2}, 1 \rangle = \langle a_0, a_1, \ldots, a_{n-3}, a_{n-2} + 1 \rangle$ and apply the preceding problem to the second expansion and β. Since these expansions agree in their first $n - 2$ partial quotients and $a_{n-2} + 1 > a_{n-2}$, it follows that $\alpha > \beta$ if and only if $n - 2$ is even and hence if and only if n is even. If $a_{n-1} > 1$, then α also has the expansion $\langle a_0, a_1, \ldots, a_{n-1} - 1, 1 \rangle$. Then α and β agree in their first $n - 1$ partial quotients. By Problem 9-15, since $a_{n-1} - 1 < a_{n-1}$, $\alpha > \beta$ if and only if $n - 1$ is odd, that is, if and only if n is even.

9-17. In each case, determine if the value of the first continued fraction is greater than the value of the second:
(a) $\langle 1, 2, 3, 4, 5 \rangle$, $\langle 1, 2, 5, 4, 3 \rangle$;
(b) $\langle 1, 2, 3, 4, 5 \rangle$, $\langle 1, 2, 3, 5, 4, \ldots \rangle$;
(c) $\langle 1, 2, 3, 4, 5, 6, \ldots \rangle$, $\langle 1, 2, 3, 4 \rangle$;
(d) $\langle 292, 1, 15, 7 \rangle$, $\langle 292, 1, 1, \ldots \rangle$;
(e) $\langle 292, 1, 15, 7, \ldots \rangle$, $\langle 292, 1, 15 \rangle$.

Solution. Let α and β denote the first and second continued fractions, respectively. If we apply Problems 9-15 and 9-16, it follows that $\alpha > \beta$ in parts (b), (d), and (e) and $\beta > \alpha$ in parts (a) and (c).

9-18. Suppose that x is a real number greater than 1 and that $x = \langle a_0, a_1, a_2, \ldots \rangle$. (This could be a finite expansion.) Prove that $1/x = \langle 0, a_0, a_1, a_2, \ldots \rangle$.

Solution. Apply (9.14) to conclude that $\langle 0, a_0, a_1, a_2, \ldots \rangle = 0 + 1/\langle a_0, a_1, \ldots \rangle = 1/x$.

9-19. *Use (9.17) to find the first nine partial quotients of the continued fraction expansion of π.*

Solution. Applying the algorithm described in (9.17), we have $a_0 = [\pi] = 3$. Then $\alpha_1 = 1/(\alpha_0 - a_0) = 1/(\pi - 3)$, and so $a_1 = [\alpha_1] = 7$. Proceeding this way, we get $\pi = \langle 3, 7, 15, 1, 292, 1, 1, 1, 2, \ldots \rangle$.

Note. The algorithm is easy to implement on a calculator. Subtract the integral part of π (namely, 3), then invert; subtract the integral part of the resulting number, then invert. Repeating the process gives successive partial quotients of π. But be aware that after a certain number of steps (depending on the accuracy of the calculator), the partial quotients computed will not be correct. Most calculators, however, should give the first nine partial quotients of π correctly.

▷ **9-20.** *(a) Let α be irrational, and suppose that $a/b < \alpha < c/d$, where a, b, c, and d are positive integers such that $bc - ad = 1$. Prove that at least one of a/b and c/d is a convergent of α.*

(b) Give an example where precisely one of a/b and c/d is a convergent, and show that it is possible for both a/b and c/d to be convergents. (Hint. Assume that the inequality in (9.39) fails for both a/b and c/d.)

Solution. (a) Note that we cannot have $|\alpha - a/b| = 1/2b^2$, since α is irrational. Thus if both $|\alpha - a/b| < 1/2b^2$ and $|\alpha - c/d| < 1/2d^2$ fail, then

$$1/bd = (bc - ad)/bd = c/d - a/b = |\alpha - a/b| + |\alpha - c/d| > 1/2b^2 + 1/2d^2.$$

Multiplying by $2b^2d^2$, we obtain $2bd > b^2 + d^2$, i.e., $(b - d)^2 < 0$, which is impossible. Thus either $|\alpha - a/b| < 1/2b^2$ or $|\alpha - c/d| < 1/2d^2$, and so, by (9.39), at least one of a/b and c/d is a convergent of α.

(b) To show that both a/b and c/d can be convergents, let α be any irrational, and let k be odd. Then by (9.12), $p_{k-1}/q_{k-1} < \alpha < p_k/q_k$, and $p_k q_{k-1} - p_{k-1} q_k = 1$, by (9.7). If $\alpha = \sqrt{6}$, then $7/3 < \sqrt{6} < 5/2$ and $3 \cdot 5 - 7 \cdot 2 = 1$; however, only $5/2$ is a convergent of $\sqrt{6}$. (The number $7/3$ is a *secondary convergent* of $\sqrt{6}$; see the last section of problems for the properties of secondary convergents.)

9-21. *Suppose α is irrational and $\alpha > 1$. Let p_k/q_k and P_k/Q_k denote, respectively, the kth convergent of α and $1/\alpha$. Prove that $P_k/Q_k = q_{k-1}/p_{k-1}$. (Hint. Use Problem 9-18.)*

Solution. By Problem 9-18, $1/\alpha = \langle 0, a_0, a_1, a_2, \ldots \rangle$. Hence the kth convergent of $1/\alpha$ is $\langle 0, a_0, a_1, \ldots, a_{k-1} \rangle = 1/\langle a_0, a_1, \ldots, a_{k-1} \rangle = q_{k-1}/p_{k-1}$, by (9.4).

Periodic Continued Fractions

9-22. *Set up a table, and use (9.24) to find the continued fraction expansion of (a) $\sqrt{47}$; (b) $(4 + \sqrt{2})/3$; (c) $(7 - \sqrt{11})/3$.*

Solution. (a) Note that $\sqrt{47} = (0 + \sqrt{47})/1$ and $1 | 47 - 0^2$. From (9.24),

k	0	1	2	3	4	5
r_k	0	6	5	5	6	6
s_k	1	11	2	11	1	11
a_k	6	1	5	1	12	

Since $r_5 = r_1$ and $s_5 = s_1$, the entries in the table repeat, beginning with $k = 5$. Thus $\sqrt{47} = \langle 6, \overline{1,5,1,12} \rangle$.

(b) Since 3 does not divide $2 - 4^2$, multiply numerator and denominator by $|-3|$ to get $(12 + \sqrt{18})/9$ (see the proof of (9.23)); here, $d = 18$, $r_0 = 12$, $s_0 = 9$. Construct the following table:

k	0	1	2	3	4
r_k	12	-3	4	4	4
s_k	9	1	2	1	2
a_k	1	1	4	8	

Since the column entries for $k = 4$ coincide with the column entries for $k = 2$, we have $(4 + \sqrt{2})/3 = \langle 1, 1, \overline{4, 8} \rangle$.

(c) Write $(7 - \sqrt{11})/3$ as $(-7 + \sqrt{11})/(-3)$; since -3 does not divide $11 - 7^2$, multiply numerator and denominator by 3 to get $(-21 + \sqrt{99})/(-9)$; then $d = 99$, $r_0 = -21$, $s_0 = -9$. Compute the following table:

k	0	1	2	3	4	5	6
r_k	-21	12	8	6	3	7	8
s_k	-9	5	7	9	10	5	7
a_k	1	4	2	1	1	3	

Since $r_6 = r_2$ and $s_6 = s_2$, we have $(7 - \sqrt{11})/3 = \langle 1, 4, \overline{2, 1, 1, 3} \rangle$.

9-23. (a) Set up a table as in (9.25) to find the continued fraction expansion of $\sqrt{130}$.
(b) Use this table to determine the continued fraction expansion of $(7 + \sqrt{130})/9$. (*Hint.* Use (9.24) and (9.19).)

Solution. (a) Set up the following table:

k	0	1	2	3	4
r_k	0	11	7	11	11
s_k	1	9	9	1	9
a_k	11	2	2	22	

Thus $\sqrt{130} = \langle 11, \overline{2,2,22} \rangle$.

(b) According to (9.24.iii), $(7+\sqrt{130})/9 = (r_2+\sqrt{130})/s_2 = \alpha_2$, and $\alpha_2 = \langle a_2, a_3, \ldots \rangle$, by (9.19). Thus $(7 + \sqrt{130})/9 = \langle 2, 22, 2, 2, 22, 2, \ldots \rangle = \langle \overline{2, 22, 2} \rangle$.

294　　CHAPTER 9: CONTINUED FRACTIONS

9-24. Prove that $\sqrt{n^2+1} = \langle n, \overline{2n} \rangle$ and $\sqrt{n^2+2} = \langle n, \overline{n, 2n} \rangle$ for $n \geq 1$.

Solution. Use (9.24) to construct the following tables for $\sqrt{n^2+1}$ and $\sqrt{n^2+2}$, respectively:

k	0	1	2
r_k	0	n	n
s_k	1	1	1
a_k	n	$2n$	

k	0	1	2	3
r_k	0	n	n	n
s_k	1	2	1	2
a_k	n	n	$2n$	

In the first table, $r_2 = r_1$ and $s_2 = s_1$, so the partial quotients repeat, beginning with $a_1 = 2n$; hence $\sqrt{n^2+1} = \langle n, \overline{2n} \rangle$. In the second table, $r_3 = r_1$ and $s_3 = s_1$, and thus the partial quotients repeat, beginning with $a_1 = n$; therefore $\sqrt{n^2+2} = \langle n, \overline{n, 2n} \rangle$.

9-25. Prove that (a) $\sqrt{n^2-1} = \langle n-1, \overline{1, 2n-2} \rangle$ for $n \geq 2$;
(b) $\sqrt{n^2-n} = \langle n-1, \overline{2, 2n-2} \rangle$ for $n \geq 2$;
(c) $\sqrt{n^2-2} = \langle n-1, \overline{1, n-2, 1, 2n-2} \rangle$ for $n \geq 3$.

Solution. Proceed as in Problem 9-24, using the following tables for $\sqrt{n^2-1}$, $\sqrt{n^2-n}$, and $\sqrt{n^2-2}$, respectively:

k	0	1	2	3
r_k	0	$n-1$	$n-1$	$n-1$
s_k	1	$2n-2$	1	$2n-2$
a_k	$n-1$	1	$2n-2$	

k	0	1	2	3
r_k	0	$n-1$	$n-1$	$n-1$
s_k	1	$n-1$	1	$n-1$
a_k	$n-1$	2	$2n-2$	

k	0	1	2	3	4	5
r_k	0	$n-1$	$n-2$	$n-2$	$n-1$	$n-1$
s_k	1	$2n-3$	2	$2n-3$	1	$2n-3$
a_k	$n-1$	1	$n-2$	1	$2n-2$	

9-26. Find the continued fraction expansion of $\sqrt{n^2+1}+n^2$, where $n \geq 1$. (Hint. Use Problem 9-24.)

Solution. By Problem 9-24, we have $\sqrt{n^2+1}+n^2 = n^2 + \langle n, \overline{2n} \rangle = n^2+n+1/\langle \overline{2n} \rangle = \langle n^2+n, \overline{2n} \rangle$.

Another proof: Use (9.18). Let $\alpha_0 = \sqrt{n^2+1}+n^2$; then $a_0 = [\alpha_0] = n+n^2$. Set $\alpha_1 = 1/(\alpha_0 - a_0) = 1/(\sqrt{n^2+1}-n) = \sqrt{n^2+1}+n$; then $a_1 = [\alpha_1] = 2n$. Now note that $\alpha_2 = 1/(\alpha_1 - a_1) = 1/(\sqrt{n^2+1}-n) = \alpha_1$, and so $a_2 = a_1$. It is then clear that $a_k = a_1$ for every $k \geq 2$, and hence $\sqrt{n^2+1}+n^2 = \langle n^2+n, \overline{2n} \rangle$.

9-27. Let d be a positive integer that is not a perfect square, and suppose a, b, r, and s are integers, with b and s not zero. If $(a+\sqrt{d})/b = (r+\sqrt{d})/s$, prove that $a = r$ and $b = s$.

Solution. The above equation implies that $(s-b)\sqrt{d} = br - as$. If $b \neq s$, then $\sqrt{d} = (br-as)/(s-b)$, contradicting the fact that $\sqrt{d}$ is irrational. Thus $b = s$, and therefore $a = r$.

9-28. *In Theorem 9.24, s_{k+1} was defined by the formula $s_{k+1} = (d - r_{k+1}^2)/s_k$. Show that $s_{k+1} = s_{k-1} + a_k(r_k - r_{k+1})$.*

Solution. We have $s_k s_{k+1} = d - r_{k+1}^2$ and $s_{k-1} s_k = d - r_k^2$. Subtracting, we obtain

$$s_k(s_{k+1} - s_{k-1}) = r_k^2 - r_{k+1}^2 = (r_k - r_{k+1})(r_k + r_{k+1}).$$

But $r_k + r_{k+1} = a_k s_k$ by the definition of r_{k+1}. Substituting and dividing by s_k, we obtain $s_{k+1} - s_{k-1} = a_k(r_k - r_{k+1})$, which yields the desired formula.

Note. When d is very large, the expression for s_{k+1} given in this problem can usually be evaluated faster than the expression $(d - r_{k+1}^2)/s_k$, since division is ordinarily significantly slower than multiplication, which in turn is much slower than addition. The expression $(d - r_{k+1}^2)/s_k$ is more convenient when we are working with a calculator.

9-29. *Let $\alpha = u_a + v_a\sqrt{d}$ and $\beta = u_b + v_b\sqrt{d}$, where u_a, v_a, u_b, and v_b are rational numbers. Show that $(\alpha + \beta)' = \alpha' + \beta'$, $(\alpha\beta)' = \alpha'\beta'$, and, if $\beta \neq 0$, $(\alpha/\beta)' = \alpha'/\beta'$. (See (9.26) for a definition of the conjugate α' of α.)*

Solution. We have $(\alpha+\beta)' = (u_a + u_b + (v_a + v_b)\sqrt{d})' = (u_a - v_a\sqrt{d}) + (u_b - v_b\sqrt{d}) = \alpha' + \beta'$. Similarly, $(\alpha\beta)' = u_a u_b + dv_a v_b - (u_a v_b + u_b v_a)\sqrt{d} = \alpha'\beta'$.

For division, we need only verify that $(1/\beta)' = 1/\beta'$. But $1/\beta = \beta'/(\beta\beta') = (u_b - v_b\sqrt{d})/(u_b^2 - dv_b^2)$. Taking the conjugate, we obtain $(u_b + v_b\sqrt{d})/(u_b^2 - dv_b^2)$. It is easy to check that this is $1/\beta'$.

9-30. *Prove the converse of Lagrange's Theorem (9.28): The value of every periodic continued fraction is a quadratic irrational.*

Solution. Let $\alpha = \langle c_0, \ldots, c_n, \overline{a_0, \ldots, a_{m-1}} \rangle$ and let $\beta = \langle \overline{a_0, \ldots, a_{m-1}} \rangle$. Note that $\alpha = \langle c_0, \ldots, c_n, \beta \rangle$; hence by (9.4), $\alpha = (\beta p_n + p_{n-1})/(\beta q_n + q_{n-1})$, where p_k/q_k denote the kth convergent of $\langle c_0, \ldots, c_n \rangle$. We can solve for β in terms of α, obtaining

$$\beta = -(\alpha q_{n-1} - p_{n-1})/(\alpha q_n - p_n). \tag{1}$$

But by (9.4), we have

$$\beta = \langle a_0, \ldots, a_{m-1}, \beta \rangle = (\beta p'_{m-1} + p'_{m-2})/(\beta q'_{m-1} + q'_{m-2}), \tag{2}$$

where p'_k/q'_k are the convergents of $\langle a_0, a_1, \ldots, a_{m-1} \rangle$. Now use (1) to substitute for β in (2). After some simplification, we find that α satisfies a quadratic equation with integer coefficients. Since α is irrational by (9.16), it follows that α is a quadratic irrational.

Purely Periodic Continued Fractions

9-31. *Let* $\alpha = \langle \overline{1,2,3} \rangle$. *Prove that* $\alpha = (4+\sqrt{37})/7$. *Using (9.24), set up a table to check your answer.*

Solution. By (9.14), we have $\alpha = 1+1/(2+1/(3+1/\alpha)) = (10\alpha+3)/(7\alpha+2)$. This gives the equation $7\alpha^2 - 8\alpha - 3 = 0$, and hence $\alpha = (4+\sqrt{37})/7$, since α is clearly positive. (Note that $\alpha > 1$ and its conjugate $\alpha' = (4-\sqrt{37})/7$ satisfies $-1 < \alpha' < 0$; thus α is reduced, and so its continued fraction expansion should be purely periodic, by (9.31).)

We can check our answer by computing the following table for $(4+\sqrt{37})/7$ (there is no need to rewrite this expression, since $7 | 37 - 4^2$):

k	0	1	2	3
r_k	4	3	5	4
s_k	7	4	3	7
a_k	1	2	3	

Since the entries for $k=3$ are the same as for $k=0$, we conclude that $(4+\sqrt{37})/7 = \langle \overline{1,2,3} \rangle$.

9-32. *Evaluate* (a) $\langle 1,1,1,\ldots \rangle$; (b) $\langle 2,3,1,1,1,\ldots \rangle$; (c) $\langle \overline{1,2} \rangle$; (d) $\langle 1,3,\overline{1,2} \rangle$.

Solution. Denote these four continued fractions by α, β, γ, and δ, respectively. (a) By (9.14), we have $\alpha = 1+1/\alpha$, and hence $\alpha^2 - \alpha - 1 = 0$. Since α is clearly positive, we use the positive root to conclude that $a = (1+\sqrt{5})/2$.

(b) Using (9.14), we have $\beta = 2+1/(3+1/\alpha) = (7\alpha+2)/(3\alpha+1)$, where $\alpha = (1+\sqrt{5})/2$. Thus $\beta = (25-\sqrt{5})/10$.

(c) Since $\gamma = 1+1/(2+1/\gamma) = (3\gamma+1)/(2\gamma+1)$, we obtain the equation $2\gamma^2 - 2\gamma - 1 = 0$, and hence $\gamma = (1+\sqrt{3})/2$.

(d) We have $\delta = 1+1/(3+1/\gamma) = (4\gamma+1)/(3\gamma+1) = 3-\sqrt{3}$.

9-33. *Suppose that a and b are positive integers and $a \mid b$. If $b = ac$, prove that* $\langle \overline{b,a} \rangle = (b+\sqrt{b^2+4c})/2$.

Solution. Let $\alpha = \langle \overline{b,a} \rangle$; then (9.14) implies that $\alpha = b+1/(a+1/\alpha) = ((ab+1)\alpha + b)/(a\alpha+1)$. This leads to the quadratic equation $a\alpha^2 - ab\alpha - b = 0$. Clearly, α is positive, and hence we use the positive root of this equation. Thus $\alpha = (ab+\sqrt{a^2b^2+4ab})/2a = (ab+\sqrt{a^2b^2+4a^2c})/2a = (b+\sqrt{b^2+4c})/2$.

9-34. *Find the value of the purely periodic continued fraction* $\langle n,n,n,\ldots \rangle$.

Solution. Let $\alpha = \langle n,n,n,\ldots \rangle$. Then (9.14) implies that $\alpha = n+1/\langle n,n,n,\ldots \rangle = n+1/\alpha$. This leads to the quadratic equation $\alpha^2 - n\alpha - 1 = 0$, whose roots are $(n \pm \sqrt{n^2+4})/2$. Since α is positive, we have $\alpha = (n+\sqrt{n^2+4})/2$.

Another proof: Use the preceding problem, with $a = b = n$ and $c = 1$.

9-35. *Determine which of the following numbers have a purely periodic continued fraction expansion:* (a) $(1 + \sqrt{3})/2$; (b) $(3 + \sqrt{13})/2$; (c) $5 + \sqrt{32}$; (d) $(7 - \sqrt{11})/5$.

Solution. In view of (9.31), we must decide in each case if the given quadratic irrational is reduced, that is, if $\alpha = (r + \sqrt{d})/s$ and its conjugate $\alpha' = (r - \sqrt{d})/s$ satisfy $\alpha > 1$ and $-1 < \alpha' < 0$. It is easily checked that both conditions hold for the first three numbers, so each of them is reduced, and hence each has a purely periodic expansion. Since $(7 - \sqrt{11})/5 < 1$, the last number is not reduced and therefore does not have a purely periodic expansion.

9-36. (a) *Let* $\alpha = (2 + \sqrt{10})/2$. *Using* (9.24), *set up a table to compute* r_k, s_k, *and* a_k *for* $0 \le k \le 4$. *Find the continued fraction expansion of* α.

(b) *Using only part* (a), *express* $\langle \overline{1,1,2} \rangle$ *and* $\langle \overline{1,2,1} \rangle$ *as* $(r + \sqrt{d})/s$, *where* $s \mid d - r^2$.

Solution. (a) If $\alpha = (r + \sqrt{d})/s$, then $d = 10$ and $r = s = 2$; thus $s \mid d - r^2$. (We therefore do not need to modify α by multiplying the numerator and denominator by $|s|$.) Using the formulas in (9.24), we obtain

k	0	1	2	3	4
r_k	2	2	1	2	2
s_k	2	3	3	2	3
a_k	2	1	1	2	

(Note that the table values repeat, beginning with $k = 3$.) Hence $\alpha = \langle \overline{2,1,1} \rangle$.

(b) According to (9.19) and (9.24.iii), $\alpha_k = \langle a_k, a_{k+1}, \ldots \rangle = (r_k + \sqrt{d})/s_k$. Thus $\langle \overline{1,1,2} \rangle = \langle a_1, a_2, a_3, \ldots \rangle = \alpha_1 = (r_1 + \sqrt{d})/s_1 = (2 + \sqrt{10})/3$. Similarly, $\langle \overline{1,2,1} \rangle = \langle a_2, a_3, a_4, \ldots \rangle = \alpha_2 = (r_2 + \sqrt{d})/s_2 = (1 + \sqrt{10})/3$.

9-37. *By definition, a quadratic irrational α is reduced if $\alpha > 1$ and $-1 < \alpha' < 0$, where α' is the conjugate of α (see (9.30)). Is it possible for α to be less than 1 and still have $-1 < \alpha' < 0$?*

Solution. Yes. Consider, for example, $\alpha = (1 + \sqrt{5})/4$ or $\alpha = (3 + \sqrt{11})/7$. In each case, it is easy to check that $\alpha < 1$ and $-1 < \alpha' < 0$.

9-38. *Suppose $\alpha = (r + \sqrt{d})/s$ is a quadratic irrational. Prove that α is reduced if and only if $0 < r < \sqrt{d}$ and $\sqrt{d} - r < s < \sqrt{d} + r$. (In particular, the last expression implies that $0 < s < 2\sqrt{d}$.)*

Solution. First suppose that α is reduced; thus $\alpha > 1$ and $-1 < \alpha' < 0$, where $\alpha' = (r - \sqrt{d})/s$. Since $1 < \alpha - \alpha' = 2\sqrt{d}/s$, we have $s > 0$ and $s < 2\sqrt{d}$. Also, $0 < \alpha + \alpha' = 2r/s$ implies that $r > 0$, and $\alpha' < 0$ yields $r < \sqrt{d}$. From $\alpha > 1$, it follows at once that $s < r + \sqrt{d}$, and since $\alpha' > -1$, we have $(r - \sqrt{d})/s > -1$, i.e., $\sqrt{d} - r < s$.

Now suppose that $0 < r < \sqrt{d}$ and $\sqrt{d} - r < s < \sqrt{d} + r$. Then $s > 0$ and hence $\alpha = (r + \sqrt{d})/s > 1$. Also, $r < \sqrt{d}$ implies that $\alpha' < 0$. Since $\sqrt{d} - r < s$, it follows that $-s < r - \sqrt{d}$ and so $-1 < (r - \sqrt{d})/s = \alpha'$. Thus α is reduced.

Rational Approximations to Irrational Numbers

9-39. *Find the best rational approximation a/b to $\sqrt{2}$ with denominator not exceeding 70. Without calculating $\sqrt{2} - a/b$, explain why $|\sqrt{2} - a/b| < .000085$.*

Solution. We will use (9.37). First check that $\sqrt{2} = \langle 1, \overline{2} \rangle$. Now set up the following table to determine the first five convergents of $\sqrt{2}$:

k	-1	0	1	2	3	4	5	6
a_k			1	2	2	2	2	2
p_k	1	1	3	7	17	41	99	239
q_k	0	1	2	5	12	29	70	169

It then follows from (9.37) that $99/70$ is the best rational approximation to $\sqrt{2}$ among all fractions whose denominators do not exceed 70. Also, by (9.33), we have $|\sqrt{2} - 99/70| < 1/(70 \cdot 169) = .00008453 \cdots < .000085$.

9-40. *Assume that $3 - \sqrt{3} = \langle 1, 3, \overline{1, 2} \rangle$.*
(a) *Calculate r_k and s_k for $0 \leq k \leq 4$.*
(b) *Using only part (a), express $\langle \overline{1, 2} \rangle$ as a quadratic irrational.*
(c) *Use a result on convergents to decide if $3 - \sqrt{3} < 52/41$.*
(d) *Using a result on approximations, determine if $|40(3 - \sqrt{3}) - 51| < |15(3 - \sqrt{3}) - 19|$.*

Solution. (a) To apply (9.24) to calculate r_k and s_k, we must first write $3 - \sqrt{3}$ in the form $(r_0 + \sqrt{d})/s_0$, where $s \mid d - r^2$. In this case, we have $3 - \sqrt{3} = (-3 + \sqrt{3})/(-1)$, so $r_0 = -3$ and $s_0 = -1$. Hence

k	0	1	2	3	4
r_k	-3	2	1	1	1
s_k	-1	1	2	1	2
a_k	1	3	1	2	

Thus $3 - \sqrt{3} = \langle 1, 3, \overline{1, 2} \rangle$.

(b) In view of (9.19) and (9.24), $\langle \overline{1, 2} \rangle = \langle a_2, a_3, \ldots \rangle = \alpha_2 = (r_2 + \sqrt{d})/s_2 = (1 + \sqrt{3})/2$.

(c) Set up the following table:

k	-1	0	1	2	3	4	5
a_k		1	3	1	2	1	2
p_k	1	1	4	5	14	19	52
q_k	0	1	3	4	11	15	41

In view of (9.12), we have $3 - \sqrt{3} < 52/41 = c_5$.

(d) The inequality does not hold, since otherwise (9.36) would imply that $40 \geq q_5$, i.e., $40 \geq 41$.

9-41. Assume that $\sqrt{18} = \langle 4, \overline{4,8} \rangle$. Use a result on approximations to find all rational numbers a/b in lowest terms, with $10 < b < 1000$, such that $|\sqrt{18} - a/b| < 1/5b^2$.

Solution. Set up the following table for $\sqrt{18}$:

k	-1	0	1	2	3	4
a_k		4	4	8	4	8
p_k	1	4	17	140	577	4756
q_k	0	1	4	33	136	1121

If $|\sqrt{18} - a/b| < 1/5b^2 < 1/2b^2$, (9.39) implies that a/b is a convergent of $\sqrt{18}$. Thus the only possible values for a/b, with $10 < b < 1000$, are $140/33$ and $577/136$. Check (on a calculator) that $a/b = 140/33$ does not satisfy the given inequality, but $a/b = 577/136$ does.

▷ **9-42.** Suppose that $|\sqrt{18} - r/s| < |\sqrt{18} - 140/33|$. Prove that $s \geq 48$. (Hint. First assume that $|\sqrt{18} - r/s| < 1/2s^2$.)

Solution. If $|\sqrt{18} - r/s| < 1/2s^2$, then r/s is a convergent, by (9.39). Since $s < 136$, we must have either $r/s = 140/33$, $17/4$, or $4/1$. (See the table in the preceding problem.) Clearly, $140/33$ is not possible, and the other values are impossible in view of (9.34). We therefore conclude that $1/2s^2 \leq |\sqrt{18} - r/s| < |\sqrt{18} - 140/33| < 1/(33 \cdot 136)$, using (9.33). It follows that $s > \sqrt{33 \cdot 136/2} = 47.37\ldots$, and hence $s \geq 48$.

9-43. Is $22/9$ one of the convergents of $\sqrt{6}$? Explain without computing convergents.

Solution. Yes; $|\sqrt{6} - 22/9| = .005045 \cdots < 1/(2 \cdot 9^2) = .006172\ldots$. Now apply (9.39) to conclude that $22/9$ must be a convergent of $\sqrt{6}$.

9-44. Find the best rational approximation to $(\sqrt{6} - 4)/2$ with denominator not exceeding 396.

Solution. Write $(\sqrt{6} - 4)/2$ in the form $(r_0 + \sqrt{d})/s_0$; thus $d = 6$, $r_0 = -4$, $s_0 = 2$. Since $s \mid d - r^2$, we can apply (9.24) to fill in the following tables:

k	0	1	2	3
r_k	-4	2	2	2
s_k	2	1	2	1
a_k	-1	4	2	

k	-1	0	1	2	3	4	5
a_k		-1	4	2	4	2	4
p_k	1	-1	-3	-7	-31	-69	-307
q_k	0	1	4	9	40	89	396

Now apply (9.37) to conclude that $-307/396$ is the best rational approximation to $(\sqrt{6} - 4)/2$ among all rationals with denominators less than or equal to 396.

9-45. Suppose that a and b are positive integers and $\sqrt{2} < a/b < 99/70$. Show that $b > 169$.

Solution. Using (9.33) and the table in Problem 9-39, we have $0 < |99/70 - a/b| < |\sqrt{2} - 99/70| < 1/(70 \cdot 169)$. Multiply by $70b$ to get $0 < |99b - 70a| < b/169$. Since $|99b - 70a|$ is clearly a positive integer, we must have $b/169 > 1$, that is, $b > 169$.

9-46. Let α be irrational, and suppose that a and b are integers, with b positive, such that $|\alpha - a/b| < |\alpha - p_k/q_k|$. Prove that $b > q_{k+1}/2$.

Solution. By the triangle inequality, $0 < |a/b - p_k/q_k| \leq |a/b - \alpha| + |\alpha - p_k/q_k| < 2|\alpha - p_k/q_k| < 2/q_k q_{k+1}$. Multiply by bq_k to get $0 < |aq_k - bp_k| < 2b/q_{k+1}$. Since $|aq_k - bp_k|$ is a positive integer, we have $2b/q_{k+1} > 1$, and thus $b > q_{k+1}/2$.

Note. The usual result, (9.37), implies that $b > q_k$ if $|\alpha - a/b| < |\alpha - p_k/q_k|$. Hence if $q_{k+1}/2 < q_k$, the conclusion of this problem is weaker. For example, if $\alpha = \pi$, then $p_2/q_2 = 333/106$, $p_3/q_3 = 355/113$, and $p_4/q_4 = 103993/33102$; thus for $k = 2$, $b > q_k$ implies that $b > 106$, whereas $b > q_{k+1}/2$ implies only that $b > 56$. But the conclusion of this problem can give a much stronger result: When $k = 3$, $b > q_k$ implies that $b > 113$, but $b > q_{k+1}/2$ implies that $b > 16551$!

9-47. Assume that $|\sqrt{2} - r/s| < |\sqrt{2} - 41/29|$, where $s > 0$. Show that $s \geq 36$. (Hint. Use the preceding problem.)

Solution. Refer to the table in Problem 9-39. Since $|\sqrt{2} - r/s| < |\sqrt{2} - p_4/q_4|$, it follows from the preceding problem that $s > q_5/2$, that is, $s \geq 36$.

9-48. Let p_k/q_k be the kth convergent of the irrational number α. If a and b are integers, with $b \geq 1$, and $\alpha < a/b < p_k/q_k$ (or $p_k/q_k < a/b < \alpha$), prove that $b > q_{k+1}$.

Solution. The argument is essentially the same as the solution to Problem 9-45. Using (9.33), we have $0 < |a/b - p_k/q_k| < |\alpha - p_k/q_k| < 1/q_k q_{k+1}$, and multiplying by bq_k gives $0 < |aq_k - bp_k| < b/q_{k+1}$. Since $|aq_k - bp_k|$ is a positive integer, it follows that $b/q_{k+1} > 1$, that is, $b > q_{k+1}$.

9-49. Let p_k/q_k denote the kth convergent of the irrational α. Prove that $1/2q_k q_{k+1} < |\alpha - p_k/q_k| < 1/q_k q_{k+1}$ for every $k \geq 0$.

Solution. The second inequality is proved in (9.33). To prove the first inequality, note that α lies between $c_k = p_k/q_k$ and $c_{k+1} = p_{k+1}/q_{k+1}$ (see the Note following (9.12)). Also, α is closer to c_{k+1} than to c_k, by (9.34). Therefore the distance between α and c_k is greater than half the distance between c_k and c_{k+1}. Using (9.7), it is easy to see that the distance between c_k and c_{k+1} is $1/q_k q_{k+1}$, which gives the first inequality.

9-50. Suppose α is irrational and $|s\alpha - r| \geq 1/s$, where $(r, s) = 1$ and $s \geq 1$. Prove that r/s cannot be one of the convergents of α.

Solution. The given inequality implies that $|\alpha - r/s| \geq 1/s^2$, which contradicts (9.33) if r/s is a convergent.

9-51. If p_k/q_k is the kth convergent of the irrational number α, prove that $|\alpha - p_k/q_k| > 1/q_k(q_k + q_{k+1})$.

Solution. It was shown in (9.18) that $\alpha - p_k/q_k = (-1)^k/q_k(\alpha_{k+1}q_k + q_{k-1})$. But $\alpha_{k+1} < a_{k+1} + 1$ (since $a_{k+1} = [\alpha_{k+1}]$). Therefore $\alpha_{k+1}q_k + q_{k-1} < a_{k+1}q_k + q_{k-1} + q_k = q_{k+1} + q_k$, and the result follows.

▷ **9-52.** Let α be irrational. Prove that at least one of every two successive convergents of the continued fraction expansion of α satisfies $|\alpha - p/q| < 1/2q^2$. (Hint. Use the identity $|\alpha - p_k/q_k| + |p_{k+1}/q_{k+1} - \alpha| = 1/q_k q_{k+1}$.)

Solution. The identity of the hint is an immediate consequence of the fact that α lies between p_k/q_k and p_{k+1}/q_{k+1} and that $p_{k+1}q_k - p_k q_{k+1} = \pm 1$. Suppose now that $|\alpha - p_k/q_k| \geq 1/2q_k^2$ and $|\alpha - p_{k+1}/q_{k+1}| \geq 1/2q_{k+1}^2$. Since α is irrational, equality is not possible. Using the identity, we therefore have $1/q_k q_{k+1} > 1/2q_k^2 + 1/2q_{k+1}^2$. Multiplying this inequality through by $2q_k^2 q_{k+1}^2$ and simplifying, we obtain $(q_k - q_{k+1})^2 < 0$, which is impossible.

▷ **9-53.** Prove Theorem 9.39: If α is irrational and c/d is a rational number, with $d \geq 1$, such that $|\alpha - c/d| < 1/2d^2$, then c/d is one of the convergents of the continued fraction expansion of α. (Hint. Use (9.35).)

Solution. Let u and v be integers, with v positive, such that $u/v \neq c/d$ and $|v\alpha - u| \leq |d\alpha - c|$. By the triangle inequality, we have $|u/v - c/d| \leq |\alpha - u/v| + |\alpha - c/d|$. Since $|d\alpha - c| < 1/2d$, it follows that $|u/v - c/d| < 1/2dv + 1/2d^2$. Multiplying by dv, we obtain the inequality $|du - cv| < 1/2 + v/2d$. But $du - cv$ is a nonzero integer, and hence $v > d$. The result now follows from (9.35).

Rational Approximations to π

9-54. Suppose that r/s is a rational number such that $\pi < r/s < 22/7$. Prove that $s > 106$. (Hint. See the solution of Problem 9-45.)

Solution. By (9.33), we have $0 < |22/7 - r/s| < |\pi - 22/7| < 1/(7 \cdot 106)$, since $22/7$ is a convergent of π. If we multiply by $7s$, we obtain $0 < |22s - 7r| < s/106$. Since $|22s - 7r|$ is a positive integer, we must have $s/106 > 1$, i.e., $s > 106$.

▷ **9-55.** If r and s are positive integers and $\pi < r/s < 22/7$, prove that $s \geq 113$. (Hint. Consider separately the cases $r/s < 355/113$ and $r/s \geq 355/113$. See (9.38).)

Solution. First suppose that $r/s < 355/113$ (the third convergent of π); then $|\pi - r/s| < |\pi - 355/113|$, and so $s > 113$ by (9.37). Now suppose that $\pi < 355/113 \leq r/s < 22/7$. If $r/s = 355/113$, then $s \geq 113$ (in fact, $s = 113$ if $(r, s) = 1$). If $r/s > 355/113$, then $0 < |r/s - 22/7| < |22/7 - 355/113| = 1/(7 \cdot 113)$ (by (9.8) or by direct calculation). Multiply by $7s$ to get $0 < |7r - 22s| < s/113$. Since $7r - 22s$ is a nonzero integer, we have $|7r - 22s| \geq 1$, and therefore $s/113 > 1$, i.e., $s > 113$. Thus in each case, $s \geq 113$.

9-56. (a) Let r and s be positive integers such that $|\pi - r/s| < |\pi - 355/113|$. Show that $s \geq 16552$. (Hint. Use a problem from the previous section.)

(b) Find the best rational approximation to π among all rationals with denominator not exceeding 16551.

Solution. (a) From the table in (9.38), we have $p_3/q_3 = 355/113$ and $q_4 = 33102$. Thus it follows from Problem 9-46 that $s > q_4/2 = 16551$.

(b) In view of part (a), $355/113$ is the best rational approximation to π with denominator not exceeding 16551.

9-57. Find the best rational approximation to π with denominator less than 130000. (Hint. Extend the table in (9.38) to $k = 8$, and use Problem 9-46.)

Solution. Compute the following table (see (9.38)):

k	-1	0	1	2	3	4	5	6	7	8
a_k		3	7	15	1	292	1	1	1	2
p_k	1	3	22	333	355	103993	104348	208341	312689	833719
q_k	0	1	7	106	113	33102	33215	66317	99532	265381

By (9.37), p_7/q_7 is the best rational approximation to π among all rationals with denominators less than or equal to $q_7 = 99532$. Now suppose that $|\pi - r/s| < |\pi - p_7/q_7|$. Using Problem 9-46, it follows that $s > q_8/2 > 130000$. Thus we conclude that there is *no* rational r/s with $s < 130000$ that approximates π better than $p_7/q_7 = 312689/99532$.

9-58. Let r/s be a rational number such that $r/s \neq 22/7$. If $0 < s \leq 53$, prove that $|\pi - 22/7| < |\pi - r/s|$. In other words, show that $22/7$ approximates π better than any other rational with denominator less than or equal to 53.

Solution. Use Problem 9-46. Note that $22/7 = p_1/q_1$ and $333/106 = p_2/q_2$ are convergents of π (see (9.38)). If $|\pi - r/s| < |\pi - 22/7|$, Problem 9-46 implies that $s > q_2/2 = 106/2 = 53$. Since $s \leq 53$, it follows that $|\pi - 22/7| < |\pi - r/s|$. (Note that $|\pi - r/s|$ cannot equal $|\pi - 22/7|$; otherwise, since $r/s \neq 22/7$, we have $r/s = 2\pi - 22/7$, which is not rational.)

Note. The conclusion is true if $s \leq 56$ but not for $s = 57$; in fact, the rational $179/57$ approximates π better than $22/7$. The number $179/57$ is a *secondary convergent* of π; see the following section of problems for the properties of secondary convergents. For example, it is true that the best rational approximation to π with denominator less than or equal to 56 must be either a convergent or a secondary convergent of π (see Problem 9-66). It is easy to check that there is no secondary convergent of π with denominator between 53 and 56, and hence the conclusion of Problem 9-58 holds for $s \leq 56$. This shows that while $|\alpha - r/s| < |\alpha - p_k/q_k|$ implies that $s > q_k$ (by (9.37)), it does *not* imply that $s \geq q_{k+1}$. (For π, $q_1 = 7$ and $q_2 = 106$, but there are (exactly seven) rationals with denominators between 7 and 106 that approximate π better than $22/7$. See Problem 9-69.)

▷ **9-59.** *Find a rational number r/s such that $|\pi - r/s| < |\pi - 22/7|$ and $s < 106$. (Hint. Approximate π by $333/106 = \langle 3,7,15 \rangle$, and note that $\langle 3,7,15 \rangle = \langle 3,7,14,1 \rangle$. Then apply (9.33) and Problem 9-49.)*

Solution. Find the convergents of $\langle 3,7,14,1 \rangle$ from the following table:

k	-1	0	1	2	3
a_k			3	7	14	1
p_k	1	3	22	311	333
q_k	0	1	7	99	106

Note that $22/7$ and $333/106$ are consecutive convergents of π ($311/99$ is not a convergent of π, but it is a *secondary convergent*; see the following section). Now use (9.7), (9.33), and the triangle inequality to get

$$|\pi - 311/99| \leq |\pi - 333/106| + |333/106 - 311/99|$$
$$< 1/106^2 + |333 \cdot 99 - 311 \cdot 106|/(99 \cdot 106)$$
$$= 1/106^2 + 1/(99 \cdot 106) < 1/(2 \cdot 7 \cdot 106).$$

However, Problem 9-49 implies that $1/(2 \cdot 7 \cdot 106) < |\pi - 22/7|$, and thus we conclude that $|\pi - 311/99| < |\pi - 22/7|$.

Secondary Convergents

Suppose α is irrational, and let p_k/q_k denote the kth convergent of α. For $k \geq 2$, consider the sequence

$$\frac{p_{k-2}}{q_{k-2}}, \frac{p_{k-2} + p_{k-1}}{q_{k-2} + q_{k-1}}, \frac{p_{k-2} + 2p_{k-1}}{q_{k-2} + 2q_{k-1}}, \ldots, \frac{p_{k-2} + a_k p_{k-1}}{q_{k-2} + a_k q_{k-1}} = \frac{p_k}{q_k}.$$

Except for the first term p_{k-2}/q_{k-2} and the last term p_k/q_k, the numbers in this sequence are called *secondary convergents* of α. To simplify notation, let $c_{k,t} = p_{k,t}/q_{k,t} = (p_{k-2} + tp_{k-1})/(q_{k-2} + tq_{k-1})$, where $1 \leq t \leq a_k - 1$. It is obvious that if $a_k = 1$, then there will be no secondary convergents between p_{k-2}/q_{k-2} and p_k/q_k.

Note that the secondary convergents $c_{k,t}$ all lie between a pair of convergents whose subscripts differ by 2 and which are therefore (by (9.12)) either both greater than α (if k is odd) or both less than α (if k is even). We will show below that every secondary convergent of α is in lowest terms and that the secondary convergents between p_{k-2}/q_{k-2} and p_k/q_k are strictly increasing when k is even and strictly decreasing when k is odd; it follows that they are all on the same side of α. And like convergents, they satisfy the same type of equation given in (9.7): If r/s and r'/s' are consecutive secondary convergents, then $rs' - r's = \pm 1$.

These properties allow us to answer completely the question raised in the Note following (9.37): *If s is a positive integer, what is the best approximation to the irrational α among all rational numbers with denominators less than or equal to s?* As we will show in Problem 9-66, for any given s, the best such rational approximation is always either a convergent or a secondary convergent.

For clarity, we consider the convergents of $\pi = \langle 3, 7, 15, 1, \ldots \rangle$; thus $p_0/q_0 = 3/1$, $p_1/q_1 = \langle 3, 7 \rangle = 22/7$, and $p_2/q_2 = \langle 3, 7, 15 \rangle = 333/106$. By definition, the secondary convergents between $3/1$ and $333/106$ are the numbers $(3+22t)/(1+7t)$, with $1 \leq t \leq a_2 - 1 = 14$; thus the first secondary convergent is $25/8$ and the last is $311/99$. These secondary convergents can then be characterized in the following useful way: *They are precisely the values of $\langle 3, 7, t \rangle$ for $1 \leq t \leq 14$.* That this is generally true is proved in the next result.

9-60. Suppose α is irrational, with $\alpha = \langle a_0, a_1, a_2, \ldots \rangle$, and let $p_{k,t}/q_{k,t}$ be a secondary convergent between p_{k-2}/q_{k-2} and p_k/q_k. Prove that $p_{k,t}/q_{k,t} = \langle a_0, a_1, \ldots, a_{k-1}, t \rangle$.

Solution. By (9.4), we have $\langle a_0, a_1, \ldots, a_{k-1}, t \rangle = (tp_{k-1} + p_{k-2})/(tq_{k-1} + q_{k-2}) = p_{k,t}/q_{k,t}$.

The next few problems establish some of the basic properties of secondary convergents.

9-61. If r/s and r'/s' are any two consecutive secondary convergents between p_{k-2}/q_{k-2} and p_k/q_k, prove that $rs' - r's = (-1)^k$.

Solution. Suppose $r/s = c_{k,t}$ and $r'/s' = c_{k,t+1}$. It is then easy to check from the definition that for *any* value of t, $rs' - r's = p_{k-1}q_k - p_k q_{k-1} = (-1)^k$, using (9.7). (The terms in t cancel.)

9-62. If $p_{k,t}/q_{k,t}$ is a secondary convergent of α, prove that $(p_{k,t}, q_{k,t}) = 1$.

Solution. This follows directly from the preceding problem, for if d is a common divisor of $p_{k,t}$ and $q_{k,t}$, then $d \mid (-1)^k$ and hence $d = 1$. The result is also an easy consequence of Problem 9-60: Since $p_{k,t}/q_{k,t} = \langle a_0, a_1, \ldots, a_{k-1}, t \rangle$, (9.4) implies that $p_{k,t}/q_{k,t}$ is just the last convergent of this continued fraction. Now apply (9.7).

9-63. Prove that the secondary convergents between p_{k-2}/q_{k-2} and p_k/q_k form a strictly increasing sequence if k is even and a strictly decreasing sequence if k is odd. (*Hint.* This follows easily from Problems 9-15 and 9-60.)

Solution. This can be established directly from the definition of secondary convergents by showing that $c_{k,t} - c_{k,t+1} = (-1)^k/(q_{k,t} q_{k,t+1})$, from which the conclusion follows. An easier way to prove the result is to use Problem 9-60 to write $c_{k,t} = \langle a_0, a_1, \ldots, a_{k-1}, t \rangle$ and $c_{k,t+1} = \langle a_0, a_1, \ldots, a_{k-1}, t+1 \rangle$. The conclusion then follows easily from Problem 9-15.

The following results establish the approximation properties of secondary convergents.

▷ **9-64.** Let α be irrational, and suppose that $|\alpha - c/d| < |\alpha - p_{k,t}/q_{k,t}|$. Prove that either $c/d = p_{k-1}/q_{k-1}$ or $d > q_{k,t}$.

Solution. As usual, let $c_{k-1} = p_{k-1}/q_{k-1}$. Use the triangle inequality to write $|c/d - c_{k-1}| \le |c/d - \alpha| + |\alpha - c_{k-1}| \le |\alpha - c_{k,t}| + |\alpha - c_{k-1}| = |c_{k-1} - c_{k,t}|$. (The last equality holds since $\alpha - c_{k-1}$ and $\alpha - c_{k,t}$ are of opposite sign: $c_{k,t}$ is on the same side of α as c_k and c_{k-2}, whereas c_{k-1} is on the other side of α, by (9.12).) If the first inequality above is actually an equality, then c/d and c_{k-1} must lie on opposite sides of α, and thus α is between c/d and c_{k-1}. If equality holds in the second inequality above, then c/d and $c_{k,t}$ are the same distance from α, and hence if $c/d \ne c_{k,t}$, then c/d is on the *same* side of α as c_{k-1}, a contradiction. Therefore at least one of these two inequalities is a strict inequality.

Note that $p_{k-1}q_{k,t} - q_{k-1}p_{k,t} = p_{k-1}q_{k-2} - p_{k-2}q_{k-1} = (-1)^k$, by (9.7). Thus if $c/d \ne c_{k-1}$, we have $0 < |c/d - c_{k-1}| < |c_{k-1} - c_{k,t}| = 1/q_{k-1}q_{k,t}$. Multiplying by dq_{k-1} then gives $1 \le |cq_{k-1} - dp_{k-1}| < d/q_{k,t}$, and hence $d > q_{k,t}$.

9-65. Let p/q be a convergent or a secondary convergent of the irrational number α. Prove that if c/d is a rational number that lies between α and p/q, then $d > q$. (Hint. Use the preceding problem.)

Solution. If p/q is a convergent, the result follows from (9.37). Now suppose that p/q is a secondary convergent, say $c_{k,t}$. Note that the convergent c_{k-1} cannot lie between α and $c_{k,t}$, since c_{k-1} and $c_{k,t}$ are on opposite sides of α; thus c/d cannot be equal to c_{k-1}. It follows from the preceding problem that $d > q$.

Here is the main result about the approximating properties of convergents and secondary convergents.

▷ **9-66.** Suppose α is irrational, and let s be a positive integer. Prove that the best approximation to α among all rational numbers with denominators less than or equal to s is either a convergent or a secondary convergent of α. Equivalently, prove that if r/s is such that every rational number between r/s and α has a denominator greater than s, then r/s is either a convergent or a secondary convergent of α.

Solution. We will prove the second form of the result. Let S denote the increasing sequence consisting of the even-numbered convergents of α, together with all of the secondary convergents between them. Similarly, let T denote the decreasing sequence of odd-numbered convergents of α and all of the secondary convergents between them. Suppose that r/s is neither a convergent nor a secondary convergent of α. We consider two cases. If $r/s < \alpha$, then r/s is on the same side of α as the even-numbered convergents, and thus either r/s lies between two consecutive terms of S or $r/s < p_0/q_0 = a_0$. Similarly, if $r/s > \alpha$, then r/s is on the same side of α as the odd-numbered convergents, and therefore either r/s lies between two consecutive terms of T or $r/s > (p_{-1} + p_0)/(q_{-1} + q_0) = a_0 + 1$ (the only secondary convergent of α that is greater than p_1/q_1). If either $r/s < a_0 = [\alpha] < \alpha$ or $r/s > a_0 + 1 > \alpha$ (since $a_0 \le \alpha < a_0 + 1$), then r/s does *not* satisfy the hypothesis stated in the problem, since there is a rational number between α and r/s with denominator 1, namely, a_0

or a_0+1, respectively. (Note that the denominator is supposed to be greater than s, and $s \geq 1$.)

We may therefore suppose that r/s lies between consecutive terms of S or T. Denote these terms by $c = p/q$ and $c' = p'/q'$; thus either $c' < r/s < c < \alpha$ or $\alpha < c < r/s < c'$. In each case, c lies between r/s and α; also, $0 < |r/s - c'| < |c - c'| = 1/qq'$ (using Problem 9-61). Multiplying by sq', we get $0 < |rq' - sp'| < s/q$, and since $|rq' - sp'|$ is a positive integer, it follows that $s > q$. Thus, between r/s and α, there is a rational p/q with denominator *less than* s, contradicting our hypothesis. We conclude therefore that r/s is either a convergent or a secondary convergent of α.

Of the secondary convergents between p_{k-2}/q_{k-2} and p_k/q_k, some will approximate α better than the convergent p_{k-1}/q_{k-1}, and some will not. The next problem describes precisely which secondary convergents are better approximations.

▷ **9-67.** Let α be irrational. Prove that $|\alpha - p_{k,t}/q_{k,t}| < |\alpha - p_{k-1}/q_{k-1}|$ if and only if either (i) $t > a_k/2$ or (ii) $t = a_k/2$ and $\langle a_k, a_{k-1}, \ldots, a_1\rangle > \langle a_k, a_{k+1}, \ldots\rangle$. In particular, if $t < a_k/2$, then $p_{k,t}/q_{k,t}$ is never a better approximation to α than p_{k-1}/q_{k-1}.

Solution. From the definition of $p_{k,t}$ and $q_{k,t}$, we have

$$|\alpha - p_{k,t}/q_{k,t}| < |\alpha - p_{k-1}/q_{k-1}|$$

if and only if

$$|t(\alpha q_{k-1} - p_{k-1}) + \alpha q_{k-2} - p_{k-2}|/q_{k,t} < |\alpha q_{k-1} - p_{k-1}|/q_{k-1}.$$

By (9.4) and (9.19), $\alpha = \langle a_0, a_1, \ldots, a_{k-1}, \alpha_k\rangle = (\alpha_k p_{k-1} + p_{k-2})/(\alpha_k q_{k-1} + q_{k-2})$, and hence $\alpha_k = -(\alpha q_{k-2} - p_{k-2})/(\alpha q_{k-1} - p_{k-1})$. Thus the last inequality holds if and only if $|(t - \alpha_k)(\alpha q_{k-1} - p_{k-1})|/q_{k,t} < |\alpha q_{k-1} - p_{k-1}|/q_{k-1}$, i.e., if and only if $|\alpha_k - t|q_{k-1} < q_{k,t} = q_{k-2} + tq_{k-1}$. Since $t < a_k < \alpha_k$, $\alpha_k - t$ is positive, and thus the last inequality is equivalent to $2tq_{k-1} + q_{k-2} > \alpha_k q_{k-1}$. This condition is satisfied when $2t \geq a_k + 1$, since then $2tq_{k-1} + q_{k-2} \geq (a_k + 1)q_{k-1} > \alpha_k q_{k-1}$, and is *not* satisfied if $2t \leq a_k - 1$, for then

$$2tq_{k-1} + q_{k-2} \leq (a_k - 1)q_{k-1} + q_{k-1} = a_k q_{k-1} < \alpha_k q_{k-1}.$$

Finally, if $2t = a_k$, the condition holds if and only if $a_k q_{k-1} + q_{k-2} > \alpha_k q_{k-1}$, i.e., $q_k/q_{k-1} > \alpha_k$, and by Problem 9-8, the inequality $q_k/q_{k-1} > \alpha_k$ is equivalent to $\langle a_k, a_{k-1}, \ldots, a_1\rangle > \langle a_k, a_{k+1}, \ldots\rangle$.

Note. This problem shows that for the secondary convergents between p_{k-2}/q_{k-2} and p_k/q_k, only those in the *second* half of the sequence are closer to α than is p_{k-1}/q_{k-1} (which is on the other side of α). If a_k is *even*, there will be a "middle" secondary convergent, and this must be checked separately according to the condition given above.

9-68. Suppose α is irrational and $\alpha = \langle a_0, a_1, \ldots \rangle$. It follows from the preceding problem that if $a_k \geq 3$, then $|\alpha - (p_k - p_{k-1})/(q_k - q_{k-1})| < |\alpha - p_{k-1}/q_{k-1}|$. (Note that $(p_k - p_{k-1})/(q_k - q_{k-1})$ is the last secondary convergent between p_{k-2}/q_{k-2} and p_k/q_k.) Give an example to show that the above inequality need not hold if $a_k = 2$. (Hint. Consider $\alpha = (1 + \sqrt{3})/2$.)

Solution. Check that $\alpha = (1 + \sqrt{3})/2 = \langle \overline{1, 2} \rangle$; then $a_3 = 2$, $p_1/q_1 = 3/2$, $p_2/q_2 = 4/3$, and $p_3/q_3 = 11/8$. The only secondary convergent between $3/2$ and $11/8$ is $(3+4)/(2+3) = 7/5$, but $|\alpha - 4/3| < |\alpha - 7/5|$.

Finally, we apply the preceding results to the secondary convergents of π.

9-69. Use Problem 9-68 to determine which secondary convergents of π between 3 and $333/106$ approximate π better than $22/7$. What is the best approximation to π among all rational numbers with denominators not exceeding 100?

Solution. Refer to Example 9.38: $p_0/q_0 = 3/1$, $p_1/q_1 = 22/7$, and $p_2/q_2 = 333/106$. Thus the secondary convergents are given by $(3 + 22t)/(1 + 7t)$ for $1 \leq t \leq 14$:

$$25/8, \quad 47/15, \quad 69/22, \quad \ldots, \quad 157/50, \quad 179/57, \quad \ldots, \quad 289/92, \quad 311/99.$$

By the preceding problem, the secondary convergents for $1 \leq t \leq 7$ (that is, the convergents up to and including $157/50$) give a *worse* approximation to π than $22/7$, while those for $8 \leq t \leq 14$ ($179/57$ to $311/99$) give successively better approximations than $22/7$. Since the next convergent or secondary convergent after $311/99$ is $333/106$, it follows from Problem 9-66 that $311/99$ is the best approximation to π among all rationals with denominators less than or equal to 100.

Note. Because $a_2 = 15$, there is no critical middle value $t = a_k/2$ to check separately. (See the following problem.)

9-70. Prove that if r/s is a better approximation to π than $355/113$, then $s \geq 16604$. (Compare Problem 9-56.) Thus, $355/113$ is the best approximation to π among all rationals with denominators not exceeding 16603. (Hint. Use Problems 9-66, 9-67, and 9-15.)

Solution. By Problem 9-66, an approximation to π better than $p_3/q_3 = 355/113$ must be either a convergent or a secondary convergent. Since the next convergent of π after $355/113$ is $p_4/q_4 = 103993/33102$ (see (9.38)), we must look at the secondary convergents of π between $333/106$ and $103993/33102$, namely, rational numbers of the form $(333 + 355t)/(106 + 113t)$, where $1 \leq t \leq 291$. According to Problem 9-67, the secondary convergents for $147 \leq t \leq 291$ give (successively) better approximations than $355/113$. But since 292 is even, the secondary convergent for $t = 146$ must be checked separately by determining whether or not $\langle 292, 1, 15, 7 \rangle > \langle 292, 1, 1, 1 \ldots \rangle$. (See the table in Problem 9-57.) By Problem 9-15, this inequality is true (here, $n = 2$ and $15 > 1$), and hence we conclude that the next best approximation to π after $355/113$ is the secondary convergent $p_{4,146}/q_{4,146} = 52163/16604$.

EXERCISES FOR CHAPTER 9

1. Use continued fractions to find a solution of
   (a) $136x + 49y = 1$;
   (b) $247x - 776y = -2$.

2. Find a solution of the following equations using continued fractions:
   (a) $79x + 212y = -3$;
   (b) $-85x + 19y = 7$.

3. Determine the two simple continued fraction expansions of $397/121$ and $-7/22$.

4. Find the continued fraction expansions of the rational numbers 3.09 and 9.115.

5. What rational number does the continued fraction $\langle 1,2,3,4,5,6 \rangle$ represent?

6. Determine the rational numbers represented by the following simple continued fractions: (a) $\langle 3,1,2,1,4 \rangle$; (b) $\langle -5,4,3,2,1 \rangle$.

7. Find the convergents of the finite continued fraction $\langle 3,6,1,7,2,1 \rangle$.

8. Calculate the convergents of $\langle -2,5,1,4,9 \rangle$.

9. Determine the first four convergents of the continued fraction expansion of 2.0867.

10. Express $739/34$ and $-739/34$ as finite simple continued fractions.

11. Determine the simple continued fraction expansion of (a) $\sqrt{98}$; (b) $(1+\sqrt{5})/2$; (c) $(12+\sqrt{11})/5$.

12. Find the first six partial quotients of the infinite continued fraction expansion of (a) $\sqrt[3]{2}$; (b) e; (c) π^2.

13. Calculate the first six convergents of the infinite continued fraction expansion of (a) e; (b) π^2.

14. In each case, determine if the value of the first continued fraction is greater than the value of the second:
    (a) $\langle 6,7,5,1,3 \rangle$, $\langle 6,7,3,1,5 \rangle$;
    (b) $\langle 8,3,1,7,4 \rangle$, $\langle 8,3,1,4,7,\ldots \rangle$;
    (c) $\langle 4,3,7,9,6,\ldots \rangle$, $\langle 4,3,7,9 \rangle$;
    (d) $\langle 11,7,5,63 \rangle$, $\langle 11,7,4,\ldots \rangle$;
    (e) $\langle 5,19,45,38,\ldots \rangle$, $\langle 5,19,45 \rangle$.

15. Evaluate (a) $\langle \overline{1,2,1} \rangle$; (b) $\langle 5,\overline{1,2,1} \rangle$; (c) $\langle \overline{4,5} \rangle$.

16. Find the continued fraction expansion of $\sqrt{311}$.

17. Prove that $\sqrt{9n^2+6} = \langle 3n, \overline{n, 6n} \rangle$.

18. Determine if the following irrational numbers have a purely periodic continued fraction expansion: (a) $(1+\sqrt{11})/3$; (b) $(2+\sqrt{6})/2$; (c) $(7+\sqrt{101})/3$.

EXERCISES

19. (a) Find the continued fraction expansion of $6 - \sqrt{6}$ by constructing a table for r_k and s_k.
    (b) Use part (a) to express $\langle \overline{2, 4} \rangle$ as a quadratic irrational.

20. Find the irrational number whose continued fraction expansion is $\langle 1, \overline{2, 3} \rangle$.

21. Determine the value of the periodic continued fraction $\langle 1, 1, \overline{3, 5, 7} \rangle$.

22. (a) Calculate the continued fraction expansion of $(3 - \sqrt{10})/2$ by setting up a table to compute r_k, s_k, and a_k.
    (b) Use part (a) to express $\langle \overline{12, 3} \rangle$ as a quadratic irrational.

23. (a) Find the infinite continued fraction expansion of $(2 + \sqrt{21})/3$ by setting up a table to compute r_k and s_k.
    (b) Use the table constructed in part (a) to determine the continued fraction expansion of $(13 + \sqrt{189})/4$.

24. Find the best approximation to $\sqrt{2}$ among all rational numbers with denominator not exceeding 985.

25. Determine the best rational approximation to $\sqrt[3]{2}$ with denominator less than or equal to 50.

26. What is the best rational approximation to e with denominator less than 8?

27. Find the best approximation to $\sqrt[3]{9}$ among all rational numbers with denominator not exceeding (a) 487; (b) 300. (Hint. For (b), look at the secondary convergents.)

28. What is the best rational approximation to $\sqrt{2}$ with denominator not exceeding 100? (Hint. Use secondary convergents.)

29. The rational $961/462$ is a convergent of $\sqrt[3]{9}$. Does it follow that $|\sqrt[3]{9} - 961/462| < 1/(2 \cdot 462^2)$?

30. Without computing the continued fraction expansion of $\sqrt[3]{50}$, prove that $70/19$ is a convergent.

31. (a) Calculate the first five partial quotients in the continued fraction expansion of 2π.
    (b) What is the best rational approximation to 2π with denominator not exceeding 7? With denominator not exceeding 53?

32. Without computing the continued fraction expansion of $\alpha = (1 + \sqrt{3})/2$, decide if $153/112$ and $571/418$ are consecutive convergents of α.

33. Find the best rational approximation to π with denominator not exceeding 25000.

34. Use a result on approximations to prove that $52/25$ is a convergent of $\sqrt[3]{9}$.

35. Suppose that $\sqrt[3]{9} > a/b > 52/25$, where a and b are positive integers. Prove that $b > 462$.

310   CHAPTER 9: CONTINUED FRACTIONS

36. Assume that $\sqrt{41} = \langle 6, \overline{2, 2, 12} \rangle$. Use a result on approximations to find (if any exist) all rational numbers a/b in lowest terms, with $50 < b < 250$, such that $|\sqrt{41} - a/b| < 1/3b^2$.

NOTES FOR CHAPTER 9

1. Most of the basic *formulas* of this chapter were known to Euler, but essentially all of the *theory* was developed by Lagrange. In particular, Lagrange proved that all quadratic irrationals have a periodic continued fraction expansion. He also investigated the approximation of real numbers by continued fractions, including the properties of secondary convergents.

2. There is an intimate connection between continued fractions and the Euclidean Algorithm. In particular, the Extended Euclidean Algorithm, which was described briefly in the Note after (1.23), is essentially the same as the recursive procedure for calculating the convergents of the rational number a/b (see (9.10)). There are only minor notational differences: The alternation of signs, which is explicit in (9.10), is hidden in the usual description of the Extended Euclidean Algorithm. The Extended Euclidean Algorithm, or, equivalently, the algorithm for computing the p_k and q_k, gives a very efficient procedure for expressing the greatest common divisor of a and b as a linear combination of a and b. It is simple to implement on a computer or programmable calculator and also works well for calculations by hand. For hand calculations, it is best to set up a table as in Example 9.6. For small numbers a and b, the back substitution procedure described after (1.23) is perhaps more natural.

If α is an irrational number, the procedure for computing the partial quotients $a_0, a_1, \ldots$ can be viewed as a kind of Euclidean Algorithm. Let x and y be real numbers, with $y > 0$, and let z be the smallest positive real number of the form $x - ny$, where n ranges over the *integers*. Then $x = qy + z$ for some integer q, and it is easy to see that $q = [x/y]$. By analogy with the integers, we can call q the *quotient* and z the *remainder* when x is divided by y. With this understanding of quotient and remainder, apply the ordinary Euclidean Algorithm to the numbers α and 1. It is easy to see that the successive "quotients" we obtain are just the partial quotients $a_0, a_1, \ldots$ in the continued fraction expansion of α.

A closely related geometric idea is used in Book X of Euclid's *Elements* to prove the equality of ratios. Greek mathematicians even had standard technical terms, *anthypharesis* and *antanairesis*, for this generalization of the Euclidean Algorithm. If we apply the algorithm to the diagonal and side of a square (that is, find the continued fraction expansion of $\sqrt{2}$), then we observe the periodicity quite quickly. This can be turned into an elegant geometric proof of the irrationality of $\sqrt{2}$. There has been speculation that this was known to Greek mathematicians by the fourth century B.C.

If the Extended Euclidean Algorithm is applied to the real numbers a and b, where b/a is irrational (for example, $a = 1$ and $b = \alpha$), we can express successive remainders in the form $r_n = ax_n + by_n$. The sequence (r_n) has limit 0, and therefore the sequence $(-x_n/y_n)$ converges to b/a. This gives an elementary way to obtain rational approximations to b/a, a technique that is essentially equivalent to the method using continued fractions.

3. Theorem 9.28 gives a complete characterization of the continued fraction expansion of quadratic irrationals, that is, irrational numbers which are roots of quadratic polynomials with integer coefficients. However, no such characterization is known for real numbers (other than rationals and quadratic irrationals) that are roots of polynomials of degree greater than two. In fact, the continued fraction expansion is not known for *any* algebraic number of higher degree. It is not even known, for example, whether the partial quotients of $\sqrt[3]{2}$ are bounded.

In contrast, the continued fraction expansion of certain *transcendental* numbers is known. Perhaps the most interesting one was found in 1737 by Euler. Let e denote the base for natural logarithms. Then

$$e - 1 = \langle 1,1,2,1,1,4,1,1,6,1,1,8,\ldots\rangle \quad \text{and} \quad \frac{e+1}{e-1} = \langle 2,6,10,14,\ldots\rangle.$$

4. Consider the infinite series

$$\frac{p_0}{q_0} + \left(\frac{p_1}{q_1} - \frac{p_0}{q_0}\right) + \left(\frac{p_2}{q_2} - \frac{p_1}{q_1}\right) + \cdots + \left(\frac{p_{k+1}}{q_{k+1}} - \frac{p_k}{q_k}\right) + \cdots,$$

where p_k/q_k denotes the kth convergent of the irrational α. In view of (9.9), the terms of this series (beginning with the second) are alternately positive and negative. Let $b_0 = p_0/q_0 = [\alpha]$ and $b_{k+1} = p_{k+1}/q_{k+1} - p_k/q_k$ for $k \geq 0$; it follows from (9.7) that $b_{k+1} = (-1)^k/q_k q_{k+1}$, and since $q_k \geq k$, the sequence $(|b_k|)$, after the first term, decreases monotonically to 0.

Thus $\sum_{k=0}^{\infty} b_k$ is an alternating series whose nth partial sum $s_n = b_0 + b_1 + \cdots + b_n$ is clearly p_n/q_n. Therefore the sum of this infinite series is, by definition, $\lim_{n\to\infty} p_n/q_n$, namely, α (see (9.13)). We therefore have

$$\alpha = [\alpha] + \sum_{k=0}^{\infty} \frac{(-1)^k}{q_k q_{k+1}} = [\alpha] + \frac{1}{q_0 q_1} - \frac{1}{q_1 q_2} + \frac{1}{q_2 q_3} - \cdots.$$

It is worth remarking that *this expression for α depends only on the values of q_k*, not on the values of p_k. Hence *every* convergent of α can be found using only $[\alpha]$ and the sequence (q_k):

$$\frac{p_n}{q_n} = s_n = [\alpha] + \sum_{k=0}^{n-1} \frac{(-1)^k}{q_k q_{k+1}}.$$

From the usual results on alternating series, we can easily derive the following properties for the continued fraction expansion of α:

(i) The subsequence (s_{2n}) is strictly increasing, with limit α, and (s_{2n+1}) is strictly increasing, with limit α; in terms of continued fractions, the even-numbered convergents are less than α and successively closer to α, and the odd-numbered convergents are greater than α and successively closer to α (see (9.9) and (9.12)).

(ii) $|b_{n+2}| < |\alpha - s_n| < |b_{n+1}|$; in terms of convergents, $|\alpha - p_n/q_n| < 1/q_n q_{n+1}$ (which is (9.33)), and $|\alpha - p_{n+1}/q_{n+1}| < |\alpha - p_n/q_n|$, that is, the convergents of α are successively closer to α (see (9.34).

The above discussion applies equally well (with only minor changes) when α is rational.

BIOGRAPHICAL SKETCHES

Joseph Louis Lagrange was born in 1736 in Italy and studied mathematics at the University of Turin. He quickly became well known, making basic contributions to analysis, including the calculus of variations, the theory of differential equations, and mechanics. In 1766, when Euler left Berlin for St. Petersburg, Lagrange gave up his position at the Royal School of Artillery in Turin to become head of the mathematical section of the Berlin Academy. After the death of Euler in 1783, Lagrange was regarded as the most important mathematician in Europe. In 1787, he accepted an invitation from Louis XVI to join the Académie Française.

Lagrange did important work in many areas of mathematics, most notably mechanics, algebra, and number theory. Most of Lagrange's number-theoretic work was done during the first half of his Berlin period, from 1766 to 1777. His contributions to number theory include the first published proof of Wilson's Theorem and the proof that every positive integer can be represented as the sum of four squares. (This result had been conjectured by Bachet and Fermat, and progress toward a proof had been made by Euler.) Lagrange developed the basic theory of continued fractions and used it to prove that Pell's Equation always has a solution. In addition, Lagrange made major contributions to the analysis of quadratic forms. His contributions in algebra include results in the theory of equations, group theory, and the theory of polynomial congruences.

Lagrange did almost no mathematics during the turbulent years of the French Revolution. In 1793, he was appointed head of a committee charged with standardizing French weights and measures, and Lagrange is regarded as the chief architect of the metric system. He was made a Senator of the Empire by Napoleon.

Lagrange died in 1813 and was buried in the Pantheon.

REFERENCES

Claude Brezinski, *History of Continued Fractions and Padé Approximants*, Springer-Verlag, New York, 1991.

> This history deals not only with the number-theoretic uses of continued fractions, but also with the many applications to differential equations, probability theory, and the approximation of functions. The book quotes at length from sources that would otherwise be difficult to locate, and it has a massive bibliography.

C.D. Olds, *Continued Fractions*, Random House, New York, 1963.

> This is a highly readable book, written in leisurely style, with many detailed numerical examples. It contains most of the important results of this chapter, together with a basic treatment of Pell's Equation. *Continued Fractions* is the most accessible book on the subject available in English and is highly recommended as a supplement to our more condensed presentation.

CHAPTER TEN
Pell's Equation

As noted in the previous chapter, finding good rational approximations to an irrational number is closely connected with the continued fraction expansion of the irrational. Around 400 B.C., Indian and Greek mathematicians gave the approximations $17/12$ and $577/408$ for $\sqrt{2}$. (These are, respectively, the third and seventh convergents of the continued fraction expansion of $\sqrt{2}$.) In the third century B.C., Archimedes approximated $\sqrt{3}$ by $265/153$ (the eighth convergent) and $1351/780$ (the eleventh convergent).

The problem of obtaining rational approximations to $\sqrt{d}$, where d is a positive integer that is not a perfect square, leads in a natural way to the Diophantine equation $x^2 - dy^2 = 1$, for if the pair (x, y) is a solution of this equation with $y \neq 0$, then $(x/y)^2 = d + 1/y^2$, and hence when y is large, x/y is a good approximation to $\sqrt{d}$. (The second-century Greek mathematician Theon of Smyrna was able to generate solutions of the equations $x^2 - 2y^2 = \pm 1$ using a recurrence relation.) Because of a possibly mistaken reference by Euler, the equation $x^2 - dy^2 = 1$ is known as *Pell's Equation*, although the English mathematician John Pell (1611–1685) had little to do with it. One of the earliest and most famous occurrences of Pell's Equation appears in the *Cattle Problem of Archimedes*, which leads to the equation $x^2 - 4729494y^2 = 1$. The least positive solution, found in 1880, has a y-value that is 41 digits long.

A detailed treatment of Pell's Equation appears in the works of the Indian mathematicians Brahmagupta (seventh century) and Bhaskara (twelfth century). The first European mathematician to make a serious investigation of Pell's Equation was Fermat. In 1657, Fermat challenged John Wallis (1616–1703), the most prominent English mathematician before Isaac Newton, and his patron Lord William Brouncker (1620–1684) to solve the equations $x^2 - 61y^2 = 1$ and $x^2 - 109y^2 = 1$. In his reply to Fermat, Brouncker gave a very efficient method for solving Pell's Equation, which yields, in the case of

$x^2 - 109y^2 = 1$, a solution with an x-value that has 15 digits. As Euler noted in 1759, Brouncker's method is the same as the algorithm for determining the continued fraction expansion of $\sqrt{d}$. (In fact, an essentially equivalent procedure appears some five centuries earlier in the work of Bhaskara.) Euler used the continued fraction expansion of $\sqrt{d}$ to find the least positive solution of $x^2 - dy^2 = 1$ and showed how additional solutions can be generated from a given solution by a recurrence relation.

Both Wallis and Fermat had conjectured, correctly, that Pell's Equation always has a solution, in fact infinitely many solutions. In 1768, Lagrange published the first complete proof of this result and showed that every solution comes from the continued fraction expansion of $\sqrt{d}$. Although Euler had seen the connection almost a decade earlier, Lagrange was the first to give a rigorous treatment of Pell's Equation using continued fractions, in a series of three papers presented to the Berlin Academy from 1768 to 1770.

RESULTS FOR CHAPTER 10

Pell's Equation $x^2 - dy^2 = 1$

Throughout this chapter, d represents a positive integer that is not a perfect square. As we have already noted, the technique for solving $x^2 - dy^2 = 1$ used by Bhaskara, Brouncker, and Lagrange is equivalent to the method for determining the continued fraction expansion of $\sqrt{d}$. In this section we will show that Pell's Equation has infinitely many solutions in integers for any choice of d, that all positive solutions are found among the convergents of $\sqrt{d}$, and that all positive solutions can be generated directly from the *least positive solution*.

By (9.31), a quadratic irrational α has a purely periodic continued fraction if and only if it is reduced, that is, if $\alpha > 1$ and its conjugate α' satisfies $-1 < \alpha' < 0$. Although $\sqrt{d}$ is not reduced, it is easily checked that the quadratic irrational $[\sqrt{d}] + \sqrt{d}$ is reduced. (Its conjugate is simply $[\sqrt{d}] - \sqrt{d}$.) This fact can be used to describe the continued fraction expansion of $\sqrt{d}$.

(10.1) Theorem. *The continued fraction expansion of $\sqrt{d}$ is of the form*

$$\langle a_0, \overline{a_1, a_2, a_3, \ldots, a_3, a_2, a_1, 2a_0} \rangle,$$

where $a_0 = [\sqrt{d}]$.

Proof. Let $a_0 = [\sqrt{d}]$. If $\alpha = [\sqrt{d}] + \sqrt{d}$, then α has a purely periodic expansion, and its first partial quotient equals $[\alpha] = 2[\sqrt{d}] = 2a_0$. Thus

$$\alpha = [\sqrt{d}] + \sqrt{d} = \overline{\langle 2a_0, a_1, a_2, \ldots, a_n \rangle} = \langle 2a_0, a_1, a_2, \ldots, a_n, \overline{2a_0, a_1, a_2, \ldots, a_n} \rangle.$$

Hence if we subtract $[\sqrt{d}] = a_0$ from each side, we get

$$\sqrt{d} = \langle a_0, \overline{a_1, a_2, \ldots, a_n, 2a_0} \rangle.$$

We show finally that the partial quotients $a_1, a_2, \ldots, a_n$ are "symmetric," that is, $a_1 = a_n$, $a_2 = a_{n-1}, \ldots, a_n = a_1$. By (9.31), we have

$$-1/\alpha' = \frac{1}{\sqrt{d} - [\sqrt{d}]} = \overline{\langle a_n, a_{n-1}, \ldots, a_2, a_1, 2a_0 \rangle}.$$

It is also clear that $\sqrt{d} - [\sqrt{d}] = \langle 0, \overline{a_1, a_2, \ldots, a_n, 2a_0} \rangle$, from which it easily follows that $1/(\sqrt{d} - [\sqrt{d}]) = \overline{\langle a_1, a_2, \ldots, a_n, 2a_0 \rangle}$. Therefore

$$\overline{\langle a_1, a_2, \ldots, a_n, 2a_0 \rangle} = \overline{\langle a_n, a_{n-1}, \ldots, a_2, a_1, 2a_0 \rangle},$$

and hence $a_1 = a_n$, $a_2 = a_{n-1}, \ldots$.

Notes. 1. The period of $\sqrt{d}$ begins after the first term, and the partial quotients (except for the first and last) form a symmetric sequence where *the innermost term repeats precisely when the length of the period is odd*. Thus for example, $\sqrt{29} = \langle 5, \overline{2, 1, 1, 2, 10} \rangle$ and $\sqrt{33} = \langle 5, \overline{1, 2, 1, 10} \rangle$.

2. *The only partial quotients of $\sqrt{d}$ that equal $2a_0 = 2[\sqrt{d}]$ are the last terms in a period.* (See Problem 10-6.) Thus if we are calculating the continued fraction expansion of $\sqrt{29}$, the partial quotients a_i need only be computed until a value of $2a_0 = 2[\sqrt{29}] = 10$ is reached.

3. A continued fraction of the form $\langle a_0, \overline{a_1, a_2, \ldots, a_2, a_1, 2a_0} \rangle$ is not necessarily the expansion of $\sqrt{d}$ for some positive integer d. However, it is true that any continued fraction of this form is $\sqrt{r}$ for some *rational* number $r > 1$. For example, $\langle 2, \overline{1, 1, 4} \rangle = \sqrt{13/2}$ and $\langle 4, \overline{3, 8} \rangle = \sqrt{56/3}$. (See Problem 10-3.)

Conversely, if $r > 1$ *is rational and r is not the square of a rational, then $\sqrt{r}$ is of the form* $\langle a_0, \overline{a_1, a_2, \ldots, a_2, a_1, 2a_0} \rangle$. (Express $\sqrt{r}$ as $\sqrt{d}/s$, and replace $\sqrt{d}$ by $\sqrt{d}/s$ in the proof of (10.1).)

If (r, s) and (u, v) are solutions of $x^2 - dy^2 = 1$ with r, s, u, and v positive, then $r < u$ if and only if $s < v$ (since $u^2 - r^2 = d(v^2 - s^2)$). Thus if solutions in positive integers exist, there is such a solution with a *least* positive x-value. In fact, we will show in (10.6) that $x^2 - dy^2 = 1$ always has a solution in positive integers.

PELL'S EQUATION $x^2 - dy^2 = 1$ 317

(10.2) Definition. A solution (u, v) of $x^2 - dy^2 = 1$ is *positive* if $u > 0$ and $v > 0$. The *least positive solution*, or *fundamental solution*, of $x^2 - dy^2 = 1$ is the positive solution (x_1, y_1) such that $x_1 < u$ and $y_1 < v$ for every other positive solution (u, v).

More generally, if the positive solutions are ordered so that the x-values (or y-values) form an increasing sequence, then the second smallest x-value defines the *second positive solution*, the third smallest x-value the *third positive solution*, and so forth.

Note. For small values of d, a reasonable way to find the least positive solution of $x^2 - dy^2 = 1$ is simply to try successive positive values for y until the smallest value is found for which $dy^2 + 1$ is a perfect square.

To find all solutions in integers of Pell's Equation, it is clearly enough to consider only positive solutions. The next result shows that every positive solution of $x^2 - dy^2 = 1$ is found among the convergents of $\sqrt{d}$.

(10.3) Theorem. *Suppose (p, q) is a positive solution of $x^2 - dy^2 = 1$. Then p/q is one of the convergents of the continued fraction expansion of $\sqrt{d}$.*

Proof. Write $p - q\sqrt{d} = 1/(p + q\sqrt{d})$; then $p/q - \sqrt{d} = 1/q(p + q\sqrt{d})$. Since clearly $p > q\sqrt{d}$, we have $p + q\sqrt{d} > 2q\sqrt{d}$, and hence

$$0 < \frac{p}{q} - \sqrt{d} < \frac{1}{2q^2\sqrt{d}} < \frac{1}{2q^2}.$$

Thus it follows from (9.39) that p/q is a convergent of $\sqrt{d}$.

Note. If (p, q) is a positive solution of $x^2 - dy^2 = 1$, then $p^2 = dq^2 + 1$, and therefore $p/q > \sqrt{d}$. Thus by (9.9), p/q must be an *odd*-numbered convergent of $\sqrt{d}$.

The converse of (10.3) is false: *Not every convergent of $\sqrt{d}$ generates a solution of $x^2 - dy^2 = 1$.* Which convergents *do* give solutions? The answer can be given in terms of the *length* of the period of the continued fraction of $\sqrt{d}$.

(10.4) Lemma. *Let p_k/q_k denote the kth convergent of $\sqrt{d}$, and let r_k and s_k be as in (9.24). Then for any $k \geq 0$,*

$$\frac{p_k + q_k\sqrt{d}}{p_{k-1} + q_{k-1}\sqrt{d}} = \frac{r_{k+1} + \sqrt{d}}{s_k}.$$

Proof. We prove the result by induction on k. For $k = 0$, note that $r_0 = 0$ and $s_0 = 1$, since $\sqrt{d} = (0 + \sqrt{d})/1$; thus $r_1 = a_0 s_0 - r_0 = a_0$. Therefore the right side is $a_0 + \sqrt{d}$, and the left side is $(p_0 + q_0\sqrt{d})/(p_{-1} + q_{-1}\sqrt{d}) = a_0 + \sqrt{d}$.

Now suppose the result holds for the positive integer k. Since $p_{k+1} = a_{k+1}p_k + p_{k-1}$ and $q_{k+1} = a_{k+1}q_k + q_{k-1}$, it follows that

$$p_{k+1} + q_{k+1}\sqrt{d} = a_{k+1}(p_k + q_k\sqrt{d}) + p_{k-1} + q_{k-1}\sqrt{d}.$$

Dividing both sides by $p_k + q_k\sqrt{d}$ and using the induction hypothesis, we obtain

$$\frac{p_{k+1} + q_{k+1}\sqrt{d}}{p_k + q_k\sqrt{d}} = a_{k+1} + \frac{p_{k-1} + q_{k-1}\sqrt{d}}{p_k + q_k\sqrt{d}} = a_{k+1} + \frac{s_k}{r_{k+1} + \sqrt{d}}$$

$$= a_{k+1} + s_k \frac{\sqrt{d} - r_{k+1}}{d - r_{k+1}^2} = a_{k+1} + \frac{\sqrt{d} - r_{k+1}}{s_{k+1}}$$

$$= \frac{a_{k+1}s_{k+1} - r_{k+1} + \sqrt{d}}{s_{k+1}} = \frac{r_{k+2} + \sqrt{d}}{s_{k+1}},$$

which is the result with k replaced by $k+1$.

(10.5) Theorem. *Suppose that the continued fraction expansion of $\sqrt{d}$ has a period of length m. Let p_k/q_k denote the kth convergent of $\sqrt{d}$, and define r_k, s_k as in (9.24). Then*

(i) $p_{k-1}^2 - dq_{k-1}^2 = (-1)^k s_k$ *for every* $k \geq 0$;

(ii) $s_k > 0$ *for every* $k \geq 0$;

(iii) $s_k = 1$ *if and only if* $m | k$.

Proof. (i) Using the preceding lemma and taking conjugates, we find that

$$\frac{p_i + q_i\sqrt{d}}{p_{i-1} + q_{i-1}\sqrt{d}} = \frac{r_{i+1} + \sqrt{d}}{s_i} \quad \text{and} \quad \frac{p_i - q_i\sqrt{d}}{p_{i-1} - q_{i-1}\sqrt{d}} = \frac{r_{i+1} - \sqrt{d}}{s_i}$$

for all $i \geq 0$. Multiplying the two expressions together and using the fact that $d - r_{i+1}^2 = s_i s_{i+1}$, we get

$$\frac{p_i^2 - dq_i^2}{p_{i-1}^2 - dq_{i-1}^2} = -\frac{s_{i+1}}{s_i}.$$

If we take the product of the above expressions from $i = 0$ to $i = k-1$, noting the cancellations and using the fact that $p_{-1} = 1$, $q_{-1} = 0$, and $s_0 = 1$, we obtain $p_{k-1}^2 - dq_{k-1}^2 = (-1)^k s_k$.

(ii) It follows from (9.9) and the definition of an infinite continued fraction that every odd convergent of $\sqrt{d}$ is greater than $\sqrt{d}$ and every even convergent is less than $\sqrt{d}$. Thus $(p_i^2 - dq_i^2)/(p_{i-1}^2 - dq_{i-1}^2)$ is always negative, and hence s_{i+1}/s_i is always positive. Since $s_0 = 1$, it follows that $s_k > 0$ for every $k \geq 0$.

(iii) Let m be the length of the period of $\sqrt{d}$. By (10.1), $\sqrt{d}$ has a continued fraction expansion of the form

$$\sqrt{d} = \langle a_0, \overline{a_1, \ldots, a_{m-1}, 2a_0} \rangle.$$

If we define α_k as in (9.17), then $[\sqrt{d}] + \sqrt{d} = \alpha_m = \alpha_{2m} = \alpha_{3m} = \ldots$. By (9.24.iii), $\alpha_k = (r_k + \sqrt{d})/s_k$, and therefore $s_{jm} = 1$ for every $j \geq 1$.

Now suppose $s_k = 1$ for some $k \geq 1$. Since α_k has a purely periodic continued fraction expansion, (9.31) implies that α_k is reduced, and therefore $-1 < \alpha_k' < 0$. But $\alpha_k' = (r_k - \sqrt{d})/s_k = r_k - \sqrt{d}$, and so $0 < \sqrt{d} - r_k < 1$, that is, $r_k = [\sqrt{d}]$. Thus $\alpha_k = [\sqrt{d}] + \sqrt{d}$, and hence k must be a multiple of the period of $[\sqrt{d}] + \sqrt{d}$, that is, a multiple of m. This completes the proof.

Notes. 1. Since $\sqrt{d}$ has a periodic continued fraction expansion, the partial quotients a_k repeat after $m+1$ terms, where m is the length of the period of $\sqrt{d}$, and hence the integers r_k and s_k defined in (9.24) also repeat after $m+1$ terms. (See the Computational Note following (9.24).) Thus $r_i = r_{m+i}$ and $s_i = s_{m+i}$ for $i = 1, 2, 3, \ldots$.

2. For $\sqrt{d}$, the integers r_k and s_k satisfy the inequalities $0 < r_k < \sqrt{d}$ and $0 < s_k < 2\sqrt{d}$. (See Problem 10-1.)

(10.6) Theorem. *Suppose the continued fraction expansion of $\sqrt{d}$ has a period of length m.*

(i) *If m is even, then all positive solutions of $x^2 - dy^2 = 1$ are given by $x = p_{jm-1}, y = q_{jm-1}$ for $j = 1, 2, 3, \ldots$.*

(ii) *If m is odd, then all positive solutions of $x^2 - dy^2 = 1$ are given by $x = p_{2jm-1}, y = q_{2jm-1}$ for $j = 1, 2, 3, \ldots$.*

Proof. By (10.3), all positive solutions of $x^2 - dy^2 = 1$ arise from a convergent of $\sqrt{d}$. If (p_k, q_k) is any positive solution, it follows from (10.5) that k is odd and $m \mid k+1$, that is, $k = jm - 1$ for some $j \geq 1$. If m is even, any value of j is permissible, while if m is odd, then j must be even.

(10.7) Corollary. *The equation $x^2 - dy^2 = 1$ has infinitely many solutions in positive integers. If m is the length of the period of $\sqrt{d}$, then the fundamental solution is (p_{m-1}, q_{m-1}) if m is even and (p_{2m-1}, q_{2m-1}) if m is odd.*

(10.8) Example. To find the fundamental solution of $x^2 - 29y^2 = 1$, first set up the following table to find the continued fraction of $\sqrt{29}$:

k	0	1	2	3	4	5
r_k	0	5	3	2	3	5
s_k	1	4	5	5	4	1
a_k	5	2	1	1	2	10

Thus $\sqrt{29} = \langle 5, \overline{2, 1, 1, 2, 10} \rangle$. Here, the length of the period is $m = 5$, so by (10.7), the fundamental solution is (p_9, q_9). To find the fundamental solution, we construct the following table:

k	−1	0	1	2	3	4	5	6	7	8	9
a_k		5	2	1	1	2	10	2	1	1	2
p_k	1	5	11	16	27	70	727	1524	2251	3775	9801
q_k	0	1	2	3	5	13	135	283	418	701	1820

Thus the smallest positive solution of $x^2 - 29y^2 = 1$ is $x = p_9 = 9801$, $y = q_9 = 1820$.

We turn now to the question of finding further solutions once the fundamental solution is known. In the above example, the next positive solution is (p_{19}, q_{19}), by (10.6.ii). We could extend the table and compute additional convergents until we reach p_{19} and q_{19}, but there is a simpler method for finding additional solutions, which will be described in (10.10). We require the following lemma.

(10.9) Lemma. *Let (a_1, b_1) and (a_2, b_2) be arbitrary (not necessarily positive) solutions of $x^2 - dy^2 = 1$. Define integers r and s by*

$$(a_1 + b_1\sqrt{d})(a_2 + b_2\sqrt{d}) = r + s\sqrt{d}.$$

Then (r, s) is also a solution of $x^2 - dy^2 = 1$. If (a_1, b_1) and (a_2, b_2) are positive solutions, then (r, s) is itself a positive solution.

Proof. It is easy to check that $r = a_1 a_2 + b_1 b_2 d$ and $s = a_1 b_2 + a_2 b_1$; hence $(a_1 - b_1\sqrt{d})(a_2 - b_2\sqrt{d}) = r - s\sqrt{d}$. Thus

$$r^2 - ds^2 = (r + s\sqrt{d})(r - s\sqrt{d})$$
$$= (a_1 + b_1\sqrt{d})(a_2 + b_2\sqrt{d})(a_1 - b_1\sqrt{d})(a_2 - b_2\sqrt{d})$$
$$= (a_1^2 - db_1^2)(a_2^2 - db_2^2) = 1.$$

Note. The above proof is easily adapted to give a more general result: *If (a_1, b_1) is a solution of $x^2 - dy^2 = M$ and (a_2, b_2) is a solution of $x^2 - dy^2 = N$, then (r, s) is a solution of $x^2 - dy^2 = MN$.* This can also be derived from Brahmagupta's identity

$$(r^2 - ds^2)(u^2 - dv^2) = (ru \pm svd)^2 - d(rv \pm su)^2,$$

which appears in the seventh century in his study of solutions of $x^2 - dy^2 = N$.

PELL'S EQUATION $x^2 - dy^2 = 1$

(10.10) Theorem. *Let (x_1, y_1) be the fundamental solution of $x^2 - dy^2 = 1$. Then all positive solutions are given by (x_n, y_n), where the integers x_n and y_n are defined by $x_n + y_n\sqrt{d} = (x_1 + y_1\sqrt{d})^n$ for $n \geq 1$. Also, (x_n, y_n) is the nth positive solution (in the sense of (10.2)).*

Proof. It follows from (10.9) that (x_n, y_n) is a positive solution. Now suppose (r, s) is any positive solution. To simplify notation, let $\beta = r + s\sqrt{d}$, $\alpha_n = x_n + y_n\sqrt{d}$, and $\alpha'_n = x_n - y_n\sqrt{d}$. Note that $\alpha_n \alpha'_n = x_n^2 - dy_n^2 = 1$, and so $\alpha'_n = 1/\alpha_n > 0$. Also, $\alpha_n < \alpha_{n+1}$ for every n, since $\alpha_n = \alpha_1^n$ and $\alpha_1 > 1$; hence there exists $n \geq 1$ such that $\alpha_n < \beta \leq \alpha_{n+1}$. Multiply this inequality by α'_n to get $1 < \beta\alpha'_n \leq \alpha_{n+1}\alpha'_n = (\alpha_1\alpha_n)\alpha'_n = \alpha_1$.

Let $\tau = \beta\alpha'_n$; then clearly, $\tau = a + b\sqrt{d}$ for integers a and b (see the proof of (10.9)). Since $(x_n, -y_n)$ is a solution of $x^2 - dy^2 = 1$, the preceding lemma implies that (a, b) is also a solution.

Since $\tau > 1$ and $\tau\tau' = a^2 - db^2 = 1$, we have $0 < \tau' < 1$; therefore $1 < a + b\sqrt{d}$ and $0 < a - b\sqrt{d} < 1$. By adding and subtracting these two inequalities, we find that a and b are positive. Thus (a, b) is a positive solution such that $a + b\sqrt{d} \leq x_1 + y_1\sqrt{d}$, and therefore, since (x_1, y_1) is the *least* positive solution, we must have $a = x_1$ and $b = y_1$. Hence $\beta = \alpha^{n+1}$.

Finally, it is clear from the definition of x_n and y_n that $x_1 < x_2 < x_3 < \cdots$ and $y_1 < y_2 < y_3 < \cdots$. Since *all* positive solutions are given by (x_n, y_n) for $n \geq 1$, it follows from Definition 10.2 that (x_n, y_n) is the nth positive solution.

Notes. 1. Taking n to be negative (or zero) in (10.10) yields nonpositive solutions. In fact, *any* solution of $x^2 - dy^2 = 1$ is of the form (x_n, y_n) or $(-x_n, -y_n)$ for a suitable integer n. (See Problem 10-29.)

2. If we use (10.6), it follows that $(x_n, y_n) = (p_{nm-1}, q_{nm-1})$ or $(x_n, y_n) = (p_{2nm-1}, q_{2nm-1})$, depending on whether m (the length of the period of $\sqrt{d}$) is even or odd.

(10.11) Corollary. *The integers x_n and y_n defined in (10.10) satisfy the following recurrence relations:*

$$x_{n+1} = x_1 x_n + y_1 y_n d, \qquad y_{n+1} = x_1 y_n + y_1 x_n.$$

Proof. These follow at once from (10.10) by observing that $x_{n+1} + y_{n+1}\sqrt{d} = (x_1 + y_1\sqrt{d})(x_1 + y_1\sqrt{d})^n = (x_1 + y_1\sqrt{d})(x_n + y_n\sqrt{d}) = (x_1 x_n + y_1 y_n d) + (x_1 y_n + y_1 x_n)\sqrt{d}$.

(10.12) Example. In (10.8), we showed that the fundamental solution of $x^2 - 29y^2 = 1$ is $(p_9, q_9) = (9801, 1820)$. To find the next positive solution, use the recurrence relations in (10.11), with $n = 1$, to get

$$x_2 = 9801 \cdot 9801 + 1820 \cdot 1820 \cdot 29 = 192119201$$
$$y_2 = 9801 \cdot 1820 + 1820 \cdot 9801 = 35675640.$$

In particular, it follows that there is no positive solution of $x^2 - 29y^2 = 1$ with x between 9801 and 192119201. By (10.6), the next solution after (p_9, q_9) is (p_{19}, q_{19}), and hence $p_{19} = 192119201$ and $q_{19} = 35675640$.

The Equation $x^2 - dy^2 = -1$

In this section, we study the related equation $x^2 - dy^2 = -1$. An argument that is virtually identical to the proof of (10.3) shows that any solution of this equation must again be of the form (p, q), where p/q is a convergent of $\sqrt{d}$ (see Problem 10-33). It follows from (10.5) that (p_k, q_k) is a solution if and only if $m \mid k + 1$. Thus $k = jm - 1$ for some $j \geq 1$, and so m must be odd. Combined with (10.6), we therefore have the following result.

(10.13) Theorem. *Let m be the length of the period of $\sqrt{d}$, and let p_k/q_k denote the kth convergent of $\sqrt{d}$.*

(i) If m is even, the equation $x^2 - dy^2 = -1$ has no solutions.

(ii) If m is odd, then all positive solutions of $x^2 - dy^2 = -1$ are given by $x = p_{jm-1}$, $y = q_{jm-1}$ for $j = 1, 3, 5, 7, \ldots$, and all positive solutions of $x^2 - dy^2 = 1$ are given by $x = p_{jm-1}$, $y = q_{jm-1}$ for $j = 2, 4, 6, 8, \ldots$. The least positive solution of $x^2 - dy^2 = -1$ is (p_{m-1}, q_{m-1}), and the least positive solution of $x^2 - dy^2 = 1$ is (p_{2m-1}, q_{2m-1}).

Just as in the case of $x^2 - dy^2 = 1$, further solutions of the equation $x^2 - dy^2 = -1$ can be found from its least positive solution without computing additional convergents of $\sqrt{d}$. In fact, the solutions to both equations can be characterized in the following way.

(10.14) Theorem. *Suppose that $x^2 - dy^2 = -1$ has a solution, and let (r, s) denote the smallest positive solution. For $n \geq 1$, define positive integers x_n and y_n by $x_n + y_n\sqrt{d} = (r + s\sqrt{d})^n$. Then all positive solutions of $x^2 - dy^2 = -1$ are given by (x_n, y_n) with n odd, and all positive solutions of $x^2 - dy^2 = 1$ are given by (x_n, y_n) with n even. In particular, the fundamental solution of $x^2 - dy^2 = 1$ is (x_2, y_2).*

Proof. We first show that (x_2, y_2) is the fundamental solution of $x^2 - dy^2 = 1$. Let (g, h) denote the least positive solution of $x^2 - dy^2 = 1$; then in view of (10.13), we have $r+s\sqrt{d} < g+h\sqrt{d}$. It is clear from Brahmagupta's identity that (x_2, y_2) is a positive solution of $x^2 - dy^2 = 1$, and so, by (10.10), $x_2 + y_2\sqrt{d} = (g+h\sqrt{d})^n$ for some $n \geq 1$. If n is even, say $n = 2k$, then $(r+s\sqrt{d})^2 = (g+h\sqrt{d})^{2k}$ and hence $r+s\sqrt{d} = (g+h\sqrt{d})^k$. Because $r+s\sqrt{d} < g+h\sqrt{d}$, we must have $k=0$, that is, $r+s\sqrt{d}=1$, which is impossible. Thus n must be odd. If $n = 2k+1$, then $(r+s\sqrt{d})^2 = (g+h\sqrt{d})^{2k+1}$, and since $r+s\sqrt{d} < g+h\sqrt{d}$, it follows that $(g+h\sqrt{d})^{2k} < r+s\sqrt{d} < g+h\sqrt{d}$. Therefore $k=0$ and so $x_2 + y_2\sqrt{d} = g+h\sqrt{d}$, from which we conclude that $x_2 = g$ and $y_2 = h$.

Since (x_2, y_2) is the fundamental solution of $x^2 - dy^2 = 1$, (10.10) implies that *all* positive solutions of $x^2 - dy^2 = 1$ are given by (x_n, y_n) with n even.

If n is odd, it follows from the Note after (10.9) that (x_n, y_n) is a positive solution of $x^2 - dy^2 = -1$. We now show that any positive solution (u, v) of $x^2 - dy^2 = -1$ is of the form (x_n, y_n), where n is odd. Write $(r+s\sqrt{d})(u+v\sqrt{d}) = a+b\sqrt{d}$; then (a, b) is a positive solution of $x^2 - dy^2 = 1$, by Brahmagupta's identity. Therefore $a + b\sqrt{d} = (r+s\sqrt{d})^k$ for some even integer $k \geq 2$, and so $u + v\sqrt{d} = (r+s\sqrt{d})^{k-1} = x_{k-1} + y_{k-1}\sqrt{d}$, with $k-1$ odd. It follows that $u = x_{k-1}$ and $v = y_{k-1}$, which completes the proof.

It is generally rather time-consuming to express $(r+s\sqrt{d})^n$ in the form $u+v\sqrt{d}$. The following result provides an easy method to use on a calculator to find further positive solutions of $x^2 - dy^2 = 1$ (and $x^2 - dy^2 = -1$ if this equation has solutions) once the least positive solution has been determined. For a proof, see Problem 10-37.

(10.15) Theorem. *Let (r, s) be the least positive solution of $x^2 - dy^2 = -1$ if solutions exist; otherwise, let (r, s) be the least positive solution of $x^2 - dy^2 = 1$. For $n \geq 1$, define positive integers x_n and y_n by $x_n + y_n\sqrt{d} = (r+s\sqrt{d})^n$. Then x_n is the nearest integer to $(r+s\sqrt{d})^n/2$, and y_n is the nearest integer to $x_n/\sqrt{d}$.*

(10.16) Example. Refer to (10.8). Since $\sqrt{29}$ has a period of length 5, it follows from (10.13) that the least positive solution of $x^2 - 29y^2 = -1$ is $(p_4, q_4) = (70, 13)$; thus in the notation of (10.15), we have $(r, s) = (70, 13)$. It follows from the preceding two results that the least positive solution of $x^2 - 29y^2 = 1$ is $(x_2, y_2) = (9801, 1820)$, since x_2 is the nearest integer to $(70 + 13\sqrt{29})^2/2$ – namely, 9801 – and $y_2 = 1820$, the nearest integer to $x_2/\sqrt{29} = 9801/\sqrt{29}$. (By (10.13.ii) or (10.6.ii), it follows that $p_9 = 9801$, $q_9 = 1820$.) Similarly, the next positive solution of $x^2 - 29y^2 = -1$ after $(70, 13)$ is $(x_3, y_3) = (1372210, 254813)$, since the nearest integer to $(70+13\sqrt{29})^3/2$ is 1372210 and the nearest integer to $1372210/\sqrt{29}$ is 254813. (Thus, $p_{14} = x_3 = 1372210$ and $q_{14} = y_3 = 254813$.)

Finally, check that the nearest integer to $(70+13\sqrt{29})^4/2$ is 192119201, and the nearest integer to $192119201/\sqrt{29}$ is 35675640. Hence the next positive solution of $x^2 - 29y^2 = 1$ after $(9801, 1820)$ is $(x_4, y_4) = (192119201, 35675640)$. In particular, we have $p_{19} = x_4 = 192119201$ and $q_{19} = y_4 = 35675640$. (Incidentally, in view of (9.33), it follows that the convergent p_{19}/q_{19} approximates $\sqrt{29}$ to better than 15 decimal places.)

The Equation $x^2 - dy^2 = N$

Let N be a fixed nonzero integer. If d is negative, then $x^2 - dy^2 = N$ can have only a finite number of solutions, since $|x| \le \sqrt{N}$ and $|y| \le \sqrt{N/|d|}$. If d is a perfect square, say $d = k^2$, then we have $(x+ky)(x-ky) = N$, and there are again only a finite number of solutions, since there are only a finite number of ways to factor N. We will therefore suppose that d is a positive integer that is not a perfect square.

The equation $x^2 - dy^2 = \pm N$, for a given positive N, appears in early Greek mathematics, in relation to finding rational approximations to $\sqrt{d}$ when d is not a perfect square. In the seventh century, Brahmagupta considered the equations $x^2 - dy^2 = \pm 1$ and stated the identity

$$(x^2 - dy^2)(z^2 - dt^2) = (xz \pm dyt)^2 - d(xt \pm yz)^2,$$

which shows, in particular, that a solution of $x^2 - dy^2 = N$ multiplied (as above) by a solution of $x^2 - dy^2 = 1$ yields another solution of $x^2 - dy^2 = N$. Since Pell's Equation $x^2 - dy^2 = 1$ has infinitely many solutions, there will also be infinitely many solutions of $x^2 - dy^2 = N$ if there are any.

(10.17) Theorem. *If the equation $x^2 - dy^2 = N$ has a solution in positive integers, then it has infinitely many.*

Deciding when $x^2 - dy^2 = N$ has a solution is much more complicated in the general case than for $N = 1$. For example, for a given value of N, a solution need not arise from a convergent of $\sqrt{d}$: Simply take N to be $a^2 - db^2$, where a and b are *any* given integers. For certain values of N, however, solutions must come from a convergent of $\sqrt{d}$. (See Problem 10-58 for a proof of the following result.)

(10.18) Theorem. *If $|N| < \sqrt{d}$ and (r, s) is a positive solution of $x^2 - dy^2 = N$, then r/s is one of the convergents of the continued fraction expansion of $\sqrt{d}$.*

Notes. 1. The integers r and s need not be relatively prime. If they are not, let $e = (r, s)$; then $r/e = p_k$ and $s/e = q_k$ for some $k \geq 0$, where p_k/q_k is a convergent of $\sqrt{d}$. For example, $(10, 2)$ is a solution of $x^2 - 24y^2 = 4$, and it is easily checked that $5/1$ is the convergent p_1/q_1 of $\sqrt{24}$.

2. There can be values of N with $|N| > \sqrt{d}$ for which the conclusion of the preceding theorem also holds. For example, $x^2 - 23y^2 = -7$ has the solution $(19, 4)$, and $19/4$ is the fourth convergent of $\sqrt{23}$.

We end this section with a comment about generating further solutions of the equation $x^2 - dy^2 = N$ from a given solution. We showed at the beginning of this section that a solution of $x^2 - dy^2 = 1$ and a solution of $x^2 - dy^2 = N$ together generate another solution of $x^2 - dy^2 = N$. In particular, if (x_1, y_1) is the fundamental solution of $x^2 - dy^2 = 1$ and (u, v) is the least positive solution of $x^2 - dy^2 = N$, then the positive integers x_n and y_n defined by

$$x_n + y_n\sqrt{d} = (u + v\sqrt{d})(x_1 + y_1\sqrt{d})^n$$

give another positive solution of $x^2 - dy^2 = N$, but (x_n, y_n) *may not give every positive solution of* $x^2 - dy^2 = N$. An example of this is given in Problem 10-62.

One difficulty is that within a period of $\sqrt{d}$, a certain value of s_k may be taken on several times, and hence there may be several solutions of $x^2 - dy^2 = N$ within a single period. In fact, for $n = 2$, the formula above gives a solution (x_2, y_2) of $x^2 - dy^2 = N$ which is the first solution in the *next* period, and thus it will skip over other solutions in its own period. But if there is only one solution within a period (that is, $N = (-1)^k s_k$ for just one value of k), then the above formula *will* generate all positive solutions of $x^2 - dy^2 = N$.

It can be shown that for any nonzero N, there is a *finite* collection of solutions of $x^2 - dy^2 = N$ such that *any* solution can be obtained from one of these by "multiplying" by a solution of $x^2 - dy^2 = 1$. (See Problem 10-63.)

Pell's Equation and Sums of Two Squares

The theory developed for finding solutions to Pell's Equation can be used to obtain primitive representations as a sum of two squares of certain integers, namely, integers d for which the equation $x^2 - dy^2 = -1$ has a nontrivial solution. In particular, the method described below will produce primitive representations of any prime of the form $4k + 1$. We first prove the following result.

(10.19) Theorem. *Let p be a prime of the form $4k + 1$. Then the equation $x^2 - py^2 = -1$ has a solution; equivalently, the length of the period of $\sqrt{p}$*

326   CHAPTER 10: PELL'S EQUATION

is odd. More generally, the length of the period of $\sqrt{p^n}$ is odd for any odd integer n.

Proof. Let (u,v) be the *smallest* solution of $x^2 - py^2 = 1$ in positive integers. Since $u^2 - pv^2 = 1$ and $p \equiv 1 \pmod 4$, we have $u^2 - v^2 \equiv 1 \pmod 4$, and hence u is odd and v is even. Now $u^2 - pv^2 = 1$ if and only if $(u+1)(u-1) = pv^2$. Note that $u+1$ and $u-1$ are even and differ by 2, and therefore $(u+1, u-1) = 2$. Let $v = 2w$ and $u+1 = 2t$; then $t(t-1) = pw^2$. There are two cases to consider: Either t is of the form r^2 and $t-1$ is of the form ps^2, or $t = ps^2$ and $t-1 = r^2$. In the first case, we have $r^2 - ps^2 = t - (t-1) = 1$; but $r < u$, which contradicts the fact that (u,v) is the least positive solution of $u^2 - pv^2 = 1$. Thus we must have $t = ps^2$ and $t - 1 = r^2$; then $r^2 - ps^2 = (t-1) - t = -1$. Essentially the same argument works for p^n.

Note. The solution (r,s) obtained in the preceding proof is in fact the *least* positive solution of $x^2 - py^2 = -1$. To see this, note that by (10.14), $r + s\sqrt{d} = (p_{m-1} + q_{m-1}\sqrt{d})^n$ for an odd positive integer n, and $u + v\sqrt{d} = (p_{m-1} + q_{m-1}\sqrt{d})^2$. Since $r + s\sqrt{d} < u + v\sqrt{d}$, it follows that $n = 1$.

The technique described next is the method used by Legendre in 1808 to write a prime of the form $4k+1$ as a sum of two squares. It applies equally well, however, to any positive integer d for which $\sqrt{d}$ has a period of odd length. (An example using this technique is given in Problem 10-67.)

(10.20) Theorem. *Let d be a positive integer that is not a perfect square, and suppose that the length of the period of $\sqrt{d}$ is odd. Then d has a primitive representation as a sum of two relatively prime squares. In particular, every prime of the form $4k+1$ is a sum of two relatively prime squares.*

Proof. Since the length of the period of $\sqrt{d}$ is odd, the continued fraction expansion of $\sqrt{d}$ has the form $\langle a_0, \overline{a_1, a_2, \ldots, a_k, a_k, \ldots, a_2, a_1, 2a_0} \rangle$, where the middle term in the period repeats (see Note 1 following (10.1)). By (9.19), $\alpha_{k+1} = \langle \overline{a_k, a_{k-1}, \ldots, a_1, 2a_0, a_1, \ldots, a_{k-1}, a_k} \rangle$. Using (9.24), let $\alpha = \alpha_{k+1}$, $r = r_{k+1}$, and $s = s_{k+1}$; then the conjugate of α is $\alpha' = (r - \sqrt{d})/s$, and hence $\alpha\alpha' = (r^2 - d)/s^2$.

If β is the irrational number obtained from α by reversing the terms in the period of α, then clearly, $\beta = \alpha$, and so, by (9.31), $\beta = -1/\alpha'$. Thus $\alpha\alpha' = -1$ and therefore $(r^2 - d)/s^2 = -1$, that is, $d = r^2 + s^2$.

We prove finally that $(r,s) = 1$. Since $d = r^2 + s^2$ and $s_k s_{k+1} = d - r_{k+1}^2 = s_{k+1}^2$, we have $s_k = s_{k+1}$. If p is prime and divides both r and s, then p also divides d. Since $p \mid d$ and $p \mid s_k$, (10.5.i) implies that $p \mid p_{k-1}^2$, and hence $p \mid p_{k-1}$; similarly, $p \mid d$ and $p \mid s_{k+1}$ imply that $p \mid p_k$. But by (9.7), $p_k q_{k-1} - p_{k-1} q_k = (-1)^{k-1}$, and hence p must divide ± 1, a contradiction. Thus $(r, s) = 1$.

Note. The converse of Theorem 10.20 is not true: If a positive integer d has primitive representations, $\sqrt{d}$ need not have a period of odd length. (See Problem 10-68.)

An Application: Factoring Large Numbers

If p_{k-1}/q_{k-1} is a convergent of the continued fraction expansion of $\sqrt{d}$, (10.5.i) implies that $p_{k-1}^2 - dq_{k-1}^2 = (-1)^k s_k$. It often happens that s_k is a perfect square for an even value of k, and this is the basis of an efficient algorithm for factoring large numbers, which we describe next.

(10.21) Discussion. Let N be an odd positive integer. If N can be written as $a^2 - b^2$ with $a - b \neq 1$, then N is composite and can be factored as $(a-b)(a+b)$. This technique, known as *Fermat factorization*, is not very efficient since it requires us to find a square of the form $a^2 - N$, and this may involve checking up to $N/6$ values of a if N is composite.

There is a weaker condition, however, that also leads to a factorization of N. Suppose there are positive integers a and b such that

$$a^2 \equiv b^2 \pmod{N} \quad \text{with} \quad b < a < N \quad \text{and} \quad a + b \neq N.$$

Then $a^2 - b^2 = (a-b)(a+b)$ is divisible by N, and the restrictions on a and b imply that neither $a - b$ nor $a + b$ is divisible by N. Now apply the Euclidean Algorithm to find $d_1 = (a - b, N)$ and $d_2 = (a + b, N)$. It is clear that d_1 and d_2 are then proper divisors of N, that is, $1 < d_1 < N$ and $1 < d_2 < N$. (See Problem 10-72.)

The congruence $x^2 \equiv y^2 \pmod{N}$ is known as *Legendre's congruence*. Thus the problem of factoring N reduces to finding solutions of Legendre's congruence that produce nontrivial divisors of N. The method we will use to find them, called *Legendre's Factoring Method*, utilizes the continued fraction expansion of $\sqrt{N}$. (Legendre's Method is one of a family of factorization algorithms that use continued fractions. See the Notes at the end of the chapter.)

(10.22) Legendre's Factoring Method. Let N be an odd positive integer that is not a perfect square. Let $d = N$ and define the integers r_k and s_k as in (9.24). Then by (10.5.i),

$$p_{k-1}^2 - Nq_{k-1}^2 = (-1)^k s_k,$$

that is,
$$p_{k-1}^2 \equiv (-1)^k s_k \pmod{N}.$$

If k is even and s_k is a square, say $s_k = c^2$, then

$$x \equiv p_{k-1} \pmod{N} \quad \text{and} \quad y \equiv c \pmod{N}$$

is a solution of Legendre's congruence. If this is a trivial solution, that is, if $p_{k-1} \equiv \pm c \pmod{N}$, then the method outlined above produces only the obvious factors 1 and N. But if $p_{k-1} \not\equiv \pm c \pmod{N}$, then this method yields proper divisors of N. (See Problems 10-72 and 10-78.)

Legendre's Factoring Method can be summarized as follows.

1. Let N be an odd positive integer that is not a perfect square. First compute r_k, s_k, a_k, and p_k. This can be done in *one* table. (See the following example.) Note that q_k need not be calculated. The values of p_k should be reduced modulo N as soon as they exceed N to keep the numbers of manageable size.

2. When an s_k with k even is generated, test whether s_k is a perfect square. (Ignore all of the s_k with k odd.)

3. If $s_k = c^2$ and $p_{k-1} \not\equiv \pm c \pmod{N}$, let $a = p_{k-1}$ and $b = c$. (If $p_{k-1} \equiv \pm c \pmod{N}$, look for the next s_k that is a square.)

4. Use the Euclidean Algorithm to find $d_1 = (a-b, N)$ and $d_2 = (a+b, N)$. Then d_1 and d_2 are proper divisors of N. In fact, $d_1 d_2 = N$ (see Problem 10-77); thus the Euclidean Algorithm need only be applied once to find, say, d_1; then d_2 is simply N/d_1.

5. If either d_1 or d_2 is composite, the process can be repeated to find additional factors of N.

Note. It can happen that the period of $\sqrt{N}$ is very short; for example, $\sqrt{n^2+1} = \langle n, \overline{2n} \rangle$ for every $n \geq 1$. When this occurs, it is less likely that we will find usable squares among the s_k. In this case, N can be multiplied by another positive integer, called a *doping factor*, to extend the length of the period. See Problems 10-81 to 10-83.

This technique can also be used even if the period of $\sqrt{N}$ is long but the factoring algorithm produces no usable squares for, say, the first 500 values of s_k. If this occurs, we can use a doping factor and start the process again.

(10.23) Example. We illustrate Legendre's Factoring Method for the relatively small integer $N = 76183$; a much larger value of N is used in Problem 10-76. (The number 76183 was chosen because the length of its period is only 20. The period of the integer used in Problem 10-76 has 1178 terms!) Construct the following table:

k	0	1	2	3	4	5	6
r_k	0	276	270	199	269	270	266
s_k	1	7	469	78	49	67	81
a_k	276	78	1	6	11	8	6
p_k	276	21529	21805	(76176)	(21728)	(21451)	

The numbers in parentheses give the least positive residues of p_k modulo 76183; for example, $76176 \equiv 6 \cdot 21805 + 21529 \pmod{76183}$. We still apply the formula $p_{k+1} = a_{k+1}p_k + p_{k-1}$, but whenever possible, we use the least positive residue of a number.

The first s_k that is a square is $s_4 = 49$; therefore $p_3^2 \equiv s_4 \pmod{76183}$, that is, $76176^2 \equiv 7^2 \pmod{76183}$. However, since $76176 \equiv -7 \pmod{76183}$, Legendre's Factoring Method does not produce proper divisors of 76183.

The next s_k that is a square is $s_6 = 81$; thus $p_5^2 \equiv s_6 \pmod{76183}$, which implies that $21451^2 \equiv 9^2 \pmod{76183}$. Since $(21451 - 9, 76183) = (21442, 76183) = 71$ and $(21451 + 9, 76183) = (21460, 76183) = 1073$, it follows that 71 *and* 1073 *are divisors of* 76183. In fact, $76183 = 71 \cdot 1073$. (The number 1073 is not prime, however; $1073 = 29 \cdot 37$.)

Note. If m is the length of the period of $\sqrt{N}$, then $s_m = 1$. In this example, since $m = 20$ is even, we have $p_{19}^2 \equiv s_{20} \pmod{76183}$, that is, $6178^2 \equiv 1 \pmod{76183}$, which implies that 37 *and* 2059 *are divisors of* 76183. In general, however, it is not practical to use $s_m = 1$ to look for factors of N, since s_m is the *last* entry in a period which could be very long and thus require a great deal of computation to find p_{m-1}.

We mention finally that even *nonsquare* values of s_k can be used to factor N. If $s_i = s_j = a$, with i and j of the same parity, then $(p_i p_j)^2 \equiv (-1)^{i+j} s_i^2 = a^2 \pmod{N}$, and so Legendre's Factoring Method can be applied. When m is even, we have $s_{m-k} = s_k$, and hence we can use this technique beginning with the innermost pair of s_k values, then the next symmetric pair, and so on.

This method works well when the period is not very long, since in this case, there may not be many (or any) usable squares to work with. An example using this approach is given in Problem 10-84.

PROBLEMS AND SOLUTIONS

The Continued Fraction Expansion of $\sqrt{d}$

10-1. Let $\alpha = \sqrt{d}$ (where d is not a perfect square), and define α_k, r_k, and s_k as in Theorem 9.24. Prove that $0 < r_k < \sqrt{d}$ and $0 < s_k < 2\sqrt{d}$ for every $k \geq 1$. (Hint. Use (10.5).)

Solution. By (10.5.ii), $s_k > 0$ for all k, and $0 < s_k = (d - r_k^2)/s_{k-1}$ implies that $r_k < \sqrt{d}$. Now suppose that $r_k \le 0$. Then $s_{k-1} \le a_{k-1}s_{k-1} \le r_{k-1} < \sqrt{d}$, since $r_k = a_{k-1}s_{k-1} - r_{k-1}$. From (9.24), $s_k = (d - r_k^2)/s_{k-1}$; thus $1 > \alpha_{k-1} - a_{k-1} = 1/\alpha_k = s_k/(\sqrt{d} + r_k) = (\sqrt{d} - r_k)/s_{k-1} \ge \sqrt{d}/s_{k-1} > 1$, a contradiction. Hence $r_k > 0$.

Finally, $r_{k+1} = a_k s_k - r_k$ implies that $s_k = (r_k + r_{k+1})/a_k \le r_k + r_{k+1} < 2\sqrt{d}$, and hence $s_k < 2\sqrt{d}$.

10-2. Apply the preceding problem to get an upper bound for the length of the period of $\sqrt{d}$. (Hint. Use (9.24.iii).)

Solution. There are at most $[\sqrt{d}]$ values for r_k and $[2\sqrt{d}]$ values for s_k; thus there are at most $[\sqrt{d}][2\sqrt{d}] < 2d$ distinct pairs r_k, s_k. Therefore there are positive integers i and j, with $i < j \le 2d$, such that $r_i = r_j$ and $s_i = s_j$. It follows from (9.24.iii) that $\alpha_i = \alpha_j$, and hence $\alpha_{i+t} = \alpha_{j+t}$ for every $t \ge 0$. Consequently, the partial quotients of $\sqrt{d}$ repeat, and the length of the period of $\sqrt{d}$ must be less than $2d$.

Note. In fact, the length of the period of $\sqrt{d}$ is less than $.72\sqrt{d} \log d$ and is often much shorter.

10-3. Let $\alpha = \langle a_0, \overline{a_1, a_2, \ldots, a_2, a_1, 2a_0} \rangle$, where the sequence $a_1, a_2, \ldots, a_2, a_1$ of partial quotients is symmetric. Show that $\alpha = \sqrt{r}$, where r is a rational number greater than 1. (Hint. Use (9.31).)

Solution. Since α has a periodic continued fraction expansion, α is a quadratic irrational. Let $\alpha = (a + \sqrt{d})/b$; then $\alpha = a_0 + 1/\beta$, where $\beta = \langle \overline{a_1, a_2, \ldots, a_2, a_1, 2a_0} \rangle$, and so $-1/\beta' = a_0 - \alpha'$. But by (9.31), $-1/\beta' = \langle \overline{2a_0, a_1, a_2, \ldots, a_2, a_1} \rangle = a_0 + \alpha$. Hence $a_0 + \alpha = a_0 - \alpha'$, i.e., $\alpha + \alpha' = 0$. It follows immediately that $a = 0$ and $\alpha = \sqrt{d}/b = \sqrt{d/b^2}$. Let $r = d/b^2$. Since $2a_0$ must be positive, we have $a_0 \ge 1$, and therefore $r > 1$.

▷ **10-4.** Let $\sqrt{d} = \langle a_0, \overline{a_1, \ldots, a_{m-1}, 2a_0} \rangle$, and define r_k and s_k as in (9.24). Prove that the values of r_k and s_k are symmetric within each period. In particular, if m is the length of the period of $\sqrt{d}$, prove that $s_i = s_{m-i}$ ($i = 0, 1, 2, \ldots, m$) and $r_{i+1} = r_{m-i}$ ($i = 0, 1, 2, \ldots, m-1$). (More generally, for any $t \ge 0$, it is true that $s_{tm+i} = s_{tm+m-i}$ and $r_{tm+i+1} = r_{tm+m-i}$ for the above values of i.)

Solution. The proof is by induction on i. In view of (10.1), we may write $\sqrt{d} = \langle a_0, \overline{a_1, a_2, \ldots, a_1, 2a_0} \rangle$, where $a_1 = a_{m-1}$, $a_2 = a_{m-2}$, ... If $\alpha_k = \langle a_k, a_{k+1}, \ldots \rangle$, then $\alpha_k = (r_k + \sqrt{d})/s_k$, by (9.24.iii). By periodicity, $\alpha_1 = \alpha_{m+1}$ and hence $r_1 = r_{m+1}$, $s_1 = s_{m+1}$. Using (10.5.iii), we have $s_0 = s_m = 1$.

To show $r_1 = r_m$, note that $a_0 s_0 - r_0 = r_1$, and hence $r_1 = a_0$ since $r_0 = 0$. By periodicity, we have $\alpha_1 = \alpha_{m+1}$, i.e., $(r_1 + \sqrt{d})/s_1 = (r_{m+1} + \sqrt{d})/s_{m+1}$. Thus $(s_{m+1} - s_1)\sqrt{d} = s_1 r_{m+1} - s_{m+1} r_1$, and since $\sqrt{d}$ is irrational, we must have $s_1 = s_{m+1}$, and therefore $r_1 = r_{m+1}$. By definition, $a_m s_m - r_m = r_{m+1}$; since $a_m = 2a_0 = 2r_1$ and $r_{m+1} = r_1$, it follows that $2r_1 - r_m = r_1$, i.e., $r_1 = r_m$. Thus both equations hold for $i = 0$.

Now suppose we have shown that $s_i = s_{m-i}$ and $r_{i+1} = r_{m-i}$ for some $i \geq 0$. By definition (see (9.24)), we have $(d - r_{i+1}^2)/s_i = s_{i+1}$, and hence $(d - r_{m-i}^2)/s_{m-i} = s_{i+1}$. Since $(d - r_{m-i}^2)/s_{m-i} = s_{m-i-1}$ (see (9.24)), it follows that $s_{i+1} = s_{m-(i+1)}$, and hence the first equation is established for $i+1$.

We now establish the second equation for $i+1$, namely, $r_{i+2} = r_{m-(i+1)}$. By definition, $r_{i+2} = a_{i+1}s_{i+1} - r_{i+1}$. Since $s_{i+1} = s_{m-(i+1)}$, $a_{i+1} = a_{m-(i+1)}$, and $r_{i+1} = r_{m-i}$ (the induction hypothesis), it follows that $r_{i+2} = a_{m-(i+1)}s_{m-(i+1)} - r_{m-i} = r_{m-(i+1)}$.

Finally, the more general result follows from the above equations and the periodicity of the continued fraction expansion of $\sqrt{d}$. Since $\alpha_{tm+i} = \alpha_i$ and $\alpha_{tm+m-i} = \alpha_{m-i}$, we have $s_{tm+i} = s_i = s_{m-i} = s_{tm+m-i}$; similarly, $\alpha_{tm+i+1} = \alpha_{i+1}$ and $\alpha_{tm+m-i} = \alpha_{m-i}$ imply that $r_{tm+i+1} = r_{i+1} = r_{m-i} = r_{tm+m-i}$.

10-5. Suppose $\sqrt{d} = \langle a_0, \overline{a_1, \ldots, a_{m-1}, 2a_0} \rangle$. Prove that $a_k \leq a_0$ if $k \leq m-1$. (Hint. Use (9.24) and Problem 10-1.)

Solution. We may suppose that $k \geq 1$. Define α_k, r_k, and s_k as in (9.24); then $\alpha_k = (r_k + \sqrt{d})/s_k$ and $a_k = [\alpha_k]$. By (10.5), $s_k \geq 2$ since $k < m$ (m is the length of the period of $\sqrt{d}$). Also, Problem 10-1 implies that $0 < r_k < \sqrt{d}$. Hence $\alpha_k < (2\sqrt{d})/2 = \sqrt{d}$, and so $a_k = [\alpha_k] \leq [\sqrt{d}] = a_0$.

10-6. Suppose $\sqrt{d} = \langle a_0, \overline{a_1, \ldots, a_{m-1}, 2a_0} \rangle$. If $a_k = 2a_0$ for some k, must k be a multiple of m, the length of the period of $\sqrt{d}$? (Use the preceding problem.)

Solution. Yes; this follows directly from the preceding problem, since every term in the period, except the last, must be less than or equal to a_0. Thus $a_k = 2a_0$ if and only if a_k is the last term in a period, i.e., if and only if k is a multiple of m.

▷ **10-7.** *Prove that the continued fraction expansion of $\sqrt{d}$ has a period of length 1 if and only if $d = n^2 + 1$ for some $n \geq 1$.*

Solution. By Problem 9-24, if $d = n^2 + 1$, then $\sqrt{d} = \langle n, \overline{2n} \rangle$. Now suppose that the continued fraction expansion of $\sqrt{d}$ has a period of length 1; then $\sqrt{d} = \langle n, \overline{2n} \rangle$ by (10.1). Let $\alpha = \langle \overline{2n} \rangle$; thus $\alpha = 2n + 1/\alpha$, i.e., $\alpha^2 - 2n\alpha - 1 = 0$. Hence $\alpha = n + \sqrt{n^2 + 1}$, and therefore $\sqrt{d} = n + 1/\alpha = n + (\sqrt{n^2+1} - n) = \sqrt{n^2+1}$.

▷ **10-8.** *Show that the period of $\sqrt{d}$ can be arbitrarily long by proving the following result: If m is a positive integer, there is a positive integer d, not a perfect square, such that the length of the period of $\sqrt{d}$ is m.*

Solution. We illustrate the technique by producing a positive integer d such that $\sqrt{d}$ has period of length 5. Let n be a positive integer, and define $\alpha = \langle n, \overline{2, 2, 2, 2, 2n} \rangle$; then $\alpha = n + 1/\langle 2, 2, 2, 2, n + \alpha \rangle$, by (9.14). It follows from (9.4) that $\alpha - n = ((n+\alpha)q_3 + q_2)/((n+\alpha)p_3 + p_2)$, where p_i/q_i are the convergents of $\langle 2, 2, 2, 2 \rangle$. It is easily checked that $q_i = p_{i-1}$ for $i = 1, 2, 3, 4$. The above equation for $\alpha - n$ can be rewritten as

$$(\alpha - n)(n + \alpha)p_3 + (\alpha - n)p_2 = (n + \alpha)q_3 + q_2;$$

since $p_2 = q_3$ and $p_3 = q_4$, this becomes

$$(\alpha^2 - n^2)q_4 = 2nq_3 + q_2. \quad (1)$$

Let $n = q_4 + 1$; then

$$2nq_3 + q_2 = 2(q_4 + 1)q_3 + q_2 = 2q_4q_3 + 2q_3 + q_2 = q_4(2q_3 + 1).$$

Thus (1) becomes $\alpha^2 = (q_4 + 1)^2 + 2q_3 + 1$. Set $d = (q_4 + 1)^2 + 2q_3 + 1$; then $\sqrt{d} = \alpha = \langle q_4 + 1, \overline{2, 2, 2, 2, 2(q_4 + 1)} \rangle$.

In general, to obtain d such that $\sqrt{d}$ has a period of length m, follow the above procedure, replacing $\langle 2, 2, 2, 2 \rangle$ by $\langle 2, 2, 2, \ldots, 2 \rangle$ (m 2's). Then q_4 is replaced by q_{m-1}, and $\sqrt{d} = \langle q_{m-1} + 1, \overline{2, 2, \ldots, 2, 2(q_{m-1} + 1)} \rangle$ has a period of length m.

10-9. Let p_k/q_k denote the kth convergent of $\sqrt{2}$. Prove that p_k is never a square if $k \geq 1$. (Hint. Use Problem 8-22.)

Solution. Since $\sqrt{2} = \langle 1, \overline{2} \rangle$ has a period of length 1, it follows from (10.13) that $p_k^2 - 2q_k^2 = \pm 1$ for every $k \geq 0$. Now suppose that $p_k = x^2$ for some k, and let $y = q_k$; then $x^4 - 2y^2 = \pm 1$. But by Problem 8-22, the equation $x^4 - 2y^2 = 1$ has no solution in positive integers, and $x^4 - 2y^2 = -1$ has only one positive solution, namely, $x = y = 1$. In this case, we must have $p_k = q_k = 1$, and hence $k = 0$.

Pell's Equation $x^2 - dy^2 = 1$

10-10. Find the first three positive solutions of $x^2 - 13y^2 = 1$.

Solution. Since $\sqrt{13} = \langle 3, \overline{1, 1, 1, 1, 6} \rangle$, (10.6) implies that the fundamental solution is $(p_9, q_9) = (649, 180)$. Now use (10.10) to generate additional solutions. Since $(649 + 180\sqrt{13})^2 = 842401 + 233640\sqrt{13}$ and $(649 + 180\sqrt{13})^3 = (649 + 180\sqrt{13})^2(649 + 180\sqrt{13}) = (842401 + 233640\sqrt{13})(649 + 180\sqrt{13}) = 1093435849 + 303264540\sqrt{13}$, the next two positive solutions are $(842401, 233640)$ and $(1093435849, 303264540)$.

10-11. Find the general form of a positive solution of $x^2 - 21y^2 = 1$ in terms of the convergents of $\sqrt{21}$, and determine the fundamental solution.

Solution. Check that $\sqrt{21} = \langle 4, \overline{1, 1, 2, 1, 1, 8} \rangle$; hence the period has length 6. By (10.6), all solutions are given by $x = p_{6j-1}$, $y = q_{6j-1}$ for $j \geq 1$, where p_k/q_k is the kth convergent of $\sqrt{21}$. Thus the fundamental solution is $(p_5, q_5) = (55, 12)$.

10-12. Let (u, v) be a positive solution of $x^2 - 2y^2 = 1$. Prove that u is not divisible by any prime of the form $8k + 5$ or $8k + 7$.

Solution. Since $u^2 = 2v^2 + 1$ is odd, any prime divisor of u must be odd. If p is prime and $p | u$, then $-2v^2 \equiv 1 \pmod{p}$, and hence $1 = (-2v^2/p) = (-2/p)$. Thus, by (5.13), p must be of the form $8k + 1$ or $8k + 3$.

PROBLEMS AND SOLUTIONS   333

10-13. *Suppose $\sqrt{d}$ has a period of odd length. Prove that $\sqrt{k^2 d}$ has a period of length 1 for infinitely many values of k. (Hint. Note that $\sqrt{n^2+1} = \langle n, \overline{2n} \rangle$, by Problem 9-24.)*

Solution. Since $\sqrt{n^2+1} = \langle n, \overline{2n} \rangle$, it suffices to show that $k^2 d = n^2 + 1$ for infinitely many values of k and n. But this is simply the equation $n^2 - dk^2 = -1$, which has infinitely many solutions, by (10.13).

10-14. *Suppose d is not a perfect square. Prove that $\sqrt{k^2 d}$ has a period of length 2 for infinitely many values of k. (Hint. Use Problem 9-25.)*

Solution. By Problem 9-25, $\sqrt{n^2-1} = \langle n-1, \overline{1, 2n-2} \rangle$. Thus it is enough to show that $k^2 d = n^2 - 1$ for infinitely many values of k and n, i.e., $n^2 - dk^2 = 1$. This equation will always have infinitely many solutions for any d, by (10.7).

10-15. *Let k be a positive integer. Prove that there exist an infinite number of solutions of $x^2 - dy^2 = 1$ for which y is a multiple of k.*

Solution. Let $D = k^2 d$; then (10.7) implies that $x^2 - Dy^2 = 1$ has infinitely many solutions. Let (a,b) be any solution of $x^2 - Dy^2 = 1$. Since $x^2 - Dy^2 = 1$ can be written as $x^2 - d(ky)^2 = 1$, it follows that (a, kb) is a solution of $x^2 - dy^2 = 1$, and the result follows.

10-16. *Let p_i/q_i be the ith convergent of $\sqrt{d}$. If k is any positive integer, use the preceding problem to show that k divides q_i for infinitely many values of i.*

Solution. Problem 10-15 shows that k divides infinitely many y-values in solutions of $x^2 - dy^2 = 1$. But by (10.3), every solution is of the form (p_i, q_i), and so the result follows.

10-17. *Let (x_1, y_1) be the fundamental solution of $x^2 - dy^2 = 1$, and define x_n and y_n by $x_n + y_n \sqrt{d} = (x_1 + y_1 \sqrt{d})^n$. Prove that $y_n \mid y_{tn}$ for every positive integer t.*

Solution. Note that $x_{tn} + y_{tn} \sqrt{d} = (x_1 + y_1 \sqrt{d})^{tn} = ((x_1 + y_1 \sqrt{d})^n)^t = (x_n + y_n \sqrt{d})^t$. If we expand $(x_n + y_n \sqrt{d})^t$ using the Binomial Theorem, then every term except for the first has a factor y_n, and hence $y_n \mid y_{tn}$.

▷ **10-18.** *By (10.10), every positive solution of $x^2 - dy^2 = 1$ is of the form (x_n, y_n), where $x_n + y_n \sqrt{d} = (x_1 + y_1 \sqrt{d})^n$ and (x_1, y_1) is the fundamental solution. Prove that there are infinitely many primes p such that p divides x_n for infinitely many values of n. (Hint. Use Problem 10-15.)*

Solution. We show first that for any finite collection $\{p_1, p_2, \ldots, p_j\}$ of primes, there is a prime p not in the collection and an integer n such that $p \mid x_n$. Let $P = p_1 p_2 \cdots p_j$ and define $D = dP^2$. By Problem 10-15, there is a y_m such that $P \mid y_m$. Since x_m and

y_m are relatively prime, it follows that none of the primes $p_1, p_2, \ldots, p_j$ can divide x_m, and so there is some prime p not in the collection which divides x_m.

We now prove that if the prime p divides x_m, then p divides x_i for infinitely many values of i. By Problem 10-15, there are infinitely many values of n for which $p \mid y_n$. But $(x_m + y_m\sqrt{d})(x_n + y_n\sqrt{d}) = x_{m+n} + y_{m+n}\sqrt{d}$, where $x_{m+n} = x_m x_n + y_m y_n d$. Thus if $p \mid y_n$, it follows that $p \mid x_{m+n}$. (In fact, in view of Problem 10-17, if $p \mid y_n$, then $p \mid x_{m+tn}$ for every $t \geq 1$.) Therefore any prime divisor of x_m divides infinitely many of the x_i.

The following problem shows that *all* values of p_k and q_k can be computed once their values in the first period are known.

10-19. Let p_i/q_i denote the ith convergent of $\sqrt{d}$, and let m be the length of the period of $\sqrt{d}$. Prove that $p_{i+m} + q_{i+m}\sqrt{d} = (p_i + q_i\sqrt{d})(p_{m-1} + q_{m-1}\sqrt{d})$ for any $i \geq -1$. (Hint. Use (10.4).)

Solution. We will show that $(p_{i+m} + q_{i+m}\sqrt{d})/(p_i + q_i\sqrt{d})$ is constant; setting $i = -1$ then implies that the ratio is $p_{m-1} + q_{m-1}\sqrt{d}$, since $p_{-1} = 1$ and $q_{-1} = 0$.

By periodicity, $r_{i+m+1} = r_{i+1}$ and $s_{i+m} = s_i$. Let $\beta_k = p_k + q_k\sqrt{d}$. Using (10.4), we obtain

$$\frac{\beta_{i+m}}{\beta_{i+m-1}} = \frac{r_{i+m+1} + \sqrt{d}}{s_{i+m}} = \frac{r_{i+1} + \sqrt{d}}{s_i} = \frac{\beta_i}{\beta_{i-1}}.$$

Therefore $\beta_{i+m}/\beta_i = \beta_{i+m-1}/\beta_{i-1}$, and the result follows.

10-20. Use the preceding problem to show that if the period of $\sqrt{d}$ has length m, then $p_{km-1} + q_{km-1}\sqrt{d} = (p_{m-1} + q_{m-1}\sqrt{d})^k$ for any positive integer k.

Solution. The result is obviously true for $k = 1$. It is therefore enough to show that for any positive integer j,

$$p_{(j+1)m-1} + q_{(j+1)m-1}\sqrt{d} = (p_{jm-1} + q_{jm-1}\sqrt{d})(p_{m-1} + q_{m-1}\sqrt{d}).$$

This follows immediately from the preceding problem by setting $i = jm - 1$.

▷ **10-21.** Let p_i/q_i denote the ith convergent of $\sqrt{d}$. Prove that for any given p_i, there are infinitely many values of j such that p_j is a multiple of p_i. (Hint. Use Problem 10-19.)

Solution. Let (x_n, y_n) denote the nth positive solution of the equation $x^2 - dy^2 = 1$. Then for any $i \geq 1$, Problem 10-19, (10.6) and (10.10) show that $p_{i+tm} + q_{i+tm}\sqrt{d} = (p_i + q_i\sqrt{d})(x_t + y_t\sqrt{d})$, and thus $p_{i+tm} = x_t p_i + d y_t q_i$. By the result of Problem 10-15, there are infinitely many t such that $p_i \mid y_t$, and for any such t, we have $p_i \mid p_{i+tm}$.

10-22. (a) Prove that the product $(r+s\sqrt{d})(u+v\sqrt{d})$ is also of the form $x+y\sqrt{d}$.
(b) Recall that the conjugate of $\gamma = x + y\sqrt{d}$ is the number $\gamma' = x - y\sqrt{d}$. Prove that $(\alpha\beta)' = \alpha'\beta'$ (i.e., the conjugate of a product is the product of the conjugates). In particular, show that the conjugate of $(r+s\sqrt{d})^n$ is $(r-s\sqrt{d})^n$.

Solution. (a) $(r + s\sqrt{d})(u + v\sqrt{d}) = (ru + svd) + (rv + su)\sqrt{d}$.

PROBLEMS AND SOLUTIONS   335

(b) Let $\alpha = r + s\sqrt{d}$ and $\beta = u + v\sqrt{d}$. By (a), we have $(\alpha\beta)' = (ru + svd) - (rv + su)\sqrt{d} = (r - s\sqrt{d})(u - v\sqrt{d}) = \alpha'\beta'$. An easy induction argument shows that the conjugate of a product of n factors is the product of the conjugates. In particular, the conjugate of $(r + s\sqrt{d})^n$ is $(r - s\sqrt{d})^n$.

10-23. Let (x_{10}, y_{10}) be the tenth positive solution of $x^2 - 3y^2 = 1$. (a) Determine if $x_{10} + y_{10}\sqrt{3}$ is between 500000 and 600000. (b) Use (10.15) to find x_{10} and y_{10}.

Solution. (a) Since $\sqrt{3} = \langle 1, \overline{1, 2} \rangle$, the period has length $m = 2$, and so, by (10.6), the fundamental solution is $(p_1, q_1) = (2, 1)$. Using (10.10) and Note 2 following it, we have $x_{10} + y_{10}\sqrt{3} = (2 + \sqrt{3})^{10} = 524174$ (to the nearest integer).

By (10.15), x_{10} is the nearest integer to $(2 + \sqrt{3})^{10}/2$, namely, 262087. Thus $y_{10} = 151316$, the nearest integer to $x_{10}/\sqrt{3}$.

10-24. Use (10.15) to find the seventh convergent p_7/q_7 of $\sqrt{20}$.

Solution. Note that $\sqrt{20} = \langle 4, \overline{2, 8} \rangle$; Hence $(p_1, q_1) = (9, 2)$ is the fundamental solution of $x^2 - 20y^2 = 1$. By (10.6), (p_7, q_7) is the fourth positive solution, namely, (x_4, y_4). Hence, by (10.15), x_4 is 551841, the nearest integer to $(9 + 2\sqrt{20})^4/2$; thus $y_4 = 11592$, the nearest integer to $x_4/\sqrt{20}$. Therefore $p_7/q_7 = 51841/11592$.

▷ **10-25.** Let (a, b) be the fundamental solution of $x^2 - dy^2 = 1$.
  (a) Prove that $0 < a - b\sqrt{d} < \sqrt{2} - 1$.
  (b) Suppose $\sqrt{d}$ has a period of length m. Prove that $0 < a - b\sqrt{d} < 1/m$ if m is even and $0 < a - b\sqrt{d} < 1/2m$ if m is odd. (Hint. Use (9.33).)

Solution. (a) Since $a > b$, we have $a + b\sqrt{d} > 1 + \sqrt{d} \geq 1 + \sqrt{2}$. Thus $a - b\sqrt{d} = 1/(a + b\sqrt{d}) < 1/(1 + \sqrt{2}) = \sqrt{2} - 1$. Clearly, $a - b\sqrt{d}$ must be positive, since $(a + b\sqrt{d})(a - b\sqrt{d}) = a^2 - db^2 = 1$.

(b) By (10.6), $a = p_{m-1}$, $b = q_{m-1}$ or $a = p_{2m-1}$, $b = q_{2m-1}$, according as m is even or odd. First suppose m is even. Using (9.33), we have $a - b\sqrt{d} = b|\sqrt{d} - a/b| < q_{m-1} \cdot 1/(q_{m-1}q_m) = 1/q_m \leq 1/m$, since $q_m \geq m$ (see the Note following (9.3)). Similarly, if m is odd, then $a - b\sqrt{d} < 1/q_{2m} \leq 1/2m$.

10-26. Prove that the sum of the first n integers is a perfect square for infinitely many even values of n and infinitely many odd values of n. Find the first six such n.

Solution. Note that $1 + 2 + \cdots + n = n(n + 1)/2$; thus we want $n(n + 1) = 2a^2$. Since $(n, n + 1) = 1$, if n is even, we look for r and s such that $n = 2r^2$ and $n + 1 = s^2$. Hence $s^2 = n + 1 = 2r^2 + 1$, and so we need r and s such that $s^2 - 2r^2 = 1$, which, by (10.7), has infinitely many solutions. Similarly, if n is odd, we look for r and s such that $n + 1 = 2r^2$ and $n = s^2$, i.e., r and s such that $s^2 - 2r^2 = -1$, which, by (10.13), also has infinitely many solutions.

By (10.13), the first three solutions of $x^2 - 2y^2 = -1$ are $(1,1)$, $(7,5)$, and $(41,29)$, giving $n = 1, 49$, and 1681. Similarly, the first three solutions of $x^2 - 2y^2 = 1$ are $(3,2)$, $(17,12)$, and $(99,70)$, which give $n = 8, 288$, and 9800.

10-27. *Let (a,b) be any solution of $x^2 - dy^2 = 1$. Prove that (a,b) is a positive solution if and only if $a + b\sqrt{d} > 1$.*

Solution. If (a,b) is a positive solution, then $a + b\sqrt{d} \geq 1 + \sqrt{d} > 1$. Now suppose that $a + b\sqrt{d} > 1$. Clearly, a and b cannot be both negative. If exactly one of a and b is less than or equal to 0, then $|a - b\sqrt{d}| \geq |a + b\sqrt{d}|$, and thus $|a - b\sqrt{d}||a + b\sqrt{d}| > 1$, contradicting the fact that $a^2 - db^2 = 1$. It follows that a and b must be positive.

10-28. *Does there exist a positive solution of $x^2 - 23y^2 = 1$ with $24 < x < 1151$?*

Solution. No. Since $\sqrt{23} = \langle 4, \overline{1,3,1,8} \rangle$, (10.6) implies that the fundamental solution is $(p_3, q_3) = (24, 5)$. In view of (10.10), since $(24 + 5\sqrt{23})^2 = 1151 + 240\sqrt{23}$, the next positive solution is $(1151, 240)$.

10-29. *Let n be any integer (positive, negative, or zero), and define x_n and y_n as in (10.10). Prove that every solution of $x^2 - dy^2 = 1$ (allowing for all variations of signs) is of the form (x_n, y_n) or $(-x_n, -y_n)$.*

Solution. For $n = 0$, we have $x_0 + y_0\sqrt{d} = 1$, and hence $x_0 = 1$, $y_0 = 0$. Thus the trivial solutions $(1,0)$ and $(-1,0)$ are precisely (x_0, y_0) and $(-x_0, -y_0)$, respectively. Now let (r,s) be any positive solution of $x^2 - dy^2 = 1$. Then, by (10.10), $r = x_n$ and $s = y_n$ for some positive integer n, and the solution $(-r, -s)$ is just $(-x_n, -y_n)$.

Since $(x_1 + y_1\sqrt{d})(x_1 - y_1\sqrt{d}) = 1$, we have $x_1 - y_1\sqrt{d} = (x_1 + y_1\sqrt{d})^{-1}$, and therefore $x_n - y_n\sqrt{d} = (x_1 + y_1\sqrt{d})^{-n} = (x_{-n}, y_{-n})$. Thus the solution $(r, -s)$ is (x_{-n}, y_{-n}), and the solution $(-r, s)$ is $(-x_{-n}, -y_{-n})$.

When we solve the Diophantine equation $x^2 - dy^2 = 1$, we are finding *lattice points* (that is, points with integer coordinates) on the hyperbola $x^2 - dy^2 = 1$. Note the close connection between the formulas obtained in the next problem and the usual formulas for $\cosh(a+b)$ and $\sinh(a+b)$, as well as the formulas for $\cos(a+b)$ and $\sin(a+b)$.

10-30. *Let m and n be positive integers, and for any k, define x_k, y_k as in (10.10). Show that $x_{m+n} = x_m x_n + y_m y_n d$ and $y_{m+n} = x_m y_n + y_m x_n$.*

Solution. We have

$$x_{m+n} + y_{m+n}\sqrt{d} = (x_1 + y_1\sqrt{d})^{m+n} = (x_1 + y_1\sqrt{d})^m (x_1 + y_1\sqrt{d})^n$$
$$= (x_m + y_m\sqrt{d})(x_n + y_n\sqrt{d})$$
$$= x_m x_n + y_m y_n d + (x_m y_n + y_m x_n)\sqrt{d},$$

and the result follows.

PROBLEMS AND SOLUTIONS   337

Note. Once we know x_1 and y_1, this problem can be used to calculate x_k and y_k quickly for large values of k by using a variant of the repeated squaring method for finding powers. For example, to find x_{130} and y_{130}, we find x_2, y_2, then x_4, y_4 using $m = n = 2$, then $x_8, y_8, \ldots$, and finally x_{130}, y_{130} using $m = 128$ and $n = 2$.

10-31. Let (x_k, y_k) be the kth positive solution of $x^2 - dy^2 = 1$. Show that $x_{2k}/y_{2k} = (x_k/y_k + dy_k/x_k)/2$. (Hint. See the preceding problem.)

Solution. By Problem 10-30, $x_{2k} = x_k^2 + y_k^2 d$ and $y_{2k} = 2x_k y_k$. The result now follows by dividing and simplifying.

Note. Define the rational numbers r_n by $r_0 = x_1/y_1$, $r_1 = x_2/y_2$, $r_2 = x_4/y_4$, $r_3 = x_8/y_8$, and so on. The above problem shows that in general, $r_{n+1} = (r_n + d/r_n)/2$. Note that the r_i are precisely the approximations to $\sqrt{d}$ that we obtain if we use *Newton's Method*, a technique from first-year calculus, starting with the estimate $\sqrt{d} \approx x_1/y_1$. For square roots, the method was fully described by Cataldi in 1613.

10-32. *Prove that there exist infinitely many triples of consecutive integers each of which is a sum of two squares. Find three such triples.*

Solution. By (10.7), the equation $x^2 - 2y^2 = 1$ has infinitely many positive solutions. If (a, b) is a positive solution, let $n = a^2$. Then $n + 1 = a^2 + 1$ is clearly a sum of two squares, as is $n - 1 = 2b^2$. The first three positive solutions are $(3, 2)$, $(17, 12)$, and $(99, 70)$, which give, respectively, the triples 8, 9, 10; 288, 289, 290; and 9800, 9801, 9802.

The Equation $x^2 - dy^2 = -1$

10-33. *Suppose (p, q) is a positive solution of $x^2 - dy^2 = -1$. Prove that p/q is one of the convergents of $\sqrt{d}$.*

Solution. Since $p^2 - dq^2 = -1$, we have $p - q\sqrt{d} = -1/(p + q\sqrt{d})$; thus $p/q - \sqrt{d} = -1/q(p + q\sqrt{d})$. Clearly $p \geq q$, and so $p + q\sqrt{d} > 2q$. Hence $|p/q - \sqrt{d}| < 1/q(2q) = 1/2q^2$. It now follows from (9.39) that p/q is a convergent of $\sqrt{d}$.

10-34. *Let r and s be integers, and define x_n and y_n by $x_n + y_n\sqrt{d} = (r + s\sqrt{d})^n$. Prove that $x_n = ((r + s\sqrt{d})^n + (r - s\sqrt{d})^n)/2$ and $y_n = ((r + s\sqrt{d})^n - (r - s\sqrt{d})^n)/2\sqrt{d}$. (Hint. Use Problem 10-22.)*

Solution. By definition, the conjugate of $x_n + y_n\sqrt{d}$ is $x_n - y_n\sqrt{d}$, so by part (b) of Problem 10-22, we have $x_n - y_n\sqrt{d} = (r - s\sqrt{d})^n$. The identities now follow directly.

10-35. *Use the recurrence relations given in (10.11) to find the first three positive solutions of $x^2 - 10y^2 = -1$ and $x^2 - 10y^2 = 1$.*

Solution. Check that the least positive solution of $x^2 - 10y^2 = -1$ is $(3, 1)$; thus by (10.11), $x_{n+1} = 3x_n + 10y_n$ and $y_{n+1} = x_n + 3y_n$. In view of (10.14), the first three positive solutions of $x^2 - 10y^2 = -1$ are given by (x_n, y_n) for $n = 1, 3$, and 5, and the first three positive solutions of $x^2 - 10y^2 = 1$ correspond to $n = 2, 4$, and 6. Using the above recurrence relations, we get $(x_2, y_2) = (19, 6)$; $(x_3, y_3) = (117, 37)$; $(x_4, y_4) = (721, 228)$; $(x_5, y_5) = (4443, 1405)$; and $(x_6, y_6) = (27379, 8658)$.

10-36. (a) Use (10.15) to find the first two positive solutions of $x^2 - 41y^2 = 1$ and $x^2 - 41y^2 = -1$.
 (b) Use part (a) to determine the value of k such that $q_k = 1311360$.
 (c) Use (10.13), (10.14), and (10.15) to find p_{14}/q_{14}. (Do not compute this convergent directly.)

Solution. (a) Check that $\sqrt{41} = \langle 6, \overline{2, 2, 12} \rangle$; hence the period of $\sqrt{41}$ has length $m = 3$. The smallest positive solution of $x^2 - 41y^2 = -1$ is therefore $(p_2, q_2) = (32, 5) = (x_1, y_1)$. Define x_n and y_n by $x_n + y_n\sqrt{41} = \gamma^n$, where $\gamma = 32 + 5\sqrt{41}$; then by (10.13), all solutions of $x^2 - 41y^2 = -1$ are given by (x_n, y_n) with n odd, and all solutions of $x^2 - 41y^2 = 1$ are given by (x_n, y_n) with n even. Thus if we use (10.15), the second positive solution of $x^2 - 41y^2 = -1$ is $(x_3, y_3) = (131168, 20485)$, since 131168 is the nearest integer to $\gamma^3/2$ and 20485 is the nearest integer to $131168/\sqrt{41}$.
 The least positive solution of $x^2 - 41y^2 = 1$ is $(x_2, y_2) = (2049, 320)$, since 2049 is the nearest integer to $\gamma^2/2$ and 320 is the nearest integer to $2049/\sqrt{41}$. Similarly, the nearest integer to $\gamma^4/2$ is 8396801 and the nearest integer to $8396801/\sqrt{41}$ is 1311360; thus the next positive solution is $(x_4, y_4) = (8396801, 1311360)$.
 (b) The y-value in the second positive solution of $x^2 - 41y^2 = 1$ is 1311360. But by (10.13), the second solution is also given by (p_{11}, q_{11}). Thus $q_{11} = 1311360$; hence $k = 11$.
 (c) By (10.13), (p_{14}, q_{14}) is the third positive solution of $x^2 - 41y^2 = -1$, which, by (10.14), is (x_5, y_5). Using (10.15), we find that $x_5 = 537526432$ and $y_5 = 83947525$. Thus $p_{14}/q_{14} = 537526432/83947525$ (which, incidentally, appproximates $\sqrt{41}$ to better than 16 decimal places).

10-37. *Prove Theorem 10.15: Let (r, s) be the least positive solution of $x^2 - dy^2 = -1$ if solutions exist; otherwise, let (r, s) be the least positive solution of $x^2 - dy^2 = 1$. For $n \geq 1$, define positive integers x_n and y_n by $x_n + y_n\sqrt{d} = (r + s\sqrt{d})^n$. Then x_n is the nearest integer to $(r + s\sqrt{d})^n/2$, and y_n is the nearest integer to $x_n/\sqrt{d}$. (Hint. Use Problem 10-34.)*

Solution. In either case, we have $|r^2 - ds^2| = 1$, and so $|r + s\sqrt{d}||r - s\sqrt{d}| = 1$. Hence $|r - s\sqrt{d}| = 1/(r + s\sqrt{d}) < 1/2$, since $r + s\sqrt{d} > 2$. Problem 10-34 shows that $x_n = ((r + s\sqrt{d})^n + (r - s\sqrt{d})^n)/2$ for every $n \geq 1$. Thus $|x_n - (r + s\sqrt{d})^n/2| = |r - s\sqrt{d}|^n/2 < 1/4$, so x_n is the nearest integer to $(r + s\sqrt{d})^n/2$.
 Since $|x_n - y_n\sqrt{d}| = |r - s\sqrt{d}|^n < 1/2$, we have $|y_n - x_n/\sqrt{d}| < 1/2\sqrt{2} < 1/2$, and hence y_n is the nearest integer to $x_n/\sqrt{d}$.

Note. To find successive solutions on a programmable calculator using the preceding result, note that the nearest integer to x is simply $[x + 1/2]$.

10-38. (a) *Find a positive solution of $x^2 - 99y^2 = 1$ by inspection.*

(b) *Without finding the continued fraction expansion of $\sqrt{99}$, prove that $x^2 - 99y^2 = -1$ has no solution in positive integers.*

Solution. It is easy to see that $(10, 1)$ is a solution of $x^2 - 99y^2 = 1$. In view of (10.14), since (clearly) $x_1 < x_2 < x_3 < \cdots$ and $y_1 < y_2 < y_3 < \cdots$, any positive solution (a, b) of $x^2 - 99y^2 = -1$ must have $a < 10$ and $b < 1$, which is impossible. (The same conclusion also follows from Problem 10-33 and the fact that $q_1 < q_2 < q_3 < \cdots$. See the Note following (9.3).)

Alternatively, we can see that $x^2 - 99y^2 = -1$ does not have a solution by noting that if (x, y) is a solution, then $x^2 \equiv -1 \pmod{3}$, which is impossible.

10-39. *Use Problem 10-20 to prove Theorems 10.10 and 10.13.*

Solution. Let (p, q) be a positive solution of $x^2 - dy^2 = \pm 1$; then by (10.3) and Problem 10-33, p/q is a convergent of the continued fraction expansion of $\sqrt{d}$. If m is the period of this expansion, then $p = p_{jm-1}$ and $q = q_{jm-1}$ for some integer j, by (10.5). It follows from Problem 10-20 that $p + q\sqrt{d} = (p_{m-1} + q_{m-1}\sqrt{d})^j$. Taking conjugates and multiplying, we find that $p^2 - dq^2 = (p_{m-1}^2 - dq_{m-1}^2)^j$.

If m is odd, then $p_{m-1}^2 - dq_{m-1}^2 = -1$ by (10.5.i); thus (p, q) is a solution of $x^2 - dy^2 = -1$ if j is odd and a solution of $x^2 - dy^2 = 1$ if j is even. In particular, if (p, q) is a solution of $x^2 - dy^2 = 1$, then $p + q\sqrt{d}$ is a power of $(p_{m-1} + q_{m-1}\sqrt{d})^2$. If m is even, then by the same reasoning, we conclude that $p + q\sqrt{d}$ is a power of $p_{m-1} + q_{m-1}\sqrt{d}$.

10-40. *Prove that there are infinitely many even and infinitely many odd positive integers n for which $n^2 + (n+1)^2$ is a perfect square. Find the first five such n. (Hint. First show that $n^2 + (n+1)^2 = k^2$ is equivalent to $(2n+1)^2 - 2k^2 = -1$.)*

Solution. If we multiply the equation $n^2 + (n+1)^2 = k^2$ by 2 and complete the square, we obtain the equivalent equation $(2n+1)^2 - 2k^2 = -1$. Since $\sqrt{2} = \langle 1, \overline{2} \rangle$ has a period of odd length, this equation has infinitely many solutions.

Since $(1, 1)$ is the least positive solution, (10.14) implies that all positive solutions are given by $(1+\sqrt{2})^t$, where t is an odd positive integer, or, equivalently, by $(1+\sqrt{2})(3+2\sqrt{2})^s$ for any nonnegative integer s. Thus if (a, b) is a solution, then $(3a + 4b, 2a + 3b)$ is the next solution. Modulo 4, the first component is $3a$, so starting with $a = 1$ and noting that $a = 2n + 1$ and $b = k$, we get successively $2n + 1 \equiv 1, 3, 1, 3, \ldots$ modulo 4. Hence the corresponding values of n are alternately even and odd.

Now apply (10.15): Since all positive solutions are given by $(1 + \sqrt{2})^t$, where t is odd, the first five positive values of n correspond to $t = 3, 5, 7, 9,$ and 11, which gives $n = 3, 20, 119, 696,$ and 4059. (These values can also be computed by taking $(a, b) = (1, 1)$ and calculating successive values of $(3a + 4b, 2a + 3b)$.)

Note. This problem shows that there are infinitely many primitive Pythagorean triples (x, y, z) where x and y are consecutive integers and y is even. Compare Problem 8-13.

10-41. Let $k \geq 2$. Prove that the continued fraction expansion of $\sqrt{k^2 - 1}$ has an even period by showing that the equation $x^2 - (k^2 - 1)y^2 = -1$ is not solvable.

Solution. Clearly, $(k, 1)$ is a positive solution of $x^2 - (k^2 - 1)y^2 = 1$. Arguing as in Problem 10-38, if (a, b) is the least positive solution of $x^2 - (k^2 - 1)y^2 = -1$, then $a < k$ and $b < 1$, which is impossible. Thus there are no solutions, and hence, by (10.13), the length of the period of $\sqrt{k^2 - 1}$ is even.

Another solution: Since $k^2 - 1 \equiv 0$ or $-1 \pmod 4$, $x^2 - (k^2 - 1)y^2$ is congruent to either 0, 1, or 2 modulo 4 and in particular can never be -1.

10-42. Argue as in the solution of the preceding problem to show that the length of the period of $\sqrt{k^2 + 1}$ is odd.

Solution. Since $(k, 1)$ is obviously a solution of $x^2 - (k^2 + 1)y^2 = -1$, the conclusion follows from (10.13).

10-43. The fundamental solution of $x^2 - 880y^2 = 1$ is $(89, 3)$. Without calculating the continued fraction expansion of $\sqrt{880}$, determine if $x^2 - 880y^2 = -1$ has any solutions in positive integers. Decide if the length of the period of $\sqrt{880}$ is odd or even.

Solution. If (a, b) is a solution of $x^2 - 880y^2 = -1$, then $a < 89$ and $b < 3$ (see the solution of Problem 10-38). But neither $y = 1$ nor $y = 2$ gives a solution. Thus $x^2 - 880y^2 = -1$ is not solvable. In view of (10.13), this implies that the length of the period of $\sqrt{880}$ is even.

Alternatively, we can use a congruential argument to show that $x^2 - 880y^2 = -1$ is not solvable. In general, $x^2 - 4dy^2 = -1$ is not solvable, since the congruence $x^2 \equiv -1 \pmod 4$ does not have a solution.

10-44. If d has a prime divisor of the form $4k + 3$, prove that the length of the period of $\sqrt{d}$ is even.

Solution. Suppose $q \mid d$, where q is prime and of the form $4k + 3$. Then $x^2 - dy^2 = -1$ implies that $x^2 \equiv -1 \pmod q$. Hence -1 is a quadratic residue of q, which contradicts (5.11). Since $x^2 - dy^2 = -1$ is not solvable, it follows from (10.13) that $\sqrt{d}$ has a period of even length.

10-45. Without calculating the continued fraction expansion of $\sqrt{9943}$, determine if the length of its period is odd or even.

Solution. Since $9943 \equiv 3 \pmod 4$, the previous problem implies that the equation $x^2 - 9943y^2 = -1$ is not solvable. Thus by (10.13), the period of $\sqrt{9943}$ must be even.

10-46. If $x^2 - dy^2 = -1$ is solvable and (r, s) is the least positive solution, prove that the fundamental solution of $x^2 - dy^2 = 1$ is $(2r^2 + 1, 2rs)$.

Solution. If a and b are defined by $(r+s\sqrt{d})^2 = a+b\sqrt{d}$, it follows from (10.14) that (a,b) is the fundamental solution of $x^2 - dy^2 = 1$. Since $r^2 - ds^2 = -1$, we have $a = r^2 + ds^2 = r^2 + (r^2+1) = 2r^2+1$ and $b = 2rs$.

10-47. Use (8.14) to show that if $x^2 - dy^2 = -1$ has a solution, then d has a primitive representation as a sum of two squares.

Solution. Suppose that $s^2 - dt^2 = -1$; then $s^2 \equiv -1 \pmod{d}$. Therefore by (8.14), there are relatively prime integers a and b such that $sa \equiv b \pmod{d}$ and $d = a^2 + b^2$.

Note. The same result is proved in Theorem 10.20, using properties of the continued fraction expansion of $\sqrt{d}$. In that theorem, we obtain an explicit representation of d as a sum of two squares. For another approach, also using continued fractions, see Problem 9-10.

10-48. If p is prime, prove that $x^4 - 361y^4 = -p$ has no solution in positive integers.

Solution. If (a,b) is a solution, then $(a^2 - 19b^2)(a^2 + 19b^2) = -p$, and hence $a^2 + 19b^2 = p$, $a^2 - 19b^2 = -1$. But the latter equation has no solution, since by (5.11), we cannot have $a^2 \equiv -1 \pmod{19}$.

10-49. Let (u,v) be the fundamental solution of $x^2 - dy^2 = 1$, and suppose there exist positive integers r and s such that $(r+s\sqrt{d})^2 = u+v\sqrt{d}$. Prove that (r,s) is a positive solution of $x^2 - dy^2 = -1$ and is, in fact, the least positive solution. (Hint. First show that $r^2 - ds^2 = \pm 1$. Then use (10.14).)

Solution. Expand $(r+s\sqrt{d})^2$ to conclude that $u = r^2 + ds^2 > r^2 \geq r$. It is clear that $(r^2 - ds^2)^2 = (r+s\sqrt{d})^2(r-s\sqrt{d})^2 = (u+v\sqrt{d})(u-v\sqrt{d}) = 1$; thus $r^2 - ds^2 = \pm 1$. We cannot have $r^2 - ds^2 = 1$, since $r < u$ and (u,v) is the least positive solution of $x^2 - dy^2 = 1$. Hence $r^2 - ds^2 = -1$.

Now suppose that (a,b) is the smallest positive solution of $x^2 - dy^2 = -1$. By (10.14), $u + v\sqrt{d} = (a+b\sqrt{d})^2$, and hence $(r+s\sqrt{d})^2 = (a+b\sqrt{d})^2$. Thus $r+s\sqrt{d} = a+b\sqrt{d}$, since both numbers are positive, and so $r = a$ and $s = b$. We conclude therefore that (r,s) is the least positive solution of $x^2 - dy^2 = -1$.

The Equation $x^2 - dy^2 = N$

10-50. Let N be a nonzero integer. Prove that the equation $x^2 - dy^2 = N$ has only a finite number of solutions if d is negative or if d is a perfect square.

Solution. Suppose $d < 0$. If $N < 0$, there are clearly no solutions, while if $N > 0$, we must have $|x| \leq \sqrt{N}$ and $|y| \leq \sqrt{N/|d|}$. Now assume that d is a perfect square, say, $d = m^2$; then $x^2 - dy^2 = (x+my)(x-my) = N$. Thus $x+my = r$ and $x-my = s$, where $N = rs$. Since there are only a finite number of ways to factor N, it follows that $x^2 - dy^2 = N$ has only finitely many solutions.

10-51. Suppose (r, s) is a positive solution of $x^2 - dy^2 = M$ and (u, v) is a positive solution of $x^2 - dy^2 = N$. Define (a, b) by $a + b\sqrt{d} = (r + s\sqrt{d})(u + v\sqrt{d})$; thus $a = ru + svd$ and $b = rv + su$. Prove that (a, b) is a positive solution of $x^2 - dy^2 = MN$.

Solution. It is easy to check that $a - b\sqrt{d} = (r - s\sqrt{d})(u - v\sqrt{d})$. Thus $a^2 - db^2 = (a + b\sqrt{d})(a - b\sqrt{d}) = (r^2 - ds^2)(u^2 - dv^2) = MN$, and hence (a, b) is a solution of $x^2 - dy^2 = MN$. (This is just *Brahmagupta's identity*.)

10-52. Use the preceding problem to find three positive solutions of $x^2 - 10y^2 = 31$.

Solution. By inspection, $(3, 2)$ is a solution of $x^2 - 10y^2 = -31$. Since $\sqrt{10} = \langle 3, \overline{6} \rangle$, the period of $\sqrt{10}$ is odd, and so $x^2 - 10y^2 = -1$ is solvable. By (10.13) (or by inspection), the least positive solution is $(p_0, q_0) = (3, 1)$. For convenience, define the "product" $(r, s)(u, v)$ to be $(ru + svd, rv + su)$. Then by (10.14), the next two positive solutions are $(x_2, y_2) = (3, 1)(3, 1) = (19, 6)$ and $(x_3, y_3) = (3, 1)(19, 6) = (117, 37)$. (These solutions could also be found by computing the convergents of $\sqrt{10}$, since $(x_2, y_2) = (p_1, q_1)$ and $(x_3, y_3) = (p_2, q_2)$.)

If we multiply each of these solutions of $x^2 - 10y^2 = -1$ by the solution $(3, 2)$ of $x^2 - 10y^2 = -31$, we obtain the solutions $(29, 9)$, $(177, 56)$, and $(1091, 345)$ of $x^2 - 10y^2 = 31$.

10-53. (a) Given that $(13, 2)$ is a solution of $x^2 - 41y^2 = 5$, find another solution. (*Hint.* Use Problems 10-36 and 10-51.)
(b) Find two solutions of $x^2 - 41y^2 = -5$.

Solution. (a) It is enough to "multiply" a given solution of $x^2 - 41y^2 = 5$ by a solution of $x^2 - 41y^2 = 1$, according to the formula given in Problem 10-51. By Problem 10-36, $(2049, 320)$ is a solution of $x^2 - 41y^2 = 1$. Thus $(13, 2)(2049, 320) = (13 \cdot 2049 + 2 \cdot 320 \cdot 41, 13 \cdot 320 + 2 \cdot 2049) = (52877, 8258)$, and so $(52877, 8258)$ is also a solution of $x^2 - 41y^2 = 5$.

(b) Using the approach in part (a), "multiply" $(13, 2)$ by the solutions $(32, 5)$ and $(131168, 20485)$ of $x^2 - 41y^2 = -1$ (see Problem 10-36). This gives the solutions $(13, 2)(32, 5) = (826, 129)$ and $(13, 2)(131168, 20485) = (3384954, 528641)$ of $x^2 - 41y^2 = -5$.

10-54. Use Problem 10-51 to find three positive solutions of $x^2 - 29y^2 = 5$, with each solution having $y < 40000$. (*Hint.* See Example 10.16.)

Solution. By inspection, $(11, 2)$ is one solution (in fact, the least positive solution) of $x^2 - 29y^2 = 5$. By Example 10.16, $(9801, 1820)$ is a solution of $x^2 - 29y^2 = 1$; thus if we proceed as in the solution of the previous problem, it follows that $(11, 2)(9801, 1820) = (213371, 39622)$ is a solution of $x^2 - 29y^2 = 5$.

To find a third solution having $y < 40000$, note that a solution of $x^2 - 29y^2 = -1$ multiplied by a solution of $x^2 - 29y^2 = -5$ also gives a solution of $x^2 - 29y^2 = 5$. By Example 10.8, $(70, 13)$ solves $x^2 - 29y^2 = -1$, and since $|-5| < \sqrt{29}$, (10.18) implies that solutions of $x^2 - 29y^2 = -5$ are found among the convergents of $\sqrt{29}$. It is easy to check that $(16, 3)$ is a solution of $x^2 - 29y^2 = -5$, and hence $(70, 13)(16, 3) = (2251, 418)$ is also a solution of $x^2 - 29y^2 = 5$.

10-55. *Suppose that $x^2 - dy^2 = N$ has a solution in positive integers. If the length of the period of $\sqrt{d}$ is odd, prove that $x^2 - dy^2 = -N$ also has a solution in positive integers.*

Solution. Since the period of $\sqrt{d}$ is odd, $x^2 - dy^2 = -1$ has a solution (u, v), by (10.13). Let (a, b) be a solution of $x^2 - dy^2 = N$, and define r, s by $r = au + bvd$ and $s = av + bu$. Then by Brahmagupta's identity, (r, s) is a solution of $x^2 - dy^2 = -N$.

10-56. *Prove or disprove: If $x^2 - dy^2 = N$ is not solvable, then $x^2 - dy^2 = -N$ is solvable.*

Solution. This is false. For example, if $\sqrt{d}$ has a period of odd length, the preceding problem shows that one equation has solutions if and only if the other does.

10-57. *Prove or disprove: If $1 < |N| < \sqrt{d}$ and $x^2 - dy^2 = N$ is solvable, then $x^2 - dy^2 = -N$ is also solvable.*

Solution. This is false. For example, $x^2 - 21y^2 = -3$ has the solution $(9, 2)$, but $x^2 - 21y^2 = 3$ has no solution, for if $a^2 - 21b^2 = 3$, then $a^2 \equiv 3 \pmod 7$, a contradiction since $(3/7) = -1$.

▷ **10-58.** *Prove Theorem 10.18: If $|N| < \sqrt{d}$ and (r, s) is a positive solution of $x^2 - dy^2 = N$, then r/s is one of the convergents of the continued fraction expansion of $\sqrt{d}$. (Hint. For $N > 0$, start as in the proof of (10.3) and use (9.39). For $N < 0$, replace d by $1/d$ and use Problem 9-21.)*

Solution. Suppose first that $N > 0$. Then $r/s - \sqrt{d} = N/(s(r + s\sqrt{d}))$. Since N is positive, $r - s\sqrt{d} > 0$, and thus $r > s\sqrt{d}$. Therefore $0 < r/s - \sqrt{d} < \sqrt{d}/s(2s\sqrt{d}) = 1/2s^2$. It follows from (9.39), that r/s is a convergent of $\sqrt{d}$.

Suppose now that $N < 0$. Divide the equation $r^2 - ds^2 = N$ by $-1/d$ to obtain $s^2 - (1/d)r^2 = -N/d$, where $-N/d > 0$. Arguing as above, we find that $s/r - \sqrt{1/d} = -N/dr(s + r\sqrt{1/d})$, and since $s > r\sqrt{1/d}$, it follows that $0 < s/r - \sqrt{1/d} < 1/2r^2$. Therefore s/r is a convergent of $1/\sqrt{d}$, and hence by Problem 9-21, r/s is a convergent of $\sqrt{d}$.

10-59. *Let d be a positive integer that is not a perfect square, and suppose d is divisible by a prime q of the form $4k + 3$. If $(N, d) = 1$ and $x^2 - dy^2 = N$ has a solution in integers, prove that $x^2 - dy^2 = -N$ does not.*

Solution. Suppose a and b are integers such that $a^2 - db^2 = N$; then $a^2 \equiv N \pmod q$, and since $(N, q) = 1$, N is a quadratic residue of q, i.e., $(N/q) = 1$. If there exist integers r and s such that $r^2 - ds^2 = -N$, then $r^2 \equiv -N \pmod q$ and hence $(-N/q) = 1$. But then $1 = (-N/q) = (-1/q)(N/q) = (-1/q)$, a contradiction, since -1 is not a quadratic residue of primes of the form $4k + 3$ (see (5.11)).

10-60. If p_k/q_k is a convergent of $\sqrt{d}$ and (p_k, q_k) is a solution of $x^2 - dy^2 = N$, prove that $|N| < 2\sqrt{d}$. (Thus, if (a, b) is a positive solution of $x^2 - dy^2 = N$ and $N > 2\sqrt{d}$, then a/b cannot be a convergent of $\sqrt{d}$.) (Hint. Use (10.5) and Problem 10-1.)

Solution. In view of (10.5), $|N| = s_k$ for some $k \geq 0$, and since $s_k < 2\sqrt{d}$ by Problem 10-1, it follows that $|N| < 2\sqrt{d}$.

10-61. By considering $d = 153$, show that it is possible for $x^2 - dy^2 = N$ and $x^2 - dy^2 = -N$ to be both solvable, where $0 < N < \sqrt{d}$, although $x^2 - dy^2 = -1$ is not solvable.

Solution. It is easy to check that $\sqrt{153} = \langle 12, \overline{2,1,2,2,2,1,2,24}\rangle$. Thus the length of the period is 8, and so, by (10.13), $x^2 - dy^2 = -1$ is not solvable. However, $s_1 = s_4 = 9$, and since $p_{k-1}^2 - q_{k-1}^2 = (-1)^k s_k$, it follows that $x^2 - dy^2 = 9$ and $x^2 - dy^2 = -9$ are both solvable.

More generally, if $\sqrt{d}$ has a period of even length and there are indices i and j of opposite parity such that $s_i = s_j$, then, taking $N = s_i$, we find that $x^2 - dy^2 = N$ and $x^2 - dy^2 = -N$ are both solvable but $x^2 - dy^2 = -1$ is not.

Note. The first three odd values of d with this property are 153, 261, and 369; the first three even values are 212, 234, and 244. If a computer program is written to find such integers d, the initial data would seem to indicate that the corresponding value of N is a square (for example, $N = 9$ when $d = 153$). This is not always the case, however; the first even d with N not a square is $d = 466$ ($s_3 = s_6 = 15$), and the first odd such d is $d = 657$ ($s_3 = s_6 = 27$).

10-62. Let (x_1, y_1) be the fundamental solution of $x^2 - dy^2 = 1$, and let (u, v) be the least positive solution of $x^2 - dy^2 = N$. Define u_n and v_n by

$$u_n + v_n\sqrt{d} = (x_1 + y_1\sqrt{d})^n(u + v\sqrt{d}).$$

It follows from Problem 10-51 that (u_n, v_n) is also a positive solution of $x^2 - dy^2 = N$. Give an example to show that not all positive solutions need to be of the form (u_n, v_n); in particular, show that (u_1, v_1) need not be the next positive solution after (u, v). (Hint. Consider $x^2 - 23y^2 = -7$.)

Solution. If we compute s_k for $\sqrt{23}$ (where s_k is defined as in (9.24)), we find that $s_k = 7$ for $k = 1, 3, 5, \ldots$. Using (10.5), it follows that $(p_0, q_0) = (4, 1)$, $(p_2, q_2) = (19, 4)$, and $(p_4, q_4) = (211, 44), \ldots$ are all solutions of $x^2 - 23y^2 = -7$. The fundamental solution of $x^2 - 23y^2 = 1$ is easily found to be $(24, 5)$, but since $(24 + 5\sqrt{23})(4 + \sqrt{23}) = 211 + 44\sqrt{23}$, it is clear that the solution $(19, 4)$ has been skipped and cannot be generated by this method.

A solution (a, b) of the equation $x^2 - dy^2 = N$ is called *primitive* if a and b are relatively prime. If (a, b) is a primitive solution, then (a, b) is said to belong to the class C_s if $a \equiv sb \pmod{N}$. (Without loss of generality, we may assume that $0 \leq s < |N|$.) The next problem shows that solutions in the same class can be obtained in a simple way from each other.

▷ **10-63.** Let (a_1, b_1) and (a_2, b_2) be primitive solutions of $x^2 - dy^2 = N$ that both belong to the class C_s. Then there is a solution (u, v) of Pell's Equation $x^2 - dy^2 = 1$ such that $a_2 + b_2\sqrt{d} = (a_1 + b_1\sqrt{d})(u + v\sqrt{d})$. (Hint. Divide $a_2 + b_2\sqrt{d}$ by $a_1 + b_1\sqrt{d}$.)

Solution. By rationalizing the denominator, we obtain

$$\frac{a_2 + b_2\sqrt{d}}{a_1 + b_1\sqrt{d}} = \frac{(a_2 + b_2\sqrt{d})(a_1 - b_1\sqrt{d})}{a_1^2 - db_1^2} = \frac{a_1a_2 - db_1b_2}{N} + \frac{a_1b_2 - b_1a_2}{N}\sqrt{d}.$$

Let $u = (a_1a_2 - db_1b_2)/N$ and $v = (a_1b_2 - b_1a_2)/N$; then $a_2 + b_2\sqrt{d} = (a_1 + b_1\sqrt{d})(u + v\sqrt{d})$. Taking conjugates and multiplying, we obtain $a_2^2 - db_2^2 = (a_1^2 - db_1^2)(u^2 - dv^2)$, and therefore $N = N(u^2 - dv^2)$, i.e., $u^2 - dv^2 = 1$. It remains to show that u and v are *integers*.

Using the fact that $a_1 \equiv sb_1 \pmod{N}$ and $a_1^2 - db_1^2 = N$, we find that $b_1^2(s^2 - d) \equiv 0 \pmod{N}$. Since the solution (a_1, b_1) is primitive, b_1 and N are relatively prime, and so $s^2 - d \equiv 0 \pmod{N}$. Thus $a_1a_2 - db_1b_2 \equiv b_1b_2(s^2 - d) \equiv 0 \pmod{N}$, and therefore N divides $a_1a_2 - db_1b_2$, i.e., u is an integer. Since $a_1b_2 - b_1a_2 \equiv sb_1b_2 - b_1sb_2 = 0 \pmod{N}$, v is also an integer. This completes the proof.

Note. For any N, there are at most $|N|$ classes of primitive solutions of $x^2 - dy^2 = N$. If we find a solution in each class, then *all* primitive solutions can be obtained by "multiplying" by a solution of $x^2 - dy^2 = 1$.

10-64. Show that any solution of $x^2 - dy^2 = N$ is of the form (mA, mB), where (A, B) is a primitive solution of $x^2 - dy^2 = N/m^2$.

Solution. Let (a, b) be a solution of $x^2 - dy^2 = N$, and let m be the greatest common divisor of a and b. Set $a = mA$ and $b = mB$; clearly, A and B are relatively prime. Substituting in the equation, we find that $m^2A^2 - dm^2B^2 = N$. Thus $m^2 | N$, and (A, B) is a primitive solution of $x^2 - dy^2 = N/m^2$.

10-65. Prove that $x^2 - 1999y^2 = -2$ has no solution in positive integers. (Hint. Use (5.13.i), noting that 1999 is a prime.)

Solution. Any positive solution (a, b) must satisfy $a^2 \equiv -2 \pmod{1999}$, which is impossible, since 1999 is of the form $8k - 1$ and -2 is a quadratic nonresidue of such primes, by (5.13.i).

10-66. Prove that $x^4 - 9y^4 = N$ is not solvable in positive integers for $N = \pm 1, \pm 2, \pm 3, \pm 4$.

Solution. Factor $x^4 - 9y^4$ as $(x^2 - 3y^2)(x^2 + 3y^2)$ and note that $x^2 + 3y^2 \geq 4$ if x and y are positive. Since $x^2 - 3y^2$ must be an integer, it follows that there can be no solutions when $N = \pm 1, \pm 2, \pm 3$. If $(x^2 - 3y^2)(x^2 + 3y^2) = \pm 4$, we must have $x^2 + 3y^2 = 4$ and hence $x = y = 1$. But then $x^2 - 3y^2 = -2$, giving $x^4 - 9y^4 = -8$.

Pell's Equation and Sums of Two Squares

10-67. *Show that 115202 is a sum of two relatively prime squares, and use the technique described in the proof of (10.20) to find a primitive representation of 115202.*

Solution. First check that $\sqrt{115202} = \langle 339, \overline{2,2,2,2,2,2,678} \rangle$, using the technique outlined in Chapter 9. Thus the length of the period is odd, and so (10.20) implies that 115202 has primitive representations. We now follow the method used in the proof of (10.20). Let $\alpha_4 = \langle \overline{2,2,2,678,2,2,2} \rangle$. Since $\alpha_0 = \sqrt{115202} = (0 + \sqrt{115202})/1$ and $a_0 = [\sqrt{115202}] = 339$, we can use (9.24) to construct the following table:

k	0	1	2	3	4
r_k	0	339	223	243	239
s_k	1	281	233	241	241
a_k	339	2	2	2	

Hence by (9.24.iii), we have $\alpha_4 = (239 + \sqrt{115202})/241$, and therefore $115202 = 239^2 + 241^2$.

▷ **10-68.** *Show that the converse of (10.20) is not true by finding a positive integer d (not a square) such that d is the sum of two relatively prime squares but $\sqrt{d}$ has a period of even length.*

Solution. Some examples: $\sqrt{34} = \langle 5, \overline{1,4,1,10} \rangle$ and $34 = 3^2 + 5^2$. Also, $\sqrt{650} = \langle 25, \overline{2,50} \rangle$; since $650 = 2 \cdot 5^2 \cdot 13$, (8.15) implies that 650 has primitive representations (namely, $11^2 + 23^2$ and $17^2 + 19^2$). Finally, check that $\sqrt{9490} = \langle 97, \overline{2,2,2,194} \rangle$; since $9490 = 2 \cdot 5 \cdot 13 \cdot 73$, it follows from (8.15) that 9490 has a primitive representation.

10-69. *Suppose d has no primitive representations; equivalently, suppose d is divisible either by 4 or by a $4k+3$ prime (see (8.15)). Then it follows from (10.20) that $\sqrt{d}$ has a period of even length. Use this to show that $\sqrt{539784}$ and $\sqrt{16725138}$ each have periods of even length.*

Solution. Clearly, 539784 is divisible by 4; hence (8.15) implies that there are no primitive representations. (But there are nonprimitive representations. Why?) Thus by (10.20), the period of $\sqrt{539784}$ is even.

Because the sum of the digits of 16725138 divisible by 3 but not by 9, it follows that 16725138 is divisible by 3 but not by 9. Thus (8.9) implies that there are no representations at all. Hence $\sqrt{16725138}$ has a period of even length.

10-70. *Let p be a prime of the form $8k+5$. Show that the equation $x^2 - 2py^2 = -1$ has a solution (equivalently, the length of the period of $\sqrt{2p}$ is odd). (Hint. Imitate the proof of (10.19), and use (5.12) and (5.13).)*

Solution. Let (u,v) be the smallest solution of $x^2 - 2py^2 = 1$ in positive integers. Clearly, u is odd and v is even. Now $u^2 - 2pv^2 = 1$ if and only if $(u+1)(u-1) = 2pv^2$.

Note that $u+1$ and $u-1$ are even and differ by 2, and therefore $(u+1, u-1) = 2$. Let $v = 2w$ and $u+1 = 2t$; then $t(t-1) = 2pw^2$. There are four cases to consider: (i) t is of the form $2r^2$ and $t-1$ is of the form ps^2; (ii) $t = ps^2$ and $t-1 = 2r^2$; (iii) $t = r^2$ and $t-1 = 2ps^2$; and (iv) $t = 2ps^2$ and $t-1 = r^2$.

In case (i), we have $2r^2 - ps^2 = t - (t-1) = 1$. It follows from this that 2 is a quadratic residue of p, which contradicts (5.12). In case (ii), we have $ps^2 - 2r^2 = 1$, and thus -2 is a quadratic residue of p, contradicting (5.13.i). In case (iii), we obtain $r^2 - 2ps^2 = 1$; but $r < u$, which contradicts the fact that (u, v) is the least positive solution of $u^2 - 2pv^2 = 1$. Thus we must have $t = 2ps^2$ and $t - 1 = r^2$ (case (iv)), and therefore $r^2 - 2ps^2 = (t-1) - t = -1$.

Note. If p is of the form $8k+1$, the length of the period of $\sqrt{2p}$ can be even. This happens, for example, when $p = 17$ (see Problem 10-68).

The next problem proves part of the Law of Quadratic Reciprocity using Pell's Equation.

10-71. *Let p and q be primes of the form $4k + 3$. By considering the least positive solution of $x^2 - pqy^2 = 1$, show that p is a quadratic residue of q if and only if q is not a quadratic residue of p. (Hint. Imitate the proof of (10.19).)*

Solution. Let (u, v) be the smallest positive solution of $x^2 - pqy^2 = 1$. Using the same argument as in the proof of (10.19), we arrive at the equation $t(t-1) = pqw^2$. There are four cases to consider: (i) t is of the form r^2 and $t-1$ is of the form pqs^2; (ii) $t = pqs^2$ and $t-1 = r^2$; (iii) $t = pr^2$ and $t-1 = qs^2$; and (iv) $t = qs^2$ and $t-1 = pr^2$.

In case (i), we obtain $r^2 - pqs^2 = 1$; since $r < u$, this contradicts the fact that (u, v) is the least positive solution of $u^2 - pqv^2 = 1$. In case (ii), we have $r^2 - pqs^2 = -1$. This implies that -1 is a quadratic residue of p, contradicting (5.11). In case (iii), we have $pr^2 - qs^2 = 1$; thus p is a quadratic residue of q. Since $-qs^2 \equiv 1 \pmod{p}$ and -1 is a quadratic nonresidue of p, it follows that q is a quadratic nonresidue of p. In the same way, we can show, in case (iv), that q is a residue of p and p is a nonresidue of q. Since only cases (iii) and (iv) are possible, it follows that p is a residue of q if and only if q is a nonresidue of p.

An Application: Factoring Large Numbers

10-72. *(a) Suppose $a^2 \equiv b^2 \pmod{N}$, with $0 < b < a < N$ and $a+b \neq N$. Let $d_1 = (a-b, N)$ and $d_2 = (a+b, N)$. Prove that d_1 and d_2 are proper divisors of N, that is, $1 < d_1 < N$ and $1 < d_2 < N$.*

(b) If $a + b = N$ in part (a), will d_1 and d_2 always give the trivial divisors 1 and N of N?

Solution. (a) Since $1 < a - b < N$, we clearly cannot have $d_1 = N$. Also, since $2 < a + b < 2N$ and $a + b \neq N$, we cannot have $d_2 = N$. It remains to show that we cannot have $d_1 = 1$ or $d_2 = 1$. If $d_1 = 1$, then $a + b \equiv 0 \pmod{N}$, contradicting the fact that $d_2 \neq N$. Similarly, if $d_2 = 1$, then $a - b \equiv 0 \pmod{N}$, contradicting the fact that $d_1 \neq N$.

(b) Not necessarily. For example, $12^2 \equiv 3^2 \pmod{15}$ gives $d_1 = (9, 15) = 3$.

10-73. *Use Legendre's Factoring Method, described in (10.22), to find a non-trivial divisor of 15925.*

Solution. Let $N = 15925$; then check that $\sqrt{N} = \langle 126, \overline{5, 6, 1, 4, 3, \ldots} \rangle$ (see (9.24)). Set up the following table:

k	0	1	2	3	4
r_k	0	126	119	97	84
s_k	1	49	36	181	49
a_k	126	5	6	1	4
p_k	126	631	3912	4543	

We now look for values of s_k, with k even, that are perfect squares. Since $s_2 = 6^2$, the congruence $p_{k-1}^2 \equiv (-1)^k s_k \pmod{N}$ becomes $631^2 \equiv 6^2 \pmod{N}$. Using the Euclidean Algorithm, we obtain $d_1 = (631 - 6, N) = 25$ and $d_2 = (631 + 6, N) = 637$. In fact, $15925 = 25 \cdot 637$ (see Problem 10-77). Note that in this case, neither d_1 nor d_2 is prime ($637 = 49 \cdot 13$).

We could also apply this technique to $s_4 = 7^2$. In this case, we get $d_1 = (4543 - 7, N) = 7$ and $d_2 = (4543 + 7, N) = 2275$.

10-74. *Use Legendre's Factoring Method to factor 22223.*

Solution. Let $N = 22223$. Using the algorithm described in (9.24), we obtain $\sqrt{N} = \langle 149, \overline{13, 1, 1, 4, 1, 1, 1, 1, 1, 4, 1, 1, 13, 298} \rangle$. Now set up the following table, reducing the values of p_k modulo N to keep the size manageable:

k	0	1	2	3	4	5	6	7	8
r_k	0	149	137	20	119	113	50	71	71
s_k	1	22	157	139	58	163	121	142	121
a_k	149	13	1	1	4	1	1	1	1
p_k	149	1938	2087	4025	18187	22212	18176	18165	

The first s_k, with k even, that is a square is $s_6 = 11^2$, and $p_5 \equiv 22212 \pmod{N}$; thus $22212^2 \equiv 11^2 \pmod{N}$. But $22212 + 11 = N$, so the method gives only the trivial factors $d_1 = (22212 + 11, N) = N$ and $d_2 = (22212 - 11, N) = 1$.

We next check $s_8 = 11^2$; then $p_7 \equiv 18165 \pmod{N}$, and since $18165 + 11 \neq N$, we expect to get proper divisors of N. In fact, $d_1 = (18154, N) = 313$ and therefore $d_2 = (18176, N) = N/313 = 71$ (see Problem 10-77). Note that in this example, the divisors 313 and 71 are both primes.

The next two problems should be done with the aid of a computer, since the calculation of a_k, r_k, s_k, and p_k cannot reasonably be done otherwise. A simple computer program can quickly check if s_k is a square and then rapidly calculate the greatest common divisors d_1 and d_2.

PROBLEMS AND SOLUTIONS 349

10-75. *Use Legendre's Factoring Method to factor 623809.*

Solution. Let $N = 623809$. The first even k for which s_k is a square is $s_{14} = 27^2$, with $p_{13} \equiv 623782$. (All of the p-values have been reduced modulo N.) However, since $p_{13} + 27 \equiv 0 \pmod{N}$, we obtain only the trivial factors 1 and N (see Problem 10-78). We therefore check the next s_k, with k even, that is a square, namely, $s_{30} = 4^2$, with $p_{29} \equiv 466366$. This produces the proper divisors $d_1 = (466362, N) = 1993$ and $d_2 = (466370, N) = 313$. Both of these divisors are prime, and in fact, $N = 313 \cdot 1993$. (That $N = d_1 d_2$ for N odd is proved in Problem 10-77.)

The length of the period of $\sqrt{N}$ is 726. It is interesting to note that among the first 727 values of s_k, there are 29 usable squares, 15 of which give a nontrivial factorization of N.

10-76. *Factor 2633383 using Legendre's Factoring Method.*

Solution. Let $N = 2633383$; then the length of the period of $\sqrt{N}$ is 1178, and so we expect to find many value of s_k that are perfect squares. However, this example shows that even though $\sqrt{N}$ produces many usable square s_k (with k even), the first four such s_k, namely, s_{24}, s_{40}, s_{92}, and s_{144}, give only trivial factors of N. The values are as follows (p_{k-1} has been reduced modulo N): $s_{24} = 53^2$, $p_{23} \equiv 53$; $s_{40} = 11^2$, $p_{39} \equiv 2633373$; $s_{92} = 33^2$, $p_{91} \equiv 33$; and $s_{144} = 3^2$, $p_{143} \equiv 2633380$. The next usable square, however, does produce a factorization: $s_{174} = 39^2$, $p_{173} \equiv 364311$ gives $d_1 = (364272, N) = 7589$ and $d_2 = (364350, N) = 347$. Both 347 and 7589 are prime, and in fact, $2633383 = 347 \cdot 7589$. (Actually, it is always true that $d_1 d_2 = N$ when N is odd; see Problem 10-77.)

▷ **10-77.** *(See (10.22).) Let N be odd, and suppose $p_{k-1}^2 - N q_{k-1}^2 = c^2$.*

(a) If $d_1 = (p_{k-1} - c, N)$ and $d_2 = (p_{k-1} + c, N)$, prove that $d_1 d_2 = N$.

(b) Let p^* be the least nonnegative residue of p_{k-1} modulo N; then $p^* \geq 1$. If $d_1^* = (p^* - c, N)$ and $d_2^* = (p^* + c, N)$, prove that $d_1^* d_2^* = N$.

Solution. Write $p = p_{k-1}, q = q_{k-1}$; then $(p-c)(p+c) = Nq^2$. Let $N = \pi_1^{n_1} \cdots \pi_r^{n_r}$ be the prime factorization of N; thus for each i, $\pi_i^{n_i}$ divides $(p-c)(p+c)$, and therefore there are at least n_i factors of π_i distributed between $p-c$ and $p+c$. (There are *at least* n_i factors, since q^2 could conceivably also contain factors of π_i.). Thus, in forming d_1 and d_2, there are at least n_i factors of π_i distributed between d_1 and d_2, and so $d_1 d_2$ is divisible by N.

We now show that N is divisible by $d_1 d_2$. Let π be a prime such that $\pi | d_1$ and $\pi | d_2$, let π^a be the largest power of π that divides $d_1 d_2$, and let P be the product of all such π^a. Each π^a divides $(p-c)(p+c)$. (To see this, note that $d_1 | p-c$ and $d_2 | p+c$, and so $d_1 d_2 | (p-c)(p+c)$; since $\pi^a | d_1 d_2$, the result follows.) Thus each π^a divides Nq^2. Also, $\pi | 2p$ and $\pi | 2c$; since π also divides N and N is odd, π must be odd, and therefore $\pi | p$. But $(p, q) = 1$ by (9.7), and so π cannot divide q. Hence we conclude that π^a must divide N. Thus *all* factors of π in $d_1 d_2$ are in N. In view of our definition of P, it follows that $P | N$.

Now let Q be the product of π^b, where π ranges over all primes dividing d_1 or d_2 but not both and π^b is the largest power of π dividing $d_1 d_2$. Clearly, P and Q are relatively prime, and $d_1d_2 = PQ$. Since $d_1 | N$ and $d_2 | N$, Q must divide N. (Write $Q = Q_1Q_2$, where $(Q_1, Q_2) = 1$ and $Q_1 | d_1$, $Q_2 | d_2$. Then $Q_1 | N$ and $Q_2 | N$, and therefore N is divisible by $Q = Q_1Q_2$.)

Thus both P and Q divide N, and since $(P, Q) = 1$, it follows that $PQ | N$, that is, $d_1d_2 | N$. Since we have already shown that $N | d_1d_2$, we conclude that $d_1d_2 = N$.

(b) Let $p^* = p + Nk$. Then, using (1.22), we have $d_1^* = (p^* - c, N) = (p + Nk - c, N) = (p - c, N) = d_1$. Similarly, $d_2^* = d_2$ and thus, by part (a), $d_1^* d_2^* = N$.

10-78. *Refer to Legendre's Factoring Method (10.22). Suppose N is an odd positive integer. If k is even, $s_k = c^2$, and $p_{k-1} \equiv \pm c \pmod{N}$, prove that $d_1 = (p_{k-1} - c, N)$ and $d_2 = (p_{k-1} + c, N)$ are (in some order) the two trivial divisors 1 and N of N.*

Solution. The easiest way to prove this is to use the preceding problem. Let $p = p_{k-1}$; then $p \equiv \pm c \pmod{N}$ implies that either $p - c$ or $p + c$ is a multiple of N. Hence either $d_1 = N$ or $d_2 = N$, and so, by Problem 10-77, $d_2 = 1$ or $d_1 = 1$, respectively.

The result can also be proved directly, as follows. By (10.5.i), we have $p^2 - Nq^2 = c^2$, i.e., $(p - c)(p + c) = Nq^2$. If $p \equiv c \pmod{N}$, then $d_1 = (p - c, N) = N$. We now show that $d_2 = 1$. If the prime π divides $p + c$ and N, then, because $N | p - c$, π divides $2p$ and hence π divides p, since N is odd. Let $p - c = mN$; then $mN(p + c) = Nq^2$, and therefore $m(p + c) = q^2$. Since $\pi | p + c$, it follows that $\pi | q$, contradicting the fact that $(p, q) = 1$ (see (9.7)). Thus no prime π divides both $p + c$ and N, i.e., $d_2 = 1$. In essentially the same way, we can show that if $p \equiv -c \pmod{N}$, then $d_2 = N$ and $d_1 = 1$.

10-79. *Find an even integer N such that d_1d_2 does not equal N, where d_1 and d_2 are the divisors of N obtained from Legendre's Factoring Method. Can $d_1d_2 = N$ when N is even? (Hint. Consider $N = 78$ and $N = 48$.)*

Solution. Let $N = 78$. Then $\sqrt{78} = \langle 8, \overline{1, 4, 1, 16} \rangle$, and it is easy to check that $s_4 = 1$, $p_3 = 53$ is a solution of Legendre's congruence. This gives $d_1 = (52, 78) = 26$ and $d_2 = (54, 78) = 6$, so that $d_1d_2 = 2N$.

It can also happen that $d_1d_2 = N$ when N is even. If $N = 48$, then $s_1 = 1$, $p_0 = 6$ gives $d_1 = (6, 48) = 6$ and $d_2 = (8, 48) = 8$. In fact, for the same N, it is possible that $d_1d_2 = N$ or $2N$, depending on the values of s_k and p_{k-1}. For example, if $N = 88$, then $s_2 = 9$, $p_1 = 19$ gives $d_1d_2 = 8 \cdot 22 = 2N$, whereas $s_4 = 9$, $p_3 = 47$ yields $d_1d_2 = 44 \cdot 2 = N$.

10-80. *Give an example to show that if N is odd and $a^2 \equiv b^2 \pmod{N}$, with $0 < b < a < N$ and $a + b \neq N$, then d_1d_2 need not equal N, where $d_1 = (a - b, N)$ and $d_2 = (a + b, N)$. Why doesn't the proof of Problem 10-77 apply here?*

Solution. Consider, for example, $21^2 \equiv 6^2 \pmod{135}$. Then $d_1 = (15, 135) = 15$ and $d_2 = (27, 135) = 27$, but d_1d_2 does not equal 135. The argument used in the solution of Problem 10-77 fails here, since in Problem 10-77, we have $p^2 - a^2 = Nq^2$, where p and q are *relatively prime*. (Thus we have made particular use of the fact that if p/q

is a convergent of $\sqrt{N}$, then $(p,q) = 1$.) In this example, however, $21^2 - 6^2 = 135 \cdot 3$, and 3 is not relatively prime to either 21 or 6.

Doping Factors. The period of $\sqrt{N}$ may be too short to produce an adequate number of values of s_k that are squares; for example, $\sqrt{n^2 + 1} = \langle n, \overline{2n} \rangle$ for any $n \geq 1$. In other cases, the period may be long but may contain no squares s_k leading to proper factors of N in, say, the first 500 values tested. In both situations, we can multiply N by a *doping factor* D and then apply Legendre's Factoring Method again to $\sqrt{DN}$. The doping factor need not be large to expand the period of $\sqrt{N}$ by a considerable amount; in fact, $D = 3$ or $D = 5$ will usually produce the desired factorization. If we take D to be odd, Problem 10-77 shows that only d_1 need be calculated for DN, since then $d_2 = DN/d_1$.

10-81. *Use Legendre's Factoring Method, with a doping factor of 3, to factor 1267877. Repeat the process with a doping factor of 5.*

Solution. Let $N = 1267877$; then $\sqrt{N} = \langle 1126, \overline{2252} \rangle$. (In fact, $N = 1126^2 + 1$.) Since the period has length 1, the algorithm produces no proper divisors of N. If we now use a doping factor of 3, then $\sqrt{3N}$ has a period of length 80, and Legendre's Factoring Method (with the values of p_k reduced modulo $3N$) gives $s_6 = 841 = 29^2$, $p_5 \equiv 195029$. Since $p_5 \not\equiv \pm 29 \pmod{3N}$, we obtain proper divisors of $3N$ by calculating $d_1 = (195000, 3N) = 39$ and hence $d_2 = 3N//d_1 = 97529$. Thus N factors as $13 \cdot 97529$. (Note that 97529 is not prime: $97529 = 17 \cdot 5737$.)

If we use a doping factor of 5, then $\sqrt{5N}$ has a period of length 55, and the algorithm gives $s_{12} = 16^2$, $p_{11} = 4876466$. Then $d_1 = (4876466 - 16, 5N) = 487645$ and so $d_2 = 5N/d_1 = 13$. Thus N is the product of $d_1/5 = 97529$ and 13.

10-82. *Suppose Legendre's Factoring Method is used with a doping factor D that is an odd prime, where we may suppose that the integer N to be factored is not divisible by D. (For example, if $D = 3$, simply check that the sum of the digits of N is not a multiple of 3.) Let P and C be p_{k-1} and $\sqrt{s_k}$ values for DN such that $p_{k-1} \not\equiv \pm c \pmod{DN}$. If $d_1 = (P - C, N)$ and $d_2 = (P+C, N)$, prove that $d_1 d_2 = N$.*

Solution. Let $D_1 = (P - C, DN)$ and $D_2 = (P + C, DN)$; then $D_1 D_2 = DN$, by Problem 10-77. Since there is only one factor of D on the right side, exactly one of D_1 and D_2 is divisible by D. Suppose $D \mid D_1$; then $D \nmid D_2$ and so $D \nmid P + C$. Thus $D_2 = (P + C, DN) = (P + C, N) = d_2$. Also, since $D \mid P - C$ and $D \nmid N$, we have $D_1 = (P - C, DN) = D(P - C, N) = Dd_1$. Therefore $DN = D_1 D_2 = Dd_1 d_2$, and hence $N = d_1 d_2$. Similarly, if $D \mid D_2$, then $D_1 = d_1$ and $D_2 = Dd_2$, and hence $N = d_1 d_2$.

Note. Suppose $P \not\equiv \pm C \pmod{DN}$. The preceding problem shows that if the doping factor is an odd prime that does not divide N, then, to find divisors of N, it is enough to calculate $(P - C, N)$ or $(P + C, N)$ directly, instead of first determining factors of DN by computing $(P - C, DN)$ or $(P + C, DN)$. Using this fact in Problem

10-81, we obtain, for a doping factor of 3, $d_1 = 13$ and $d_2 = 97529$; for a doping factor of 5, we get $d_1 = 97529$ and so $d_2 = 13$.

10-83. *The use of a doping factor D may, strictly speaking, produce proper factors of DN, but these factors might simply be D or N and thus yield only the trivial factors 1 and N of N. Prove this by applying Legendre's Factoring Method to $N = 4097$ with a doping factor $D = 3$.*

Solution. Check that $\sqrt{N}$ has a period of length 1 (since $4097 = 64^2 + 1$), but the length of the period of $\sqrt{3N}$ is 34. Applying Legendre's Factoring Method to $3N$ yields $s_6 = 25$ and $p_5 = 4102$. Thus, for $3N$, we obtain the divisors $d_1 = (4097, 3N) = N$ and $d_2 = 3N/d_1 = 3$, which gives only the factors 1 and N of N.

However, the next usable pair for $3N$ is $s_{10} = 121$ and $p_9 = 9169$, which gives $d_1 = 241$ and $d_2 = 51$. This yields the nontrivial factorization $N = 241 \cdot 17$.

10-84. *Use nonsquare values of s_k to factor 34579. (See the end of the section on Legendre's Factoring Method.)*

Solution. Let $N = 34579$; then $\sqrt{N} = \langle 185, \overline{1, 20, 1, 7, 3, 4, 2, 4, 3, 7, 1, 20, 1, 370} \rangle$, and so the length of the period is 14. If we go to the middle of the period, we find that $s_6 = s_8 = 78$; modulo N, we then have $p_5 = 32559$ and $p_7 = 12325$. Since $p_5^2 \equiv s_6$ and $p_7^2 \equiv s_8 \pmod{N}$, it follows that $(p_5 p_7)^2 \equiv s_6 s_8 \pmod{N}$, i.e., $380^2 \equiv 78^2 \pmod{N}$. Hence $d_1 = (380 - 78, N) = (302, N) = 151$, and therefore $d_2 = N/151 = 229$. Thus 34579 factors as $151 \cdot 229$.

Note. The only usable *square* s_k in the first period is the *last* value, namely, $s_{14} = 1$. Since $p_{13} \equiv 7098 \pmod{N}$, we have $d_1 = (7097, N) = 151$ and hence $d_2 = N/d_1 = 229$.

EXERCISES FOR CHAPTER 10

1. Find the general form of a positive solution of $x^2 - 11y^2 = 1$ in terms of the convergents of $\sqrt{11}$, and determine the least positive solution.
2. Find the unique positive solution (a, b) of $x^2 - 14y^2 = 1$ such that $10^6 < a + b\sqrt{14} < 10^8$.
3. Determine the fourth positive solution of $x^2 - 7y^2 = 1$.
4. (a) Find the least positive solution of $x^2 - 74y^2 = 1$ and $x^2 - 74y^2 = -1$.
   (b) Determine the next positive solution of each equation.
5. How many positive solutions (a, b) of $x^2 - 13y^2 = -1$ have the property that $10^5 < a + b\sqrt{13} < 10^{15}$?
6. What is the third positive solution of $x^2 - 8y^2 = 1$?
7. The fundamental solution of $x^2 - 5y^2 = -1$ is $(2, 1)$. Use this to find the seventh convergent p_7/q_7 of $\sqrt{5}$.

8. If (a, b) is a positive solution of $x^2 - 3y^2 = 1$, prove that a is not divisible by any prime of the form $6k + 5$. (Hint. Use (5.10) and (5.13).)

9. Prove or disprove: If $x^2 - dy^2 = -1$ is solvable, then $x^2 - d^n y^2 = -1$ is solvable for every odd $n \geq 1$.

10. Calculate the continued fraction expansion of $\sqrt{29}$, and prove that $k\sqrt{29}$ has a period of length 1 for infinitely many values of k. (Hint. Note that the length of the period of $\sqrt{n^2 + 1}$ is 1, by Problem 9-24.)

11. There are infinitely many triples of consecutive integers each of which is a sum of two squares. Find all such triples where each member is between 100,000 and 400,000.

12. (a) Determine, by inspection, the least positive solution of $x^2 - 899y^2 = 1$.
    (b) Without calculating the continued fraction expansion of $\sqrt{899}$, explain why $x^2 - 899y^2 = -1$ is not solvable. Is the length of the period of $\sqrt{899}$ odd or even?

13. Prove that there are infinitely many positive integers n such that $n/3 + 1$ and $n + 1$ are both perfect squares. Find the first two such values of n.

14. By inspection, find the least positive solution of $x^2 - 143y^2 = 1$. Use this to decide if $x^2 - 143y^2 = -1$ is solvable.

15. Show that $x^2 - 1995y^2 = -1$ has no solutions in positive integers. Is the length of the period of $\sqrt{1995}$ odd or even?

16. Prove or disprove: The length of the period of $\sqrt{50000007}$ is even. (Do not calculate the continued fraction of $\sqrt{50000007}$!)

17. Prove that the equation $x^4 - 121y^4 = -45$ has no solution in positive integers. (Hint. First show that the length of the period of $\sqrt{11}$ is even.)

▷ 18. Show that $x^4 - 441y^4 = -5991$ has no solution in positive integers.

19. Determine, by inspection, a positive solution of $x^2 - 12y^2 = 1$ and $x^2 - 12y^2 = 13$. Use these solutions to find another positive solution of $x^2 - 12y^2 = 13$.

▷ 20. Find five positive solutions of $x^2 - 17y^2 = -8$, each solution having $y < 4000$.

21. Prove that $x^2 - 311y^2 = -3$ has no solution in positive integers. (Hint. Use (5.13).)

22. Determine if the equation $x^2 - 313y^2 = 5$ is solvable.

23. Prove or disprove: If (r, s) is a positive solution of $x^2 - 519y^2 = -14$, then r/s is a convergent of $\sqrt{519}$.

24. Use the technique described in the proof of (10.20) to find a primitive representation of 21437 as a sum of two squares.

25. Use (10.20) to find a representation of 84922 as a sum of two relatively prime squares.

26. Without calculating their continued fraction expansions, show that $\sqrt{793268}$ and $\sqrt{1759251}$ each have periods of even length.
27. Use Legendre's Factoring Method to determine the prime factorization of 49387.
28. Calculate a proper divisor of 24569 using Legendre's Factoring Method.
29. Using Legendre's Factoring Method, find the prime factorization of 83731.

NOTES FOR CHAPTER 10

1. Additional solutions of Pell's Equation can be found either by using a recursive formula (see Theorem 10.11 and Problem 10-30) or by determining the nearest integer to $(r + s\sqrt{d})^n$, where (r, s) is the least positive solution (Theorem 10.15). For computational purposes, the "nearest integer" algorithm will quickly exceed the accuracy of the calculator or computer being used. The recursive definitions, however, do not have this limitation, since all of the calculations are done in integer arithmetic.

On a computer (unlike a calculator), there is no real difference in speed between the two methods, but the recursive approach is preferable, since floating-point arithmetic is inherently less accurate than integer arithmetic. However, the usefulness of (10.15) is evident if a *calculator* is used to find additional solutions (as on an exam).

2. Theorems 10.13 and 10.14 show that the values of p_{jm-1} ($j \geq 1$) can be computed once p_{m-1} and q_{m-1} are known, according to the relation $(p_{m-1} + q_{m-1})^j = p_{jm-1} + q_{jm-1}\sqrt{d}$. In fact, *all* values of p_k and q_k can be determined if their values in the first period are known:

$$p_{i+m} + q_{i+m}\sqrt{d} = (p_i + q_i\sqrt{d})(p_{m-1} + q_{m-1}\sqrt{d}).$$

For a proof of this, see Problem 10-19.

3. Factoring Large Numbers. Legendre's Factoring Method is actually a very old technique, dating back to the second half of the eighteenth century. However, it was not really a practical algorithm to use before the advent of fast computers, since for large N, there are, in general, a great number of calculations necessary to find the s_k that are perfect squares.

In addition to Legendre's Factoring Method (10.22), there are a number of other important factoring algorithms that use similar ideas, among them Fermat's Method, Euler's Method, and Gauss's Method (see the book by H. Riesel listed in the References). A particularly efficient implementation of the continued fraction algorithm is embodied in Shank's Method. This is also discussed in Riesel's book, where a PASCAL program for employing the algorithm is given. Finally, we mention a factorization algorithm due to Morrison and Brillhart, also based on continued fractions, which is very efficient. Unlike Legendre's Method and Shank's Method, in which successive values of s_k are computed until a square is found, this algorithm tries to combine various s_k to form a square; thus it may require much less time to produce a factorization.

A fairly short QuickBasic program that implements Legendre's Factoring Method and utilizes doping factors is given in "A Continued Fraction Approach for Factoring Large Numbers" by Robert A. Coury in the *Pi Mu Epsilon Journal* (Vol. 9, No. 1, 1989). This paper also contains a table of large numbers, their factors, the number of square s_k the program checks until a proper factorization is found, and the subscript k that produces this factorization. The program can be run on any IBM-compatible personal computer.

BIOGRAPHICAL SKETCHES

Bhaskara, also known as Bhaskara Acharya, is the most famous Indian astronomer-mathematician of the premodern period. He was born in 1114, in the Indian state of Mysore. Bhaskara's most famous work, the *Siddanta Siromani*, was written in 1150. The first two books, called the *Leelivati* and the *Bijaganita*, are textbooks of arithmetic and algebra, and the rest concerns itself with mathematical astronomy. As was common in that era, all of the work is written in verse.

The *Leelivati* contains basic arithmetic and geometry, as well as a large collection of recreational problems, written in beautifully fanciful language. (There is a story, not necessarily true, that Bhaskara had a daughter named Leelivati, who was widowed early. He taught her mathematics to assuage her grief and composed *Leelivati* for her.) The book ends with a section on linear Diophantine equations. All of the mathematics of *Leelivati* was probably known to Brahmagupta, some five hundred years earlier.

The *Bijaganita* is mathematically much more sophisticated. In the main, it concerns itself with Pell's Equation $x^2 - dy^2 = 1$ and related Diophantine problems. The book presents an efficient algorithm for solving Pell's Equation, closely related to the procedure Brouncker was to discover five hundred years later. Like Brouncker, Bhaskara did not prove that his method always yields a solution. One of the worked examples in the *Bijaganita* is the equation $x^2 - 61y^2 = 1$. In only a few steps, Bhaskara produces the solution $x = 226,153,980$. Fermat challenged Frenicle in 1657 with precisely this problem!

In studying Pell's Equation, Bhaskara was working within a long-standing Indian mathematical tradition. Brahmagupta, for example, knew how to generate additional solutions from given solutions. It is clear that Bhaskara had algebraic knowledge that would not be matched in Europe until the seventeenth century.

Bhaskara died in approximately 1185.

REFERENCES

Leonhard Euler, *Elements of Algebra*, translated by John Hewlett, Springer-Verlag, New York, 1984.

This is probably the most widely read of all books about mathematics other than Euclid's *Elements*, having been printed more than 30 times in six languages. Most texts in elementary algebra are direct descendants of this book, for Euler's influence ensured that his notation, terminology, and choice of topics became standard.

The first few sections are elementary, containing material that is still covered in high school algebra courses. The final third of the book deals with Diophantine equations. It begins with integer solutions of linear equations and goes on to study Pell's Equation, sums of two squares, and related problems. Euler then discusses at length the equations $x^4 + y^4 = z^2$ and $x^3 + y^3 = z^3$, as well as a number of problems of the kind Diophantus studied, where rational solutions are sought. The book is always very clear, and each technique is profusely illustrated with examples.

Euler's *Algebra* ends with a lengthy supplement by Lagrange in which he lays out the theory of continued fractions, including a full theoretical analysis of Pell's Equation. Lagrange then gives a procedure for finding all integer solutions of arbitrary Diophantine equations of the second degree in two variables. Lagrange's addition, and indeed the entire book, is still very much worth reading.

Hans Riesel, *Prime Numbers and Computer Methods for Factorization*, Birkhäuser, Boston, 1987.

Riesel describes some of the basic number-theoretic algorithms, focusing on ideas that are needed for primality testing and factorization. In particular, he sketches the parts of the theory of Pell's Equation needed to understand Legendre's Factoring Method and related techniques. The book is easy to read and gives good insight about computational issues. An interesting feature is a collection of computer programs, including a complete multiple-precision package for the elementary arithmetical operations.

CHAPTER ELEVEN

The Gaussian Integers and Other Quadratic Extensions

We consider first a simple problem: For which nonzero integers n does the Diophantine equation $x^2 - y^2 = n$ have a solution, and how many solutions are there? To arrive at an answer, we *factor* the polynomial $x^2 - y^2$ and obtain the equivalent equation $(x - y)(x + y) = n$.

If n is odd, let d be a divisor (positive or negative) of n. Set $x + y = d$ and $x - y = n/d$. Then $x = (d + n/d)/2$ and $y = (d - n/d)/2$ are clearly integers, and $x^2 - y^2 = n$. Moreover, all pairs (x, y) of integers satisfying the equation arise in this way. Recalling that $\tau(k)$ is the number of positive divisors of k, we see that there are $2\tau(|n|)$ choices for d, and thus $2\tau(|n|)$ solutions in integers (positive, negative, or zero) of the equation $x^2 - y^2 = n$.

The key step in the analysis of the solutions of the Diophantine equation $x^2 - y^2 = n$ is the factorization $x^2 - y^2 = (x - y)(x + y)$. It is tempting to apply a similar idea to the equation $x^2 + y^2 = n$. Although the polynomial $x^2 + y^2$ does not factor over the real numbers, we do have the factorization $x^2 + y^2 = (x - yi)(x + yi)$, where $i^2 = -1$. This leads to an examination of the number-theoretic properties of complex numbers of the form $a + bi$, where a and b are integers. We will see later in this chapter how the use of such complex numbers, called *Gaussian integers*, makes the theory of representations as a sum of two squares quite transparent.

Euler seems to have been the first to use such extensions of the ordinary integers to prove a number-theoretic result. In attempting to prove that $x^3 + y^3 = z^3$ has no integer solutions with $xyz \neq 0$, he was led to look at the factorization $z^3 - y^3 = (z - y)(z - y\omega)(z - y\omega^2)$, where $\omega = (-1 + \sqrt{-3})/2$ is a cube root of 1. Without explicitly saying so, he assumed that numbers of the form $a + b\omega$, where a and b are integers, have factorization properties much like those of the ordinary integers. Euler then used a descent argument that settled the case $n = 3$ of Fermat's Last Theorem.

The next significant use of such extensions of the integers was by Gauss, in a series of papers on biquadratic reciprocity. The simple fact that $x^2 - 1 = (x-1)(x+1)$ is important for the analysis of quadratic residues, and the factorization $x^4 - 1 = (x-1)(x+1)(x-i)(x+i)$ is helpful in dealing with fourth power residues. To use this factorization effectively, one first needs to establish the basic properties of Gaussian integers, and this Gauss proceeded to do.

During the nineteenth century, investigation of properties of "integers" of a more general kind became important in the study of higher reciprocity laws and also of Diophantine equations, particularly Fermat's Last Theorem and quadratic forms. Some of the most important mathematicians of the era made contributions, including Dirichlet, Kummer, Eisenstein, Jacobi, Kronecker, and Dedekind. The field that they developed, *algebraic number theory*, is one of the basic areas of modern number theory.

RESULTS FOR CHAPTER 11

The Gaussian Integers

(11.1) Definition. A *Gaussian integer* is a complex number of the form $a + bi$, where a and b are integers and $i = \sqrt{-1}$.

It is clear that every ordinary integer is a Gaussian integer and that the sum, difference, and product of Gaussian integers is a Gaussian integer. We will generally use lowercase Greek letters to denote Gaussian integers. In this chapter, ordinary integers will often be called *rational integers*.

(11.2) Definitions. (i) Let α and β be Gaussian integers. We say that α *divides* β if there is a Gaussian integer γ such that $\beta = \gamma\alpha$. In this case, we write $\alpha | \beta$.

(ii) The Gaussian integer ϵ is called a *unit* if $\epsilon | 1$.

(iii) If ϵ is a unit, then $\epsilon\alpha$ is called an *associate* of α.

(iv) The Gaussian integer π is a *Gaussian prime* if π is not a unit, and for any Gaussian integers α and β such that $\pi = \alpha\beta$, either α or β is a unit.

(v) If $\alpha = x + yi$, where x and y are rational integers, then $\bar{\alpha}$, the *conjugate* of α, is the Gaussian integer $x - yi$.

(vi) If $\alpha = x + yi$, then the *norm* of α is

$$N(\alpha) = \alpha\bar{\alpha} = (x + yi)(x - yi) = x^2 + y^2.$$

The following fundamental property of the norm will be used repeatedly.

(11.3) Lemma. *Let α and β be Gaussian integers. Then $N(\alpha\beta) = N(\alpha)N(\beta)$.*

Proof. Let $\alpha = u + vi$ and $\beta = x + yi$. Then $\alpha\beta = (u+vi)(x+yi) = (ux - vy) + (uy + vx)i$. It follows that

$$N(\alpha\beta) = (ux - vy)^2 + (uy + vx)^2 = (u^2 + v^2)(x^2 + y^2) = N(\alpha)N(\beta).$$

Note. The identity $(ux - vy)^2 + (uy + vx)^2 = (u^2 + v^2)(x^2 + y^2)$ is familiar; we have already seen it in Chapter 8, where it played an important role in expressing an integer as the sum of two squares. In that chapter, however, it appeared as an isolated fact. Here, we see the identity as expressing a fundamental *algebraic* property of the norm.

Since $N(x + yi) = x^2 + y^2$, it is clear that the norm of a Gaussian integer is a nonnegative rational integer and that 0 is the only Gaussian integer with norm 0. There is also a simple characterization of units in terms of their norm.

(11.4) Lemma. *The Gaussian integer α is a unit if and only if $N(\alpha) = 1$. The only Gaussian integers which are units are ± 1 and $\pm i$.*

Proof. If $N(\alpha) = 1$, then $\alpha\bar{\alpha} = 1$; in particular, α divides 1 and so α is a unit. Conversely, if α is a unit, then $\alpha\beta = 1$ for some Gaussian integer β, and hence $N(\alpha)N(\beta) = N(\alpha\beta) = N(1) = 1$. Since the norms of α and β are nonnegative integers, it follows that $N(\alpha) = 1$.

Let $\alpha = x + yi$; then $N(\alpha) = x^2 + y^2$. Clearly, $N(\alpha) = 1$ if and only if $y = 0$ and $x = \pm 1$, or $x = 0$ and $y = \pm 1$; hence the only Gaussian units are ± 1 and $\pm i$.

The next result plays the same role in the theoretical development as the Division Algorithm did in Chapter 1. Note the close similarity between the next few results and the basic theorems of Chapter 1.

(11.5) Lemma (Division Algorithm). *Let α and β be Gaussian integers, with α nonzero. Then there exist Gaussian integers γ and ρ such that $\beta = \gamma\alpha + \rho$ and $N(\rho) < N(\alpha)$.*

Proof. Let $\beta/\alpha = x + yi$, where x and y are real numbers, and let u and v be, respectively, integers nearest to x and y. (These may not be uniquely determined.) Define $\gamma = u + vi$, and set $\rho = \beta - \gamma\alpha$; clearly, $\beta = \gamma\alpha + \rho$. It remains to verify that $N(\rho) < N(\alpha)$. Since $\beta = (x + yi)\alpha$, we have $\rho = ((x - u) + (y - v)i)\alpha$. But $|x - u| \leq 1/2$ and $|y - v| \leq 1/2$, and hence $N((x - u) + (y - v)i) \leq 1/4 + 1/4 = 1/2$. Therefore $N(\rho) \leq \frac{1}{2}N(\alpha)$, and the result follows.

We now develop, as in Chapter 1, a theory of greatest common divisors. The appropriate definition is as follows.

(11.6) Definition. The Gaussian integer δ is a *greatest common divisor* of α and β if (i) δ divides α and β and (ii) δ is divisible by every common divisor of α and β.

We prove next that greatest common divisors as defined above exist and can be represented as a linear combination of α and β.

(11.7) Theorem. *If α and β are Gaussian integers, not both zero, then α and β have a greatest common divisor δ, which can be represented as $\delta = \lambda\alpha + \mu\beta$, where λ and μ are Gaussian integers.*

Proof. Let I be the set of all numbers of the form $\phi\alpha + \psi\beta$, where ϕ and ψ range over the Gaussian integers. Let δ be an element of I of smallest positive norm, and suppose that $\delta = \lambda\alpha + \mu\beta$. We show that δ is a greatest common divisor of α and β.

We first prove that $\delta \mid \alpha$. By (11.5), we have $\alpha = \kappa\delta + \rho$, where $N(\rho) < N(\delta)$. Then
$$\rho = \alpha - \kappa\delta = \alpha - \kappa(\lambda\alpha + \mu\beta) = (1 - \kappa\lambda)\alpha + (-\kappa\mu)\beta,$$
and so we have expressed ρ as a linear combination of α and β. Since $N(\rho) < N(\delta)$, this contradicts the definition of δ unless the remainder ρ is 0. Thus we conclude that $\delta \mid \alpha$; similarly, $\delta \mid \beta$. It is obvious that if $\gamma \mid \alpha$ and $\gamma \mid \beta$, then $\gamma \mid \lambda\alpha + \mu\beta$ and therefore $\gamma \mid \delta$. Thus δ is a greatest common divisor of α and β.

Note. Suppose that δ and δ' are both greatest common divisors of α and β. Since δ' is a divisor of α and of β, it follows from the definition of greatest common divisor that $\delta \mid \delta'$. Similarly, we have $\delta' \mid \delta$. Thus $N(\delta) \leq N(\delta')$ and $N(\delta') \leq N(\delta)$, and hence δ and δ' have the same norm. Since $\delta \mid \delta'$, it follows that $\delta' = \epsilon\delta$, where ϵ is a unit, and therefore δ' is an associate of δ.

Conversely, if δ is a greatest common divisor of α and β and δ' is an associate of δ, then δ' is also a greatest common divisor of α and β. Moreover, if we have expressed δ as a linear combination of α and β, it is easy to express δ' as a linear combination of α and β.

The following result extends to Gaussian primes a familiar property of ordinary primes and uses essentially the same proof.

(11.8) Lemma. *Let π be a Gaussian prime. If $\pi \mid \alpha\beta$, then $\pi \mid \alpha$ or $\pi \mid \beta$.*

Proof. If $\pi \nmid \alpha$, then 1 is a greatest common divisor of π and α. Hence there exist Gaussian integers λ and μ such that $\lambda\pi + \mu\alpha = 1$. Multiplying both sides of the equation by β, we obtain $\lambda\pi\beta + \mu\alpha\beta = \beta$. Since π divides both $\lambda\pi\beta$ and $\mu\alpha\beta$, it follows that π divides β.

Unique Factorization for Gaussian Integers

We show first that Gaussian integers can be decomposed into a product of primes.

(11.9) Theorem. *If α is a Gaussian integer other than 0 or a unit, then α can be expressed as a product of Gaussian primes.*

Proof. The proof is by induction on the norm of α. Suppose that the result is true for all Gaussian integers of norm less than n; we show that the result must then hold for Gaussian integers α of norm n. If α is a Gaussian prime, there is nothing to prove. Otherwise, there exist Gaussian integers β and γ, neither of which is a unit, such that $\alpha = \beta\gamma$. But since $N(\alpha) = N(\beta)N(\gamma)$ and neither β nor γ is a unit, we must have $N(\beta) < n$ and $N(\gamma) < n$. Thus by the induction hypothesis, both β and γ can be expressed as a product of Gaussian primes, and therefore α is also a product of Gaussian primes.

The next result shows that an analogue of the Fundamental Theorem of Arithmetic holds for Gaussian integers.

(11.10) Unique Factorization Theorem. *Suppose that $\alpha_1 \alpha_2 \cdots \alpha_r = \epsilon \beta_1 \beta_2 \cdots \beta_s$, where the α_i and β_i are Gaussian primes and ϵ is a unit. Then $r = s$, and the β_i can be rearranged so that for all i, β_i is an associate of α_i.*

Proof. Let $\gamma = \alpha_1 \alpha_2 \cdots \alpha_r$. The result is clear when γ is a Gaussian prime, for then $r = s = 1$ and $\alpha_1 = \epsilon \beta_1$. We prove the result in general by induction on r. Thus suppose that the unique factorization result holds for all Gaussian integers that have at least one factorization as a product of $r-1$ (not necessarily distinct) prime factors; we will show that it must hold for all Gaussian integers that have an expression as the product of r prime factors.

Suppose that $\alpha_1 \alpha_2 \cdots \alpha_r = \epsilon \beta_1 \beta_2 \cdots \beta_s$. Since α_r divides the product of the β_i, it must divide at least one of the β_i. By rearranging the factors if necessary, we may assume that $\alpha_r | \beta_s$. Since β_s is prime, it must be an associate of α_r; thus $\beta_s = \epsilon' \alpha_r$, where ϵ' is a unit. Cancelling α_r from both sides of the equation

$$\alpha_1 \alpha_2 \cdots \alpha_r = \epsilon \epsilon' \beta_1 \beta_2 \cdots \beta_{s-1} \alpha_r,$$

we obtain

$$\alpha_1 \cdots \alpha_{r-1} = \epsilon \epsilon' \beta_1 \cdots \beta_{s-1}.$$

By the induction hypothesis, we therefore have $r - 1 = s - 1$, and hence $r = s$. It also follows from the induction hypothesis that the numbers $\beta_1, \ldots, \beta_{r-1}$ can be rearranged so that they are associates of $\alpha_1, \ldots, \alpha_{r-1}$, which proves the theorem.

Note. The unique factorization theorem for Gaussian integers can be reworded so as to look more like the ordinary unique factorization theorem. If $a + bi$ is a nonzero Gaussian integer, we will temporarily call $a + bi$ *positive* if $a > 0$ and $b \geq 0$. It is easy to verify that any nonzero Gaussian integer has a unique positive associate. Then any nonzero Gaussian integer has (apart from the order of the factors) a unique representation of the form $\epsilon \pi_1^{a_1} \pi_2^{a_2} \cdots \pi_r^{a_r}$, where ϵ is a unit, the π_i are distinct positive Gaussian primes, and the a_i are positive integers.

The Gaussian Primes

Note first that if $N(\alpha)$ is a rational prime, then α is a Gaussian prime. For suppose that $\alpha = \beta\gamma$; then $N(\alpha) = N(\beta)N(\gamma)$. Thus if $N(\alpha)$ is a rational prime, then $N(\beta) = 1$ or $N(\gamma) = 1$, and therefore by (11.4), either β or γ is a unit. It follows from the definition that α is a Gaussian prime.

We next observe that any Gaussian prime divides some rational prime. To see this, let π be a Gaussian prime, and let $n = N(\pi)$. Since $N(\pi) = \pi\bar{\pi}$, we have $\pi | n$. Now express n as a product of rational primes, say, $n = p_1 p_2 \cdots p_k$. Since $\pi | n$, it follows from (11.8) that $\pi | p_i$ for some i. Thus in looking for Gaussian primes, it is enough to examine factors of rational primes.

In the Gaussian integers, the rational prime 2 has the factorization $2 = (1 - i)(1 + i)$. Since $1 - i$ has norm 2, which is prime, it follows that $1 - i$ and its associates $1 + i$, $-1 + i$, and $-1 - i$ are Gaussian primes.

Rational primes p of the form $4k + 3$ are also Gaussian primes. For suppose that $p = \alpha\beta$; then $N(p) = p^2 = N(\alpha)N(\beta)$. But p cannot be the norm of a Gaussian integer, since no number of the form $4k + 3$ is the sum of two squares. Therefore either $N(\alpha)$ or $N(\beta)$ must be 1, and hence either α or β is a unit. It follows that p is a Gaussian prime. Thus the only Gaussian primes that divide p are the four associates of p.

It remains to deal with rational primes p of the form $4k + 1$. By (8.8), any such prime can be expressed as a sum of two squares. If $p = u^2 + v^2$, then $p = (u + vi)(u - vi)$. Since the norms of $u + vi$ and $u - vi$ are prime, each of these numbers is a Gaussian prime. Hence $u + vi$, $u - vi$, and their associates are the only Gaussian primes that divide p.

The discussion above can be summarized as follows.

(11.11) Theorem. *The Gaussian integer π is a Gaussian prime if and only if one of the following holds:*

(i) π is $1 - i$ or an associate;

(ii) π is a rational prime of the form $4k + 3$ or an associate;

(iii) $N(\pi) = p$, where p is a rational prime of the form $4k + 1$.

An Application: Gaussian Integers and Sums of Two Squares

Gaussian integers can be used to give alternative proofs for all of the results about representations as a sum of two squares that were proved in Chapter 8.

As our first example, we give another proof of the fact that every prime p of the form $4k+1$ is a sum of two squares. We show first that p is *not* a Gaussian prime. Since -1 is a quadratic residue of p, there is a rational integer x such that $x^2 \equiv -1 \pmod{p}$. Thus $p \mid x^2 + 1$ and therefore $p \mid (x-i)(x+i)$. If p were a Gaussian prime, it would follow from (11.8) that $p \mid x - i$ or $p \mid x + i$, but plainly neither is the case.

Since p is not a Gaussian prime, there exist Gaussian integers α and β, neither of which is a unit, such that $p = \alpha\beta$. Since $N(p) = p^2 = N(\alpha)N(\beta)$ and neither $N(\alpha)$ nor $N(\beta)$ is equal to 1, it follows that $N(\alpha) = N(\beta) = p$. Thus if $\alpha = u + iv$, then $p = u^2 + v^2$, and hence p is a sum of two squares.

The unique factorization theorem for Gaussian integers can also be used to count the number $N(n)$ of representations of n as a sum of two squares. (Do not confuse $N(n)$, the number of representations, with $N(\alpha)$, the norm of α.)

(11.12) Theorem. *Let $n = 2^a \prod p_j^{a_j} \prod q_j^{b_j}$, where the p_j are $4k+1$ primes, the q_j are $4k+3$ primes, and each b_j is even. Then*

$$N(n) = 4 \prod (a_j + 1).$$

(A product of no terms is interpreted to be 1.)

Proof. We need to count the number of ordered pairs (u, v) of integers such $u^2 + v^2 = n$, that is, $(u + vi)(u - vi) = n$. For each p_j, let c_j and d_j be integers such that $p_j = c_j^2 + d_j^2$. Then, in the Gaussian integers, n has the factorization

$$n = i^a (1-i)^{2a} \prod (c_j + d_j i)^{a_j} (c_j - d_j i)^{a_j} \prod q_j^{b_j},$$

where $1 - i$, $c_j \pm d_j i$, and q_j are all Gaussian primes.

Any factor $u + vi$ of n has the form

$$\epsilon (1-i)^t \prod (c_j + d_j i)^{u_j} (c_j - d_j i)^{v_j} \prod q_j^{w_j}, \tag{1}$$

where ϵ is a unit, $0 \le t \le 2a$, $0 \le u_j, v_j \le a_j$, and $0 \le w_j \le b_j$. Moreover, by the Unique Factorization Theorem, all distinct expressions of form (1) lead to distinct factors of n. By taking conjugates, we find that $u - vi$ has the representation

$$\bar{\epsilon}(1+i)^t \prod (c_j - d_j i)^{u_j} (c_j + d_j i)^{v_j} \prod q_j^{w_j}. \tag{2}$$

The condition $(u+vi)(u-vi) = n$ then yields $n = 2^t \prod p_j^{u_j+v_j} \prod q_j^{2w_j}$; thus $t = a$, $u_j + v_j = a_j$, and $2w_j = b_j$. It follows that t and the w_j are completely determined by n, that we have for each j precisely $a_j + 1$ possible choices for u_j, and that once u_j has been chosen, v_j is determined. Since there are clearly four choices for the unit ϵ in (1), we conclude that there are $4\prod(a_j+1)$ choices for $u + vi$. This completes the proof.

Factorization in the Gaussian integers also throws additional light on representations of n as a sum of two relatively prime squares. We give a proof of (8.18) based on Gaussian integers.

(11.13) Theorem. *Suppose that $n = p_1^{a_1} \cdots p_r^{a_r}$, where $p_1, p_2, \ldots, p_r$ are distinct primes of the form $4k + 1$. Then n and $2n$ each have precisely 2^{r+2} primitive representations as a sum of two squares.*

Proof. We prove the result for n; the proof for $2n$ is similar. For each j, let $p_j = (c_j + d_j i)(c_j - d_j i)$. The proof of (11.12) shows that the representations of n as $u^2 + v^2$ are obtained by setting $u + iv = \epsilon \prod (c_j + d_j i)^{u_j} (c_j - d_j i)^{v_j}$, where ϵ is a unit and $u_j + v_j = a_j$. If $0 < u_j < a_j$, then $(c_j + d_j i)(c_j - d_j i)$ divides $u + vi$, so p_j divides $u + vi$. Taking conjugates, we find that p_j divides $2u$ and $2v$, and since p_j is odd, p_j divides u and v. Hence we do not get a primitive representation.

Thus for each j, we must choose $u_j = 0$ or $u_j = a_j$, which then produces a primitive representation of n. For suppose, to the contrary, that p is a prime dividing u and v; then p divides n, and hence $p = p_j$ for some j. Thus $c_j + d_j i$ and $c_j - d_j i$ each divide $u + vi$, which is impossible since if $u_j = 0$, then $c_j + d_j i$ does not divide $u + vi$, and if $u_j = a_j$, then $c_j - d_j i$ does not divide $u + vi$.

It follows that for each j, we have the two options $u_j = 0$ or $u_j = a_j$, giving a total of 2^r choices. Since there are four possibilities for ϵ, we conclude that the total number of primitive representations is 2^{r+2}.

An Application of Gaussian Integers to Diophantine Equations

We have already used Gaussian integers to analyze the Diophantine equation $x^2 + y^2 = n$. In this section, we present another example. The main tool is the following analogue of a familiar fact about the integers.

(11.14) Theorem. *Suppose $\alpha\beta = \gamma^n$, where α, β, and γ are Gaussian integers and α and β are relatively prime. Then there exists a unit ϵ and a Gaussian integer δ such that $\alpha = \epsilon\delta^n$.*

Proof. We use the Unique Factorization Theorem, in the form that states that (apart from the order of the terms) any nonzero Gaussian integer can be uniquely expressed as a unit times a product of "positive" Gaussian primes. (See the Note following (11.10).) Let

$$\alpha = \epsilon_a \prod \pi_j^{a_j}, \quad \beta = \epsilon_b \prod \pi_j^{b_j}, \quad \gamma = \epsilon_c \prod \pi_j^{c_j}.$$

Since $\alpha\beta = \gamma^n$, we have $a_j + b_j = nc_j$ for every j. Also, since α and β are relatively prime, no Gaussian prime can occur in the factorizations of both α and β; thus for any j, we have $a_j = 0$ or $a_j = nc_j$. It follows that every a_j is a multiple of n, that is, α is a unit times an nth power.

Example. We use properties of the Gaussian integers to find all solutions of the Diophantine equation $x^2 + 1 = y^3$ or, equivalently, $(x+i)(x-i) = y^3$. First we show that if x, y is a solution, then $x+i$ and $x-i$ are relatively prime. If they are not, there is a Gaussian prime π which divides $x+i$ and $x-i$; then $\pi | 2$, since π divides $(x+i) - (x-i)$. It is easy to see that x must be even, for if x is odd, then $x^2 + 1 \equiv 2 \pmod{4}$, and no cube can be congruent to 2 modulo 4. Since x is even, we have $\pi | x$; since $\pi | x + i$ by assumption, it follows that $\pi | i$, which is impossible.

By (11.14), $x+i$ can be expressed in the form $\epsilon\delta^3$, where ϵ is a unit. It is easy to check that for any unit ϵ, we have $\epsilon^3 = \epsilon$, so in fact $x+i = (\epsilon\delta)^3$. Let $x + i = (u+vi)^3 = (u^3 - 3uv^2) + (3u^2v - v^3)i$; then $x = u^3 - 3uv^2$ and $1 = 3u^2v - v^3$.

Since $v(3u^2 - v^2) = 1$, it follows that $v = \pm 1$. If $v = 1$, then $3u^2 - v^2 = 1$ and hence $3u^2 = 2$, which is impossible. If $v = -1$, then $3u^2 - v^2 = -1$, giving $u = 0$. Thus $x = 0$, and therefore $y = 1$. We conclude that the only solution in integers of the Diophantine equation $x^2 + 1 = y^3$ is $x = 0$, $y = 1$. (For more examples, see Problems 11-30 and 11-31.)

The Integers of $Q(\sqrt{d})$

(11.15) Definition. *Suppose $d \neq 1$, and assume that d is square-free. Then $Q(\sqrt{d})$ is the set of all numbers of the form $x + y\sqrt{d}$, where x and y range over the rationals. The number $x + y\sqrt{d}$ is said to be an* integer *of $Q(\sqrt{d})$ if either (i) $d \equiv 2$ or $3 \pmod{4}$ and x and y are ordinary integers or (ii) $d \equiv 1 \pmod{4}$ and x and y are both integers or each is half an odd integer.*

Notes. **1.** When $d \equiv 1 \pmod{4}$, the definition of an integer of $Q(\sqrt{d})$ can be motivated as follows. Call a complex number α an *algebraic integer* if α is a root of a polynomial with integer coefficients and leading coefficient equal to 1. It is not difficult to show that if $\alpha \in Q(\sqrt{d})$, then α is an algebraic integer if and only if α is an integer of $Q(\sqrt{d})$ in the sense of the preceding definition (see Problems 11-41 and 11-42).

2. If $d \equiv 1 \pmod{4}$, the integers of $Q(\sqrt{d})$ are precisely the complex numbers of the form $x + y\omega$, where x and y are ordinary integers and $\omega = (1 + \sqrt{d})/2$.

In what follows, we will need to know that if d is not the square of an integer, then $\sqrt{d}$ is irrational. It is just as easy to prove a more general result.

(11.16) Theorem. *Let $P(x)$ be a polynomial with integer coefficients and leading coefficient 1. If z is a rational number such that $P(z) = 0$, then z is an integer. In particular, if the integer d is not a perfect square, then $\sqrt{d}$ is irrational.*

Proof. Let $P(x) = x^k + a_{k-1}x^{k-1} + \cdots + a_1 x + a_0$, and suppose that $P(z) = 0$, where z is rational. Then $z = s/t$, where s and t are integers. Without loss of generality, we may assume that t is positive and $(s, t) = 1$. We will show that $t = 1$.

Substituting s/t for z and multiplying by t^k, we obtain

$$a_{k-1}s^{k-1}t + a_{k-2}s^{k-2}t^2 + \cdots + a_1 s t^{k-1} + a_0 t^k = -s^k.$$

The left side is clearly divisible by t, and so $t \mid s^k$. Since $(s, t) = 1$, it follows that $t = 1$.

Finally, to show that $\sqrt{d}$ is not rational unless d is a perfect square, note that $\sqrt{d}$ is a root of the polynomial equation $x^2 - d = 0$.

Note. If α is an element of $Q(\sqrt{d})$, then there exist *uniquely determined* rational numbers x, y such that $\alpha = x + y\sqrt{d}$. This is an immediate consequence of (11.16).

Elementary Arithmetical Operations. It is easy to see that if α and β are in $Q(\sqrt{d})$, then $\alpha \pm \beta$ and $\alpha\beta$ are also in $Q(\sqrt{d})$. It is also true that if $\beta \neq 0$, then α/β is in $Q(\sqrt{d})$. To prove this, it is enough to show that $1/\beta$ is in $Q(\sqrt{d})$.

Let $\beta = s + t\sqrt{d}$, where s and t are rational. If $t = 0$, then β is rational and the result holds. If $t \neq 0$, note that $s - t\sqrt{d} \neq 0$: for if $s - t\sqrt{d} = 0$, then

$\sqrt{d} = s/t$, contradicting (11.16). Therefore

$$\frac{1}{s+t\sqrt{d}} = \frac{s-t\sqrt{d}}{(s+t\sqrt{d})(s-t\sqrt{d})} = \frac{s}{s^2-dt^2} + \frac{-t}{s^2-dt^2}\sqrt{d},$$

and thus $1/\beta$ is in $Q(\sqrt{d})$.

We have already examined the case $d = -1$ in detail; the integers of $Q(\sqrt{-1})$ are just the Gaussian integers. We now sketch the basic arithmetic of $Q(\sqrt{d})$. The fundamental definitions are close analogues of the corresponding definitions for the Gaussian integers.

(11.17) Definitions. (i) Let α, β be integers of $Q(\sqrt{d})$. Then α *divides* β if there is an integer γ of $Q(\sqrt{d})$ such that $\beta = \gamma\alpha$.

(ii) An integer ϵ of $Q(\sqrt{d})$ is called a *unit* if ϵ divides 1. If $\beta = \epsilon\alpha$ for some unit ϵ, then β is called an *associate* of α.

(iii) An integer π of $Q(\sqrt{d})$ is *prime* if π is not a unit, and for any integers α and β of $Q(\sqrt{d})$ such that $\pi = \alpha\beta$, either α or β is a unit.

(iv) If x and y are rational numbers, then $x - y\sqrt{d}$ is the *conjugate* of $x + y\sqrt{d}$, and $x^2 - dy^2$ is the *norm* of $x + y\sqrt{d}$. We will use the notation $\bar{\alpha}$ for the conjugate of α, and $N(x + y\sqrt{d})$ for the norm of $x + y\sqrt{d}$.

Note. The conjugate of α was defined, for positive d, in Chapter 9; there, it was denoted by α'. We use $\bar{\alpha}$ here to emphasize the fact that this is a generalization of the familiar notion of complex conjugate.

We next state some results about units, associates, conjugates, and norm. The proofs are all straightforward.

(11.18) Theorem.

(i) If ϵ and ϵ' are units, then so are $\epsilon\epsilon'$ and $1/\epsilon$.

(ii) If α and β are elements of $Q(\sqrt{d})$, then $N(\alpha\beta) = N(\alpha)N(\beta)$.

(iii) If α is an integer of $Q(\sqrt{d})$, then α is a unit if and only if $N(\alpha) = \pm 1$.

(iv) If α, β are in $Q(\sqrt{d})$, then $\overline{\alpha\beta} = \bar{\alpha}\bar{\beta}$.

(v) If $\alpha|\beta$, then $N(\alpha)|N(\beta)$.

(vi) If α is an integer of $Q(\sqrt{d})$ and $N(\alpha) = \pm p$, where p is a rational prime, then α is prime.

Up to this point, the results mirror closely standard properties of the Gaussian integers. It is also easy to show that any integer of $Q(\sqrt{d})$ which is not 0 or a unit can be expressed as a product of primes. The proof is almost identical to the proof of (11.9); the only significant difference is that since norms can be negative in $Q(\sqrt{d})$, we prove the result by induction on the *absolute value* of the norm.

As the following example shows, however, *we do not in general have unique factorization for the integers of* $Q(\sqrt{d})$.

Example. We show that unique factorization fails for the integers of $Q(\sqrt{-6})$. First note that the only units are ± 1. This is straightforward, since by (11.18.iii), $x + y\sqrt{-6}$ is a unit if and only if $x^2 + 6y^2 = 1$, and the only integer solutions of this equation are $x = \pm 1$, $y = 0$.

Next observe that $10 = 2 \cdot 5 = (2 + \sqrt{-6})(2 - \sqrt{-6})$; clearly, neither $2 + \sqrt{-6}$ nor $2 - \sqrt{-6}$ is an associate of 2 or 5. We will show that 2, 5, and $2 \pm \sqrt{-6}$ are all primes, and thus 10 has two essentially different representations as a product of primes.

Neither 2 nor 5 can be the norm of an integer of $Q(\sqrt{-6})$, since $N(x + y\sqrt{-6}) = x^2 + 6y^2$ and $x^2 + 6y^2$ cannot take on the values 2 or 5. We now show that 2 is prime. If $2 = \alpha\beta$, then by taking norms, we find that $4 = N(\alpha)N(\beta)$; thus $N(\alpha) \mid 4$. Since $N(\alpha) \neq 2$, we must have $N(\alpha) = 1$ or $N(\alpha) = 4$, and therefore either α or β is a unit. In essentially the same way, we can show that 5 and $2 \pm \sqrt{-6}$ are prime.

It is also true that integers of $Q(\sqrt{-6})$ need not have a greatest common divisor. For example, let $\alpha = 10$ and $\beta = 5\sqrt{-6}$. If α and β have a greatest common divisor δ, then $N(\delta)$ is a common divisor of $N(\alpha)$ and $N(\beta)$, and hence $N(\delta) \mid 50$. Note that $2 + \sqrt{-6}$ is a common divisor of α and β, since $(2+\sqrt{-6})(2+\sqrt{-6}) = 10$ and $(2+\sqrt{-6})(3+\sqrt{-6}) = 5\sqrt{-6}$. Also, 5 is obviously a common divisor of α and β. Thus $N(5)$ and $N(2 + \sqrt{-6})$ must each divide $N(\delta)$, and therefore $50 \mid N(\delta)$. We conclude that $N(\delta) = 50$. But this is impossible, for it is easily checked that the equation $x^2 + 6y^2 = 50$ does not have a solution in integers.

Note. Unique factorization also fails for the integers of $Q(\sqrt{-5})$; see Problem 11-48.

In a few cases, we can prove a Unique Factorization Theorem for the integers of $Q(\sqrt{d})$ using essentially the same proof as for Gaussian integers.

(11.19) Definition. $Q(\sqrt{d})$ is *Euclidean* if for any integers α and β of $Q(\sqrt{d})$, with $\alpha \neq 0$, there exist integers γ and ρ of $Q(\sqrt{d})$ such that $\beta = \gamma\alpha + \rho$ and $0 \leq N(\rho) < N(\alpha)$.

The ordinary Division Algorithm served as a foundation for Chapter 1, and a close analogue (Lemma 11.5) was used in developing the factorization properties of Gaussian integers. Essentially the same Division Algorithm is built into the above *definition* of "Euclidean."

The basic results about Gaussian integers can be stated and proved in *exactly the same way* for the integers of $Q(\sqrt{d})$, as long as $Q(\sqrt{d})$ is Euclidean.

(11.20) Unique Factorization Theorem for $Q(\sqrt{d})$. *Assume that $Q(\sqrt{d})$ is Euclidean. Suppose that $\alpha_1 \alpha_2 \cdots \alpha_r = \epsilon \beta_1 \beta_2 \cdots \beta_s$, where the α_i and β_i are primes of $Q(\sqrt{d})$ and ϵ is a unit. Then $r = s$, and the β_i can be rearranged so that for all i, β_i is an associate of α_i.*

In view of (11.20), it is natural to try to show that $Q(\sqrt{d})$ is Euclidean for as many d as possible. Unfortunately, $Q(\sqrt{d})$ is rarely Euclidean. The proof of (11.5), with minor modifications, can be used to show that $Q(\sqrt{d})$ is Euclidean for $d = \pm 2$ and ± 3 (see Problem 11-56). In fact, it is not hard to prove that the only negative square-free d for which $Q(\sqrt{d})$ is Euclidean are $d = -1$, -2, -3, -7, and -11. It is also known that the only positive square-free d for which $Q(\sqrt{d})$ is Euclidean are $d = 1$ (the ordinary integers), 2, 3, 5, 6, 7, 11, 13, 17, 19, 21, 29, 33, 37, 41, 57, and 73.

Notes. 1. Unique factorization holds for the integers of $Q(\sqrt{-3})$ (see Theorem 11.20 and Problem 11-56), but not for the set of numbers of the form $x + y\sqrt{-3}$, where x and y are ordinary integers (Problem 11-47).

2. The integers of $Q(\sqrt{d})$ can have unique factorization even when $Q(\sqrt{d})$ is not Euclidean; see the Chapter Notes for details.

Primes of $Q(\sqrt{d})$ and Diophantine Equations

We look next at the problem of which primes p can be represented in the form $p = x^2 + 2y^2$, where x and y are integers. Fermat knew the answer to this question, but it is not clear that he had a proof. Obviously, 2 has such a representation, and thus we may assume that p is odd.

Suppose that $p = x^2 + 2y^2$. Since x must be odd, we have $x^2 \equiv 1 \pmod 8$. Also, since $2y^2 \equiv 0$ or $2 \pmod 8$, it follows that $p \equiv 1$ or $p \equiv 3 \pmod 8$. Therefore primes of the form $8k + 5$ or $8k + 7$ *cannot* be represented in the form $x^2 + 2y^2$. We will show that every prime of the form $8k + 1$ or $8k + 3$ *can* be represented as $x^2 + 2y^2$.

Let p be a (rational) prime of the form $8k + 1$ or $8k + 3$. By (5.13.i), -2 is a quadratic residue of p, and hence $x^2 \equiv -2 \pmod p$ for some integer x, that is, p divides $(x - \sqrt{-2})(x + \sqrt{-2})$. Since $Q(\sqrt{-2})$ is Euclidean, the analogue of (11.8) holds. Thus, if p were prime in $Q(\sqrt{-2})$, we would have $p \mid x - \sqrt{-2}$ or $p \mid x + \sqrt{-2}$. But clearly, p does not divide $x \pm \sqrt{-2}$, so p is not prime.

Thus there exist integers α and β of $Q(\sqrt{-2})$, neither of which is a unit, such that $p = \alpha\beta$. Taking norms, we find that $p^2 = N(\alpha)N(\beta)$, and therefore

$N(\alpha) = p$. If $\alpha = x + y\sqrt{-2}$, then we have $N(\alpha) = x^2 + 2y^2 = p$, and so p is representable in the form $x^2 + 2y^2$. We have therefore proved the following result.

(11.21) Theorem. *The prime p is representable in the form $p = x^2 + 2y^2$ if and only if $p = 2$ or $p \equiv 1$ or $3 \pmod 8$.*

We could also characterize all integers that have a representation of the form $x^2 + 2y^2$, find the number of representations, and so on. The arguments are essentially the same as the corresponding arguments for $x^2 + y^2$ given in (11.12) and (11.13).

(11.22) Theorem. *The positive integer n is representable in the form $x^2 + 2y^2$ if and only if n is of the form $n = 2^a \prod p_j^{a_j} \prod q_j^{b_j}$, where each p_j is a prime of the form $8k + 1$ or $8k + 3$, each q_j is a prime of the form $8k + 5$ or $8k + 7$, and each b_j is even. If n is representable, then n has $2 \prod (a_j + 1)$ representations.*

An essentially identical argument works for the form $x^2 + 3y^2$. Similar ideas can be used to analyze representations by $x^2 - dy^2$, where d is any square-free integer such that the integers of $Q(\sqrt{d})$ have unique factorization. If we do not have unique factorization, the problem is much more difficult.

Other applications of the arithmetic of $Q(\sqrt{d})$ to Diophantine equations can be found in Problems 11-57 and 11-58.

Units of $Q(\sqrt{d})$

In this section, we examine the units of $Q(\sqrt{d})$, where d is square-free. The case $d < 0$ is easy to deal with, but the situation is more complicated when d is positive.

Let d be negative. Since d is square-free, d is congruent to 1, 2, or 3 modulo 4. If $d \not\equiv 1 \pmod 4$, then any integer α of $Q(\sqrt{d})$ is of the form $x + y\sqrt{d}$, where x and y are integers. Note that $N(\alpha) = x^2 + |d|y^2$. If $d \leq -2$, then $x^2 + |d|y^2 \geq x^2 + 2y^2$. Since $x^2 + 2y^2 \geq 2$ unless $y = 0$, it follows that the only elements of norm 1 are ± 1. If $d = -1$, then the units are ± 1 and $\pm i$.

Suppose now that d is congruent to 1 modulo 4. Just as in the preceding paragraph, we can show that ± 1 are the only units of the form $x + y\sqrt{d}$, where x and y are integers. We look next for units of the form $\alpha = (x + y\sqrt{d})/2$, where x and y are *odd*. Taking norms, we find that such an α is a unit if and only if $x^2 + |d|y^2 = 4$. If $d \leq -7$, then $x^2 + |d|y^2 > 4$ unless $y = 0$; in particular, y is not odd. Finally, let $d = -3$; we are then looking for the odd solutions of the Diophantine equation $x^2 + 3y^2 = 4$. These are clearly $x = \pm 1$, $y = \pm 1$. We can summarize the situation as follows.

(11.23) Theorem. *Let $d < 0$ be square-free. The units of $Q(\sqrt{d})$ are ± 1 unless $d = -1$ or $d = -3$. If $d = -1$, the units are ± 1 and $\pm i$, and if $d = -3$, the units are ± 1, $\pm(1+\sqrt{-3})/2$, and $\pm(1-\sqrt{-3})/2$.*

We have found that if $d < 0$, then $Q(\sqrt{d})$ has only a finite number of units, usually two. The situation is entirely different when d is positive. By (11.18.iii), if $x^2 - dy^2 = \pm 1$, then $x + y\sqrt{d}$ is a unit of $Q(\sqrt{d})$. But (10.7) shows that Pell's Equation $x^2 - dy^2 = 1$ has infinitely many integer solutions, and therefore $Q(\sqrt{d})$ has infinitely many units.

The next result provides a complete description of the units in the case $d > 1$.

(11.24) Theorem. *If $d > 1$, then there is a unique unit $\epsilon > 1$ of $Q(\sqrt{d})$ such that any unit is of the form $\pm \epsilon^n$, where n is an integer, positive, negative, or zero. (This unit ϵ is called the fundamental unit of $Q(\sqrt{d})$.)*

Proof. First we show that if λ is a unit other than ± 1, then one of λ, $1/\lambda$, $-\lambda$, or $-1/\lambda$ is of the form $u + v\sqrt{d}$, where u and v are *positive*. There are two cases to consider, according as $N(\lambda) = 1$ or $N(\lambda) = -1$.

Let $\lambda = a + b\sqrt{d}$, and suppose first that $N(\lambda) = 1$. By changing the sign of λ if necessary, we may assume that $a > 0$. If $b > 0$, then λ itself has the appropriate form. Otherwise, note that $1/\lambda = \bar{\lambda}/\lambda\bar{\lambda} = \bar{\lambda} = a - b\sqrt{d}$, and therefore $1/\lambda = a + (-b)\sqrt{d}$, where a and $-b$ are positive.

Suppose now that $N(\lambda) = -1$. By changing the sign of λ if necessary, we may assume that $b > 0$. If $a > 0$, we are finished. If not, observe that $1/\lambda = \bar{\lambda}/\lambda\bar{\lambda} = -\bar{\lambda} = -a + b\sqrt{d}$, and so $1/\lambda$ is of the right form.

It is therefore enough to examine units λ of the form $u + v\sqrt{d}$, with u and v positive. Such a unit is clearly greater than 1, and all units greater than 1 are of this form, since $-\lambda$, $1/\lambda$ and $-1/\lambda$ cannot be. That there *are* units of this form follows from (10.7).

For any real number x, there are only finitely many integers of $Q(\sqrt{d})$ of the form $u + v\sqrt{d}$, with u and v positive and $1 < u + v\sqrt{d} \leq x$. Therefore there is a *smallest* unit $\epsilon > 1$. We show that ϵ is the fundamental unit.

It is enough to show that every unit $\lambda > 1$ is a power of ϵ. Clearly, λ lies between two successive powers of ϵ. Suppose that $\epsilon^n < \lambda \leq \epsilon^{n+1}$; then λ/ϵ^n is a unit greater than 1. If $\lambda \neq \epsilon^{n+1}$, then $\lambda/\epsilon^n < \epsilon$, contradicting the fact that ϵ is the *smallest* unit greater than 1. Thus $\lambda = \epsilon^{n+1}$, and the result follows.

Notes. 1. If $d \equiv 2$ or $3 \pmod 4$, the units of $Q(\sqrt{d})$ arise precisely from the solutions of the Pell Equations $x^2 - dy^2 = \pm 1$, and therefore Theorem 11.24 is an immediate consequence of the results of Chapter 10. In particular, the fundamental unit of $Q(\sqrt{d})$ is $a+b\sqrt{d}$, where (a, b) is the least positive solution of $x^2 - dy^2 = -1$, if solvable, or $x^2 - dy^2 = 1$ otherwise.

2. When $d \equiv 1 \pmod{4}$, the situation is more complicated because integers of $Q(\sqrt{d})$ can also be of the form $(x+y\sqrt{d})/2$, where x and y are odd integers. In this case, since $N((x + y\sqrt{d})/2) = (x^2 - dy^2)/4$, the fundamental unit of $Q(\sqrt{d})$ is $(a+b\sqrt{d})/2$, where (a, b) is the least positive solution of $x^2 - dy^2 = -4$, if solvable, or $x^2 - dy^2 = 4$ otherwise. (The equation $x^2 - dy^2 = 4$ is always solvable: If (u, v) is a solution of $x^2 - dy^2 = 1$, then $(2u, 2v)$ is a solution of $x^2 - dy^2 = 4$.)

Examples. (a) Suppose $d = 3$. The equation $x^2 - 3y^2 = -1$ is not solvable, since the congruence $x^2 \equiv -1 \pmod{3}$ has no solution. The fundamental unit of $Q(\sqrt{3})$ is therefore $2 + \sqrt{3}$, since $(2, 1)$ is the fundamental solution of $x^2 - 3y^2 = 1$.

(b) If $d = 10$, then $d \equiv 2 \pmod{4}$ and the equation $x^2 - 10y^2 = -1$ has $(3, 1)$ as its least positive solution. Hence $3 + \sqrt{10}$ is the fundamental unit of $Q(\sqrt{10})$.

(c) Suppose $d = 5$. Since $d \equiv 1 \pmod{4}$, the fundamental unit of $Q(\sqrt{5})$ is given by $(a+b\sqrt{5})/2$, where (a, b) is the least positive solution of $x^2 - 5y^2 = -4$, if solvable, or $x^2 - 5y^2 = 4$ otherwise. Clearly, $x^2 - 5y^2 = -4$ has $(1, 1)$ as its least positive solution, and so the fundamental unit of $Q(\sqrt{10})$ is $(1 + \sqrt{5})/2$.

PROBLEMS AND SOLUTIONS

Gaussian Integers and Gaussian Primes

11-1. *Does 7 divide* $(8 - i)(4 + 5i)$?

Solution. No. We can either calculate the product $(8 - i)(4 + 5i)$ and see directly that 7 does not divide it or note that $(8 - i)(4 + 5i)$ has norm $65 \cdot 41$, which is not divisible by the norm of 7.

11-2. *For which Gaussian integers α is the conjugate $\bar{\alpha}$ an associate of α?*

Solution. Let $\alpha = s + ti$; then $\bar{\alpha} = s - ti$. We want $s - ti$ to be one of $s + ti$, $-t + si$, $-s - ti$, or $t - si$. For the first case, we need $t = 0$; for the second, $s = -t$; for the third, $s = 0$; and for the fourth, $s = t$. Thus α is of the form a, $a(1 - i)$, ai, or $a(1 + i)$, where a is a rational integer.

11-3. Let $\alpha = 3 + 4i$ and $\beta = 40 + 10i$. Find Gaussian integers γ and ρ such that $\beta = \gamma\alpha + \rho$ and $N(\rho) < N(\alpha)$ by carrying out the procedure described in the proof of (11.5).

Solution. We use the notation of (11.5). Let $\beta/\alpha = x + yi$. By "rationalizing the denominator," we find that $x = 160/25$ and $y = -130/25$. Thus the nearest integers to x and y are $u = 6$ and $v = -5$, and hence we choose $\gamma = 6 - 5i$. Therefore $\rho = \beta - \gamma\alpha = 2 + i$.

11-4. *True or false:* If α, β are Gaussian integers and $N(\alpha)|N(\beta)$, then $\alpha|\beta$.

Solution. This is not true in general. For example, let $\alpha = 2+i$ and $\beta = 2-i$. Each of α and β has norm 5, but $\alpha \nmid \beta$. (This can be verified directly by attempting to divide, or by noting that α and β each have prime norm, and hence are prime, but are not associates.)

11-5. Let $a + bi$ be a Gaussian integer. Show that $1 + i$ divides $a + bi$ if and only if a and b are both even or both odd.

Solution. Divide $a + bi$ by $1 + i$, by "rationalizing the denominator." We obtain

$$\frac{a+bi}{1+i} = \frac{a+bi}{1+i}\cdot\frac{1-i}{1-i} = \frac{a+b}{2} + \frac{b-a}{2}i.$$

But $a+b$ is even if and only if a and b have the same parity; the same is true of $b - a$. Thus $(a+bi)/(1+i)$ is a Gaussian integer if and only if a and b are both even or both odd.

Another proof: It is easy to see that $N(a+bi)$ is even if and only if a and b have the same parity. Since the only Gaussian primes with even norm are $1 + i$ and its associates, the result follows.

11-6. Complete the analysis of the Diophantine equation $x^2 - y^2 = n$ given at the beginning of this chapter by discussing the case where n is even.

Solution. We describe the integer solutions of $x^2 - y^2 = n$, where n is a nonzero *even* integer. Since $x^2 - y^2 = (x-y)(x+y)$ and $x-y$ and $x+y$ always have the same parity, $(x-y)(x+y)$ is even only if $x-y$ and $x+y$ are both even. Thus if $x^2 - y^2 = n$, then n must be divisible by 4. Let $n = 4m$; then each of $x - y$ and $x + y$ is twice a divisor of m. If d is a divisor (positive or negative) of m, set $x+y = 2d$ and $x-y = 2m/d$. Then $x = d+m/d$ and $y = d-m/d$ are clearly integers, and $x^2 - y^2 = n$. Moreover, all pairs (x,y) of integers satisfying the equation arise in this way. There are $2\tau(|m|)$ choices for d, and thus if n is even, then the equation $x^2 - y^2 = n$ has $2\tau(|m|)$ solutions.

11-7. *True or false:* If $a+bi$ is a Gaussian prime, then so is $a-bi$. What about $b + ai$?

Solution. The assertion is true. If $a - bi = (s+ti)(u+vi)$, where neither $s+ti$ nor $u+vi$ is a unit, then by taking conjugates, we obtain $a + bi = (s - ti)(u - vi)$, and neither $s - ti$ nor $u - vi$ is a unit.

If $a + bi$ is a Gaussian prime, then so is $a - bi$. Since $b + ai = -i(a - bi)$, $b + ai$ is an associate of a Gaussian prime and hence is a Gaussian prime.

374   CHAPTER 11: QUADRATIC EXTENSIONS

11-8. *Determine which of the following are Gaussian primes:* $3 + 4i$, $3 - 4i$, $5i$, $-11i$.

Solution. By (11.11), the Gaussian primes are $1 + i$ and its associates, the rational primes of the form $4k + 3$ and their associates, and numbers of the form $a + bi$, where $a^2 + b^2$ is a prime of the form $4k + 1$. Clearly, $3 + 4i$ is not any of these, for $3 + 4i$ has norm $25 = 5^2$, and 5 is a prime of the form $4k + 1$. It is easy to verify that in fact $3 + 4i = (2 + i)^2$. For the same reason, $3 - 4i$ is not a Gaussian prime. It is easy to see that $5i$ is not a Gaussian prime, since 5 is divisible by the prime $2 + i$. Finally, $-11i$ is a Gaussian prime, for it is an associate of 11, a rational prime of the form $4k + 3$.

11-9. *Prove that there are infinitely many Gaussian primes of the form $a + bi$ with $a \neq 0$ and $b \neq 0$.*

Solution. Let p be a rational prime of the form $4k + 1$. Then p has a a representation as a sum $a^2 + b^2$ of nonzero squares. Let $\pi = a + bi$. Distinct p obviously give rise to distinct π, and by Problem 5-23, there are infinitely many primes of the form $4k + 1$, so there are infinitely many Gaussian primes of the specified form.

11-10. *Suppose that the norm of γ is q^2, where q is a prime of the form $4k+3$. Is γ a Gaussian prime?*

Solution. Suppose that $\gamma = \alpha\beta$. If $N(\gamma) = q^2$ with q a rational prime, then $N(\alpha) = 1$, q, or q^2. But we cannot have $N(\alpha) = q$ if q is of the form $4k + 3$, since no number of that form is the sum of two squares. Thus $N(\alpha) = 1$ or $N(\alpha) = q^2$, and so either α or β is a unit. Therefore γ is a Gaussian prime.

11-11. *Express $3 + i$ and $6 + 7i$ as products of Gaussian primes.*

Solution. Since $3 + i$ has norm 10, the only possible prime divisors of $3 + i$ are prime divisors of 2 and of 5, namely, associates of $1 + i$ and $2 \pm i$. Divide $3 + i$ by $1 + i$ (it is easiest to do this by rationalizing the denominator, i.e., by multiplying numerator and denominator by $1 - i$); thus $3 + i = (1 + i)(2 - i)$.

The norm of $6 + 7i$ is $85 = 5 \cdot 17$. Since 5 and 17 are rational primes of the form $4k + 1$, the only possible Gaussian prime divisors of $6 + 7i$ are $2 \pm i$, $4 \pm i$, and their associates. Dividing $6 + 7i$ by $2 + i$, we find that the quotient is not a Gaussian integer. Now divide by $2 - i$, obtaining quotient $1 + 4i$, which has norm 17 and thus is prime. Therefore $6 + 7i = (2 - i)(1 + 4i)$ is a prime factorization of $6 + 7i$.

11-12. *Write $60 + 105i$ as a product of Gaussian primes.*

Solution. Since $60 + 105i = 15(4 + 7i)$, we need to factor 15 and $4 + 7i$. We have $15 = 3(2 + i)(2 - i)$, and each of the factors is prime. Since $N(4 + 7i) = 65 = 5 \cdot 13$, $4 + 7i$ is divisible by either $2 + i$ or $2 - i$. We find that $4 + 7i = (2 + i)(3 + 2i)$, and since each factor has prime norm, it is prime. Thus $60 + 105i = 3(2 + i)^2(2 - i)(3 + 2i)$ is one of the prime factorizations of $60 + 105i$.

11-13. *Find a prime factorization of* $239 - i$.

Solution. Since $N(239 - i)$ is even, it follows that $239 - i$ is divisible by $1 - i$. When we divide, we obtain the quotient $120 + 119i$. Since $N(120 + 119i) = 28561 = 13^4$, we look for prime factors with norm 13. The possibilities are $3 \pm 2i$ (and their associates). Note that $120 + 119i$ cannot be divisible by both $3 + 2i$ and $3 - 2i$, for otherwise, it would be divisible by 13, which is not the case. Dividing, we find that $120 + 119i = (3+2i)(46+9i) = (3+2i)^2(12-5i) = (3+2i)^3(2-3i)$. Thus $120 + 119i = -i(3+2i)^4$, and hence $239 - i = (-1 - i)(3 + 2i)^4$ is a prime factorization of $239 - i$.

11-14. *How many Gaussian integers have norm* 1800? *How many of these are rational integers?*

Solution. The norm of a rational integer n is n^2, so no rational integer has norm 1800. To determine the number of Gaussian integers with norm 1800, note that we are asking for the number of solutions of the Diophantine equation $u^2 + v^2 = 1800$. But $1800 = 2^3 \cdot 3^2 \cdot 5^2$, and so by (11.12), there are $4 \cdot (2 + 1) = 12$ Gaussian integers with norm 1800.

11-15. *Find all Gaussian integers of norm* 169.

Solution. We need to solve the equation $(u + vi)(u - vi) = 169$. Since $169 = 13^2$ and $13 = 2^2 + 3^2$, 169 has the prime factorization $169 = (2 + 3i)^2(2 - 3i)^2$. Thus $u + vi = \epsilon(2 + 3i)^k(2 - 3i)^{2-k}$, where ϵ is a unit and $0 \le k \le 2$ (see the proof of (11.12)). Taking $\epsilon = 1$, we conclude that $u + iv$ is one of $-5 - 12i$, 13, or $-5 + 12i$; taking other units yields associates of these. So the Gaussian integers of norm 169 are $-5 - 12i$, $12 - 5i$, $5 + 12i$, $-12 + 5i$, 13, $13i$, -13, $-13i$, $-5 + 12i$, $-12 - 5i$, $5 - 12i$, and $12 + 5i$.

11-16. *Make a list of all Gaussian primes with norm not exceeding* 60. *To make the list shorter, list only one of* π, $-\pi$, $i\pi$, *and* $-i\pi$.

Solution. Clearly, $1 + i$ is such a prime. We list next all rational primes p of the form $4k + 3$ such that $p^2 \le 60$; these are 3 and 7. Next we look at all primes p of the form $4k + 1$ with $p \le 60$ and express each as a sum of two squares. Each time, we obtain two primes, in conjugate pairs. These are $2 \pm i$, $3 \pm 2i$, $4 \pm i$, $5 \pm 2i$, $6 \pm i$, $5 \pm 4i$, and $7 \pm 2i$.

11-17. *Solve* $x + y + z = xyz = 1$ *in Gaussian integers.*

Solution. Since $xyz = 1$, all the variables must be units and thus must be chosen from ± 1 and $\pm i$. Also, either all of them are real or exactly one is real. If all are real, then from $x + y + z = 1$, we find that two are equal to 1 and the other is equal to -1, contradicting $xyz = 1$. Thus exactly two of x, y, and z are imaginary, and since $x + y + z = 1$, these two must be i and $-i$; $xyz = 1$ then implies that the third number is 1. Thus the solutions are the six permutations of $(1, i, -i)$.

11-18. *Find a greatest common divisor of* $4 + 6i$ *and* $4 - 6i$.

Solution. It is enough to find a greatest common divisor of $2 + 3i$ and $2 - 3i$. Each of these has norm 13, which is prime, so $2 + 3i$ and $2 - 3i$ are prime. They are clearly not associates, and hence 2 is a greatest common divisor of $4 + 6i$ and $4 - 6i$.

11-19. (a) *Show that* $10 + 3i$ *is a greatest common divisor of* $-25 + 47i$ *and* $17 + 16i$. (b) *Express* $-25 + 47i$ *and* $17 + 16i$ *as a product of Gaussian primes.*

Solution. (a) (b) We divide each of $-25 + 47i$ and $17 + 16i$ by $10 + 3i$ (a convenient way to do this is to rationalize the denominator by multiplying "top" and "bottom" by $10 - 3i$). The quotients are $-1 + 5i$ and $2 + i$, respectively. We factor $-1 + 5i$, noting that since -1 and 5 are odd, $-1 + 5i$ is divisible by $1 + i$. Dividing, we obtain $-1 + 5i = (1 + i)(2 + 3i)$, and so $-25 + 47i = (1 + i)(2 + 3i)(10 + 3i)$. The norm of each factor is prime, so we have obtained a prime factorization of $-25 + 47i$. Also, $17 + 16i = (2 + i)(10 + 3i)$, and since each factor has prime norm, we have found a prime factorization of $17 + 16i$. Since $2 + i$ is not an associate of $1 + i$ or $2 + 3i$, it follows that $10 + 3i$ is a greatest common divisor of $-25 + 47i$ and $17 + 16i$.

▷ **11-20.** *Refer to Definition 11.6. Show that γ is a greatest common divisor of α and β if and only if γ is a common divisor, and for every common divisor λ, we have* $N(\lambda) \leq N(\gamma)$.

Solution. If γ is a greatest common divisor of α and β and λ is a common divisor, then $\lambda \mid \gamma$. Thus $N(\lambda) \mid N(\gamma)$ and therefore $N(\lambda) \leq N(\gamma)$. Conversely, suppose that γ is a common divisor of α and β of maximum norm, and let δ be a greatest common divisor of α and β as defined in (11.6). Then $\gamma \mid \delta$; let $\delta = \mu\gamma$. Since γ has maximum norm among common divisors, it follows that $N(\mu) = 1$, and hence μ is a unit. Thus γ is a greatest common divisor of α and β (see the Note after (11.7)).

Note. The definition of greatest common divisor that we use in this chapter is one of the *properties* of the ordinary common divisor of two rational integers (see (1.6)). The new definition is algebraically more natural and makes sense even in settings where there is no obvious notion of size.

11-21. *Prove or disprove: If α and β are Gaussian integers and δ is a greatest common divisor of α and β, then $\bar{\delta}$ is a greatest common divisor of $\bar{\alpha}$ and $\bar{\beta}$.*

Solution. This is true. Since $\delta \mid \alpha$, there is a γ such that $\alpha = \gamma\delta$, and hence $\bar{\alpha} = \bar{\gamma}\bar{\delta}$. It follows that $\bar{\delta} \mid \bar{\alpha}$; similarly, $\bar{\delta} \mid \bar{\beta}$, so $\bar{\delta}$ is a *common divisor* of $\bar{\alpha}$ and $\bar{\beta}$. Now take any common divisor $\bar{\mu}$ of $\bar{\alpha}$ and $\bar{\beta}$. It is easy to see that μ is a common divisor of α and β, and hence $\mu \mid \delta$, since δ is a greatest common divisor of α and β. But then $\bar{\mu} \mid \bar{\delta}$, and thus $\bar{\delta}$ is a *greatest* common divisor of $\bar{\alpha}$ and $\bar{\beta}$.

Another solution: By the proof of (11.7), we know that δ is an element of smallest positive norm that can be represented as a linear combination of α and β. Taking complex conjugates, it follows that $\bar{\delta}$ is an element of minimal positive norm that can be represented as a linear combination of $\bar{\alpha}$ and $\bar{\beta}$.

11-22. *Show that if the rational integers a and b are relatively prime in the ordinary sense, then they are relatively prime as Gaussian integers.*

Solution. Since a and b are relatively prime, there exist (rational) integers u and v such that $au + bv = 1$. Therefore if the Gaussian integer λ divides a and b, then λ divides 1, and hence λ is a unit. Thus the only common Gaussian divisors of a and b are units, and so a and b are relatively prime as Gaussian integers.

11-23. *How many Gaussian integers divide n, where $n = 2^2 \cdot 3^3 \cdot 5^5$?*

Solution. We find a prime factorization of n of the form $\epsilon \prod \pi_i^{a_i}$, where ϵ is a unit and π_i is not an associate of π_j when $i \neq j$ (this is important for a correct count). In this case, we have $n = -(1+i)^4 3^3 (1+2i)^5 (1-2i)^5$. By unique factorization, any divisor of n has a unique representation as $\epsilon(1+i)^a 3^b (1+2i)^c (1-2i)^d$, where ϵ is a unit and a, b, c and d are nonnegative integers with $a \leq 4$, $b \leq 3$, $c \leq 5$, and $d \leq 5$. There are 4 choices for ϵ, 5 for a, and so on. Thus the total number of divisors is $4 \cdot 5 \cdot 4 \cdot 6^2 = 2880$.

If α, β, and μ are Gaussian integers, with $\mu \neq 0$, we say that $\alpha \equiv \beta$ (mod μ) if $\mu | \alpha - \beta$. A *complete residue system* modulo μ is a set S of Gaussian integers, incongruent to each other modulo μ, such that any Gaussian integer is congruent modulo μ to an element of S. The next six problems deal with congruences modulo a Gaussian integer.

11-24. *Solve the following linear congruences: (a) $2x \equiv 1 + 3i$ (mod 11); (b) $2x \equiv 1 + 3i$ (mod $3 + 8i$).*

Solution. (a) We multiply as usual by an (ordinary) inverse of 2 modulo 11. Multiplication by 6 gives $x \equiv 6 + 18i$ (mod 11), and so, for example, $6 + 7i$ is a solution.

(b) We would like to multiply the congruence through by a Gaussian integer α such that $2\alpha \equiv 1$ (mod $3 + 8i$). Since $(3 + 8i)(3 - 8i) = 73$, it is enough to find α such that $2\alpha \equiv 1$ (mod 73), and clearly $\alpha = 37$ works. Thus we obtain $x \equiv 37 + 111i$ (mod $3 + 8i$). But $37 + 111i = 14(3 + 8i) - (5 + i)$, so a simpler answer is $x \equiv -(5 + i)$ (mod $3 + 8i$).

11-25. *Let α and $\mu \neq 0$ be relatively prime Gaussian integers. Show that there is a Gaussian integer ξ such that $\alpha\xi \equiv 1$ (mod μ).*

Solution. Since 1 is a greatest common divisor of α and μ, (11.7) implies that there exist Gaussian integers κ and λ such that $\kappa\alpha + \lambda\mu = 1$. If we let $\xi = \kappa$, then $\alpha\xi \equiv 1$ (mod μ).

11-26. *Let μ and ν be relatively prime Gaussian integers. Show that if α and β are Gaussian integers, then the system of congruences*

$$x \equiv \alpha \pmod{\mu}, \qquad x \equiv \beta \pmod{\nu}$$

is solvable, and the solution is unique modulo $\mu\nu$. (Hint. See Problem 11-25 and the proof of the Chinese Remainder Theorem.)

Solution. By the preceding problem, there are Gaussian integers γ and δ such that $\gamma\nu \equiv 1$ (mod μ) and $\delta\mu \equiv 1$ (mod ν). It is easy to see that $x = \gamma\nu\alpha + \delta\mu\beta$ satisfies

378   CHAPTER 11: QUADRATIC EXTENSIONS

both congruences. If y also satisfies both congruences, then $x - y$ is divisible by μ and ν and hence by $\mu\nu$ (consider the prime factorization of $x - y$). Thus there is exactly one solution modulo $\mu\nu$.

11-27. *Let m be a positive integer. Show that any complete residue system modulo m has m^2 elements.*

Solution. It is clear that any two complete residue systems modulo m have the same number of elements. Thus it is enough to find one complete residue system with m^2 elements. Let S be the set of Gaussian integers of the form $u + vi$, where $0 \le u, v \le m-1$; clearly, S has m^2 elements. If $\alpha = s + ti$ is any Gaussian integer, let s_0, t_0 be, respectively, the least nonnegative residues of s and t modulo m. Then $s_0 + t_0 i$ is in S and is congruent to α modulo m. If $\alpha = u + vi$ and $\alpha' = u' + v'i$ are elements of S and m divides $\alpha - \alpha'$, then m divides *each* of $u - u'$ and $v - v'$. But since $0 \le u, u' \le m-1$, it follows that $u = u'$. Similarly, $v = v'$, and thus distinct elements of S are incongruent modulo m.

▷ **11-28.** *Let π be a Gaussian prime such that $N(\pi)$ is a prime p of the form $4k + 1$. Show that any complete residue system modulo π has p elements. (Hint. Use Problems 11-26 and 11-27.)*

Solution. It is easy to verify that π and $\bar\pi$ are relatively prime, for they are not associates and π has no proper divisors. Let P, A, and B be complete residue systems modulo p, π, and $\bar\pi$, respectively. If A has k elements, then B also has k elements. By Problem 11-26, any pair α, β, where $\alpha \in A$ and $\beta \in B$, determines a unique element γ of P such that $\gamma \equiv \alpha \pmod{\pi}$ and $\gamma \equiv \beta \pmod{\bar\pi}$. Moreover, every element of P arises in this way. Thus P has k^2 elements. By Problem 11-27, P has p^2 elements, and hence $k = p$.

Note. In essentially the same way, we could show that any complete residue system modulo π^n has p^n elements. By using the Chinese Remainder Theorem, we can then prove that if μ is a nonzero Gaussian integer, then any complete residue system modulo μ has $N(\mu)$ elements.

▷ **11-29.** *(A Gaussian analogue of Fermat's Theorem.) Let π be a Gaussian prime, and suppose that α is not divisible by π. Show that $\alpha^{N(\pi)-1} \equiv 1 \pmod{\pi}$. (Hint. See Problems 11-27 and 11-28, or use the Binomial Theorem.)*

Solution. The result is easy to verify if π is $1 + i$ or one of its associates. Thus we may assume that π is either a rational prime of the form $4k + 3$ or a prime factor of a rational prime of the form $4k + 1$. Let $\beta_0, \beta_1, \ldots, \beta_n$ be a complete residue system modulo π, with $\beta_0 = 0$. By Problems 11-27 and 11-28, $n = N(\pi) - 1$. Now consider $\alpha\beta_1, \alpha\beta_2, \ldots, \alpha\beta_n$. It is easy to verify that no two of these are congruent modulo π. Hence they are congruent modulo π, in some order, to $\beta_1, \beta_2, \ldots, \beta_n$. Multiplying, we obtain

$$(\alpha\beta_1)(\alpha\beta_2)\cdots(\alpha\beta_n) \equiv \beta_1\beta_2\cdots\beta_n \pmod{\pi}.$$

If we divide both sides of the above congruence by $\beta_1\beta_2\cdots\beta_n$, we obtain $\alpha^n \equiv 1 \pmod{\pi}$, which is the desired result.

Another proof: We prove the equivalent assertion that $\alpha^{N(\pi)} \equiv \alpha \pmod{\pi}$ for any Gaussian integer α. The result is clearly true for $\alpha = 0$. We next show that if the result holds for α, then it holds for β, where $\beta = \alpha \pm 1$ or $\beta = \alpha \pm i$. The arguments in the four cases are very similar, so let $\beta = \alpha + i$. Expanding by the Binomial Theorem and using the fact that the binomial coefficient $\binom{p}{k}$ is divisible by p if p is prime and $1 \leq k \leq p-1$, we find that $(\alpha + i)^{N(\pi)} \equiv \alpha^{N(\pi)} + i^{N(\pi)} \equiv \alpha + i^{N(\pi)} \pmod{\pi}$. It remains to show that $i^{N(\pi)} \equiv i \pmod{\pi}$. When $N(\pi) = 2$, we need to verify that $\pi \mid 1 + i$, which is clear. In all other cases, $N(\pi)$ is of the form $4k + 1$, and $i^{4k+1} = i$.

Note. It is not difficult to prove also an analogue of Wilson's Theorem.

11-30. *Use the factorization $x^2 + y^2 = (x+yi)(x-yi)$ to show that any primitive Pythagorean triple (x, y, z) with y even is of the form $x = a^2 - b^2$, $y = 2ab$, $z = a^2 + b^2$ for relatively prime integers a and b.*

Solution. Consider the factorization $z^2 = (x + yi)(x - yi)$. We show that $x + yi$ and $x - yi$ are relatively prime. Any Gaussian prime π that divides these two numbers divides their sum and difference and thus divides $2x$ and $2y$. Since no rational prime divides both x and y, we must have $\pi \mid 2$, and hence π is an associate of $1 + i$. Thus $(1+i)^2 \mid x^2 + y^2$, i.e., $2i \mid z^2$. This is impossible, since z is odd.

Thus by (11.14), $x + yi$ is of the form $\epsilon \delta^2$, where ϵ is a unit. Let $\delta = a + bi$; then $x + yi = \epsilon((a^2 - b^2) + 2abi)$. Since x is odd and y is even, we must have $a^2 - b^2 = \pm x$ and $2ab = \pm y$. Note that $(a, b) = 1$, since if $d \mid a$ and $d \mid b$, then d divides $x + yi$ and $x - yi$. The result now follows, by interchanging the roles of a and b (if necessary) and choosing appropriate signs for a and b.

11-31. *Give a "formula" for all solutions of the equation $x^2 + y^2 = z^3$, where x and y are relatively prime and of opposite parity.*

Solution. We want $(x + yi)(x - yi) = z^3$. By the same argument as in the preceding problem, we can show that $x + yi$ and $x - yi$ are relatively prime. Hence by (11.14), there is a unit ϵ and a Gaussian integer δ such that $x + yi = \epsilon \delta^3$. Since any unit is a perfect cube, we can in fact assume that $x + iy = \delta^3$. Let $\delta = a + bi$. Then, by expanding, we find that $x + yi = a^3 - 3ab^2 + (3a^2b - b^3)i$, and therefore $x = a^3 - 3ab^2$, $y = 3a^2b - b^3$. Since $z^3 = (a + bi)^3(a - bi)^3$, we find that $z = a^2 + b^2$.

The numbers a and b must be chosen to be relatively prime to make x and y relatively prime. Also, to ensure opposite parity for x and y, a and b must be chosen of opposite parity. We verify that with this choice, x and y are relatively prime. For if p is a prime that divides x and y, then $p \mid ay - 3bx$, i.e., $p \mid 8ab^3$; since p is odd, $p \mid a$ or $p \mid b$. But if $p \mid a$, then since $p \mid 3a^2b - b^3$, we have $p \mid b$, contradicting the choice of a and b. Similarly, $p \mid b$ leads to a contradiction. Thus all solutions of the equation $x^2 + y^2 = z^3$, where x and y are relatively prime and of opposite parity, are given by $x = a^3 - 3ab^2$, $y = 3a^2b - b^3$, and $z = a^2 + b^2$, where a and b are any two relatively prime integers of opposite parity.

The Arithmetic of $Q(\sqrt{d})$

11-32. *Show that if $\alpha | \beta$ and $|N(\alpha)| = |N(\beta)|$, then β is an associate of α.*

Solution. The result is obvious if $\beta = 0$. If $\beta \neq 0$, let $\beta = \gamma\alpha$. Since $N(\beta) = N(\gamma)N(\alpha)$ and $|N(\alpha)| = |N(\beta)|$, it follows that $|N(\gamma)| = 1$, and hence γ is a unit.

11-33. *Show that if β is an associate of α, then α is an associate of β.*

Solution. By definition, β is an associate of α if and only if there is a unit ϵ such that $\beta = \epsilon\alpha$. But if ϵ is a unit and $\epsilon' = 1/\epsilon$, then $\epsilon\epsilon' = 1$ and hence ϵ' is a unit. Clearly, $\alpha = \epsilon'\beta$, and thus α is an associate of β.

11-34. *Let m be a square-free rational integer, and suppose that α and β are integers in $Q(\sqrt{m})$ whose norms are equal. Is it always the case that β is an associate of α or an associate of the conjugate of α?*

Solution. No, and we can find examples in the Gaussian integers. Since $25 = 5^2 = 3^2 + 4^2$, 25 is the norm of 5 and also of $3 + 4i$. Clearly, $3 + 4i$ is neither an associate of 5 nor the conjugate of an associate of 5.

11-35. *Determine which of the following are primes in $Q(\sqrt{-5})$: (a) $3+4\sqrt{-5}$; (b) $7 - \sqrt{-5}$; (c) 1997. (Note. 1997 is a rational prime.)*

Solution. (a) Since the norm of $3 + 4\sqrt{-5}$ is 83, which is a rational prime, it follows that $3 + 4\sqrt{-5}$ is prime.

(b) The norm of $7 - \sqrt{-5}$ is 54, and hence the only possible nontrivial divisors of $7 - \sqrt{-5}$ have norm 2, 3, 6, 9, 18, or 27. If we look for a nontrivial divisor of smallest norm, we need examine only 2, 3, and 6 as possibilities for the norm. The norm of $a + b\sqrt{-5}$ is $a^2 + 5b^2$, which is obviously never 2 or 3. The integers with norm 6 are $1 + \sqrt{-5}$, $1 - \sqrt{-5}$, and their negatives. When we divide $7 - \sqrt{-5}$ by $1 - \sqrt{-5}$ (by rationalizing the denominator), we find that the quotient is an integer of $Q(\sqrt{-5})$, and thus $7 - \sqrt{-5}$ is not prime.

(c) Since $N(1997) = 1997^2$, the only possible nontrivial divisors of 1997 are numbers of norm 1997. We could verify by enumeration that no integer of $Q(\sqrt{-5})$ has norm equal to 1997, but there is an easier way. If $a^2 + 5b^2 = 1997$, then 1997 is a quadratic residue of 5. Since $(1997/5) = (2/5) = -1$, there do not exist integers a, b such that $a^2 + 5b^2 = 1997$, and hence 1997 is prime.

11-36. *Is 41 a prime in $Q(\sqrt{13})$?*

Solution. If $41 = \alpha\beta$, then $N(41) = 41^2 = N(\alpha)N(\beta)$. Therefore any nontrivial divisors of 41 must have norm ± 41, and thus we look for either integer solutions of $x^2 - 13y^2 = \pm 41$ or odd integers x and y such that $(x/2)^2 - 13(y/2)^2 = \pm 41$. Consider first the equation $x^2 - 13y^2 = \pm 41$. If this has a solution (x, y), then ± 41 is a quadratic residue of 13, for $x^2 \equiv \pm 41 \pmod{13}$. Evaluate the Legendre symbols $(\pm 41/13)$ in the usual way. In both cases, we obtain -1, and hence the equation $x^2 - 13y^2 = \pm 41$ does not have integer solutions. Essentially the same argument works for the equation $(x/2)^2 - 13(y/2)^2 = \pm 41$. Thus 41 is a prime in $Q(\sqrt{13})$.

11-37. *Express $\sqrt{14}$ as a product of two primes in $Q(\sqrt{14})$.*

Solution. Since $\sqrt{14}$ has norm -14, in looking for proper divisors of $\sqrt{14}$ we may confine our attention to numbers of norm ± 2 or ± 7. Note that $4 + \sqrt{14}$ has norm 2 and that $\sqrt{14}/(4 + \sqrt{14}) = -7 + 2\sqrt{14}$. Thus we have the factorization $\sqrt{14} = (4 + \sqrt{14})(-7 + 2\sqrt{14})$. Each of the factors is prime, since the absolute value of each norm is prime.

Note. The factorization that we have found gives rise to infinitely many inessential variants, since there are infinitely many units, and we can obtain a different-looking factorization by multiplying one of the factors by a unit ϵ and dividing the other factor by ϵ.

11-38. *Express $33 + 11\sqrt{-7}$ as a product of primes in $Q(\sqrt{-7})$. (Note that $-7 \equiv 1 \pmod 4$.)*

Solution. Write $33 + 11\sqrt{-7} = 11(3 + \sqrt{-7})$. Since $11 = 2^2 + 1^2 \cdot 7$, we have $11 = (2 + \sqrt{-7})(2 - \sqrt{-7})$. The integers $2 \pm \sqrt{-7}$ are prime, since each has prime norm. Now factor $3 + \sqrt{-7}$. Note that $3 + \sqrt{-7}$ is divisible by 2, since $(3 + \sqrt{-7})/2$ is an integer of $Q(\sqrt{-7})$. Also, 2 is the product of $(1 + \sqrt{-7})/2$ and $(1 - \sqrt{-7})/2$, both of which are prime since each has prime norm. It remains to deal with $(3 + \sqrt{-7})/2$. Since $N((3 + \sqrt{-7})/2) = 4$, any proper divisor of $(3 + \sqrt{-7})/2$ has norm 2. With the search thus limited, it is easy to verify that $(3 + \sqrt{-7})/2$ is the product of $(1 - \sqrt{-7})/2$ and $(-1 + \sqrt{-7})/2$.

11-39. *Determine if 97 is a prime in $Q(\sqrt{119})$.*

Solution. The norm of 97 is 97^2. Thus if $\alpha\beta = 97$, then $N(\alpha)N(\beta) = 97^2$. We show that the norm of one of the factors must be ± 1 (i.e., one of the factors is a unit) by showing that we cannot have $N(\gamma) = \pm 97$ for any integer γ of $Q(\sqrt{119})$. Suppose that $\gamma = u + v\sqrt{119}$, where u and v are rational integers. Then $N(\gamma) = \pm 97$ if and only if $u^2 - 119v^2 = \pm 97$. Note that $119 = 7 \cdot 17$. Thus if $u^2 - 119v^2 = 97$, then $u^2 \equiv 97 \pmod 7$, i.e., $u^2 \equiv -1 \pmod 7$, which is impossible since -1 is not a quadratic residue of 7. If $u^2 - 119v^2 = -97$, then $u^2 \equiv -97 \pmod{17}$, i.e., $u^2 \equiv 5 \pmod{17}$. This is impossible, since $(5/17) = (17/5) = (2/5) = -1$.

11-40. *Show that if $5|d$, then 2 is prime in $Q(\sqrt{d})$.*

Solution. First we look at the case $d \not\equiv 1 \pmod 4$. The integers of $Q(\sqrt{d})$ are of the form $a + b\sqrt{d}$, where a and b are integers. Since $N(2) = 4$, any proper divisor of 2 has norm ± 2. If $a^2 - db^2 = \pm 2$ and $5|d$, then $a^2 \equiv \pm 2 \pmod 5$, which is impossible.

If $d \equiv 1 \pmod 4$, there are also integers of the form $(a + b\sqrt{d})/2$ with a and b odd. If such an integer has norm ± 2, then $a^2 - db^2 = \pm 8$. If $5|d$, then we obtain $a^2 \equiv \pm 3 \pmod 5$, which is impossible.

11-41. *Let d, u, and v be integers. Show that $u + v\sqrt{d}$ is a root of an equation of the form $x^2 + bx + c = 0$, where b and c are integers. (b) Show that if $d \equiv 1$*

(mod 4) *and u and v are odd, then* $(u + v\sqrt{d})/2$ *is a root of an equation of the form* $x^2 + bx + c = 0$, *where b and c are integers.*

Solution. (a) The polynomial $P(x) = (x-(u+v\sqrt{d}))(x-(u-\sqrt{d})) = x^2 - 2ux + u^2 - dv^2$ is of the required form, and clearly, $P(u+v\sqrt{d}) = 0$. (b) Let $P(x) = (x-(u+v\sqrt{d})/2)(x-(u-\sqrt{d})/2) = x^2 - ux + (u^2 - dv^2)/4$. Since u and v are odd and $d \equiv 1 \pmod 4$, $u^2 - dv^2$ is a multiple of 4, and hence $P(x)$ has integer coefficients.

▷ **11-42.** *Let* $d \neq 1$ *be a square-free integer, and let r and* $s \neq 0$ *be rational numbers. Suppose that* $\alpha = r + s\sqrt{d}$ *is a root of the polynomial equation* $x^2 + bx + c = 0$, *where b and c are integers. Show that if* $d \equiv 2$ *or* $3 \pmod 4$, *then r and s are integers. If* $d \equiv 1 \pmod 4$, *prove that either r and s are both integers, or 2r and 2s are both odd integers.*

Solution. If $r + s\sqrt{d}$ is a root of the equation $x^2 + bx + c = 0$, then so is $r - s\sqrt{d}$, and since $s \neq 0$, these two roots are distinct. The sum $2r$ of the two roots is $-b$, while their product $r^2 - ds^2$ is c. Substituting, we find that $4ds^2 = b^2 - 4c$. Let $2s = m/n$, where m and n are relatively prime integers; then $dm^2 = n^2(b^2 - 4c)$. Since m and n are relatively prime, it follows that $n^2 | d$. Since d is square-free, we have $n^2 = 1$, and therefore $2s$ is an integer.

Let $2s = v$; then $b^2 - dv^2 = 4c$, and thus $b^2 - dv^2 \equiv 0 \pmod 4$. It is easy to check that if $d \equiv 2$ or $3 \pmod 4$, then b and v must be even, and hence r and s are integers. If $d \equiv 1 \pmod 4$ and b and v are not both even, then both must be odd, so $2r$ and $2s$ are both odd.

11-43. *Show that if d is negative, all the units of* $Q(\sqrt{d})$ *are powers of one of the units.*

Solution. The units are described in (11.23). Note that unless $d = -1$ or $d = -3$, they are the powers of -1. If $d = -1$, the units are the powers of i (also of $-i$). If $d = -3$, it is easy to verify that the units are the powers of $(1 + \sqrt{-3})/2$ (also of $(1 - \sqrt{-3})/2$).

11-44. *Are there any units in* $Q(\sqrt{2})$ *that lie between 1 and 10? If so, find them. If not, explain why there are none.*

Solution. It is easy to verify that $1 + \sqrt{2}$ is a unit. It is the fundamental unit, for by the proof of (11.24), any unit greater than 1 is of the form $a + b\sqrt{2}$, where a and b are positive. But if δ is of that form, with $1 < \delta < 1 + \sqrt{2}$, then $\delta = 2$, so δ is not a unit. We have $1 < (1 + \sqrt{2})^n < 10$ only for $n = 1$ and $n = 2$. Thus the units between 1 and 10 are $1 + \sqrt{2}$ and $3 + 2\sqrt{2}$.

11-45. *Show that if* π_1 *and* π_2 *are primes in* $Q(\sqrt{d})$ *and* $\pi_1 | \pi_2$, *then* π_1 *and* π_2 *are associates.*

Solution. Let $\pi_2 = \alpha \pi_1$. Since π_2 is prime, one of α and π_1 is a unit. Since π_1 is prime, it cannot be a unit. Thus α is a unit, and therefore π_1 and π_2 are associates.

11-46. Show that if m and n are distinct square-free rational integers, then $Q(\sqrt{m}) \neq Q(\sqrt{n})$.

Solution. Without loss of generality, we may assume that $n \neq 1$. We prove that the two fields are distinct by showing that $\sqrt{n}$ is not an element of $Q(\sqrt{m})$. Suppose, to the contrary, that $\sqrt{n} = s + t\sqrt{m}$, where s and t are rational; then $t \neq 0$, since $\sqrt{n}$ is not a rational number.

If $\sqrt{n} = s + t\sqrt{m}$, then $\sqrt{n} - t\sqrt{m} = s$. Squaring and simplifying, we obtain $\sqrt{mn} = (n + t^2 m - s^2)/2t$; thus $\sqrt{mn}$ is a rational number. We show that this cannot be the case by showing that mn is not a perfect square. If there is a prime p which divides one of m or n but not the other, then mn is not a perfect square, since m and n are square-free. If all primes that divide either m or n divide the other, then $m = -n$, since m and n are distinct, and again mn is not a perfect square.

11-47. (a) Let $\alpha = (7+\sqrt{-3})/2$ and $\beta = 1+2\sqrt{-3}$. Check that $13 = \alpha\bar{\alpha} = \beta\bar{\beta}$. Why doesn't this contradict the fact that $Q(\sqrt{-3})$ has the unique factorization property?

(b) Let W be the set of numbers of the form $x + y\sqrt{-3}$, where x and y are integers. Show that unique factorization fails in W. (Unique factorization does hold, however, for the set of all integers in $Q(\sqrt{-3})$. See (11.20) and Problem 11-56.)

Solution. (a) Since the norms of the given factors are prime, the factors are prime. But a short calculation shows that $\alpha/\beta = (1 - \sqrt{-3})/2$. Since $(1 - \sqrt{-3})/2$ is a unit, this means that α and β are associates. Taking conjugates, we see that $\bar{\alpha}$ and $\bar{\beta}$ are also associates. Thus the two given factorizations are essentially the same.

(b) Note that $2 \cdot 2 = (1+\sqrt{-3})(1-\sqrt{-3})$. Since 2 and $1 \pm \sqrt{-3}$ all have norm 4, any of their nontrivial divisors must have norm 2. But the equation $x^2 + 3y^2 = 2$ has no integer solutions, and so W has no elements of norm 2. It follows that 2 and $1 \pm \sqrt{-3}$ are prime in W. But in W, $1 + \sqrt{-3}$ is not an associate of 2 since $(1 + \sqrt{-3})/2$ is not an element of W, and therefore unique factorization fails in W.

11-48. Show that $Q(\sqrt{-5})$ does not have the unique factorization property.

Solution. Note that $6 = 2 \cdot 3 = (1+\sqrt{-5})(1-\sqrt{-5})$. We show that 2, 3, and $1 \pm \sqrt{-5}$ are prime, and since clearly no two of these are associates, this will show that we do not have unique factorization. Since 2, 3, and $1 \pm \sqrt{-5}$ have norms 4, 9, and 6, respectively, to show that they do not have nontrivial divisors, it is enough to show that there do not exist integers of norm 2 or 3. This is obvious.

11-49. Do the integers of $Q(\sqrt{-14})$ factor uniquely as a product of primes? If so, prove it; if not, explain why not.

Solution. Note that $15 = 3 \cdot 5 = (1 + \sqrt{-14})(1 - \sqrt{-14})$. We show that all of these factors are prime. Since $1+\sqrt{-14}$ is clearly not an associate of 3 or of 5, this will show that the integers of $Q(\sqrt{-14})$ do not have the unique factorization property.

Since $N(3) = 9$, to show that 3 is a prime, it is enough to show that no integer of $Q(\sqrt{-14})$ has norm 3. This is obvious, since the norm of $a + b\sqrt{-14}$ is $a^2 + 14b^2$, and

$a^2 + 14b^2 = 3$ does not have integer solutions. An almost identical argument shows that 5 is prime. Similarly, if $1 \pm \sqrt{-14} = \alpha\beta$, where neither α nor β is a unit, then the norms of α and β are 3 or 5, which is impossible.

11-50. *Show that $Q(\sqrt{15})$ does not have the unique factorization property.*

Solution. Note that 2 divides $(\sqrt{15}+1)(\sqrt{15}-1)$ but 2 divides neither $\sqrt{15}+1$ nor $\sqrt{15}-1$. This is incompatible with unique factorization if we can show that 2 is prime in $Q(\sqrt{15})$.

Any nontrivial divisor of 2 must have norm ± 2. But we cannot have $u^2 - 15v^2 = \pm 2$, since this would imply that $u^2 \equiv \pm 2 \pmod{5}$, which is impossible. Thus 2 is prime, and the result follows.

11-51. *Express $3 \cdot 31 \cdot 41$ as a product of primes of $Q(\sqrt{-2})$. Show that the factors are indeed prime.*

Solution. We can use the theory of (11.22), but instead, we proceed directly. Note that the norms of 3, 31, and 41 are 3^2, 31^2, and 41^2, and so any proper factors have norms 3, 31, and 41. It is easy to check that $3 = (1+\sqrt{-2})(1-\sqrt{-2})$, that no integer has norm 31, and that $41 = (3+4\sqrt{-2})(3-4\sqrt{-2})$. Since there is no integer of norm 31, it follows that 31 is prime, and since $1 \pm \sqrt{-2}$ and $3 \pm 4\sqrt{-2}$ all have prime norm, they are prime. Therefore we have the prime factorization $3 \cdot 31 \cdot 41 = 31(1+\sqrt{-2})(1-\sqrt{-2})(3+4\sqrt{-2})(3-4\sqrt{-2})$.

11-52. *Find all representations of $n = 2 \cdot 11 \cdot 17 \cdot 25$ of the form $x^2 + 2y^2$.*

Solution. Since 11 and 17 are primes of the form $8k+1$ or $8k+3$, while 5 is of the form $8k+5$ but occurs to an even power, it follows from (11.22) that n has representations of the form $x^2 + 2y^2$. In fact, n has $2(1+1)(1+1) = 8$ such representations. In any representation, 5 must divide both x and y: 5 is prime in $Q(\sqrt{-2})$, and thus if $5 \mid x^2 + 2y^2$, then 5 divides either $x+y\sqrt{-2}$ or $x-y\sqrt{-2}$ and hence divides both. It is therefore enough to represent $2 \cdot 11 \cdot 17$. Factoring into a product of primes of $Q(\sqrt{-2})$, we obtain

$$2 \cdot 11 \cdot 17 = -(\sqrt{-2})^2(3+\sqrt{-2})(3-\sqrt{-2})(3+2\sqrt{-2})(3-2\sqrt{-2}).$$

We wish to express this as $\alpha\bar{\alpha}$. There are essentially two ways to do this: Choose $\alpha = \sqrt{-2}(3+\sqrt{-2})(3+2\sqrt{-2})$ or $\alpha = \sqrt{-2}(3+\sqrt{-2})(3-2\sqrt{-2})$. The first produces the representation $2 \cdot 11 \cdot 17 = 18^2 + 2 \cdot 5^2$, and the second yields the representation $6^2 + 2 \cdot 13^2$. Multiplying by 5, we obtain representations $n = 90^2 + 2 \cdot 25^2$ and $n = 30^2 + 2 \cdot 65^2$. Each has four obvious variants obtained by changing signs.

11-53. *Suppose that d is square-free and $Q(\sqrt{d})$ has the unique factorization property. Let p be an odd (rational) prime such that d is a quadratic residue of p. Prove that p is not prime in $Q(\sqrt{d})$. (Hint. See the discussion preceding (11.21).)*

Solution. There is an integer x such that $x^2 \equiv d \pmod{p}$, since d is a quadratic residue of p; thus $p \mid (x + \sqrt{d})(x - \sqrt{d})$. If p were prime in $Q(\sqrt{d})$, then it would follow that $p \mid x + \sqrt{d}$ or $p \mid x - \sqrt{d}$, since we are assuming that the integers of $Q(\sqrt{d})$ have unique factorization. But it is clear that p does not divide $x \pm \sqrt{d}$.

11-54. Show that there exist integers x and y such that $x^2 - 3y^2 = 97$. *(Hint. Use Problem 11-53 and the fact that $Q(\sqrt{3})$ has the unique factorization property.)*

Solution. By an easy Legendre symbol calculation, we find that $(3/97) = 1$, and therefore 3 is a quadratic residue of 97. Thus by the preceding problem, 97 is not prime in $Q(\sqrt{3})$; let $97 = \alpha\beta$, where neither α nor β is a unit. Then since $N(97) = 97^2$, we must have $N(\alpha) = \pm 97$. If $\alpha = x + y\sqrt{3}$, then $x^2 - 3y^2 = \pm 97$. But $x^2 - 3y^2 = -97$ is impossible, by a simple calculation modulo 3. It follows that $x^2 - 3y^2 = 97$.

11-55. Let p be an odd prime that does not divide the square-free number d. Show that if α is an integer of $Q(\sqrt{d})$, then $\alpha^p \equiv \alpha \pmod{p}$ if d is a quadratic residue of p, and $\alpha^p \equiv \bar{\alpha} \pmod{p}$ if d is a nonresidue of p. *(Hint. Expand $(x + y\sqrt{d})^p$. Here, $\alpha \equiv \beta \pmod{p}$ means that $p \mid \alpha - \beta$.)*

Solution. Let $\alpha = x + y\sqrt{d}$. We deal with the case where x and y are integers; the argument is similar when x and y are each half an odd integer. By the Binomial Theorem, we have

$$(x + y\sqrt{d})^p = x^p + \binom{p}{1}x^{p-1}y\sqrt{d} + \cdots + \binom{p}{p-1}xy^{p-1}(\sqrt{d})^{p-1} + y^p(\sqrt{d})^p.$$

Using Fermat's Theorem and the fact that $\binom{p}{k} \equiv 0 \pmod{p}$ for any k other than 0 and p, we find that $(x + y\sqrt{d})^p \equiv x + yd^{(p-1)/2}\sqrt{d} \pmod{p}$. The result is now a direct consequence of Euler's Criterion.

▷ **11-56.** Let d be ± 2 or ± 3. Prove that if α and β are integers of $Q(\sqrt{d})$, with α nonzero, then there exist integers γ and ρ of $Q(\sqrt{d})$ such that $\beta = \gamma\alpha + \rho$ and $|N(\rho)| < |N(\alpha)|$. *(Hint. See (11.5).)*

Solution. The argument closely mirrors the one used in (11.5). Divide β by α to obtain $\beta/\alpha = x + y\sqrt{d}$ for some rational numbers x and y. Let u and v be, respectively, integers nearest to x and y, let $\gamma = u + vi$, and set $\rho = \beta - \gamma\alpha$. It remains to verify that $N(\gamma) < N(\alpha)$.

Since $\beta = (x + y\sqrt{d})\alpha$, we have $\rho = \left((x - u) + (y - v)\sqrt{d}\right)\alpha$. But $|x - u| \leq 1/2$ and $|y - v| \leq 1/2$. Thus $|N((x-u) + (y-v)\sqrt{d})| \leq 1/4 + |d|/4$. Therefore if $d = \pm 2$, we have $|N(\rho)| \leq \frac{3}{4}|N(\alpha)|$, and the result follows. For $d = 3$, note that we are trying to maximize the absolute value of $(x-u)^2 - 3(y-v)^2$, subject to the conditions $|x-u| \leq 1/2$ and $|y - v| \leq 1/2$. It is obvious that this maximum is reached if $|y - v| = 1/2$ and $x - u = 0$. Thus $|N(\rho)| \leq \frac{3}{4}|N(\alpha)|$, and again the result holds.

The case $d = -3$ requires some additional argument, since $N((1 + \sqrt{-3})/2) = 1$. If $|x - u| < 1/2$ or $|y - v| < 1/2$, then $N((x - u) + (y - v)\sqrt{-3}) < 1$, and hence $|N(\rho)| < |N(\alpha)|$. If x and y each differ by $1/2$ from an integer, then $(x + y\sqrt{-3})/2$ is an integer of $Q(\sqrt{-3})$, so we can let $\gamma = (x + y\sqrt{-3})/2$ and $\rho = 0$.

▷ **11-57.** *Find all solutions in integers of the Diophantine equation $x^2 + 2 = y^3$. (Hint. See the example that follows (11.14). You may assume that $Q(\sqrt{-2})$ has unique factorization.)*

Solution. Suppose that $x^2 + 2 = y^3$, and factor $x^2 + 2$ as $(x + \sqrt{-2})(x - \sqrt{-2})$. Any nontrivial common divisor of $x + \sqrt{-2}$ and $x - \sqrt{-2}$ must divide $2\sqrt{-2}$. But clearly, x is odd, for if x is even, then $x^2 + 2 \equiv 2 \pmod{4}$, and hence $x^2 + 2$ cannot be a perfect cube. Therefore each of $x \pm \sqrt{-2}$ has odd norm. Since any nontrivial divisor of $2\sqrt{-2}$ has even norm, it follows that $x + \sqrt{-2}$ and $x - \sqrt{-2}$ are relatively prime.

Since $Q(\sqrt{-2})$ has unique factorization, a proof similar to that of (11.14) shows that $x + \sqrt{-2}$ can be expressed as $\epsilon \gamma^3$, where ϵ is a unit. But the only units are ± 1, which are both perfect cubes, and so we may take $x + \sqrt{-2} = \gamma^3$. Let $\gamma = a + b\sqrt{-2}$. By cubing $a + b\sqrt{-2}$ and setting the result equal to $x + \sqrt{-2}$, we obtain $a^3 - 6ab^2 = x$ and $3a^2b - 2b^3 = 1$. Since b divides $3a^2b - 2b^3$, it follows that $b = \pm 1$. If $b = 1$, then $a = \pm 1$, while if $b = -1$, then the equation $3a^2b - 2b^3 = 1$ does not have a solution. Thus $x = a^3 - 6ab^2$ with $b = 1$ and $a = \pm 1$, i.e., $x = \pm 5$. Accordingly, the only integer solutions of $x^2 + 2 = y^3$ are $x = \pm 5$, $y = 3$.

▷ **11-58.** *Find the integer solutions of the equation $x^2 + 11 = y^3$. Assume without proof that the integers of $Q(\sqrt{-11})$ have unique factorization.*

Solution. First we note that if $x^2 + 11 = y^3$, then x must be even. For if x is odd, then $x^2 + 11 \equiv 4 \pmod 8$, and thus $x^2 + 11$ cannot be a pefect cube. Also, x cannot be divisible by 11, since if $11 | x$, then $11 | y$ and therefore $y^3 - x^2$ is divisible by 121, contradicting the fact that $y^3 - x^2 = 11$.

Factoring in $Q(\sqrt{-11})$, we obtain $(x + \sqrt{-11})(x - \sqrt{-11}) = y^3$. Note that if α is a common prime factor of $x + \sqrt{-11}$ and $x - \sqrt{-11}$, then α must divide $2\sqrt{-11}$, and since 2 and $\sqrt{-11}$ are clearly prime, α must be ± 2 or $\pm\sqrt{-11}$. But we cannot have $\alpha = \pm 2$, since x is even and 2 does not divide $\sqrt{-11}$. If $\alpha = \pm\sqrt{-11}$, then $\alpha | x$ and therefore $11 | x$, which is impossible. Thus $x + \sqrt{-11}$ and $x - \sqrt{-11}$ are relatively prime. It follows that $x + \sqrt{-11}$ and $x - \sqrt{-11}$ are both perfect cubes. Let $x + \sqrt{-11} = \lambda^3$. Since $-11 \equiv 1 \pmod 4$, either (i) $\lambda = a + b\sqrt{-11}$, where a and b are integers, or (ii) $\lambda = (c + d\sqrt{-11})/2$, where c and d are odd integers.

In case (i), we find that $x = a^3 - 33ab^2$ and $1 = 3a^2b - 11b^3$. But if $3a^2b - 11b^3 = 1$, then $b = 1$ or $b = -1$. If $b = 1$, then $a = \pm 2$; $b = -1$ does not give an integer value of a. Substituting in the expression $x = a^3 - 33ab^2$, we obtain the solution $x = \pm 58$; thus $y = 15$.

In case (ii), we have $x = (c^3 - 33cd^2)/8$ and $1 = (3c^2d - 11d^3)/8$. Consider the equation $3c^2d - 11d^3 = 8$. Then d divides 8, and since d is odd, it follows that $d = \pm 1$. Clearly, $d = 1$ does not yield a solution, and if $d = -1$, then $c = \pm 1$. Therefore we obtain the solution $x = \pm 4$, $y = 3$.

EXERCISES FOR CHAPTER 11

1. Suppose $N(\gamma) = 729 = 27^2$. Is γ a Gaussian prime?
2. Prove or disprove: If $a+bi$ is a Gaussian prime, where a and b are nonzero, then $N(a+bi)$ is a rational prime.
3. Express 390 as a product of Gaussian primes.
4. Write $60 + 180i$ as a product of Gaussian primes.
5. Find a prime factorization of $19 + 17i$.
6. Express $7 + 24i$ as a product of a unit and Gaussian primes of the form $a + bi$, where $a > 0$ and $b \geq 0$.
7. Find the greatest common divisors of $18 + i$ and $6 - 17i$.
8. Observe that 41 is the norm of a Gaussian prime. What is the next integer which is the norm of a Gaussian prime?
9. How many Gaussian integers have norm 1300? How many of these are divisible by 5?
10. Find the number of integers n less than 200 such that n is the norm of 12 or more Gaussian integers.
11. Suppose that α, β, and γ are Gaussian primes of odd norm. Can we ever have $\alpha + \beta = \gamma$? Explain.
12. How many different Gaussian integers divide 10^6?
13. Find all the Gaussian integers with norm 578.
14. Show that if α and β are Gaussian integers, then δ is a greatest common divisor of α and β if and only if δ is a greatest common divisor of α and $\alpha - \beta$.
15. Let α, β, and γ be Gaussian integers such that $\alpha | \gamma$ and $\beta | \gamma$. If α and β are relatively prime, show that $\alpha\beta | \gamma$. (Hint. See (1.10), or use unique factorization.)
16. Let α and β be Gaussian integers whose norms are relatively prime. Show that α and β are relatively prime. Does the converse of this result hold? Explain.
17. Prove (11.18.ii): If α and β are elements of $Q(\sqrt{d})$, then $N(\alpha\beta) = N(\alpha)N(\beta)$.
18. Prove (11.18.iii): If α is an integer of $Q(\sqrt{d})$, then α is a unit if and only if $N(\alpha) = \pm 1$.
19. Prove that $2 + \sqrt{-6}$ is a prime in $Q(\sqrt{-6})$.
20. Show that $1 + \sqrt{3}$ and $1 - \sqrt{3}$ are associates in $Q(\sqrt{3})$.
21. Determine which of the following are primes in $Q(\sqrt{-13})$: $1 + 2\sqrt{-13}$; $3 + \sqrt{-13}$; 41.

22. (a) Is $\alpha = 11$ prime in $Q(\sqrt{-3})$? If so, prove it. If not, express α as a product of primes. (b) Answer the same question when $\alpha = 23$. (Note that $-3 \equiv 1 \pmod{4}$.)

23. Show that $5 + \sqrt{15}$ is prime in $Q(\sqrt{15})$.

24. Show that for any square-free d, there are infinitely many integers of $Q(\sqrt{d})$ which are prime in $Q(\sqrt{d})$.

25. Show that $Q(\sqrt{-19})$ does not have the unique factorization property by examining factorizations of 20.

26. Find all representations of $8 \cdot 49 \cdot 121$ as $a^2 + 2b^2$, where a and b are integers.

27. Find the fundamental unit ϵ of $Q(\sqrt{101})$, and prove that ϵ is the fundamental unit.

28. Find all the units of $Q(\sqrt{3})$ between 100 and 1000.

NOTES FOR CHAPTER 11

1. Gauss knew by 1801 that the integers of $Q(\sqrt{d})$ have unique factorization when $d = -1, -2, -3, -7, -11, -19, -43, -67,$ and -163. (What he actually proved was an equivalent assertion about quadratic forms.) Gauss also conjectured that these were the only square-free negative d with this property. The conjecture remained open for more than 150 years. In 1934, H. Heilbronn and E.H. Linfoot showed that there is *at most one more* negative square-free d for which the integers of $Q(\sqrt{d})$ have unique factorization. Finally, in 1966, H.M. Stark proved that Gauss's list is indeed complete. More or less simultaneously, A. Baker also found a proof. Both arguments are exceedingly difficult.

For positive d, the question of when we have unique factorization is far from being settled. There are 37 positive square-free d under 100 for which the integers of $Q(\sqrt{d})$ have unique factorization, and $Q(\sqrt{d})$ is Euclidean for only 17 of these d. In contrast with the situation for negative d, unique factorization seems to be quite common when d is positive. However, it has not even been proved that there are infinitely many square-free positive d such that the integers of $Q(\sqrt{d})$ have unique factorization.

2. A *binary quadratic form* is a polynomial $ax^2 + bxy + cy^2$, where a, b, and c will be taken to be integers. One of the fundamental problems of the theory of quadratic forms, dating back to Bhaskara and Fermat, is to determine which integers are representable by the form $ax^2 + bxy + cy^2$. We have obtained answers to this question for the forms $x^2 + y^2$ and $x^2 + 2y^2$ (see (11.21)). By completing the square, we can see that

$$4a(ax^2 + bxy + cy^2) = (2ax + by - y\sqrt{d})(2ax + by + y\sqrt{d}),$$

where $d = b^2 - 4ac$. Thus questions of representability by quadratic forms are closely connected with factorization in $Q(\sqrt{d})$.

Consider, in particular, the form $x^2 - dy^2$. (Theorem 11.21 deals with the case $d = -2$.) We solved the problem of which integers are representable by finding the *primes p* that can be represented and then using the fact that the norm of a product is the product of the norms. Finding the representable p was a simple application of quadratic reciprocity. This approach works whenever the integers of $Q(\sqrt{d})$ have unique factorization. (When d is positive, we also need to consider representability of $-p$.) The picture is not always this simple, however. For example, 21 can be represented by the form x^2+5y^2, but neither of its prime factors can be. This phenomenon is closely connected with the fact that unique factorization fails for the integers of $Q(\sqrt{-5})$.

BIOGRAPHICAL SKETCHES

Ernst Eduard Kummer was born in Germany in 1810 and obtained his doctorate from the University of Halle. For ten years, he taught in a secondary school. Leopold Kronecker, who would himself become a well-known mathematician, was one of Kummer's pupils. In 1855, Kummer was appointed Professor of Mathematics at the University of Berlin and the Berlin War College. Until the age of 32, Kummer's work concerns itself mainly with definite integrals, differential equations, and series, most notably hypergeometric functions.

Kummer made his first major contribution to number theory through his work on higher reciprocity laws. Following ideas of Gauss and Jacobi, Kummer was led to study the arithmetic of the "integers" in the field $Q(\alpha)$ obtained by adjoining to the rationals a nontrivial root of the equation $x^n = 1$. He proved that unique factorization fails when $n = 23$ and successfully introduced "ideal numbers" to restore unique factorization. Ultimately, he would use these ideal numbers to prove a general reciprocity law. Kummer pointed out also the relevance of the arithmetic of $Q(\sqrt{d})$ to the theory of quadratic forms. Quadratic forms had been central in the number-theoretic work of Lagrange, Legendre, and Gauss. Kummer was thus the founder of algebraic number theory.

Kummer is also famous for his work on Fermat's Last Theorem. Here again, he used factorization in $Q(\alpha)$ and ultimately settled Fermat's Last Theorem for all exponents under 100. He was one of the first mathematicians to show how "abstract" algebraic concepts could be used to solve concrete problems.

Kummer died in Berlin in 1893.

REFERENCES

G.H. Hardy and E.M. Wright, *An Introduction to the Theory of Numbers* (Fourth Edition). (See Chapter 7.)

Ivan Niven, Herbert S. Zuckerman, and Hugh L. Montgomery, *An Introduction to the Theory of Numbers* (Fifth Edition), John Wiley & Sons, New York, 1991.

> Since the first edition of Niven and Zuckerman appeared in 1960, this has been a well-known textbook in elementary number theory. The material in the current edition is presented at a higher and more theoretical level than other current texts. It is, however, an excellent book to study after a basic background in number theory has been acquired, and it contains a fairly complete treatment of the topics in this chapter.

Appendix

Table 1: Primes less than 1000 and their least primitive root

Table 2: Continued fraction expansion of $\sqrt{d}$ for $d < 100$

Table 1. Primes less than 1000 and their least primitive root g

p	g	p	g	p	g	p	g
2	1	191	19	439	15	709	2
3	2	193	5	443	2	719	11
5	2	197	2	449	3	727	5
7	3	199	3	457	13	733	6
11	2	211	2	461	2	739	3
13	2	223	3	463	3	743	5
17	3	227	2	467	2	751	3
19	2	229	6	479	13	757	2
23	5	233	3	487	3	761	6
29	2	239	7	491	2	769	11
31	3	241	7	499	7	773	2
37	2	251	6	503	5	787	2
41	6	257	3	509	2	797	2
43	3	263	5	521	3	809	3
47	5	269	2	523	2	811	3
53	2	271	6	541	2	821	2
59	2	277	5	547	2	823	3
61	2	281	3	557	2	827	2
67	2	283	3	563	2	829	2
71	7	293	2	569	3	839	11
73	5	307	5	571	3	853	2
79	3	311	17	577	5	857	3
83	2	313	10	587	2	859	2
89	3	317	2	593	3	863	5
97	5	331	3	599	7	877	2
101	2	337	10	601	7	881	3
103	5	347	2	607	3	883	2
107	2	349	2	613	2	887	5
109	6	353	3	617	3	907	2
113	3	359	7	619	2	911	17
127	3	367	6	631	3	919	7
131	2	373	2	641	3	929	3
137	3	379	2	643	11	937	5
139	2	383	5	647	5	941	2
149	2	389	2	653	2	947	2
151	6	397	5	659	2	953	3
157	5	401	3	661	2	967	5
163	2	409	21	673	5	971	6
167	5	419	2	677	2	977	3
173	2	421	2	683	5	983	5
179	2	431	7	691	3	991	6
181	2	433	5	701	2	997	7

Table 2. Continued fraction expansion of $\sqrt{d}$ for $d < 100$

d	$\sqrt{d}$	d	$\sqrt{d}$
2	$\langle 1, \overline{2} \rangle$	53	$\langle 7, \overline{3, 1, 1, 3, 14} \rangle$
3	$\langle 1, \overline{1, 2} \rangle$	54	$\langle 7, \overline{2, 1, 6, 1, 2, 14} \rangle$
5	$\langle 2, \overline{4} \rangle$	55	$\langle 7, \overline{2, 2, 2, 14} \rangle$
6	$\langle 2, \overline{2, 4} \rangle$	56	$\langle 7, \overline{2, 14} \rangle$
7	$\langle 2, \overline{1, 1, 1, 4} \rangle$	57	$\langle 7, \overline{1, 1, 4, 1, 1, 14} \rangle$
8	$\langle 2, \overline{1, 4} \rangle$	58	$\langle 7, \overline{1, 1, 1, 1, 1, 1, 14} \rangle$
10	$\langle 3, \overline{6} \rangle$	59	$\langle 7, \overline{1, 2, 7, 2, 1, 14} \rangle$
11	$\langle 3, \overline{3, 6} \rangle$	60	$\langle 7, \overline{1, 2, 1, 14} \rangle$
12	$\langle 3, \overline{2, 6} \rangle$	61	$\langle 7, \overline{1, 4, 3, 1, 2, 2, 1, 3, 4, 1, 14} \rangle$
13	$\langle 3, \overline{1, 1, 1, 1, 6} \rangle$	62	$\langle 7, \overline{1, 6, 1, 14} \rangle$
14	$\langle 3, \overline{1, 2, 1, 6} \rangle$	63	$\langle 7, \overline{1, 14} \rangle$
15	$\langle 3, \overline{1, 6} \rangle$	65	$\langle 8, \overline{16} \rangle$
17	$\langle 4, \overline{8} \rangle$	66	$\langle 8, \overline{8, 16} \rangle$
18	$\langle 4, \overline{4, 8} \rangle$	67	$\langle 8, \overline{5, 2, 1, 1, 7, 1, 1, 2, 5, 16} \rangle$
19	$\langle 4, \overline{2, 1, 3, 1, 2, 8} \rangle$	68	$\langle 8, \overline{4, 16} \rangle$
20	$\langle 4, \overline{2, 8} \rangle$	69	$\langle 8, \overline{3, 3, 1, 4, 1, 3, 3, 16} \rangle$
21	$\langle 4, \overline{1, 1, 2, 1, 1, 8} \rangle$	70	$\langle 8, \overline{2, 1, 2, 1, 2, 16} \rangle$
22	$\langle 4, \overline{1, 2, 4, 2, 1, 8} \rangle$	71	$\langle 8, \overline{2, 2, 1, 7, 1, 2, 2, 16} \rangle$
23	$\langle 4, \overline{1, 3, 1, 8} \rangle$	72	$\langle 8, \overline{2, 16} \rangle$
24	$\langle 4, \overline{1, 8} \rangle$	73	$\langle 8, \overline{1, 1, 5, 5, 1, 1, 16} \rangle$
26	$\langle 5, \overline{10} \rangle$	74	$\langle 8, \overline{1, 1, 1, 1, 16} \rangle$
27	$\langle 5, \overline{5, 10} \rangle$	75	$\langle 8, \overline{1, 1, 1, 16} \rangle$
28	$\langle 5, \overline{3, 2, 3, 10} \rangle$	76	$\langle 8, \overline{1, 2, 1, 1, 5, 4, 5, 1, 1, 2, 1, 16} \rangle$
29	$\langle 5, \overline{2, 1, 1, 2, 10} \rangle$	77	$\langle 8, \overline{1, 3, 2, 3, 1, 16} \rangle$
30	$\langle 5, \overline{2, 10} \rangle$	78	$\langle 8, \overline{1, 4, 1, 16} \rangle$
31	$\langle 5, \overline{1, 1, 3, 5, 3, 1, 1, 10} \rangle$	79	$\langle 8, \overline{1, 7, 1, 16} \rangle$
32	$\langle 5, \overline{1, 1, 1, 10} \rangle$	80	$\langle 8, \overline{1, 16} \rangle$
33	$\langle 5, \overline{1, 2, 1, 10} \rangle$	82	$\langle 9, \overline{18} \rangle$
34	$\langle 5, \overline{1, 4, 1, 10} \rangle$	83	$\langle 9, \overline{9, 18} \rangle$
35	$\langle 5, \overline{1, 10} \rangle$	84	$\langle 9, \overline{6, 18} \rangle$
37	$\langle 6, \overline{12} \rangle$	85	$\langle 9, \overline{4, 1, 1, 4, 18} \rangle$
38	$\langle 6, \overline{6, 12} \rangle$	86	$\langle 9, \overline{3, 1, 1, 1, 8, 1, 1, 1, 3, 18} \rangle$
39	$\langle 6, \overline{4, 12} \rangle$	87	$\langle 9, \overline{3, 18} \rangle$
40	$\langle 6, \overline{3, 12} \rangle$	88	$\langle 9, \overline{2, 1, 1, 1, 2, 18} \rangle$
41	$\langle 6, \overline{2, 2, 12} \rangle$	89	$\langle 9, \overline{2, 3, 3, 2, 18} \rangle$
42	$\langle 6, \overline{2, 12} \rangle$	90	$\langle 9, \overline{2, 18} \rangle$
43	$\langle 6, \overline{1, 1, 3, 1, 5, 1, 3, 1, 1, 12} \rangle$	91	$\langle 9, \overline{1, 1, 5, 1, 5, 1, 1, 18} \rangle$
44	$\langle 6, \overline{1, 1, 1, 2, 1, 1, 1, 12} \rangle$	92	$\langle 9, \overline{1, 1, 2, 4, 2, 1, 1, 18} \rangle$
45	$\langle 6, \overline{1, 2, 2, 2, 1, 12} \rangle$	93	$\langle 9, \overline{1, 1, 1, 4, 6, 4, 1, 1, 1, 18} \rangle$
46	$\langle 6, \overline{1, 3, 1, 1, 2, 6, 2, 1, 1, 3, 1, 12} \rangle$	94	$\langle 9, \overline{1, 2, 3, 1, 1, 5, 1, 8, 1, 5, 1, 1, 3, 2, 1, 18} \rangle$
47	$\langle 6, \overline{1, 5, 1, 12} \rangle$	95	$\langle 9, \overline{1, 2, 1, 18} \rangle$
48	$\langle 6, \overline{1, 12} \rangle$	96	$\langle 9, \overline{1, 3, 1, 18} \rangle$
50	$\langle 7, \overline{14} \rangle$	97	$\langle 9, \overline{1, 5, 1, 1, 1, 1, 1, 1, 5, 1, 18} \rangle$
51	$\langle 7, \overline{7, 14} \rangle$	98	$\langle 9, \overline{1, 8, 1, 18} \rangle$
52	$\langle 7, \overline{4, 1, 2, 1, 4, 14} \rangle$	99	$\langle 9, \overline{1, 18} \rangle$

General References

W.W. Adams and L.J. Goldstein, *Introduction to Number Theory*, Prentice-Hall, Englewood Cliffs, New Jersey, 1976.

Tom M. Apostol, *Introduction to Analytic Number Theory*, Springer-Verlag, New York, 1976.

David M. Bressoud, *Factorization and Primality Testing*, Springer-Verlag, New York, 1989.

Harold Davenport, *The Higher Arithmetic* (Sixth Edition), Cambridge University Press, Cambridge, England, 1992.

Harold M. Edwards, *Fermat's Last Theorem, A Genetic Introduction to Algebraic Number Theory*, Springer-Verlag, New York, 1977.

Leonhard Euler, *Elements of Algebra*, translated by John Hewlett, Springer-Verlag, New York, 1984.

Carl Friedrich Gauss, *Disquisitiones Arithmeticae*, translated by Arthur A. Clarke, Yale University Press, New Haven, Connecticut, 1966.

Peter Giblin, *Primes and Programming*, Cambridge University Press, Cambridge, England, 1993.

Emil Grosswald, *Topics from the Theory of Numbers* (Second Edition), Birkhäuser, Boston, 1982.

G.H. Hardy and E.M. Wright, *An Introduction to the Theory of Numbers* (Fourth Edition), The Clarendon Press, Oxford, England, 1971.

Thomas L. Heath, *The Thirteen Books of Euclid's Elements, Volume II*, Cambridge University Press, Cambridge, England, 1926.

Thomas L. Heath, *Diophantus of Alexandria*, Dover, New York, 1964.

K. Ireland and M.I. Rosen, *A Classical Introduction to Modern Number Theory*, Springer-Verlag, New York, 1982.

Victor Klee and Stan Wagon, *Old and New Unsolved Problems in Plane Geometry and Number Theory*, Mathematical Association of America, Washington, 1991.

GENERAL REFERENCES

Donald E. Knuth, *The Art of Computer Programming, Volume 2* (Second Edition), Addison-Wesley, Reading, Massachussetts, 1981.

Neal Koblitz, *A Course in Number Theory and Cryptography*, Springer-Verlag, New York, 1987.

William J. LeVeque, *Fundamentals of Number Theory*, Addison-Wesley, Reading, Massachusetts, 1977.

L.J. Mordell, *Diophantine Equations*, Academic Press, New York, 1969.

Trygve Nagell, *Introduction to Number Theory*, Chelsea, New York, 1981 (originally published in 1951).

W. Narkiewicz, *Classical Problems in Number Theory*, Polish Scientific Publishers, Warsaw, 1986.

Ivan Niven, Herbert S. Zuckerman, and Hugh L. Montgomery, *An Introduction to the Theory of Numbers* (Fifth Edition), John Wiley & Sons, New York, 1991.

Carl D. Olds, *Continued Fractions*, Random House, New York, 1963.

Oystein Ore, *Number Theory and Its History*, McGraw-Hill, New York, 1948.

Hans Rademacher, *Lectures on Elementary Number Theory*, Blaisdell, New York, 1964.

Paulo Ribenboim, *13 Lectures on Fermat's Last Theorem*, Springer-Verlag, New York, 1979.

Paulo Ribenboim, *The Book of Prime Number Records* (Second Edition), Springer-Verlag, New York, 1989.

Hans Riesel, *Prime Numbers and Computer Methods for Factorization*, Birkhäuser, Boston, 1987.

Kenneth H. Rosen, *Elementary Number Theory and its Applications* (Third Edition), Addison-Wesley, Reading, Massachusetts, 1992.

W. Scharlau and H. Opolka, *From Fermat to Minkowski, Lectures on the Theory of Numbers and Its Historical Development*, Springer-Verlag, New York, 1985.

Jacques Sesiano, *Books IV to VII of Diophantus' Arithmetica*, Springer-Verlag, New York, 1982.

Daniel Shanks, *Solved and Unsolved Problems in Number Theory* (Third Edition), Chelsea, New York, 1985.

J.V. Uspensky and M.A. Heaslet, *Elementary Number Theory*, McGraw-Hill, New York, 1939.

I.M. Vinogradov, *Elements of Number Theory*, Dover, New York, 1954.

History

Eric T. Bell, *Men of Mathematics*, Simon & Schuster, New York, 1965.

Carl B. Boyer and Uta C. Merzbach, *A History of Mathematics* (Second Edition), John Wiley & Sons, New York, 1989.

Claude Brezinski, *History of Continued Fractions and Padé Approximants*, Springer-Verlag, New York, 1991.

David M. Burton, *The History of Mathematics* (Second Edition), Wm. C. Brown, Dubuque, Iowa, 1991.

Leonard Eugene Dickson, *History of the Theory of Numbers* (3 volumes), Chelsea, New York, 1952 (originally published in 1919).

George G. Joseph, *The Crest of the Peacock: Non-European Roots of Mathematics*, Penguin Books, New York, 1991.

Morris Kline, *Mathematical Thought from Ancient to Modern Times*, Oxford University Press, New York, 1972.

Ulrich Libbrecht, *Chinese Mathematics in the Thirteenth Century*, The MIT Press, Cambridge, Massachusetts, 1973.

James R. Newman, *The World of Mathematics* (4 volumes), Simon & Schuster, New York, 1956.

André Weil, *Number Theory: An approach through history from Hammurapi to Legendre*, Birkhäuser, Boston, 1984.

Journal Articles

Robert A. Coury, "A Continued Fraction Approach for Factoring Large Numbers," *Pi Mu Epsilon Journal*, Vol. 9, No. 1 (1989), 9–12.

David A. Cox, "Introduction to Fermat's Last Theorem," *American Mathematical Monthly*, Vol. 101, No. 1 (1994), 3–14.

Jacques Dutka, "On the Gregorian Revision of the Julian Calendar," *Mathematical Intelligencer*, Vol. 10, No. 1 (1988), 56–64.

M. Ram Murty, "Artin's Conjecture for Primitive Roots," *Mathematical Intelligencer*, Vol. 10, No. 4 (1988), 59–67.

Carl Pomerance, "The Search for Prime Numbers," *Scientific American*, Vol. 247 (1982), 136–147.

V. Frederick Rickey, "Mathematics of the Gregorian Calendar," *Mathematical Intelligencer*, Vol. 7, No. 1 (1985), 53–56.

David E. Rowe, "Gauss, Dirichlet, and the Law of Biquadratic Reciprocity," *Mathematical Intelligencer*, Vol. 10, No. 2 (1988), 13–25.

A. Seidenberg, "The Ritual Origin of Geometry," *Archive for the History of the Exact Sciences*, Vol. 1 (1963), 488–527.

Stan Wagon, "Perfect Numbers," *Mathematical Intelligencer*, Vol. 7, No. 2 (1985), 66–68.

Stan Wagon, "Primality Testing," *Mathematical Intelligencer*, Vol. 8, No. 3 (1986), 58–61.

Stan Wagon, "The Euclidean Algorithm Strikes Again," *American Mathematical Monthly*, Vol. 97, No. 2 (1990), 125–129.

Index

Algebra (Euler), 356
al-Karaji, 221
al-Khwarizmi, 7
approximations
  rational, 282
  best rational, 284
  to π, 284, 301
Arithmetica (Diophantus), 267
Aryabhata, 37
associate, 358, 367

Bachet, Claude, 6
Bertrand's Postulate, 201
Bhaskara, 355
Binary GCD Algorithm, 36
binary quadratic form, 388
Brahmagupta's identity, 320
Brouncker, William, 314

calendars, 47, 285
Carmichael function, 187
Carmichael number, 86
Chebyshev, Pavnuty, 218
Ch'in Chiu-shao, 68
Chinese Remainder Theorem, 46
complete residue system, 40
composite number, 10
congruence
  $ax \equiv b \pmod{m}$, 43
  $f(x) \equiv 0 \pmod{m}$, 101
  $f(x) \equiv 0 \pmod{p^k}$, 106

$ax^2 + bx + c \equiv 0 \pmod{p}$, 126
$x^2 \equiv a \pmod{p}$, 72
$x^2 \equiv a \pmod{p^k}$, 109
$x^2 \equiv a \pmod{m}$, 127
$x^k \equiv a \pmod{m}$, 165
congruent to, 40
conjugate
  of quadratic irrational, 280
  of Gaussian integer, 358
  of integer in $Q(\sqrt{d})$, 367
continued fraction
  convergent of, 273
  finite, 271
  infinite, 275
  period of, 278
  periodic, 278
  purely periodic, 278
  simple, 271
convergent, 273
  secondary, 303

day of the week, 47
Diophantine equation, 221
  $ax + by = c$, 15
  $x^2 + y^2 = z^2$, 222
  $x^4 + y^4 = z^2$, 225
  $x^2 - dy^2 = 1$, 314
  $x^2 - dy^2 = -1$, 322
  $x^2 - dy^2 = N$, 324
Diophantus, 267
Dirichlet, Peter Gustav Lejeune, 218
Dirichlet's Theorem, 202

INDEX

Disquisitiones Arithmeticae (Gauss), 69
divisibility tests, 42
Division Algorithm, 7, 359
Division Algorithm for Polynomials, 103
divisors, 7
  proper, 10
  number of positive, 12
  sum of positive, 12

Eisenstein, Ferdinand, 156
Elements (Euclid), 37
Euclid, 37
Euclidean Algorithm, 14
  Extended Euclidean Algorithm, 15
Euclidean quadratic field, 368
Euler, Leonhard, 99
Euler ϕ-function, 75
Euler's Criterion, 73, 129
Euler's Theorem, 76
Extended Euclidean Algorithm, 15

factoring large numbers, 347
Factoring Method, Legendre's, 327
factorization, prime, 11
Fermat, Pierre de, 98
Fermat number, 198
Fermat prime, 198
Fermat's Last Theorem, 224
Fermat's Theorem, 74
Fibonacci (Leonardo of Pisa), 69
Frenicle de Bessy, Bernard, 71
functions
  $[x]$, 13
  $N(n)$, 228
  $\pi(x)$, 200
  $\phi(m)$, 75
  $\sigma(n)$, 12
  $\tau(n)$, 12
fundamental solution, 317
Fundamental Theorem of Arithmetic, 11, 36

Gauss, Carl Friedrich, 68

Gaussian integer, 358
Gaussian prime, 358
Gauss's Lemma, 131
gcd, 8
Goldbach's Conjecture, 203
greatest common divisor, 8, 360
greatest integer function, 13
group, 67
  multiplicative, 192

index of an integer, 164
infinite descent, method of, 225
integer
  algebraic, 366
  Gaussian, 358
  of $Q(\sqrt{d})$, 365
  rational, 358
inverse modulo m, 44

Jacobi, Carl Gustav, 267
Jacobi symbol, 151

Kummer, Ernst Eduard, 389

Lagrange, Joseph Louis, 312
Lagrange's Theorem, 104
Law of Quadratic Reciprocity, 134
Least Absolute Remainder Algorithm, 35
least common multiple, 9
least universal exponent, 161, 187
Legendre, Adrien-Marie, 193
Legendre symbol, 129
Legendre's Factoring Method, 327
Liber Abaci (Fibonacci), 69
Liber Quadratorum (Fibonacci), 69
linear combination, 8
linear congruence, 43
Lucas Primality Test, 162

Mersenne, Marin, 3
Mersenne number, 197
Mersenne prime, 197
method of infinite descent, 225
modulo m, 40

modulus, 40
multiple, 7
multiplicative function, 13

nonresidue, quadratic, 128
norm
   Gaussian, 358
   in $Q(\sqrt{d})$, 367
number of positive divisors, 12
number of representations, 228, 235

order of an integer, 159

partial quotient, 271
Pell's Equation, 314
   least positive solution, 317
   $x^2 - dy^2 = 1$, 315
   $x^2 - dy^2 = -1$, 322
   $x^2 - dy^2 = N$, 324
perfect number, 196
period of a continued fraction, 278
periodic continued fraction, 278
polynomial congruence
   root, 101
   solution, 101, 102, 106
power residue, 165
prime, 10
   Gaussian, 358
   of $Q(\sqrt{d})$, 367
prime factorization, 11
Prime Number Theorem, 200
primitive representation, 229
primitive root, 160
purely periodic continued fraction, 278
Pythagorean triangle, 222
Pythagorean triple, 222
   primitive, 222

quadratic congruence, 126
quadratic extension $Q(\sqrt{d})$, 365
quadratic irrational, 278
   conjugate of, 280
   reduced, 281
quadratic nonresidue, 128

quadratic reciprocity law, 134
quadratic residue, 128
quotient, 7
   partial, 271

rational approximations, 282
   best, 284
   to π, 284, 301
reciprocity, quadratic, 134
reduced quadratic irrational, 281
reduced residue system, 75
relatively prime, 9
   in pairs, 9
remainder, 7
representation as sum of two squares, 226
   positive, 229
   primitive, 229
residue
   least nonnegative modulo m, 40
   modulo m, 40
   power, 165
   quadratic, 128
Riemann, Georg Bernhard, 219
root of polynomial congruence, 101

secondary convergent, 303
Sieve of Eratosthenes, 195
solution
   fundamental, 317
   positive, 16, 317
   $ax + by = c$, 15, 16
   $ax \equiv b \pmod{m}$, 43, 44
   $f(x) \equiv 0 \pmod{m}$, 101
   $ax^2 + bx + c \equiv 0 \pmod{p}$, 126
   $x^2 - dy^2 = 1$, 317, 319
   $x^2 - dy^2 = -1$, 322, 323
solutions, number of
   $ax + by = c$, 16
   $ax \equiv b \pmod{m}$, 43
   $f(x) \equiv 0 \pmod{m}$, 102
   $f(x) \equiv 0 \pmod{p^k}$, 106
   $x^2 \equiv a \pmod{2^k}$, 109
   $x^2 \equiv a \pmod{p^k}$, 109
   $x^2 \equiv a \pmod{m}$, 128

$x^k \equiv a \pmod{m}$, 165
sum of positive divisors, 12
sum of squares
  two squares, 228
  two squares, number of representations, 229
  two relatively prime squares, 229
  three squares, 235
  four squares, 233
  four squares, number of representations, 235

Unique Factorization Theorem, 11, 361, 369
unit
  Gaussian, 358
  of $Q(\sqrt{d})$, 367

Waring's Problem, 236
Wilson's Theorem, 73